TREE KANGAROOS

TREE KANGAROOS

Science and Conservation

Series Volume Editors

Lisa Dabek

Peter Valentine

Jacque Blessington

Karin R. Schwartz

Academic Press is an imprint of Elsevier
125 London Wall, London EC2Y 5AS, United Kingdom
525 B Street, Suite 1650, San Diego, CA 92101, United States
50 Hampshire Street, 5th Floor, Cambridge, MA 02139, United States
The Boulevard, Langford Lane, Kidlington, Oxford OX5 1GB, United Kingdom

Notices
Knowledge and best practices in this field are constantly changing. As new research and experience broaden our understanding, changes in research methods, professional practices, or medical treatment may become necessary.

Practitioners and researchers must always rely on their own experience and knowledge in evaluating and using any information, methods, compounds, or experiments described herein. In using such information or methods, they should be mindful of their own safety and the safety of others, including parties for whom they have a professional responsibility.

To the fullest extent of the law, neither the Publisher nor the authors, contributors, or editors, assume any liability for any injury and/or damage to persons or property as a matter of products liability, negligence or otherwise, or from any use or operation of any methods, products, instructions, or ideas contained in the material herein.

Library of Congress Cataloging-in-Publication Data
A catalog record for this book is available from the Library of Congress

British Library Cataloguing-in-Publication Data
A catalogue record for this book is available from the British Library

ISBN 978-0-12-814675-0

For information on all Academic Press publications
visit our website at https://www.elsevier.com/books-and-journals

Publisher: Charlotte Cockle
Acquisitions Editor: Anna Valutkevich
Editorial Project Manager: Devlin Person
Production Project Manager: Omer Mukthar
Cover Designer: Christian J. Bilbows

Typeset by SPi Global, India

Dedication

This book is dedicated to the future of all tree kangaroos and the people that live with them.

Tree Kangaroos: Science and Conservation

Cover and section photo credits

Photo Credits

Cover Image by Jonathan Byers

Section Photo Credits

I. Doria's tree kangaroo in PNG: Lisa Dabek

II. Matschie's tree kangaroo in forest in PNG: Lisa Dabek

III. Young Lumholtz's tree kangaroo in tree: Peter Valentine

IV. Manauno holding a Matschie's tree kangaroo at Dendawang field site: TKCP

V. Matschie's tree kangaroo joey in mother's pouch at Woodland Park Zoo: Ryan Hawk

VI. Researchers tracking Matschie's tree kangaroo at Wasaunon field site: Jonathan Byers

VII. Mother and young Matschie's tree kangaroo in PNG: Lisa Dabek

Contents

III

CONSERVATION SOLUTIONS: IN THE FIELD – AUSTRALIA

5. Community Conservation of Tree Kangaroos

PETER VALENTINE

6. Rehabilitation of Lumholtz's Tree Kangaroo Joeys

MARGIT CIANELLI AND KATRIN SCHMIDT

7. How an Understanding of Lumholtz's Tree Kangaroo Behavioral Ecology Can Assist Conservation

SIGRID HEISE-PAVLOV AND ELIZABETH PROCTER-GRAY

8. Tree Kangaroo Tourism as a Conservation Catalyst in Australia

ALAN GILLANDERS AND CLEVO WILSON

IV

CONSERVATION SOLUTIONS: IN THE FIELD – NEW GUINEA

9. Opportunities for Tree Kangaroo Conservation on the Island of New Guinea

BRUCE M. BEEHLER, NEVILLE KEMP, AND PHIL L. SHEARMAN

10. Creating the First Conservation Area in Papua New Guinea to Protect Tree Kangaroos

LISA DABEK AND ZACHARY WELLS

11. Land-Use Planning for a Sustainable Future in Papua New Guinea

DANNY NANE, MODI PONTIO, AND TREVOR HOLBROOK

12. A Model Tree Kangaroo Conservation Ranger Program in Papua New Guinea

DANIEL SOLOMON OKENA AND PAUL VAN NIMWEGEN

13. Community-Based Conservation on the Huon Peninsula

MIKAL EVERSOLE NOLAN, TIMMY SOWANG, AND KARAU KUNA

14. Strengthening Community Conservation Commitment Through Sustainable Livelihoods

TREVOR HOLBROOK

15. Using a One Health Model: Healthy Village-Healthy Forest

ROBERT M. LIDDELL, EMILY R. TRANSUE, JOAN CASTRO, NANCY PHILIPS, W. BLAIR BROOKS, MARTI H. LIDDELL, PETER M. RABINOWITZ, AND LISA DABEK

16. Building Conservation Leadership in Papua New Guinea for Tree Kangaroo Conservation

MIKAL EVERSOLE NOLAN, DANNY SAMANDINGKE, AND TIMMY SOWANG

Contributors

Agustina Y.S. Arobaya Biodiversity Research Centre; Faculty of Forestry, University of Papua, Manokwari, West Papua, Indonesia

Bruce M. Beehler Smithsonian Institution, Washington, DC, United States

W. Blair Brooks Dartmouth Hitchcock Medical Center, Lebanon, NH, United States

Jacque Blessington Association of Zoos and Aquariums Tree Kangaroo Species Survival Plan® Program, Kansas City, MO, United States

Simon Burchill Tree Kangaroo and Mammal Group Inc, Atherton, QLD, Australia

Jonathan Byers Perspective Solutions LLC, Missoula, MT, United States

Onnie Byers IUCN SSC Conservation Planning Specialist Group, Apple Valley, MN, United States

Joan Castro PATH Foundation Philippines, Inc., Makati, Philippines

Margit Cianelli Tree Kangaroo and Mammal Group & Tablelands Wildlife Rescue, Atherton; Lumholtz Lodge, Upper Barron, Atherton, Far North Queensland, QLD, Australia

Lisa Dabek Tree Kangaroo Conservation Program, Woodland Park Zoo, Seattle, WA, United States

Ellen Dierenfeld Association of Zoos and Aquariums Tree Kangaroo Species Survival Plan® Program, St. Louis, MO, United States

Carol Esson Ulysses Veterinary Clinic, Stratford, QLD, Australia

Claire Ford Taronga Conservation Society Australia, Mosman, NSW, Australia

Alan Gillanders Alan's Wildlife Tours, Yungaburra, QLD, Australia

Sigrid Heise-Pavlov Centre for Rainforest Studies at The School for Field Studies, Yungaburra, QLD, Australia

Margaret Highland Kansas State University, Manhattan, KS, United States

Trevor Holbrook Tree Kangaroo Conservation Program, Woodland Park Zoo, Seattle, WA, United States

Falk Huettmann EWHALE Lab – Biology & Wildlife Department, Institute of Arctic Biology, University of Alaska Fairbanks (UAF), Fairbanks, AK, United States

Razak Jaffar Wildlife Reserves Singapore, Singapore

John Kanowski Australian Wildlife Conservancy, Wondecla, QLD, Australia

Neville Kemp US AID Lestari-Indonesia, Jakarta, Indonesia

Johan F. Koibur Faculty of Animal Science; Biodiversity Research Centre, University of Papua, Manokwari, West Papua, Indonesia

Corinne Kozlowski Saint Louis Zoo, St. Louis, MO, United States

Karau Kuna Kainantu, Eastern Highlands Province, Papua New Guinea

Alana Legge Dreamworld, Gold Coast, QLD, Australia

Marti H. Liddell The Polyclinic, Seattle, WA, United States

Robert M. Liddell Center for Diagnostic Imaging Quality Institute, Seattle, WA, United States

Gayl Males South Coast Environment Centre, Goolwa Beach, SA, Australia

Thomas J. McGreevy, Jr. Department of Natural Resources Science, University of Rhode Island, Kingston, RI, United States

Philip Miller IUCN SSC Conservation Planning Specialist Group, Apple Valley, MN, United States

Sy Montgomery Nature Author, Hancock, NH, Unites States

Danny Nane Tree Kangaroo Conservation Program, Lae, Papua New Guinea

Paul van Nimwegen International Union for Conservation of Nature, Oceania Regional Office, Fiji

Mikal Eversole Nolan Lae, Papua New Guinea

Davi Ann Norsworthy Lincoln Children's Zoo, Lincoln, NE, United States

Daniel Solomon Okena Tree Kangaroo Conservation Program - Papua New Guinea, Lae, Morobe Province, Papua New Guinea

Freddy Pattiselanno Faculty of Animal Science; Biodiversity Research Centre, University of Papua, Manokwari, West Papua, Indonesia; Australasian Marsupial and Monotreme Species Specialist Group, Brisbane, QLD, Australia

Nancy Philips Dartmouth Hitchcock Medical Center, Lebanon, NH, United States

Terry M. Phillips Association of Zoos and Aquariums Tree Kangaroo Species Survival Plan® Program, Richmond, VA, United States

Modi Pontio Tree Kangaroo Conservation Program, Lae, Papua New Guinea

Gabriel Porolak University of Papua New Guinea, Port Moresby, Papua New Guinea

Elizabeth Procter-Gray Department of Medicine, University of Massachusetts Medical School, Worcester, MA, United States

Peter M. Rabinowitz Environmental and Occupational Health Sciences, University of Washington, Seattle, WA, United States

Megan Richardson Melbourne Zoo, Parkville, VIC, Australia

Danny Samandingke Lae, Papua New Guinea

Katrin Schmidt Lumholtz Lodge, Upper Barron, Atherton, Far North Queensland, QLD, Australia

Ulrich Schürer Wuppertal Zoo (retired), Wuppertal, Germany

Karin R. Schwartz Association of Zoos and Aquariums Tree Kangaroo Species Survival Plan®, Milwaukee, WI, United States

Deanna Sharpe Association of Zoos and Aquariums Tree Kangaroo Species Survival Plan® Program, Anchorage, AK, United States

Phil L. Shearman ANU College of Science, Australian National University, Canberra, ACT, Australia

Brett Smith Port Moresby Nature Park, Port Moresby, Papua New Guinea

Timmy Sowang Tree Kangaroo Conservation Program, Lae, Papua New Guinea

Jared Stabach Smithsonian National Zoo and Conservation Biology Institute, Front Royal, VA, United States

Judie Steenberg Association of Zoos and Aquariums Tree Kangaroo Species Survival Plan® Program, Maplewood, MN, United States

Emily R. Transue Washington State Health Care Authority, Olympia, WA, United States

Erika (Travis) Crook Utah's Hogle Zoo, Salt Lake City, UT, United States

Peter Valentine College of Science and Engineering, James Cook University, Townsville, QLD, Australia

Patricia Watson Redmond Fall City Animal Hospital, Redmond, WA, United States

Zachary Wells Conservation International, Crystal City, VA, United States

Clevo Wilson QUT Business School, School of Economics and Finance, Queensland University of Technology, Brisbane, QLD, Australia

Acronyms

A

AAZK American Association of Zoo Keepers
AAZPA American Association of Zoological Parks and Aquariums (now AZA)
AAZV American Association of Zoo Veterinarians
ACM Animal Care Manuals
ADT Animal Data Transfer form
AGL Above ground level
AGM Annual General Meeting
AI Artificial Insemination
ADM Anterior descending method
APDM Anterior and posterior descending method
APHIS Animal and Plant Health Inspection Service
API Application Programming Interface
ARAZPA Australasian Regional Association of Zoological Parks and Aquaria (now ZAA)
ARCS Annual Report on Conservation and Science (AZA)
ARKS Animal Records Keeping System
ASL Above sea level
ASMP Australasian Species Management Program
ASZK Australasian Society of Zoo Keeping
AWA Animal Welfare Act
AZA Association of Zoos and Aquariums (formerly AAZPA)
AZDANZ Association of Zoo Directors of Australia and New Zealand

B

BMP Birth Management Plan
BMU Bundesministerium für Umwelt, Naturschutz und nukleare Sicherheit (Federal Ministry of the Environment, Nature Conservation and Nuclear Safety – Germany)
BRD Biological Resources Division (US Geological Survey)
BRS Baiyer River Sanctuary

C

CAMC Conservation Area Management Committee (YUS)
CAMP Conservation Assessment and Management Plan
CAP Community Action Plan (for Lumholtz's tree kangaroo)
CAP Conservation Action Partnership (AZA)
CAZA Canadian Association of Zoos and Aquariums (formerly CAZPA)
CBC Complete blood count
CBSG Conservation Breeding Specialist Group (now Conservation Planning Specialist Group)
CDC Centers for Disease Control (U.S. Public Health Service)
CEPA Conservation and Environment Protection Authority (PNG)
CGIAR Consortium of International Agricultural Research Centers
CHM Canopy height model
CI Conservation International
CITES Convention on International Trade in Endangered Species of Wild Fauna and Flora
CO Conservation Officer
COMS Complementary metal oxide semiconductor
COPD Chronic Obstructive Pulmonary Disease
CPR Captive Propagation Rescue
CPM Committee for Population Management (WAZA)
CPSG Conservation Planning Specialist Group (Previously CBSG)
CRC Conservation Research Center
CSIRO Commonwealth Scientific and Industrial Research Organization (Australia)
CWS Currumbin Wildlife Sanctuary
CR Capture-Recapture
CRES Center for Reproduction of Endangered Species (ZSSD)
CSC Conservation and Sustainability Committee (WAZA)
CZA Central Zoo Authority (India)

D

DAWE Department of Agriculture, Water, and Environment (Australia)
DBH Diameter at Breast Height
DERP Display/Education/Research Population (in AZA region)
DFWP David Fleay Wildlife Park
DM Developmental milestones
DNA Deoxyribose Nucleic Acid
DS Distance Sampling

E

EAZA European Association of Zoos and Aquaria
EEP European Ex Situ Programme
EMP Ecological Monitoring Program (YUS)
ESA Ecological Society of America
ESA Endangered Species Act (USA)
EU European Union

F

FGDC Federal Geographic Data Committee
FGE Founder genome equivalent
FIC Founder importance coefficient
FWS US Fish and Wildlife Service

G

GD Gene Diversity
GEF Global Environment Facility
GIS Geographic Information System
GPS Global Positioning System
GSD Ground Sample Distance
GSMP Global Species Management Plan
GU Genome uniqueness

H

HM Harmonic mean

I

IACUC Institutional Animal Care and Use Committee
IATA International Animal Transport Association
ICP Institutional Collection Plan
IPY International Polar Year
ISB International Studbook
ISIS International Species Information System (now Species360)
ITTFFR International Trade (Fauna and Flora)
IUCN International Union for Conservation of Nature and Natural Resources
IUCN status EX – Extinct
EW – Extinct in the Wild
CR – Critically Endangered
EN – Endangered
VU – Vulnerable
NT – Near Threatened
LC – Least Concern
DD – Data Deficient
IVF In vitro fertilization
IVM In vitro maturation
IZPA Indonesian Zoological Parks Association
IZY International Zoo Yearbook

J

JAZA Japanese Association of Zoos and Aquariums
JRP Junior Ranger Program

K

KfW German Development Bank (formerly Kreditanstalt für Wiederaufbau)
KM Kernel
KV Kinship value

L

LLG Local Level Government
LNG Liquefied Natural Gas
LTK Lumholtz's tree kangaroo

M

M and M TAG Marsupial and Monotreme Taxon Advisory Group
MAC Mycobacterium Avium Complex
MCP Minimum convex polygon
METT Management Effectiveness and Tracking Tool
MHC Major histocompatibility complex
MHR Medical history report

MK	Mean kinship
MLC	Mixed lymphocyte culture
MTK	Matschie's tree kangaroo
MYA	Million Years Ago
MVP	Minimum viable population

N

NA	North American
NCBG	National Capital Botanical Gardens
NGO	Nongovernmental organization
NIH	National Institutes of Health (USA)
NHT	National Heritage Trust
NR	Necropsy Report
NSW	New South Wales
NZP	Smithsonian National Zoological Park
NZP-CRC	National Zoological Park – Conservation Research Center

P

PAAZAB	Pan-African Association of Zoos and Aquaria
PABTP	Population Analysis & Breeding and Transfer Plan (AZA)
PacLII	Pacific Islands Legal Information Institute
PAME	Protected Area Management Effectiveness
PDA	Personal Data Assistant
PDM	Posterior descending method
PHE	Population-Health-Environment
PHVA	Population and Habitat Viability Assessment
PMC	Population Management Center (AZA)
PMP	Population Management Plan (AZA)
PNG	Papua New Guinea
POMNP	Port Moresby Nature Park
PVA	Population Viability Analysis

Q

QLD	Queensland

R

RCP	Regional collection plan
RGB	Red-green-blue (imagery)

S

SAG	Scientific Advisory Group (AZA)
SAFE	Saving Animals From Extinction
SAT	Spot Assessment Technique
SCBI	Smithsonian Conservation Biology Institute (previously CRC – Conservation & Research Center (NZP))
SEAZA	Southeast Asian Zoo and Aquarium Association (previously Southeast Asian Zoo Association)
SECR	Spatially Explicit Capture Recapture
SMART	Spatial Monitoring and Reporting Tool
SPARKS	Single Population Analysis and Record Keeping System (obsolete)
SPMAG	Small Population Management Advisory Group
SSC	Species Survival Commission (IUCN)
SSP	Species Survival Plan® (AZA)
STI	Sexually transmitted infection

T

TAG	Taxon Advisory Group (AZA)
TB	Tuberculosis
TCA	Tenkile Conservation Alliance
TCR	T-cell receptor
TKCP	Tree Kangaroo Conservation Program
TKCP-PNG	Tree Kangaroo Conservation Program – Papua New Guinea
TKHM	Tree Kangaroo Husbandry Manual
TKMG	Tree-Kangaroo and Mammal Group
TKSSP	Tree Kangaroo Species Survival Plan® (AZA)
T-LoCoH	Time-Local Convex Hull
TNF	Tumor necrosis factor
TREAT	Trees for the Atherton and Evelyn Tablelands
TRI	Terrain Ruggedness Index
TRRACC	Tree Roo Rescue and Conservation Center

U

UAS	Unmanned aerial system
UK	United Kingdom
UNDP-GEF	United Nations Development Programme Global Environmental Finance
UNEP	United Nations Environmental Programme
USA	United States of America
USAID	United States Agency for International Development
USDI	United States Department of the Interior

USFWS United States Fish and Wildlife Service
USGS United States Geological Survey (USDI)

V

VHF Very high frequency

W

WAZA World Association of Zoos and Aquariums
WCMC Wildlife Conservation and Management Committee (AZA)
WCS Wildlife Conservation Society
WMA Wildlife Management Area
WPZ Woodland Park Zoo
WRS Wildlife Reserves Singapore
WTMA World Heritage Management Authority
WWF World Wide Fund for Nature (International)
WZACS World Zoo and Aquarium Conservation Strategy

Y

YFRW Yellow-footed rock wallaby
YUS Yopno-Uruwa-Som (area around three rivers)
YUS CA YUS Conservation Area
YUS CBO YUS Community Based Organization
YUS CO YUS Conservation Organization

Z

ZAA Zoo and Aquarium Association (Australasia) (formerly ARAZPA)
ZIMS Zoological Information Management System
ZOO Zoo Outreach Organisation (India)
ZPOT Zoological Park Organization of Thailand
ZV Zoos Victoria

Maps

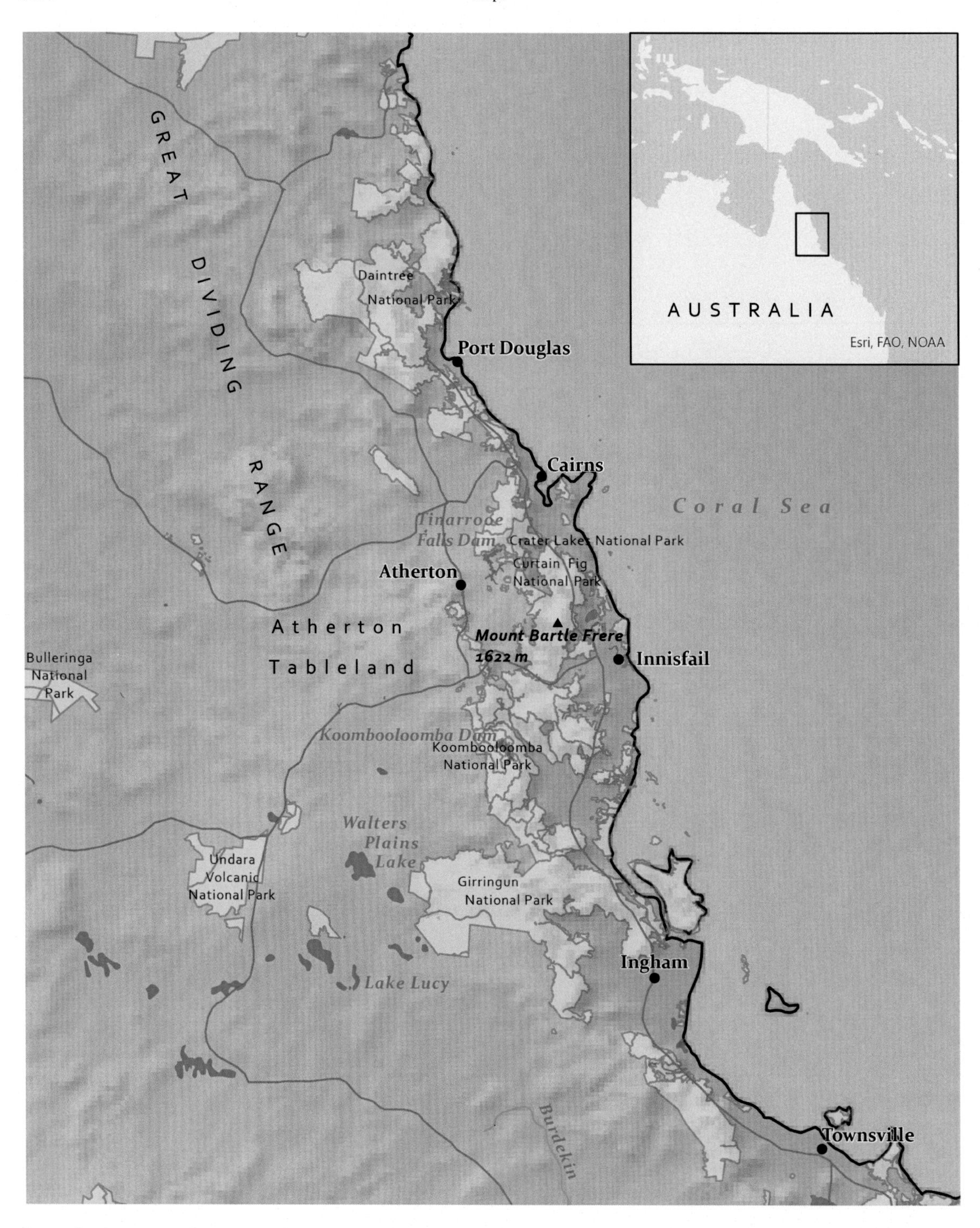

Australia *Source: Jonathan Byers.*

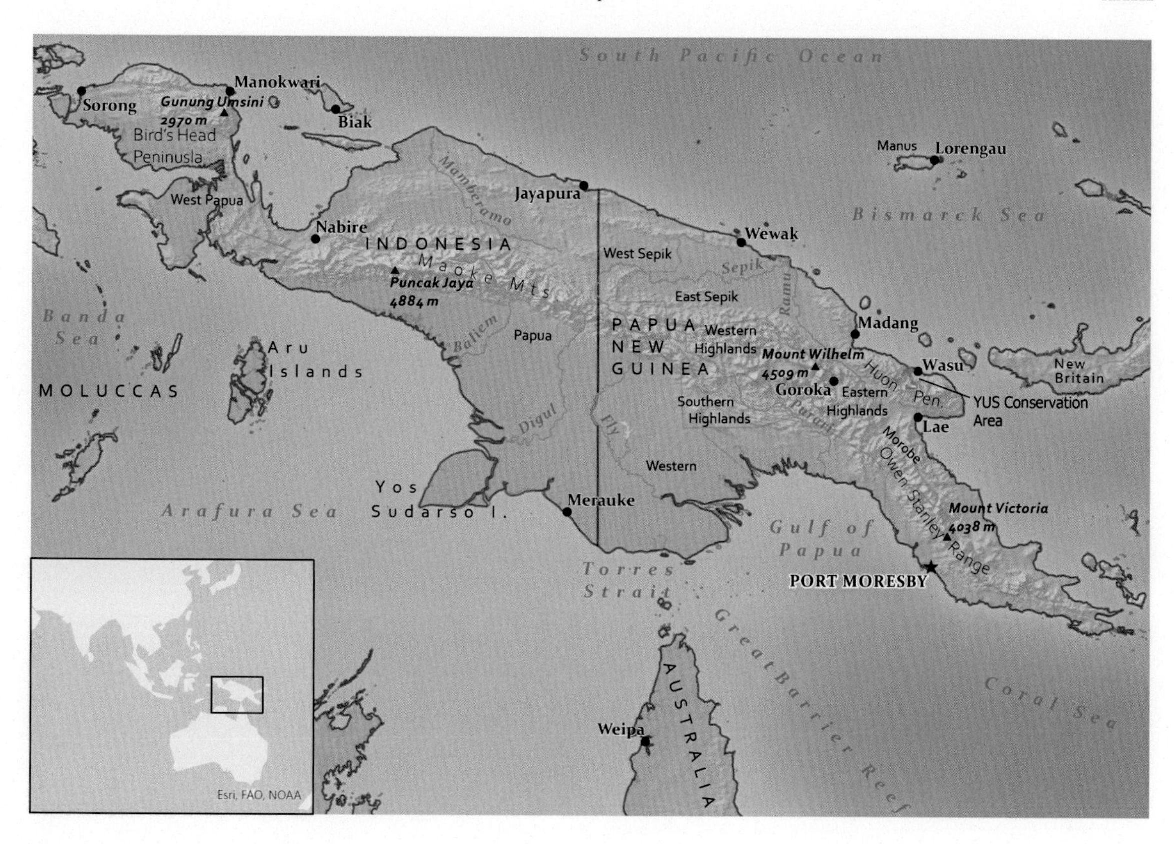

New Guinea *Source: Jonathan Byers.*

List of Other Books in the Biodiversity of the World Series

Galapagos Giant Tortoises
Edition: 1
First Published: 2020
Hardcover: 978-0-12-817554-5
eBook: 978-0-12-817555-2

Pangolins
Edition: 1
First Published: 2019
Hardcover: 978-0-12-815507-3
eBook: 978-0-12-815506-6

Whooping Cranes: Biology and Conservation
Edition: 1
First Published: 2018
Hardcover: 978-0-12-803555-9
eBook: 978-0-12-803585-6

Cheetahs: Biology and Conservation
Product Type: Book
Edition: 1
First Published: 2017
Hardcover: 978-0-12-804088-1
eBook: 978-0-12-804120-8

Snow Leopards
Edition: 1
First Published: 2016
Hardcover: 978-0-12-802213-9
eBook: 978-0-12-802496-6

Foreword: The Charisma of Tree Kangaroos

Beginning in 1964, I devoted a cumulative time of five years to studying New Guinea birds in the course of 31 expeditions to many different parts of that great island. But I never saw a tree kangaroo, until in 1981 I finally succeeded in Indonesian New Guinea's remote Foja Mountains. Those mountains are steep, forest-covered, far inland from the coast, and completely uninhabited by humans. When I was dropped near the summit by a helicopter to spend two magical weeks there, I encountered birds and animals that had never seen humans, had never been hunted, and consequently were unafraid. New Guinea's long-lost Golden-fronted Bowerbird went through its courtship display ignoring me, while I was standing only a few feet away at the opposite side of the bird's bower. I repeatedly encountered New Guinea species that elsewhere are hunted and shy, such as ground wallabies, lories, large pigeons, and New Guinea's huge Harpy Eagle. There were strange sounds, such as a hoarse rising night call like the screams of an angry cat. (It proved to be the aptly-named Feline Owlet-nightjar). Another strange sound that I didn't understand until decades later was a loud thump, as of a heavy falling object hitting the ground.

One morning, while I was descending my forest trail from my camp, I turned a corner in the trail. There, on the ground a few yards ahead of me, was my first tree kangaroo. It wasn't just ANY tree kangaroo, but an especially beautiful one, known as the Golden-mantled Tree Kangaroo. His (her?) fur was mostly rich burgundy-red and golden yellow, plus pink and white on the face, with black and yellow stripes and black and white tail rings, and about four feet long including the tail. No, I wasn't drunk or delusional; you can look up photos and paintings of that animal if you don't believe me.

When the kangaroo saw me, it jumped onto a vertical tree trunk about 10 yards ahead in the middle of my trail, grasped the trunk, and turned its head to stare at me. I stood there, holding my breath, waiting for it to run off in fear. Instead, it remained there, clutching that trunk and otherwise doing nothing, evidently as surprised at the sight of me as I was surprised at the sight of it. It stayed, and it stayed. What an unforgettable scene, and what a gorgeous animal! But eventually I bethought myself: Jared, you're here to study birds, you can't wait forever, you're not here in a contest to see whether you or the kangaroo gets bored first. I resumed walking toward the animal, and it bounded off.

That kangaroo was tame because it had never seen people and never been hunted or threatened by humans. In my subsequent 38 years in New Guinea, I've seen wild tree kangaroos at only two other locations, both nearly uninhabited or unhunted: Indonesian New Guinea's Van Rees Mountains, with only a sparse population of nomadic hunter-gatherers; and the Kikori oil fields, whose operators strictly ban hunting.

Today, tree kangaroos are New Guinea's largest native forest mammal. That makes them the most prized target of New Guinea hunters. In areas of New Guinea reachable by hunters, tree kangaroos become extirpated near settlements, occur only far from habitation, and are shy and rarely seen. But formerly, before humans arrived in New Guinea around 50,000 years ago, New Guinea supported a megafauna of large animals, as did Australia and North America and South America as well. That New Guinea megafauna included a tree kangaroo larger than its living relatives; ground kangaroos larger than any living kangaroos; and abundant large animals called diprotodonts weighing up to half a ton, as the marsupial equivalent of rhinoceroses or extinct giant American ground sloths.

The megafauna thrived in New Guinea for millions of years, surviving dozens of climate fluctuations and glacial retreats and advances during the Pleistocene. In New Guinea as in Australia, the megafauna became extinct soon after the arrival of humans around 50,000 years ago. On the face of it, that implicates humans as the cause of the megafauna's disappearance. Our courts convict human murderers (of other humans) on weaker circumstantial evidence. Surprisingly to me, though, among paleontologists the debate continues as to whether humans somehow did eliminate the megafauna in New Guinea (and in Australia, North America, and South America), or whether, instead, all those species of large animals decided to succumb to the 23rd Pleistocene climate fluctuation or just to drop dead coincidently after the arrival of those innocent vegetarian humans.

Having seen the non-reaction of that gorgeous tree kangaroo in the Foja Mountains to my appearing around the corner on my trail, I have no doubt about the answer to that debate. I could have taken all the time that I wanted to aim a spear or a bow and arrow at that animal watching me at a distance of 10 yards. The living tree kangaroos are today New Guinea's largest forest mammals, but 50,000 years ago they were the smallest members of the otherwise-vanished megafauna. As for why they survived but the rhinoceros-like diprotodonts and the giant ground kangaroos did not, the obvious explanation is that the tree kangaroos were less prime targets and more difficult prey for hunters, because they were smaller, more abundant related to their smaller size, and living mostly in trees where they were more difficult to reach and to kill.

The significance of tree kangaroos in New Guinea today far transcends their own appeal and interest. That's because conservation biologists constantly choose to talk about "charismatic megavertebrates". That phrase means large appealing vertebrates that are the species most effective at arousing public sympathy and donations. Yes, there are millions of other animal and plant species equally or more deserving of protection, because of their role in ecosystems, their biological importance and interest, and ethical reasons. But most of the public doesn't respond well to appeals for money to protect worms, bladderworts, beetles, and rats.

Instead, conservation biologists have learned to ask for donations to protect "charismatic megavertebrates": i.e., vertebrates because they are our closest relatives with which we can identify; preferably, large vertebrates close to our own size; and best of all, "charismatic ones" with a distinctive personality and appearance. That is, cute! Familiar examples of charismatic megavertebrates adopted as icons of conservation biology are the giant panda, adopted as the symbol of World Wildlife Fund, plus monkeys, zebras, and parrots. Once you've gotten people to donate money and set aside large areas of habitat to protect your chosen charismatic megavertebrate, that habitat will also protect the thousands of species of worms, bladderworts, beetles, and rats sharing that habitat with your iconic species. Conservation biologists have learned by hard experience that, when shown a picture of a worm, no matter how great the worm's value in regenerating topsoil, the public doesn't gush "How cute!" and write a big check. The public prefers pictures of pandas and parrots.

For conservation biologists working in New Guinea, tree kangaroos have no rivals as charismatic megavertebrates. They are ABC = adorable/beautiful/cute. We humans can identify with them. They can be maintained in zoos, where millions of people who will never get to observe wild tree kangaroos in New Guinea jungles will see them in enclosures and learn to love them. Tree kangaroos appeal not only to Americans and Europeans, who never heard of them before seeing one in a zoo. They also appeal to New Guineans, for whom tree kangaroos (as the largest native forest mammal) are the most rewarding hunting targets. On the Huon Peninsula, New Guineans have come to recognize that, if you want to hunt tree kangaroos, you have to begin by protecting them, otherwise you won't have any of them left to hunt. As a result, the community-based conservation area on the Huon is not only the most important area in New Guinea for tree kangaroo conservation. It's also the most important area in New Guinea for conservation of plants and animals in general, because habitats necessary to sustain tree kangaroos are also necessary to sustain worms and rats.

Besides being adorable icons for conservationists, tree kangaroos are also fascinating and important biologically. What we've already observed about their biology raises tantalizing questions of what we don't know about them. I'll conclude with two examples.

My first example starts with that strange loud thump of a heavy falling object hitting the ground that I heard in the Foja Mountains in 1981. At that time, I had no idea what caused the thump. The answer emerged only two decades later, from studies in the Huon conservation area (later gazetted as the YUS Conservation Area). It turned out that a major predator on tree kangaroos is – the New Guinea Harpy Eagle, New Guinea's largest bird of prey. Suppose that you are a tree kangaroo, perched 30 m up in the canopy of a rainforest tree, and you detect a Harpy Eagle flying straight at you and about to grab you. Think quickly: what should you do? The tree kangaroo's answer: it lets go, drops like a stone to the ground, hits the ground with a thump audible to Jared Diamond in the Foja Mountains, and bounds off into the safety of the undergrowth!

Yes, that seems like a good idea. But it's not really so simple to put into practice. Imagine yourself, my human reader, 30 m up in the canopy of a rainforest tree, watching a Harpy Eagle on close approach, and debating whether to let go and drop to the ground. For myself, being terrified of heights and not being an experienced sky jumper, I'd rather stay in the canopy and take my chances with the Harpy Eagle. To be able to count on surviving a free fall of 30 m, you need the right stuff: either the right behavior, or the right anatomy, or both.

Human paratroopers in training don't have time to devote hundreds of generations to evolving better anatomy. They're stuck with their inherited anatomy. Instead, they practice carefully how to land, roll, spread the shock, and avoid a rigid upright stance that will kill them by driving their legs up through their pelvis. Paratroopers rehearse by jumping without parachutes from heights of 5 m, which they routinely survive without injury – unlike the rest of us non-paratrooper humans.

Just as humans didn't evolve to become paratroopers but were evolutionarily predisposed to succeed at it, cats also didn't evolve to survive falls from New York skyscrapers. However, cats often do survive such falls, as an evolutionary by-product of the lifestyle of wild cats to climb trees, from which they often jump or fall. When a cat slips off a skyscraper's 100th-story balcony, starts falling, and feels the acceleration of gravity, it understandably gets worried and tenses its muscles. That's unfortunate: you'll rip yourself apart if you hit the ground with your muscles tensed. So, the falling cat's secret is that, as soon as it reaches terminal velocity and no longer experiences any further acceleration of gravity, it relaxes and hits the ground with its limbs and body flattened, thereby spreading the impact over its whole body surface.

Remember that trick if you slip from a skyscraper balcony! Of course, the trick won't do you any good if you have the misfortune to land on concrete. But the trick did help the father of a British friend of mine, who as an RAF pilot during World War Two had to jump from a damaged plane high over enemy-occupied Yugoslavia without a parachute. He landed flat on his stomach, limbs extended, in a muddy ploughed field, and survived with some broken bones. (German troops who found him there didn't want to believe his story).

We don't yet have such studies for tree kangaroos, whose jumps from trees pose many unanswered questions. Do they rotate in the air while falling, as does a cat, so as to land belly down? Do they land flat so as to spread the impact over their entire underparts, or do their limbs absorb all of the impact? Have their muscles and joints, during their millions of years of co-evolution with Harpy Eagles, evolved adaptations different from those of their non-arboreal relatives, so as to reduce the risk of injury? Will X-ray examination of captured wild tree kangaroos reveal a high frequency of broken bones? – and if so, which bones are broken most often? (The bone most often broken by cats falling from skyscrapers is the palate). Do young and old tree kangaroos jump from different heights? (I predict that young individuals will be injured less often, because they are lighter). Do tree kangaroos falling from very tall trees, especially young animals, reach terminal velocity? And – can human paratroopers learn anything of value from the answers to those questions?

My other example of unanswered questions of tree kangaroo biology crying out for study concerns their digestion of leaves that are the major component of their diet. No animal possesses its own enzymes for breaking down cellulose, which is the main component of plant cell walls. Instead, animals rely on cellulose-splitting enzymes (called cellulases) synthesized by microorganisms that live in the animal's intestine, and that split ("ferment") cellulose into sugars and other small molecules that the animal's intestine can absorb. Different herbivorous animal species have different specialized intestinal chambers to harbor their symbiotic microbes. Best known to us humans is our appendix, a small and nearly useless side-pouch of our hindgut, corresponding to the much larger, more numerous, functional caeca of geese, grouse, and many other animals. Other animals house their cellulose-splitting microbes in a chamber of their stomach or foregut, of which the rumen of cows and their relatives is well known. Rabbits increase their efficiency of utilizing cellulose by producing two types of feces: one normal type that is excreted and forgotten by the rabbit, while the other is specialized feces that the rabbit re-ingests and submits to a second round of extraction.

Where do tree kangaroos fit into this spectrum of digestive adaptation? The short answer is: we don't know. The tree kangaroo digestive tract includes a sack-like front of the stomach and a caecum side-pouch, suggesting the possibility of both foregut and hindgut fermentation. Different tree kangaroo species have different intestinal anatomies, suggesting different mixes of digestive strategies adapted to somewhat different diets. The 1996 Flannery et al. book *Tree Kangaroos* contains a section entitled "Diseases and parasites," reporting that a tree kangaroo's stomach was found to contain 132,000 worms of different sizes and shapes. We think of parasites as being almost by definition harmful, but – could those worm "parasites" in the kangaroo be beneficial? Might they serve the function of chopping up ingested leaves into small particles suitable for microbial fermentation, thereby doing for tree kangaroos what cows do for themselves by "chewing their cud," i.e. bringing up plant food from their stomach and chewing it into small fragments accessible to microbes and their enzymes? Could tree kangaroos be calling on their worms to do the hard work of synthesizing cellulose breakdown products into worm protein, which the kangaroo then digests easily? Should we think of tree kangaroos not as being herbivores dining on leaves, but instead as being carnivores dining on worms, which the kangaroo farms and grows in its stomach turned into a big intestinal compost heap?

Unanswered questions, unanswered questions! Read on in this book, to see why tree kangaroos appeal to conservationists, New Guineans, biologists, and all lovers of ABC = adorable/beautiful/cute animals!

Jared Diamond
Geography Department, UCLA, Los Angeles, CA, United States

Acknowledgments

The Editors would like to thank the many authors who contributed their knowledge and passion to this book. In addition, we would like to thank all the other colleagues who provided advice, support, information, images, stories, and knowledge about tree kangaroos. In particular, Peter Schouten generously allowed us to share his beautiful tree kangaroo illustrations in this volume. We want to make a special mention of the tree kangaroo scientists and practitioners who were unable to participate in this book. We have shared the foundational work of Tim Flannery, Roger Martin, and Mark Eldridge in several chapters and they continue to inspire us. Jim and Jean Thomas (TCA) and Karen Coombs (TRRCC) have also influenced the world of tree kangaroo conservation and science and they are mentioned in the book as well. We also thank the Publishers for recognizing the significance of tree kangaroos as a group of high conservation concern and worthy of a place in their Biodiversity of the World series.

The four Editors come from diverse disciplines and each contributed their own expertise to the knowledge and understanding of tree kangaroo science and conservation. They coordinated an equally diverse group of authors for this book who are all committed to working toward a future for tree kangaroos.

Lisa Dabek

I would like to express my deep appreciation to the other editors of this book Peter Valentine, Jacque Blessington, and Karin Schwartz. You have been amazing and I thank you so much for your dedication and passion for tree kangaroos.

I would like to thank the Administration and Board of Directors of Woodland Park Zoo and the many staff members who have worked with TKCP and supported our work, especially Trevor Holbrook. I also thank all of the staff of TKCP, past and present.

For funding and technical support, I thank Conservation International, GEF 5/United Nations Development Program, Ministry of Environment of the Government of Germany through the German Development Bank (KfW), Rainforest Trust, Shared Earth Foundation, National Geographic Society, USAID Biodiversity Project, Cardno Corporation, US Embassy of PNG, Zoos Victoria, AZA TK-SSP Institutions, AZA Marsupial and Monotreme TAG, Zoo Beauval, the Gay Jensen and Robert Plotnick Tree Kangaroo Research and Conservation Leadership Gift, and many generous individual donors. I also want to thank Tom McCarthy who mentored me in the early stages of this book process.

I want to thank the photographers who contributed their amazing images to this book including Bruce Beehler, Doug Bonham, Jonathan Byers, Ryan Hawk, and Rob Liddell.

There are many people in addition to authors who have inspired me since I first met tree kangaroos at the Woodland Park Zoo in 1987. I want to acknowledge their profound influence on my life's work with tree kangaroos: Judie Steenberg, Larry Collins, Joan Lockard, Michael Hutchins, David Towne, Sam Wasser, Margit Cianelli, Bill and Wendy Cooper, Rigel Jensen, Valerie Thompson, Jacque Blessington, Kathy Russell, Beth Carlyle-Askew, Tina Mullett, AZA TKSSP members, Megan Richardson, Wolfgang Dressen, Bruce Beehler, Russ Mittermeier, the Leahy family, Will Betz, Mambawe Manauno and family, Dono and Annie Ogate, Angie Heath, Gabriel Porolak, Daniel Okena, Nicholas Wari, Timmy Sowang, Stanley Gesang, the many YUS tree kangaroo research assistants, Andrew Krockenberger, Carol Esson, Erika Travis Crook, Trish Watson, Luis Padilla, Rick Passaro, Mikal Nolan, Zachary Wells, Ashley Brooks, Brett Smith and Michelle McGeorge, Penny LeGate and Craig Tall, Alejandro Grajal, Nik Sekhran, Karen Coombes, Terry Phillips, Kyler Abernathy, Dana Filippini, John Williar, Doug Bonham, Fred Koontz, Harriet Allen, Harmony Frazier, Carol Hosford, Betsy Dennis, Rob and Marti Liddell, Vicky Leslie, Margie Wetherald and Len Barson, Nancy Philips and Blair Brooks, Emily Transue, Joan Castro, Carolyn Marquardt, the Shannon-Dabek family, Ruth Dabek Hoffman, David Gillison and so many more friends, family, and colleagues. There are also people that have passed away who are deeply meaningful to tree kangaroo conservation and to me personally: Blair Brooks, Larry Collins, Michael Hutchins, Holly Reed, Barau Giebac, Victor Ecki and many YUS elders that I came to love and respect.

For support with writing and editing this book I thank my fiancé Bruce Ellestad for editing and providing a loving home, my beloved mom and TKCP Archivist Henny Philips, and all my family and friends. I also thank Jessie, Shelby, Jack, and Margot for their patience and canine support during all of the tree kangaroo work and writing.

I thank the National Government of PNG, CEPA, UNDP-PNG, The National Research Institute, the Morobe Provincial Government, the Madang Provincial Government, the Kabwum District Government, the Tewai Siassi District, the Wasu Local Level Government, and the YUS Local Level Government.

Finally, I would like to thank the people of YUS who are the true heroes of conservation. It has been an honor to work with you in your forests studying the tree kangaroo, Havam, Klapgaman, kapul longpela tel. You have shared with me your knowledge about tree kangaroos, forest, wildlife, culture, and so much more. May this book help support the future of tree kangaroos and the sustainable management of the YUS Living Landscape.

Peter Valentine

I am particularly grateful to the Tree Kangaroo and Mammal Group (TKMG), based on the Atherton Tablelands of far northern Queensland, the members of which have provided knowledge and inspiration about these astonishing animals. In every human society, it is crucial that caring people come together to help protect the wildlife whose habitat our own lifestyles have destroyed and who will mostly be lost without deliberate action. Many individuals have given direct support to my knowledge and awareness of tree kangaroos including past and present members of TKMG and I especially acknowledge Amy Shima, Roger Martin, Jim and Jean Thomas, John Winter, John Kanowski, Siggy Heise-Pavlov, Alan Gillanders, Martin Willis and Margit Cianelli. Margit in particular enabled me to get to know Lumholtz Tree Kangaroos in a way unimagined by me; personal and up close. I am grateful indeed to the wonderful wildlife carers who rescue individual victims of our human lifestyles and help them recover from trauma and for many, to successfully resume their lives in the wild. I also thank Martin Willis who has generously shared his photographs of tree kangaroos and to other artists including Peter Schouten and Daryl Dickson, who interpret and present the natural world in ways that bring joy to our hearts and minds. It has been a pleasure to work with the other editors and the many authors to help bring this volume to a successful conclusion. Behind that activity, I have always had support from my wife Valerie Valentine and my three children, Polly, Leonie and Kate.

In a lifetime spent with nature my principal gratitude goes to those who opened my eyes to the natural world and to those who opened my mind to science and knowledge.

Jacque Blessington

First and foremost, I would like to thank and give special recognition to Judie Steenberg. Judie has been a long-time friend and marsupial mentor (with tree kangaroo emphasis) to me after we first met at an AAZK Australasian workshop over 25 years ago. Judie was the first TKSSP Coordinator and later encouraged me to fill that same role in 2004. The TKSSP would not be what it is today without the dedication that Judie has for tree kangaroos and she is a continued wealth of knowledge as a historical advisor. Chapters 18–21 would not have been possible without Judie's proficiency in the history of tree kangaroos in captive care. It was also her hardwork and tenacity to see the project through that helped me complete this special project. Thank you, Judie, you are a very special friend. You and tree kangaroos are 'above it all'.

I would also like to send a personal thank you to Kathy Russell, Davi Ann Norsworthy and Deanna Sharpe. You have made the last several years with me on the TKSSP Management team very meaningful with all of your dedication. To all of the TKSSP Advisors and volunteers, it is your commitment to continued improvement of care that will help raise awareness and husbandry standards for this little-known species. To all of you devoted tree kangaroo keepers (you know who you are!), keep your passion flowing, it is contagious and inspiring. A special thank you to Megan Richardson, a dear friend and tree kangaroo enthusiast. Your original training has allowed people around the world to see developing joeys as they never would have seen before. To the GSMP team, thank you for your global dedication and the work that is still to come.

Lastly, I would like to thank Lisa Dabek. It was at her encouragement that I get involved with the development of this all-encompassing book on tree kangaroos. While the journey has been longer than anticipated and quite arduous at times, I am so thankful that you brought me along to help share with the world more information about such a beautiful and unique creature.

Karin R. Schwartz

I would like to thank the AZA Tree Kangaroo Species Survival Plan® (TKSSP) Steering Committee for giving me the opportunity to become involved as a data management advisor for conservation breeding management of tree kangaroos under managed care. Through the TKSSP, I became enamored with this rare and beautiful creature and learned about the scientific care and population management processes that are enacted to maintain healthy ex situ tree kangaroo populations. The TKSSP also was active in supporting the conservation of tree kangaroos in the wild and introduced me to the work of Lisa Dabek and the Tree Kangaroo Conservation Program in Papua New Guinea. Thus started a long friendship and a shared interest in connecting the valuable work in zoos with conservation of tree kangaroos in the wild.

Conservation is all about people and connections that we make to form collaborations for conservation action. Through work to determine international data standards for the development of Species360 Zoological Information Management System, and through involvement in the IUCN Species Survival Commission Conservation Planning Specialist Group (CPSG), I connected with many international colleagues that in turn, became important information sources for this book. I would like to thank my friends Chris Banks (Zoos Victoria, Australia), Peter Clark and Gert Skipper (Zoos South Australia), Biswajit Guha (Taman Safari Indonesia), Sonja Luz (Wild Reserves Singapore), Eiji Kawaguchi (previously at Zoorasia, Yokohama, Japan), and Sanjay Molur (Zoo Outreach Organization, Coimbatore, India). I thank Onnie Byers and Phil Miller and the scientists of CPSG for providing me with the philosophies important for conservation action planning and implementation.

My family turned on the support to lift me up during the writing and editing of this book. My mother Ruth Schwartz, at 97 years old, kept me entertained with her stories of renewing her connection with the piano during the pandemic lockdown and the challenges of sight reading after a hiatus of over 60 years. My daughters Lisa Newman (with fiancé Dino Vrakas) and Laura Newman (with son-in-law Rich Rhee and my two grandsons) and son David Newman provided ongoing heartwarming support of and interest in my conservation work and renewed my energy with entertaining remote virtual video chats.

I would like to thank Peter Valentine and Jacque Blessington for sharing their expertise in the editing of this book under the direction of Lisa Dabek, as well as all the authors that worked with such dedication to make this book a reality. The Woodland Park Zoo was instrumental in providing support for my participation in this project. Finally, I want to express my special gratitude to Lisa Dabek for extending her friendship and passion for tree kangaroos and inspiring me to care about the world of the tree kangaroo and the people of the Morobe Province who are the stewards of this incredible species half a world away.

SECTION I

DEFINING THE TREE KANGAROO

CHAPTER

1

What is a Tree Kangaroo? Evolutionary History, Adaptation to Life in the Trees, Taxonomy, Genetics, Biogeography, and Conservation Status

Peter Valentine[a], *Lisa Dabek*[b], *and Karin R. Schwartz*[c]

[a]College of Science and Engineering, James Cook University, Townsville, QLD, Australia
[b]Tree Kangaroo Conservation Program, Woodland Park Zoo, Seattle, WA, United States
[c]Association of Zoos and Aquariums Tree Kangaroo Species Survival Plan®, Milwaukee, WI, United States

OUTLINE

INTRODUCTION

The Macropodidae family contains the greatest diversity of all marsupial herbivores and is confined to the continent of Australia and island of New Guinea. While best known for the grassland species of kangaroos and wallabies, there are many other macropodids utilizing a wide range of habitats across the 60 or so species in the family. Despite being descended from arboreal possum-like ancestors, kangaroos subsequently evolved morphological features to enhance their terrestrial life styles and in particular their highly effective characteristic hopping locomotion. But one group returned to the trees and re-developed associated morphological

Tree Kangaroos
https://doi.org/10.1016/B978-0-12-814675-0.00010-5

elements to enable this largely arboreal life. These are the tree kangaroos; 17 forms in 14 known extant species and 3 subspecies, all in the genus *Dendrolagus* (Eldridge et al., 2018). These rarely seen animals remain poorly known at least partly because their center of diversity is in the remote mountainous rainforests of New Guinea where they feed mainly in the dense forest canopy and are particularly wary of humans, the principal predator of most tree kangaroo species. In Australia, where two endemic species occur, much more is known about one of these at least, despite extensive habitat loss from colonial clearing of rainforests. For almost every other species, sightings in the wild are extremely rare. A good illustration of this fact is the recent rediscovery by Michael Smith of a Wondiwoi tree kangaroo in New Guinea, after a period of no recorded sightings for 90 years (Pickrell, 2018). Despite the many challenges to studying tree kangaroos, there have been some valuable contributions to knowledge of this group by scientists working in New Guinea and Australia. Eight species of tree kangaroos are covered by Flannery in his monograph on the mammals of New Guinea (1995), forming an important part of the Macropodidae family. An excellent and distinctive monograph on tree kangaroos by Flannery et al. (1996) brought together scientific knowledge to that point, providing a foundation of information about early species discoveries by western scientists, some of the human-animal relationships with Indigenous peoples of New Guinea and the latest understanding of tree kangaroo taxonomy. The book also includes magnificent artwork by Peter Schouten depicting all the forms of tree kangaroos known at that time. Some of Peter Schouten's artwork is reproduced here to illustrate the species. A second more recent tree kangaroo monograph by Roger Martin (2005) provides a significant amount of additional information for many of the species. Finally, the most recent paper by Eldridge et al. (2018) provides the most up to date knowledge of the taxonomy of these elusive species based on phylogenetic research. These publications form the basis of much of the present understanding of tree kangaroos.

EVOLUTIONARY HISTORY

As the Australian continental plate drifted north following its separation from Gondwana some 50 million years ago, climatic conditions changed the vegetation from a mainly rainforest flora to a more arid-adapted vegetation community across increasing areas of the continent (Merrick et al., 2006). Eventually the once widespread rainforests became restricted to much smaller areas along the eastern coast and Tasmania, where they preserve much of the former Gondwanan biota and associated descendants. Some excellent reconstruction of the Australian mammal fauna has been possible from extensive fossil sites, particularly the Riversleigh material (Archer et al., 1991) of inland northern Queensland with its trove of mammal fossils over more than a 20 million-year period from the Miocene to the Holocene. While this site has provided an extensive representation of extinct species and ancestors of many extant species, there is no evidence of tree kangaroos.

A study of fossil remains near the central Queensland coast (Mt. Etna) however, supports the presence of several species of tree kangaroos in what was then a tropical rainforest environment until around 280,000 years ago (Hocknull et al., 2007). At least one of these now extinct species was related to the New Guinea tree kangaroo form, suggesting a secondary invasion into Australia. From about 300,000 years BP the fauna at this site changed from rainforest fauna to xeric-adapted fauna. These findings create further complexity in understanding the evolutionary biogeography of tree kangaroos. More recently it has become apparent that the capacity to climb trees may have been more widely spread in Miocene macropodoids (Den Boer et al., 2019).

Fossil material of tree kangaroos is particularly scarce in both Australia and New Guinea. One ancient extinct genus, *Bohra* was described from Pleistocene material near Wellington in New South Wales (NSW), mainly based on distinctive ankle bones (Flannery and Szalay, 1982). An older deposit further south in Victoria yielded teeth that were judged to be *Dendrolagus* (Flannery et al., 1992). Discoveries further south and west, in a Pleistocene Nullarbor Plain cave site, based on cranial material, were described as a new species of *Bohra* and the evidence was also interpreted to support a common ancestry of tree kangaroos with rock wallabies (Prideaux and Warburton, 2008). The limited fossil remains in New Guinea have added little to our understanding of the evolution of tree kangaroos (Martin, 2005). In a comprehensive osteology-based analysis, Prideaux and Warburton (2010) show morphological support for a phylogenetic alliance between rock wallabies (*Petrogale* spp.) and tree kangaroos which they then include in an expanded tribe Dendrolagini, but with a caution calling for further analysis on the possible link between tree-kangaroos and the New Guinea forest wallabies.

ADAPTATION TO LIFE IN THE TREES

Many aspects of tree kangaroo morphology illustrate how they have adapted to life in the trees. Since tree kangaroos evolved from rock wallabies it is useful to look at the key parts of the body that have allowed tree kangaroos to succeed as arboreal species. Tree kangaroos have been divided into two main groups—the more ancestral long-footed group which includes *D. bennettianus*, *D. lumholtzi*, and also *D. inustus*, and the more evolved short-footed group which includes the Goodfellow group and the Doria group as well as *D. mbaiso*. This division shows that species in the long-footed group are considered more primitive and have maintained more of the features of terrestrial kangaroos (Flannery et al., 1996; Martin, 2005).

At first glance a tree kangaroo might not look like a true kangaroo. The limbs, feet, claws, tail, ears all vary greatly from terrestrial kangaroos. The hind limbs have greatly adapted to tree climbing. The hind limbs are shorter in proportion to their body length than in terrestrial kangaroos. The ankle bones have adapted for more flexibility and lateral movement which is needed for climbing. Unlike terrestrial kangaroos, the hind limbs can move separately allowing tree kangaroos to climb as well as hop. The hind feet are shorter and broader. The claws are adapted for climbing and they have rough pads that aid them in navigating the trees. The forelimbs and hind limbs are more equal in length than in terrestrial macropodids. The forelimbs are long and more muscular for climbing. The forefeet also have long claws and rough pads that are advantageous to climbing.

Flannery et al. (1996) and Martin (2005) compare the morphology of the different species of tree kangaroos and how they represent whether each group is more terrestrial or arboreal. See Chapter 2 for further descriptions of tree kangaroo morphology.

TAXONOMY

Recent phylogenetic work (Eldridge et al., 2018) provides a much-improved appreciation of the relationships between the various taxonomic clusters of tree kangaroos and establishes a valued framework for discussion of the species. The two Australian species (*D. lumholtzi*, *D. bennettianus*) are isolated in a "strong monophyletic relationship and relatively recent early Pleistocene divergence (2.24 million years ago [MYA])". Both are long-footed in contrast to the mainly short-footed New Guinea species (but *D. inustus* is part of the long-footed group). This study also revealed five distinct New Guinea lineages comprising *D. inustus*;

D. ursinus; a Goodfellow's group (*D. goodfellowi*, *D. spadix*, and *D. matschiei*); *D. mbaiso*; and a Doria's group (containing *D. dorianus* and *D. scottae*). However, each of the three subspecies of *D. dorianus* is seen to be sufficiently divergent to support specific separation, thus giving two more species to the Doria's group, *D. notatus* and *D. stellarum*. In addition, within the Goodfellow's group the subspecies *D. g. pulcherrimus* is seen as distinct from the other identified subspecies and is accepted as likely to be a distinct species. Unfortunately, full resolution awaits a sample from the nominate Goodfellow's subspecies (*D. g. goodfellowi*) missing from this study. While acknowledging that some taxa not included in their study will inevitably lead to further adjustment, Eldridge et al. (2018) have given confidence about the 13 taxa studied. A 14th species not studied is the Wondiwoi tree-kangaroo (*D. mayri*), known only from a single specimen, but given its circumstances is very likely a distinct species (Eldridge et al., 2018). These 14 taxa are also identified in the latest conservation assessments by IUCN (see below) and therefore are the basis for the treatment in this volume.

BIOGEOGRAPHY AND CURRENT CONSERVATION STATUS OF TREE KANGAROOS

The following accounts are based mainly on the 2016 IUCN status assessment reports published in the IUCN Red List. The amount of information available for each species varies considerably and many species are very poorly known. Conservation action and protections for species in the threatened categories of the Red List (Vulnerable, Endangered, Critically Endangered) or for populations with downward population size trends may include listing within the Appendices of the Convention on International Trade in Endangered Species (CITES) (CITES, 2020), establishment of Protected Areas (PAs) in species ranges, NGO programs dedicated to tree kangaroo conservation, and local community action to preserve tree kangaroos. Each species is figured in this section using an image from the artwork of Peter Schouten.

1. **Lumholtz's Tree Kangaroo** (*Dendrolagus lumholtzi*) (Fig. 1.1).
   Conservation Status: **Near Threatened** (Woinarski and Burbidge, 2016).
   Found in the rainforests of north eastern Queensland, much of the habitat is within the Wet Tropics World Heritage Area. Significant habitat loss has occurred from clearing for agriculture, reducing prime

FIGURE 1.1 Lumholtz's Tree Kangaroo—*Dendrolagus lumholtzi. Source: PV image of artwork by Peter Schouten.*

habitat and some of the surviving population persists in relatively small pockets of rainforest along creek lines and gullies unsuitable for farmland. Threats include roadkill and domestic dogs and continuing habitat loss with some concern about the effects of climate change, especially on habitat condition through severe cyclones and thermal stress. This is probably the best studied species and has been the focus of several significant research programs and ongoing monitoring as well as strong actions for conservation and habitat restoration (Chapters 3 and 5). Conservation action includes an educational program to increase awareness through the Tree Kangaroo and Mammal Group of the Atherton Tablelands, a rescue center offering ecotourism opportunities within the range and Wildlife Carers of the Tablelands (Chapters 6 and 8), and an NGO—Tree Roo Rescue and Conservation Centre (TRRACC; Chapter 22) that takes in injured or orphaned Lumholtz's with the hope of releasing them back to the wild.

FIGURE 1.2 Bennett's Tree Kangaroo—*Dendrolagus bennettianus. Source: PV image of artwork by Peter Schouten.*

2. **Bennett's Tree Kangaroo** (*Dendrolagus bennettianus*) (Fig. 1.2).
   Conservation Status: **Near Threatened** (Winter et al., 2019).
   Confined to a limited area of rainforest in northern Queensland, mostly north of the Daintree River and Mt. Windsor Tablelands to close to Cooktown. Most of the habitat is within the Wet Tropics World Heritage Area. Upland populations appear secure. Lowland areas of forest have been disturbed by subdivision and development, but this threat has now been much eased because of the buy-back schemes in the Daintree region and better legal protection of native vegetation. There is some concern about the effects of climate change, especially on habitat condition through severe cyclones and thermal stress. Some threats are currently limited due to the cessation of traditional hunting and reduction of habitat clearing. This species has also been subject to significant research projects and study (Martin, 2005).

3. **Grizzled Tree Kangaroo** (*Dendrolagus inustus*) (Fig. 1.3).
   Conservation Status: **Vulnerable** (Leary et al., 2016a).
   This species, most likely the largest of all tree kangaroos, occurs in lowland and mid-montane tropical forests in the far western parts of the island of New Guinea (Indonesia) with another disjunct population occurring on the north coast of New Guinea (both Indonesia and Papua

FIGURE 1.3 Grizzled Tree Kangaroo—*Dendrolagus inustus. Source: PV image of artwork by Peter Schouten.*

FIGURE 1.4 Vogelkop Tree Kangaroo—*Dendrolagus ursinus. Source: PV image of artwork by Peter Schouten.*

New Guinea). The dominant threats are hunting pressures and habitat loss and degradation from agricultural expansion (including palm oil plantations). Altitudinal range is from 100 to 1500 m above sea level (asl). Listed on Appendix II of CITES, some populations may live in protected areas although the extent of their range is currently unknown. The north coast range population might be subject to positive conservation outcomes from the Tenkile Conservation Alliance (TCA) active in the region. The TCA focuses on conservation for several species including grizzled tree kangaroo, and works with local communities (Chapter 22). There remains a great deal to learn about the species.

4. **Vogelkop Tree Kangaroo [White-throated Tree Kangaroo]** (*Dendrolagus ursinus*) (Fig. 1.4).
Conservation Status: **Vulnerable** (Leary et al., 2016b).
Confined to the Vogelkop Peninsula of West Papua, Indonesian New Guinea, in four disjunct populations restricted to elevations above 1000 m asl today, despite historical records showing its range extending from sea level. A montane tropical forest species, the major threats are traditional hunting and habitat loss and degradation due to conversion of forests for agricultural development. Despite being the first tree kangaroo to receive a scientific name (in 1836), little is known about its biology, abundance or ecology. The long ears are quite different from other species. Listed on

FIGURE 1.5 Goodfellow's Tree Kangaroo—*Dendrolagus goodfellowi. Source: PV image of artwork by Peter Schouten.*

Appendix II of CITES, one population of the species is located in a protected area

5. **Goodfellow's Tree Kangaroo** (*Dendrolagus goodfellowi*) (Fig. 1.5).
   Conservation Status: **Endangered** (Leary et al., 2016c).
   This species is now only found in the central highlands of New Guinea, entirely in Papua New Guinea, a much-reduced area of occupation. The principal threats are from human hunting and from habitat loss due to agricultural development and deforestation for wood and timber. It's current distribution overlaps with high human population densities. Few studies have occurred in the wild and further information is needed on its distribution and ecology.

6. **Lowlands Tree Kangaroo** (*Dendrolagus spadix*) (Fig. 1.6).
   Conservation Status: **Vulnerable** (Leary et al., 2016d).
   One of the Goodfellow's group of species, this poorly known and rare species is found at relatively low altitude (sea level to 800 m asl), on the southern slopes and Papuan Plateau of Papua New Guinea, Eastern Highlands and Chimbu and has been associated with rugged karst landforms remote from human settlement. In recent

FIGURE 1.6 Lowlands Tree Kangaroo—*Dendrolagus spadix. Source: PV image of artwork by Peter Schouten.*

decades, much of their habitat has been subject to logging concessions (75%) and mining activity (petroleum and gas), with new roads enabling much greater access and likely population reduction for the species. Very little is known of its ecology and behavior in the wild.

7. **Huon Tree Kangaroo** [Matschie's Tree Kangaroo](*Dendrolagus matschiei*) (Fig. 1.7). Conservation Status: **Endangered** (Ziembicki and Porolak, 2016).
Confined to tropical montane and upper montane forests on the Huon Peninsula in Papua New Guinea, the species occurs between 1000 and 3300 m above sea level (asl) in naturally low population densities. The species is part of the Goodfellow's species complex. The main threat is from the growing human population with associated pressure: conversion of forest and direct hunting by local people. Other possible threats are from commercial logging and introduction of roads to the Huon Peninsula. Some concern exists about possible impacts from climate change including increased wildfire risk. This species was the focus for the establishment of the YUS Conservation Area, a community-based protected area that was established in collaboration with the Tree Kangaroo Conservation Program (TKCP, 2020; Chapters 10, 13, 22). Due to implementation of the YUS Landscape Management Plan (Chapter 11), several species are being monitored for population trends, and there are data that show that the decline of the Huon tree kangaroo may be stabilizing with a possible increase in no-take zones.

FIGURE 1.7 Matschie's Tree Kangaroo—*Dendrolagus matschiei*. Also known as Huon Tree Kangaroo. *Source: PV image of artwork by Peter Schouten.*

8. **Golden-mantled Tree Kangaroo [Weimang Tree Kangaroo]** (*Dendrolagus pulcherrimus*) (Fig. 1.8).
Conservation Status: **Critically Endangered** (Leary et al., 2016i).
This final member of the Goodfellow's group survives as a disjunct population in Indonesia (West Papua) and in a small part of the Torricelli Mountains in Papua New Guinea. This latter population is now confined to a much smaller area than it originally occupied. The Indonesian population is not protected and threats from hunting and habitat degradation continue. As a mid-elevation species (between 680 and 1700 m asl) some of the habitat is threatened by conversion to agriculture, including oil palm plantations. A major threat to the species in PNG is heavy hunting pressure and therefore the work of the Tenkile Conservation Alliance in

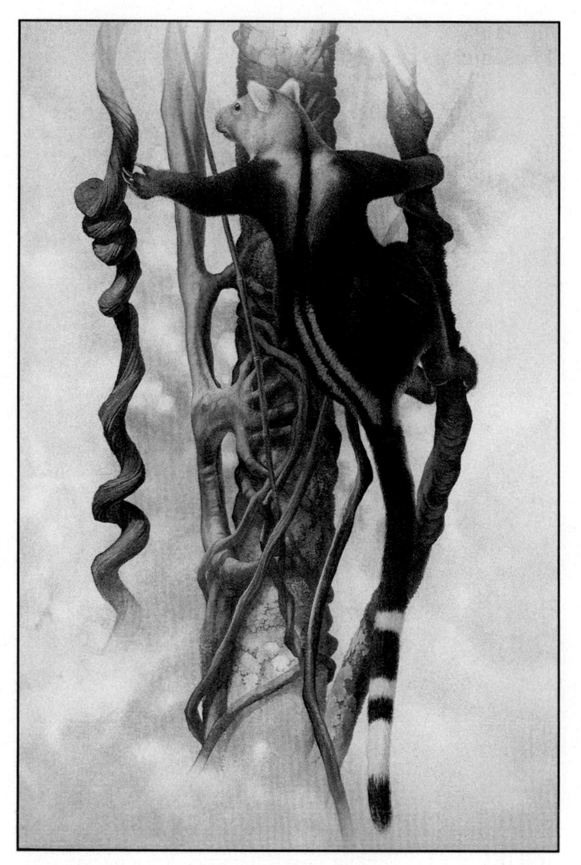

FIGURE 1.8 Golden-mantled Tree Kangaroo—*Dendrolagus pulcherrimus. Source: PV image of artwork by Peter Schouten.*

FIGURE 1.9 Dingiso Tree Kangaroo—*Dendrolagus mbaiso. Source: PV image of artwork by Peter Schouten.*

producing voluntary reductions of hunting in the Torricelli Mountains is seen as a critical part of recovery of the species (TCA, 2020; Chapter 22).

9. **Dingiso Tree Kangaroo** (*Dendrolagus mbaiso*) (Fig. 1.9).
   Conservation Status: **Endangered** (Leary et al., 2016e).
   A high elevation species, Dingiso occur in the Indonesian province of Papua in the Tembagapura and Kwiyawagi Mountains, generally between 2700 and 3500 m asl. In the last three decades, the population has dramatically declined as a result of human activities including expanded agriculture and more human hunting. Climate change also threatens the sub-alpine habitat. Traditional beliefs protect the species in the western part of its range. The species is very docile and largely terrestrial making it even more vulnerable. It remains poorly studied.

10. **Doria's Tree Kangaroo** (*Dendrolagus dorianus*) (Fig. 1.10).
    Conservation Status: **Vulnerable** (Leary et al., 2016f).
    Confined to the high elevations of the central mountains of southeastern Papua New Guinea, the population density is naturally low and is associated with mossy mid to upper primary montane tropical

FIGURE 1.10 Doria's Tree Kangaroo *Dendrolagus dorianus. Source: PV image of artwork by Peter Schouten.*

FIGURE 1.11 Tenkile Tree Kangaroo—*Dendrolagus scottae. Source: PV image of artwork by Peter Schouten.*

forest. Heavy hunting with dogs is the principal threat but increasing numbers of roads and greater access adds to the threat. Habitat loss and further degradation is likely, especially associated with the liquefied natural gas (LNG) project and other developments. The species is poorly studied in the wild.

11. **Tenkile Tree Kangaroo [Scott's Tree Kangaroo]** (*Dendrolagus scottae*) (Fig. 1.11). Conservation Status: **Critically Endangered** (Leary et al., 2019).
This is a member of the Doria's group of tree kangaroos and is very rare with an extremely restricted distribution in the Torricelli Mountain Range and a small disjunct population in the Bewani Range to the west. The cause of the dramatic decline in population over the past 30 years, leading to the critically endangered status, was heavy hunting pressure from local people. However, the community-based Tenkile Conservation Alliance (see http://www.tenkile.com), in the Torricelli Mountains, has halted the decline and produced stability over the last 5 years. There may even be a slight recovery underway as a result of voluntary restrictions on hunting by local people (Jim Thomas, pers. comm. 2019).

FIGURE 1.12 Ifola Tree Kangaroo—*Dendrolagus notatus*. *Source: PV image of artwork by Peter Schouten.*

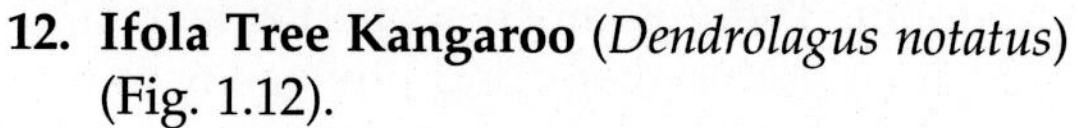

FIGURE 1.13 Seri's Tree Kangaroo—*Dendrolagus stellarum*. *Source: PV image of artwork by Peter Schouten.*

12. **Ifola Tree Kangaroo** (*Dendrolagus notatus*) (Fig. 1.12).
    Conservation Status: **Endangered** (Leary et al., 2016g).
    Another member of the Doria group, this species is found only in the central mountains of Papua New Guinea, between 900 and 3100 m asl. Generally rare it occurs with a naturally low population density. It is poorly studied in the wild and threatened by hunting for food with dogs by local people and by loss of habitat from logging and transformation for agriculture.

13. **Seri's Tree Kangaroo** (*Dendrolagus stellarum*) (Fig. 1.13).
    Conservation Status: **Vulnerable** (Leary et al., 2016h).
    This large tree kangaroo is also a member of the Doria group, confined to upper montane tropical forests of the Central Cordillera with populations extending from the Papua Province of Indonesia to the western mountains of Papua New Guinea. The species is rare and difficult to find and is subject to hunting for food in populated areas, the main threat to its survival. Additional threats might arise from climate change (increased fire, causing habitat degradation). Further studies are needed on its natural history and abundance.

FIGURE 1.14 Wondiwoi Tree Kangaroo—*Dendrolagus mayri. Source: PV image of artwork by Peter Schouten.*

14. **Wondiwoi Tree Kangaroo** (*Dendrolagus mayri*) (Fig. 1.14).
    Conservation Status: **Critically Endangered** (Leary et al., 2016i).
    Originally known only from a single specimen collected in 1928 by Ernst Mayr in the Wondiwoi Peninsula of West Papua (Indonesia) it was long thought that it might be extinct. The mossy montane forest where it was originally found is of limited extent. In 2018, the species was rediscovered in the wild for the first time in 90 years and a photograph proved its identity (Pickrell, 2018). This sighting has renewed hopes that a small population survives despite the ongoing threats, presumably from hunting. Further surveys in the remote area where it was seen are required to confirm its ongoing survival. Nothing is known of its ecology.

There are a number of potentially distinct tree kangaroo taxa that have not yet been fully described or subject to any phylogenetic analysis. It remains to be seen whether additional species may be added to the 14 covered above. The remote and rugged nature of the habitat, the ongoing threats from human hunting and development, and the low density and shy nature of these animals mitigate against field observations.

CONCLUSIONS

Despite tree kangaroos being relatively large mammals (the largest endemic mammal taxa in New Guinea), very few field studies of the 14 extant species have been completed and across the group, little is known about their ecology, behavior, social interactions, populations, and biogeography. The rarity with which most species are encountered in the wild and the fact that human hunting pressure is frequently very high means that inadequate information has been compiled about their conservation prospects, but what is known suggests their survival is in doubt. There is much scope for increased research on both ecology and phylogeny to further extend the understanding of this fascinating group of animals.

ACKNOWLEDGMENTS

The authors are particularly indebted to the pioneering work on tree kangaroos in two landmark publications. Tim Flannery is a name that will always be linked with tree kangaroos and the outstanding monograph published in 1996 with Roger Martin and Alexandra Szalay remains an excellent source of both inspiration and knowledge (Flannery et al., 1996). Sadly, it is long out of print. Roger Martin's more recent monograph (2005) has also proved critically informative along with the papers of many researchers who have brought new awareness about these remarkable animals. We also thank Peter Schouten for permission to use his amazing artwork to illustrate the species.

REFERENCES

Archer, M., Hand, S.J., Godthelp, H., 1991 (1994). Riversleigh: The Story of Ancient Rainforests in Inland Australia. Reed, Chatswood. 264 pp.

Convention on International Trade in Endangered Species (CITES), 2020. CITES Appendices. Available from: https://www.cites.org/eng/app/index.php. (29 July 2020).

Den Boer, W., Campione, N.E., Kear, B.P., 2019. Climbing adaptations, locomotory disparity and ecological convergence in ancient stem 'kangaroos'. R. Soc. Open Sci. 6, 181617. https://doi.org/10.1098/rsos.181617.

Eldridge, M.D.B., Potter, S., Helgen, K.M., Sinaga, M.H., Aplin, K.P., Flannery, T.F., Johnson, R.N., 2018. Phylogenetic analysis of the tree-kangaroos (*Dendrolagus*) reveals multiple divergent lineages within New Guinea. Mol. Phylogenet. Evol. 127, 589–599.

Flannery, T.F., 1995. Mammals of New Guinea. Reed Books, Chatswood, NSW. 568 pp.

Flannery, T.F., Szalay, F., 1982. *Bohra paulae*, a new giant fossil tree kangaroo (Marsupialia: Macropodidae) from New South Wales, Australia. Aust. Mammal. 5, 83–94.

Flannery, T.F., Rich, T.H., Turnbull, W.D., Lundelius, E.L., 1992. The Macropodoidea (Marsupialia) of the early Pliocene Hamilton local fauna, Victoria. Fieldiana Geol. 25, 1–37.

Flannery, T.F., Martin, R., Szalay, A., 1996. Tree Kangaroos: A Curious Natural History. Reed Books, Melbourne. 202 pp.

Hocknull, S.A., Zhao, J.-X., Feng, Y.-X., Webb, G.E., 2007. Responses of Quaternary rainforest vertebrates to climate change in Australia. Earth Planet. Sci. Lett. 264, 317–331.

Leary, T., Seri, L., Wright, D., Hamilton, S., Helgen, K., Singadan, R., Menzies, J., Allison, A., James, R., Dickman, C., Aplin, K., Flannery, T., Martin, R., Salas, L., 2016a. *Dendrolagus inustus*. The IUCN Red List of Threatened Species. Available from:. https://doi.org/10.2305/IUCN.UK.2016-2.RLTS.T6431A21957669.en. (24 July 2020).

Leary, T., Seri, L., Wright, D., Hamilton, S., Helgen, K., Singadan, R., Menzies, J., Allison, A., James, R., Dickman, C., Aplin, K., Salas, L., Flannery, T., Bonaccorso, F., 2016b. *Dendrolagus ursinus*. The IUCN Red List of Threatened Species. Available from: https://doi.org/10.2305/IUCN.UK.2016-2.RLTS.T6434A21956516.en. (24 July 2020).

Leary, T., Seri, L., Wright, D., Hamilton, S., Helgen, K., Singadan, R., Menzies, J., Allison, A., James, R., Dickman, C., Aplin, K., Flannery, T., Martin, R., Salas, L., 2016c. *Dendrolagus goodfellowi*. The IUCN Red List of Threatened Species. Available from: https://doi.org/10.2305/IUCN.UK.2016-2.RLTS.T6429A21957524.en. (24 July 2020).

Leary, T., Seri, L., Wright, D., Hamilton, S., Helgen, K., Singadan, R., Menzies, J., Allison, A., James, R., Dickman, C., Aplin, K., Salas, L., Flannery, T., Bonaccorso, F., 2016d. *Dendrolagus spadix*. The IUCN Red List of Threatened Species. Available from: https://doi.org/10.2305/IUCN.UK.2016-2.RLTS.T6436A21956250.en. (24 July 2020).

Leary, T., Seri, L., Wright, D., Hamilton, S., Helgen, K., Singadan, R., Menzies, J., Allison, A., James, R., Dickman, C., Aplin, K., Flannery, T., Martin, R., Salas, L., 2016e. *Dendrolagus mbaiso*. The IUCN Red List of Threatened Species. Available from: https://doi.org/10.2305/IUCN.UK.2016-2.RLTS.T6437A21956108.en. (24 July 2020).

Leary, T., Seri, L., Flannery, T., Wright, D., Hamilton, S., Helgen, K., Singadan, R., Menzies, J., Allison, A., James, R., 2016f. *Dendrolagus dorianus*. The IUCN Red List of Threatened Species. Available from: https://doi.org/10.2305/IUCN.UK.2016-2.RLTS.T6427A21957392.en. (24 July 2020).

Leary, T., Seri, L., Flannery, T., Wright, D., Hamilton, S., Helgen, K., Singadan, R., Menzies, J., Allison, A., James, R., 2016g. *Dendrolagus notatus*. The IUCN Red List of Threatened Species. Available from: https://doi.org/10.2305/IUCN.UK.2016-2.RLTS.T136732A21957010.en. (24 July 2020).

Leary, T., Seri, L., Flannery, T., Wright, D., Hamilton, S., Helgen, K., Singadan, R., Menzies, J., Allison, A., James, R., Aplin, K., Salas, L., Dickman, C., 2016h. *Dendrolagus stellarum*. The IUCN Red List of Threatened Species. Available from: https://doi.org/10.2305/IUCN.UK.2016-2.RLTS.T136812A21956889.en. (24 July 2020).

Leary, T., Seri, L., Flannery, T., Wright, D., Hamilton, S., Helgen, K., Singadan, R., Menzies, J., Allison, A., James, R., 2016i. *Dendrolagus mayri*. The IUCN Red List of Threatened Species. Available from: https://doi.org/10.2305/IUCN.UK.2016-2.RLTS.T136668A21956785.en. (24 July 2020).

Leary, T., Wright, D., Hamilton, S., Helgen, K., Singadan, R., Aplin, K., Dickman, C., Salas, L., Flannery, T., Martin, R., Seri, L., 2019. *Dendrolagus scottae*. The IUCN Red List of Threatened Species. Available from: https://doi.org/10.2305/IUCN.UK.2019-1.RLTS.T6435A21956375.en. (24 July 2020).

Martin, R., 2005. Tree-Kangaroos of Australia and New Guinea. CSIRO Publishing, Melbourne. 158 pp.

Merrick, J.R., Archer, M., Hickey, G.M., Lee, M.S.Y. (Eds.), 2006. Evolution and Biogeography of Australian Vertebrates. Ausci Publishing, Oatlands, NSW. 942 pp.

Pickrell, J., 2018. Rare Tree Kangaroo Reappears After Vanishing for 90 Years. National Geographic. September, 2018. Available from: https://www.nationalgeographic.com/animals/2018/09/rare-wondiwoi-tree-kangaroo-discovered-mammals-animals/. (25 July 2020).

Prideaux, G.J., Warburton, N.M., 2008. A new Pleistocene tree-kangaroo (Diprotodontia: Macropodidae) from the Nullarbor Plain of south-central Australia. J. Vertebr. Paleontol. 28, 463–478.

Prideaux, G.J., Warburton, N.M., 2010. An osteology-based appraisal of the phylogeny and evolution of kangaroos and wallabies (Macropodidiae: Marsulialia). Zool. J. Linnean Soc. 159, 954–987.

Tenkile Conservation Alliance (TCA), 2020. Available from: https://tenkile.com/. (24 June 2020).

Tree Kangaroo Conservation Program (TKCP), 2020. Available from: https://www.zoo.org/tkcp. (29 July 2020).

Winter, J., Burnett, S., Martin, R., 2019. *Dendrolagus bennettianus*. The IUCN Red List of Threatened Species. Available from: https://doi.org/10.2305/IUCN.UK.2019-1.RLTS.T6426A21957127.en. (24 July 2020).

Woinarski, J., Burbidge, A.A., 2016. *Dendrolagus lumholtzi*. The IUCN Red List of Threatened Species. Available from: https://doi.org/10.2305/IUCN.UK.2016-1.RLTS.T6432A21957815.en. (24 July 2020).

Ziembicki, M., Porolak, G., 2016. *Dendrolagus matschiei*. The IUCN Red List of Threatened Species. Available from: https://doi.org/10.2305/IUCN.UK.2016-2.RLTS.T6433A21956650.en. (24 July 2020).

CHAPTER

2

What is a Tree Kangaroo? Biology, Ecology, and Behavior

Lisa Dabek

Tree Kangaroo Conservation Program, Woodland Park Zoo, Seattle, WA, United States

OUTLINE

INTRODUCTION

Conservation of endangered species and the habitats in which they live relies on an understanding of many aspects of a species' biology and ecology. Information on such features as life history traits, habitat use, intra- and interspecific interactions, social behavior, feeding ecology, and reproduction is necessary to attempt to successfully restore and subsequently maintain healthy populations of endangered species.

The tropical marsupials of New Guinea and Australia, and in particular tree kangaroos and their habitat, have been the focus of increasing concern due to declining species populations. In Australia, the major threats are habitat fragmentation, roads, and dogs (Chapter 3). New Guinea is one of the last remaining large areas of pristine tropical rainforest, and the habitat is threatened by logging, mining, and land development (Chapter 4). Papua New Guinea (PNG) is one of the richest biological regions in the world, with over 6% of the world's biodiversity (Sekhran and Miller, 1996). The Macropodidae Family in New Guinea and Australia has radiated out to fill parallel niches occupied by primates, cervids, and

Tree Kangaroos
https://doi.org/10.1016/B978-0-12-814675-0.00029-4

lagomorphs in other parts of the world (Flannery, 1995; Flannery et al., 1996). There is a great deal of phyletic divergence within the macropodids and a great deal of convergence among marsupial and eutherian species in similar habitats (Chapter 1). This chapter addresses what is known currently about the biology, ecology, and behavior of tree kangaroos and serves as a foundation for the chapters on conservation of tree kangaroos.

TREE KANGAROO BIOLOGY

The genus *Dendrolagus* is the only arboreal group of kangaroos in the family Macropodidae. The genus *Dendrolagus* became secondarily arboreal from a terrestrial rock wallaby ancestor. *Dendrolagus* spp. are secondarily adapted to a lifestyle in the trees. Within the Macropodidae Family, *Dendrolagus* species have been regarded as an anomaly (Tyndale-Biscoe and Renfree, 1987). There are two main divisions of *Dendrolagus*, the more ancestral long-footed species which include the two Australian species and the more evolved short-footed species which include all but one of the New Guinea species (Eldridge et al., 2018; Flannery et al., 1996; Martin, 2005; Chapter 1). Also, based on the recent genetic work by Eldridge et al. (2018), there is now confirmation that tree kangaroos in New Guinea are divided into two main groups: Goodfellow's group (which includes *D. goodfellowi* species, *D. matschiei* and *D. spadix*) and Doria's group (which includes *D. dorianus* species and *D. scottae*). *D. mbaiso* is considered sister taxa to Doria's group but a distinct lineage although there are many ecological features that are similar. *D. bennettianus* and *D. lumholtzi* make up the separate Australian Species group.

Differences in morphology, locomotion, and feeding ecology generally have been the focus of comparisons of this genus with other macropodid species. Grand (1990) compared the body composition of *Dendrolagus matschiei* with three other macropodids and found that *D. matschiei* had less than 75% of the musculature of other macropodids, relatively more equal ratios of forelimb to hind limb bones and muscles, lower proportions of back extensor muscles, and a different gait pattern. Also, some *D. matschiei* morphological features were found to converge with another slow-moving arboreal marsupial, the ring-tailed possum, *Pseudocheirus peregrinus*. It appears that marsupials that share a common niche have a similar morphology and similar life history traits. The morphology of tree kangaroos has evolved to adapt to a life in the trees (Chapter 1). The forelimbs are longer than terrestrial kangaroos and the hind limbs are shorter. The hind limbs can move separately allowing tree kangaroos to climb as well as hop unlike terrestrial kangaroos. Warburton et al. (2012) studied the musculature of *D. lumholtzi* hind limbs and found, "…the muscles had a greater degree of internal differentiation, relatively longer fleshy bellies and very short, stout tendons of insertion. There was also a modified arrangement of muscle origins and insertions that enhance mechanical advantage." The adaptation of the hind limbs is suited for moving in trees. The fore and hind feet are adapted for tree climbing, with rough pads for gripping (Fig. 2.1) and long claws, especially on the forefeet, to facilitate climbing (Fig. 2.2). Flannery et al. (1996) describes in detail the front and hind limbs, and the front and hind feet of tree kangaroos with drawings and a comparison among the different species of tree kangaroos.

The tails of tree kangaroos are designed for balance in the trees and have lost their function as a fifth appendage as in terrestrial macropodids. The tails vary in length by group of species with the more terrestrial tree kangaroos having shorter tails (Flannery et al., 1996; Martin, 2005). Also some species have hair tufts at the end of their tails, and this has been seen in certain individual *D. matschiei* animals caught in the field (Porolak et al., 2014) (Fig. 2.3).

The markings of tree kangaroos vary from species to species. In the Goodfellow's group,

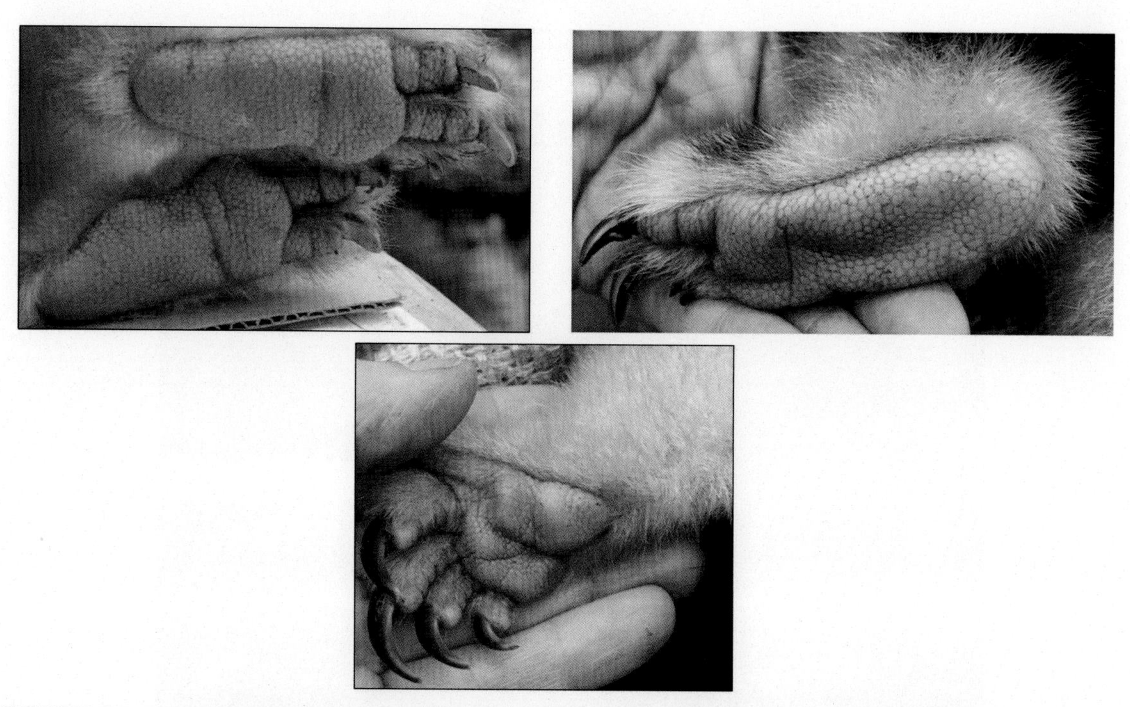

FIGURE 2.1 Rough pads on the front (five digits) and hind feet (three digits) of *Dendrolagus matschiei*. *Source: TKCP.*

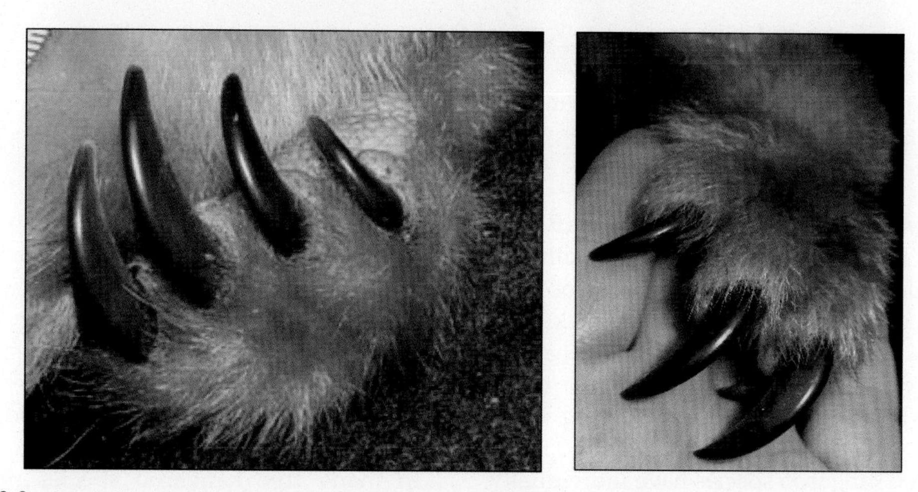

FIGURE 2.2 Long claws on the front feet of *Dendrolagus matschiei*. *Source: TKCP.*

tree kangaroos mainly have a reddish coat and some species have dark markings across the tail or down the back. In the Doria's group, the animals are brown in color and there are minimal distinct markings on the body (see Flannery et al., 1996 for a comparison of markings; Chapter 1). For *D. matschiei*, the color of the fur is the same as the reddish moss in the cloud

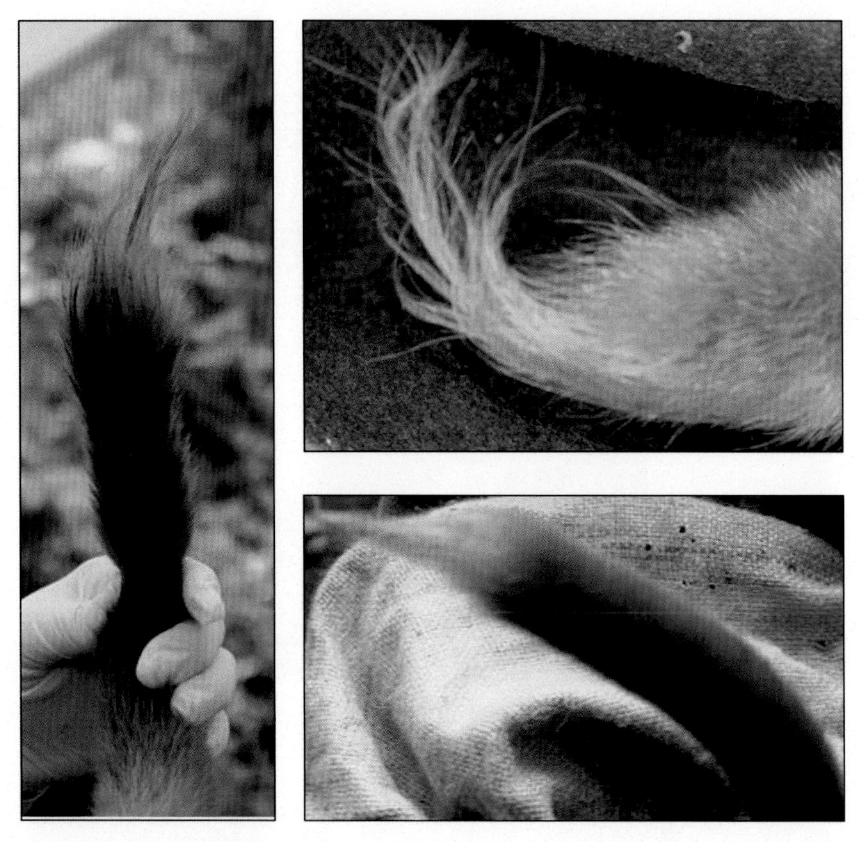

FIGURE 2.3 Tails of *Dendrolagus matschiei* with tufts of fur at the end. *Source: TKCP.*

forest trees allowing them to be camouflaged (Fig. 2.4). The fur is quite thick, especially for the high elevation species. Tree kangaroo species also have a whorl of fur on their back. This swirl allows raindrops to easily fall off the animal's back when it is curled up in the trees. This clearly is an adaption to living in a rainforest.

Some tree kangaroo species have individually distinct markings on the face. In *D. matschiei*, color patterns change from a young age to permanent markings as adults. The patterns are an excellent way to tell individual animals apart (Fig. 2.5; Chapter 19). Each individual *D. lumholtzi* has a black face mask with a unique extent and shape, allowing for individual recognition (Chapter 26). Tree kangaroo ears are smaller than other macropodid species and size varies from one species to another (Flannery et al., 1996; Martin, 2005). Many of the high elevation species like *D. matschiei* have thick-furred ears.

In many tree kangaroo species males are larger than females. There is a great deal of variation, but for *D. matschiei* the males are on average about 1 kg larger than females (Porolak et al., 2014).

FIGURE 2.4 *Dendrolagus matschiei* in a moss covered tree in YUS, Papua New Guinea. *Source: Jonathan Byers.*

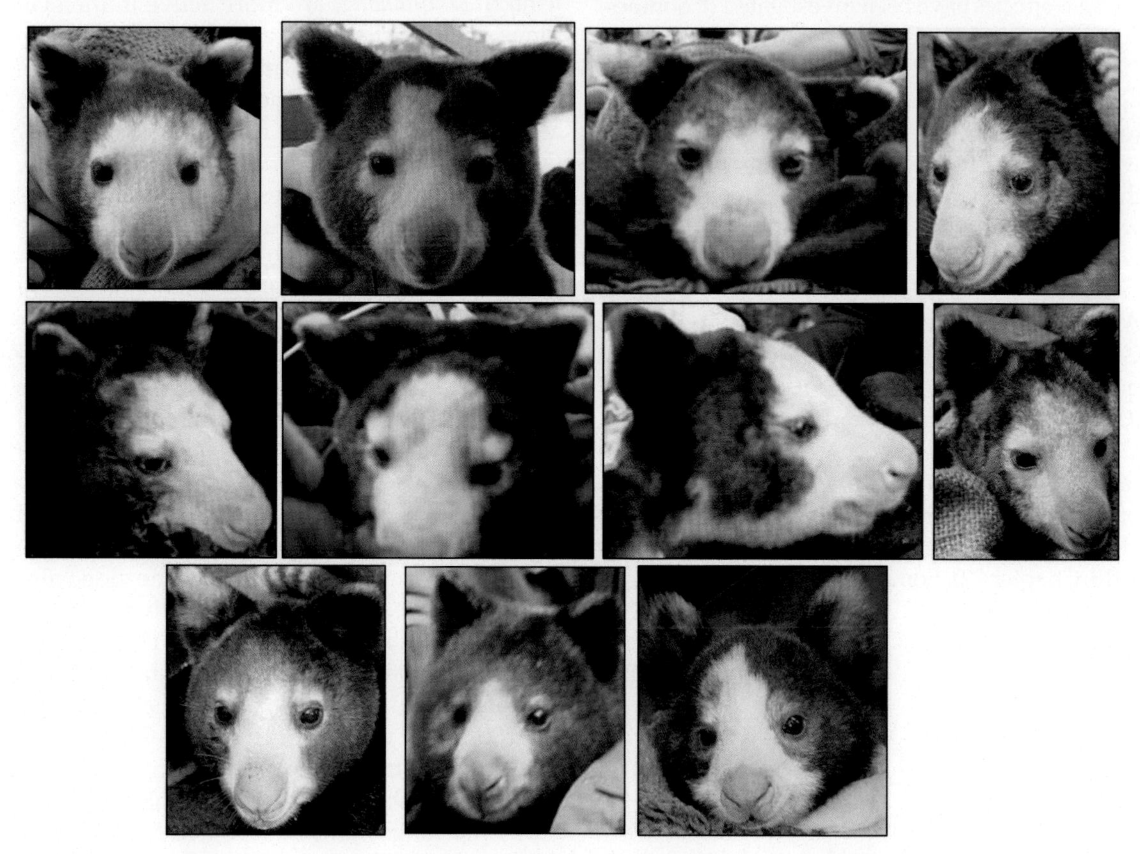

FIGURE 2.5 Individually distinct markings on the faces of wild *Dendrolagus matschiei*. Bottom row are young animals. *Source: TKCP.*

TREE KANGAROOS AND RAINFOREST ECOLOGY

There are currently 13–14 recognized species of *Dendrolagus* living in tropical rainforests ranging from northeastern Australia to the island of New Guinea (Eldridge et al., 2018; IUCN 2020; Chapter 1). Australian species live in the wet tropics in a seasonal rainforest. Most New Guinea species live in mid to high montane forests (1000–3000 m above sea level [asl]). *D. spadix*, or lowland tree kangaroo, is found at 800 m asl and below. Chapter 23 describes in detail the vegetation structure of the PNG montane forest where *D. matschiei* live.

The genus *Dendrolagus* is one of the least studied groups of macropodids. Only a few of the *Dendrolagus* species have been investigated (Coombes, 2005; Dabek, 1994; Flannery et al., 1996; Newell, 1999; Martin, 2005; Porolak et al., 2014; Procter-Gray, 1985). One reason is that most species are found in remote regions of New Guinea where wild populations have restricted distributions in the rainforest. Up until the 1980s, much of what was known about the ecology of tree kangaroos was anecdotal in nature (Husson and Rappard, 1958; Johnston and Gilles, 1918; Waite, 1894). In Australia, field studies conducted by Procter-Gray (1985), Newell (1999), and Coombes (2005) quantified ecological aspects of wild *D. lumholtzi* (Chapter 7). Ecological work has been done on *D. bennettianus* (Martin, 1992, 2005). Ecological research in PNG on *D. matschiei* is conducted as part of the Tree Kangaroo Conservation Program (TKCP) (Betz, 2001; Porolak et al., 2014; TKCP, 2018; Chapter 22). Some ecological work has been done on *D. scottae* (TCA, unpublished).

TREE KANGAROO ACTIVITY PATTERNS

Tree kangaroo species vary how much time they spend in the trees and how much time they spend on the ground. It has been documented (Groves, 1982) that certain species spend more time in the trees than others; there appears to be a correlation between the evolutionary status of the species (i.e., how primitive a *Dendrolagus* species is) and how much time they spend on the ground. Procter-Gray (1985) reported that *D. lumholtzi*, spent over 99% of their time in the trees. *D. matschiei* spend most of their time in the trees, but also go to the ground, especially to move from one tree to another. The Goodfellow's group of species (Chapter 1; Eldridge et al., 2018) is mainly arboreal and the Doria's group of species is more terrestrial.

D. matschiei are considered primarily crepuscular although some activity also occurs during the day and at night (Dabek and Byers, unpublished). *D. lumholtzi* are more active in the evenings and Procter-Gray and Ganslosser (1986) considered them "primarily nocturnal." However, individuals have been observed feeding and locomoting to some extent during the day (M. Cianelli, personal communication; Chapter 7). Other species such as *D. dorianus* are considered more diurnal (Flannery, 1995; Flannery et al., 1996). TKCP researchers are conducting research on the time budget of *D. matschiei* and where they spend their time. The studies, as of now, show that they are mainly active at dawn and dusk as well as exhibit some movement during the day and some activity at night (Chapter 23). TKCP researchers used National Geographic Critter-Cam© video cameras on tree kangaroos and documented early morning movements on the ground moving from tree to tree. The video also showed a greater amount of activity during early morning and late in the day (TKCP, 2009).

Dendrolagus spp. are mainly solitary animals (Dabek, 1994; Hutchins et al., 1991; Martin, 2005; Porolak et al., 2014; Procter-Gray, 1985). The strongest bond is between a female and her offspring. In *D. lumholtzi*, females have individual home ranges and males have home ranges that overlap with those of females and possibly other males (Chapter 7; Coombes, 2005; Procter-Gray, 1985; Newell, 1999), although sometimes small groups of animals are observed (Coombes, 2005; M. Cianelli, personal observation). In *D. bennettianus*, Martin (2005) describes separate home ranges for females and males with males overlapping several female home ranges. In *D. matschiei* females mainly have separate home ranges although there is some overlapping with home ranges of other females. Male home ranges overlap several female home ranges (Chapter 23; Porolak et al., 2014).

As rainforest prey species, tree kangaroo defense mechanisms include camouflage and being quiet. They are considered elusive and very difficult to find in the forest. Local hunters generally use hunting dogs to locate tree kangaroos, or they search for signs such as scratch marks on climbing trees or scat on the ground (Fig. 2.6). Tree kangaroos also have a distinct musky smell which humans as well as dogs can detect. Olfactory cues are important for tree kangaroo communication. They elicit only a few vocalizations such as a chuffing sound when they are upset or alerted. There are vocal communications between mother and offspring which makes sense as they are closely bonded and located near each other.

FIGURE 2.6 Scratch marks by *Dendrolagus bennettianus* on a tree trunk at Shipton's Flat, south of Cooktown, Queensland, Australia. *Source: Martin Willis.*

Only a little is known about predation of *Dendrolagus* in their habitat. The New Guinea Harpy Eagle (*Harpyopsis novaeguineae*) has been documented killing an adult *D. matschiei* (Porolak et al., 2014). In Australia, wedge-tailed eagles (*Aquila audax*) have predated on *D. lumholtzi* (M. Cianelli, personal communication). Martin (2005) has documented Amethystine or scrub python (*Simalia amethistina*) predation on *D. bennettianus* and predicts that they also predate on *D. lumholtzi*. He also discusses the potential for Boelen's python (*Simalia boelani*) in New Guinea to be a threat to tree kangaroos. In Australia, humans used to hunt tree kangaroos. Now tree kangaroos are mainly killed by dogs or cars. In New Guinea humans are the main predator, often using hunting dogs.

TREE KANGAROO FEEDING ECOLOGY

Tree kangaroos evolved from ancestral grazers to browsers. They have adapted to a diet based in the rainforest which is predominately folivorous, but also includes fruit, bark, and flowers (Dierenfeld et al., 2020; Flannery et al., 1996; Husson and Rappard, 1958; Martin, 2005; Porolak et al., 2014). The diet of *D. lumholtzi* consists of over 90% leaves including up to 33 different species (Procter-Gray, 1985). The diet of *D. matschiei* is very broad with over 90 plant species recorded in the wild that include ferns, flowers, bark, and moss, and they also ingest insects as a byproduct (Dabek and Betz, 1998; Betz, 2001; Porolak et al., 2014; TKCP, 2018). *D. bennettianus* are reported to eat creepers, ferns, and fruits in addition to leaves (Martin, 2005). There have been several reports of captive *Dendrolagus* spp. individuals killing birds (Mullett et al., 1989; Steenberg and Harke, 1984) although carnivory appears rare and they appear to be opportunists. *Dendrolagus* spp. feed while in the trees and sometimes on the ground (Porolak et al., 2014; Chapter 20).

Martin (2005) and Flannery et al. (1996) give detailed descriptions of tree kangaroo skulls and teeth and how they have adapted to being folivores. Tree kangaroos, like other taxa in the Order Diprodontia, have two protruding incisors at the front of their lower jaw (Fig. 2.7). Their teeth are specialized for grabbing and shredding leaves.

Tree kangaroos have a gastrointestinal tract that is suited for a folivorous diet. They all contain a foregut and hind gut, although the relative sizes vary among species. The long-footed group has more of a grazer-like gut and the short-footed group has more of a browser-like gut (Flannery et al., 1996; Martin, 2005). During the TKCP field research on *D. matschiei*, nematodes were documented in the foregut that appear to add in digestion (Esson and Dabek, unpublished notes; Porolak et al., 2014). There is much more to learn about the gastrointestinal physiology of tree kangaroos. As seen in other arboreal folivores, *Dendrolagus* spp. have relatively low metabolic rates (McNab, 1988). McNab also indicated that arboreality is associated with decreased levels of activity which is true in *Dendrolagus* spp. For instance, *D. lumholtzi* was found to be inactive approximately 86% of the time (Procter-Gray, 1985). McNab (1988) stated that the *D. matschiei* metabolic rate reflects a low-energy lifestyle in a thermally benign tropical rainforest environment.

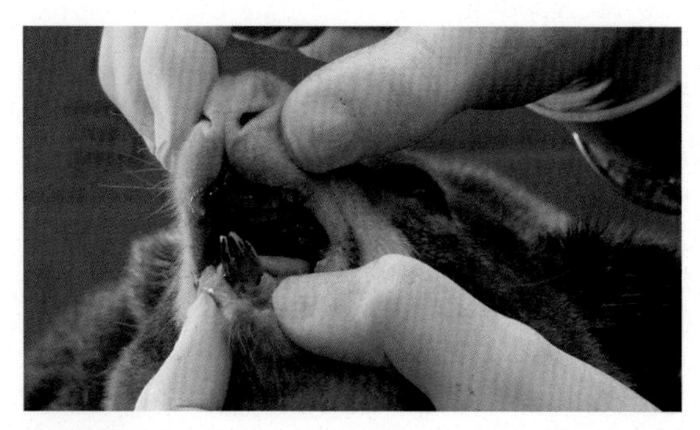

FIGURE 2.7 Teeth in *Dendrolagus matschiei* showing the two protruding incisors on the bottom jaw. *Source: TKCP.*

TREE KANGAROO REPRODUCTIVE BIOLOGY AND BEHAVIOR

All marsupials share unique features in their mode of reproduction that set them apart from eutherian mammals (Renfree, 1981). Unlike eutherian mammals that have a relatively long gestation period, marsupial young are born while still fetus-like, transport themselves into a pouch structure, and continue a majority of their development during an extensive lactation period. Maternal investment in the offspring is greatest during the lactation period (Tyndale-Biscoe, 1984). Within this general reproductive pattern, there is much variation among and within the marsupial families. Within the Macropodidae family, which includes the kangaroos, wallabies, and rat kangaroos, there is also much variation. The macropodids have many reproductive features in common; all known species are polyestrous and monovular (Tyndale-Biscoe and Renfree, 1987). In general, estrous cycles ranges from 21 to 55 days and gestation ranges from 21 to 44 days. Generally, one offspring is born at a time. In the Macropodidae Family, it has been shown that gestation length and estrous cycle length are not correlated with body size (Tyndale-Biscoe and Renfree, 1987). Therefore, comparisons can be made looking at the relationship between ecological factors and timing of reproductive events (Harvey and Read, 1988).

The habitat in which *Dendrolagus* spp. live and their folivorous diet present unique factors that appear to influence their reproductive biology and behavior (Dabek, 1991, 1994; McNab, 1988; Procter-Gray, 1985). In general, many arboreal folivores including *Dendrolagus*; *Phascolarctos* (koala); the primate genera *Lepilemur*, *Arctocebus*, and *Perodictus*; as well as many rodent species appear to live "close to the limit of their energy budget" (Charles-Dominique, 1977; Eisenberg, 1978). Due to the restrictions of their metabolic demands and arboreal lifestyles, several common features are found in these animals. They seem to be slow in their movements, have smaller home ranges, rely more on crypticity to avoid predators, and do not live in large social groups. In terms of breeding, they generally have small numbers of offspring and an overall lower reproduction potential. This last feature could include a longer developmental stage of the young, a later age at which individuals reach sexual maturity, and a longer overall estrous cycle with an extended gestation period.

An understanding of the reproductive biology of *Dendrolagus* spp. is one of the key components in the conservation of this genus. Martin (2005) stated that different *Dendrolagus* spp. show either continuous or seasonal breeding. For species living in the equatorial rainforests of New Guinea, such as *D. matschiei*, continuous breeding is appropriate because the tropical climate is maintained throughout the year. Breeding year round has been documented in wild and captive populations of *D. matschiei* (Dabek, 1991, 1994, unpublished data; Heath et al., 1990). For the two species, *D. lumholtzi* and *D. bennettianus*, that inhabit the seasonal forests of Australia, seasonal breeding has been observed. Martin (1992) found that *D. bennettianus* in Queensland, Australia tended to give birth just before the rainy season. Wildlife caregivers on the Atherton Tablelands report a higher number of offspring in June (M. Cianelli, personal communication, Chapter 6).

Due to the metabolic characterization of *Dendrolagus* spp., they have a relatively low fecundity rate in which the reproductive cycle is slowed down and extended. Most of the reproductive research has been on captive animals (Chapter 21). The gestation period is longer than any other macropodid species. *D. matschiei* has an average gestation length of 44.2 days (Heath et al., 1990).

The pouch life and development of the young is longer than most other macropodids (Dabek, 1991) (Fig. 2.8). *D. matschiei* young first emerge their heads from the pouch at about 22 weeks, the young starting to feed independently at about 26 weeks, first leave the pouch at about 28 weeks, and permanently exit the pouch at 41 weeks (Dabek, 1991; Chapter 21) (Table 2.1). Weaning takes place at about 14 months (Dabek, 1991) although young stay with their mother until at least 18 months. Most of these milestones occur later in time than for other macropodids and, therefore, set *D. matschiei* apart from many other macropodids.

The age at permanent pouch exit is significantly later for *D. matschiei* than any other macropodid. Two reasons are possible. One is that a low metabolic rate of the folivorous *D. matschiei* could cause the reproductive rate

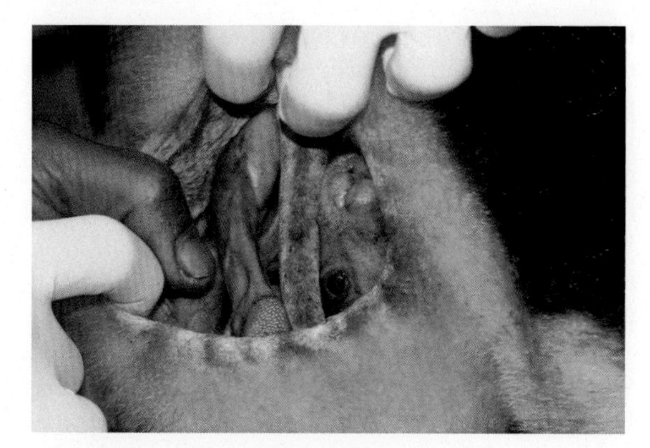

FIGURE 2.8 Joey (offspring) in the pouch of a wild female *Dendrolagus matschiei*. *Source: TKCP.*

TABLE 2.1 Mean ages of developmental milestones for *Dendrolagus matschiei* combining historical records and data from Dabek (1991).

Developmental Milestone	Mean Age (weeks)
Limb out of pouch	22.36 ± 3.26 (n = 17)
Head out of pouch	23.70 ± 3.61 (n = 28)
Fed Independently	27.6 ± 0.5 (n = 2)
1st out of pouch completely	29.04 ± 4.06 (n = 32)
Permanently out of pouch	41.33 + 1.33 (n = 7)

to be slower overall. Another reason is that the challenges of an offspring emerging out of the pouch into an arboreal and three dimensional habitat would likely require the development of more advanced locomotor skills than offspring born to terrestrial species. In general, the developmental milestones of *D. matschiei* appear to be a tight system since there is not a great deal of variation among individuals.

Females give birth to one offspring at a time which is consistent with other macropodids (Tyndale-Biscoe and Renfree, 1987). There are only two documented cases of twins, one in *D. matschiei* and one in *D. goodfellowi*, both in captivity (Lincoln Children's Zoo in Nebraska and San Diego Zoo in California, respectively) (Chapter 21). Unlike other macropodids, postpartum estrus and embryonic diapause have never been observed in *D. matschiei.*

The age at sexual maturity of *D. matschiei* is later than all other macropodid species studied thus far. The minimum age of sexual maturity for females is 2.15 years and the mean age is 3.51 years (Chapter 21). The later age at sexual maturity is similar to another arboreal folivore, the koala (*Phascolarctos cinereus*) (Martin, 1992). The reason for this difference could be based on *D. matschiei*'s ecological niche as an arboreal folivore with a lower metabolic rate (McNab, 1988). This pattern of lower fecundity is similarly found in the *P. cinereus* (Lee and Martin, 1988; Martin, 1992). It has similar timing of reproductive events to that of *D. matschiei*, including extended gestation length and development of the young, as well as longer estrous cycles.

The oldest age of female *D. matschiei* in captivity giving birth is 15 years old (although one female was documented at 17 years; Chapter 21) and the interbirth interval is about 1.39 years. Therefore, a female can have a maximum number of 8–9 offspring in her lifetime which is considerably fewer than most macropodids.

In *Dendrolagus matschiei* the patterns of steroid hormone concentrations of estrogens and progestins reveal an estrous cycle of about 55 days (Dabek, 1994; Chapter 21). *D. matschiei* has the longest estrous cycle of any known macropodid. This corroborates the idea that arboreal folivores have an overall lower reproductive potential (Charles-Dominique, 1977; Eisenberg, 1978).

In most mammals, cycling hormone levels in females are accompanied by outward indicators of estrus which include visual, olfactory/chemical, and behavioral cues. Unlike in many eutherians, estrous behaviors in macropodids are not obvious although studies on various macropodid species have documented some noticeable changes such as an initiation of pouch cleaning within a few days prior to estrus (Tyndale-Biscoe and Renfree, 1987). In the koala, females are known to increase their activity levels and decrease their food intake around the time of estrus (Lee and Martin, 1988).

In *D. matschiei*, behavioral cues of estrus are not overt although certain behaviors seem to be associated with this period. In captive studies (Dabek, 1994; Chapter 21), some females descend to the ground and approach male enclosures more often at this time. Females also are more receptive to males during estrus. If a male approaches a female, she will allow him to mount her without protest. During other times in her cycle, a female is aggressive to approaching males, swatting them with her paws and vocalizing. Males often will move away in response and avoid the female. A likely scenario for tree kangaroos is that as a female approaches estrus, an olfactory cue (pheromone) is given off which attracts the male from his broader home range that encompasses several female home ranges to the individual female's home range. During this time period the female is receptive to the male and copulation takes place. Since females are only receptive for about a 48-hour period, the cues have to be well timed.

An increase in estrogen can promote attraction in the form of sounds, odors, or visual changes (Carter, 1992). The attractiveness of female *D. matschiei* is seemingly not visual, but could be chemical, perhaps in the feces or urine. In wild *D. bennettianus* and *D. matschiei*, males approached individual females in their home ranges during the time of estrus (Martin, 1992, 2005; Porolak et al., 2014). Male and female pairs of *D. matschiei* have been observed together in trees from time to time at the TKCP field research site. Also in wild *D. bennettianus*, an increase in male activity was observed including male-male fights (Martin, 2005). It appears that *Dendrolagus* males respond to the attractiveness of the females during estrus, and the females are receptive to the males during this time period. The females observed in proximity of a male during estrus were found to be consistently receptive at that time.

All species of tree kangaroos (*Dendrolagus* spp.) are considered vulnerable or threatened in the wild. Their numbers have been reported to be adversely affected by hunting and habitat destruction in Australia and New Guinea (Flannery et al., 1996; Martin, 2005; Porolak et al., 2014). Under managed care in zoos, breeding has not sustained viable populations for most tree kangaroo species and the reproductive rates of all species are below their potential (Chapter 18).

Successful wildlife management of threatened or endangered species such as tree kangaroos depends on knowledge of an animal's reproductive biology (Lasley and Kirkpatrick, 1991). Life history information on seasonality, breeding interactions, reproductive cycles, and offspring development is necessary to closely monitor wild populations or breed endangered species in captivity.

CONCLUSIONS

This chapter has covered the complexity of a taxon that has adapted to living in the trees in some of the most remote and rugged forests in the world. These elusive animals are hard to study and yet their life histories are extraordinary and shed light on what is the largest endemic mammal of New Guinea. Studies have been done both in the wild and in zoos to unravel their mysterious lives and there is still much more to discover. Learning about the biology, ecology and behavior of tree kangaroos provides an appreciation for species that are true kangaroos and yet fill the niche of arboreal folivorous primates found elsewhere in the world. Most importantly, in order to conserve these precious species we rely on a solid understanding of their ecology and behavior. This knowledge will aid us in protecting their habitat and what they need to survive into the future.

ACKNOWLEDGMENTS

This chapter builds on the head and shoulders of the tree kangaroo scientists that have come before me. I particularly want to thank Tim Flannery and Roger Martin for their books on tree kangaroos and leading the way for learning about these extraordinary animals, as well as Judie Steenberg and Larry Collins for leading the early work on tree kangaroos in captivity. Margit Cianelli introduced me to Australian tree kangaroos and for over 20 years allowed me to interact with the tree kangaroos in her care to observe and learn about tree kangaroo biology in an intimate setting. I also want to thank all of the colleagues that have worked with me in zoos and in the field in Papua New Guinea who have helped build the knowledge of tree kangaroo biology. I thank all of the funders who have supported WPZ and TKCP's tree kangaroo research in PNG. My most sincere appreciation goes to the landowners and communities of YUS, Papua New Guinea for allowing me to study and work in the forests for which they are the stewards.

REFERENCES

Betz, W., 2001. Matschie's Tree Kangaroo (Marsupialia: Macropodidae, *Dendrolagus matschiei*) in Papua New Guinea: Estimates of Population Density and Landowner Accounts of Food Plants and Natural History (M.Phil. thesis). University of Southampton, Southampton, England.

Carter, C.S., 1992. Neuroendocrinology of sexual behavior in the female. In: Becker, J.B., Breedlove, S.M., Crews, D. (Eds.), Behavioral Endocrinology. The MIT Press, Cambridge, pp. 71–95.

Charles-Dominique, P., 1977. Ecology and Behaviour of Nocturnal Primates. Columbia University Press, New York.

Coombes, K., 2005. The Ecology and Habitat Utilisation of the Lumholtz's Tree-Kangaroo, *Dendrolagus lumholtzi* (Marsupialia: Macropodidae), on the Atherton Tablelands, Far North Queensland (Ph.D. thesis). James Cook University, Cairns.

Dabek, L., 1991. Mother-Young Relations and Development of the Young in Captive Matschie's Tree Kangaroos (*Dendrolagus matschiei*) (Master's thesis). University of Washington, Seattle, WA.

Dabek, L., 1994. The Reproductive Biology and Behavior of Captive Female Matschie's Tree Kangaroos (*Dendrolagus Matschiei*) (Ph.D. dissertation). University of Washington, Seattle, WA. Abstract in Dissertation Abstracts International 55(1 IB), 4748.

Dabek, L., Betz, W., 1998. Tree kangaroo conservation in Papua New Guinea. Endang. Spec. Update 15 (6), 114–116.

Dierenfeld, E.S., Okena, D.S., Paul, O., Dabek, L., 2020. Composition of browses consumed by Matschie's tree kangaroo (*Dendrolagus matschiei*) sampled from home ranges in Papua New Guinea. Zoo Biol., 1–5. Available from: https://doi.org/10.1002/zoo.21543.

Eisenberg, J., 1978. The evolution of arboreal herbivores in the class Mammalia. In: Montgomery, G. (Ed.), The Ecology of Arboreal Folivores. Smithsonian Institution Press, Washington, DC, pp. 135–152.

Eldridge, M.D.B., Potter, S., Helgen, K.M., Sinaga, M.H., Aplin, K.P., Flannery, T.F., Johnson, R.N., 2018. Phylogenetic analysis of the tree-kangaroos (*Dendrolagus*) reveals multiple divergent lineages within New Guinea. Mol. Phylogenet. Evol. 127, 589–599.

Flannery, T.F., 1995. Mammals of New Guinea. Robert Brown and Associates, Carina, Queensland, Australia.

Flannery, T.F., Martin, R.W., Szaley, F., 1996. Tree Kangaroos: A Curious Natural History. Reed Books, Melbourne.

Grand, T.I., 1990. Body composition and the evolution of the Macropoidae (*Potorous*, *Dendrolagus*, and *Macropus*). Anat. Embryol. 182, 85–92.

Groves, C.P., 1982. The systematics of tree kangaroos (*Dendrolagus*: Marsupialia, Macropodidae). Aust. Mammal. 5, 157–186.

Harvey, P.H., Read, A.F., 1988. How and why do mammalian life histories vary? In: Boyce, M.S. (Ed.), Evolution of Life Histories of Mammals. Yale University Press, New Haven, pp. 213–232.

Heath, A., Benner, K.S., Watson-Jones, J.B., 1990. Husbandry and management of Matschie's tree kangaroo: a case study. In: Proceedings of the AAZPA Northeastern Regional Conference, Washington D.C, pp. 518–527.

Husson, A.M., Rappard, F.W., 1958. Note on the taxonomy and habits of *Dendrolagus ursinus* Temminck and *D. leucogenys* Matschie (Mammalia: Marsupialia). Nova Guinea 9, 9–14.

Hutchins, M., Smith, G.M., Mead, D.C., Elbin, S., Steenberg, J., 1991. Social behavior of Matschie's tree kangaroos (*Dendrolagus matschiei*) and its implications for captive management. Zoo Biol. 10, 147–164.

International Union for Conservation of Nature (IUCN), 2020. IUCN Red List of Threatened Species. Available from: https://www.iucnredlist.org/. (31 July 2020).

Johnston, T.H., Gilles, C.D., 1918. Notes on records of tree kangaroos in Queensland. Aust. Zool. 1, 153–156.

Lasley, B.L., Kirkpatrick, J.F., 1991. Monitoring ovarian function in captive and free-ranging wildlife by means of urinary and fecal steroids. J. Zoo Wildl. Med. 22 (1), 23–31.

Lee, A., Martin, R., 1988. The Koala, A Natural History. New South Wales University Press, NSW.

Martin, R., 1992. An Ecological Study of Bennett's Tree-Kangaroo (*Dendrolagus bennettianus*). Project 116. World Wide Fund for Nature, Sydney, Australia.

Martin, R., 2005. Tree-Kangaroos of Australia and New Guinea. CSIRO Publishing, Melbourne.

McNab, B.K., 1988. Energy conservation in a tree kangaroo (*Dendrolagus matschiei*) and the red panda (*Ailuris fulgens*). Physiol. Zool. 61 (3), 280–292.

Mullett, T., Yoshimi, D., Steenberg, J., 1989. Tree Kangaroo Husbandry Notebook. Woodland Park Zoological Gardens, Seattle, WA.

Newell, G.R., 1999. Home range and habitat use by Lumholtz's tree-kangaroo (*Dendrolagus lumholtzi*) within a rainforest fragment in North Queensland. Wildl. Res. 26, 129–145.

Porolak, G., Dabek, L., Krockenberger, A.K., 2014. Spatial requirements of free-ranging Huon tree kangaroos, *Dendrolagus matschiei* (Macropodidae), in upper montane forest. PLoS One 9 (3), e91870. Available from: https://doi.org/10.1371/journal.pone.0091870.

Procter-Gray, E., 1985. The Behavior and Ecology of Lumholtz's Tree-Kangaroo, (*Dendrolagus lumholtzi*, Marsupialia: Macropodidae) (Dissertation). Harvard University, Cambridge, MA.

Procter-Gray, E., Ganslosser, U., 1986. The individual behaviors of Lumholtz's tree kangaroo: repertoire and taxonomic implications. J. Mammal. 67, 343–352.

Renfree, M.B., 1981. Marsupials: alternative mammals. Science 293, 100–101.

Sekhran, N., Miller, S., 1996. Papua New Guinea Department of Environment and Conservation. Papua New Guinea Country Study on Biological Diversity. Colorcraft Ltd, Hong Kong.

Steenberg, J., Harke, C., 1984. Predation on a Nicobar pigeon by a Matschie's tree kangaroo. Thylacinus 9, 14–15.

Tree Kangaroo Conservation Program (TKCP), 2009. 2009 Annual Report. TKCP, Lae, Papua New Guinea. Available from: https://www.zoo.org/document.doc?id=1723.

Tree Kangaroo Conservation Program (TKCP), 2018. 2018 Annual Report. TKCP, Lae, Papua New Guinea. Available from: https://www.zoo.org/document.doc?id=2537.

Tyndale-Biscoe, C.H., 1984. Mammal—marsupials. In: Lamming, G.E. (Ed.), Marshall's Physiology of Reproduction, fourth ed. Vol. I. Churchill Livingstone, Edinburgh (Chapter 5).

Tyndale-Biscoe, H., Renfree, M., 1987. Reproductive Physiology of Marsupials. Cambridge University Press, Cambridge.

Waite, E.R., 1894. Observations on *Dendrolagus bennettianus*. Proc. Linnean Soc. NSW 9, 571–582.

Warburton, N.M., Yakovleff, M., Malric, A., 2012. Anatomical adaptations of the hind limb musculature of tree-kangaroos for arboreal locomotion (Marsupialia: Macropodinae). Aust. J. Zool. 60, 246–258.

SECTION II

CONSERVATION CONCERNS

CHAPTER

3

Conservation of Australian Tree Kangaroos: Current Issues and Future Prospects

John Kanowski

Australian Wildlife Conservancy, Wondecla, QLD, Australia

OUTLINE

INTRODUCTION

Two species of tree kangaroo, Lumholtz's (*Dendrolagus lumholtzi*) and Bennett's (*D. bennettianus*) are endemic to Australia. Both species are restricted to the rainforests of north-east Queensland, the largest remaining area of tropical rainforest in Australia. Lumholtz's tree kangaroo inhabits rainforests between the Herbert River and the Daintree River (a north-south distance of 300 km), while Bennett's tree kangaroo inhabits rainforests north of the Daintree River to Cooktown (a north-south distance of 100 km).

Tree Kangaroos
https://doi.org/10.1016/B978-0-12-814675-0.00005-1

From an evolutionary perspective, tree kangaroos are just "hanging on" in Australia. The fossil record shows that ancestral tree kangaroos occupied ancient Australian rainforests in southern Australia (Flannery et al., 1996; Martin, 2005). As the continent became increasingly arid, rainforests contracted to wetter areas along the east coast. Tree kangaroos, in common with a number of other rainforest specialists, became restricted to the largest area of remnant rainforests, in the wet tropics (Nix and Switzer, 1991; Winter, 1997).

Tree kangaroos have a long-standing cultural significance to the Indigenous peoples who are traditional owners of the north Queensland rainforests. Their recognition in the broader community has increased in recent decades, due to the activities of the Tree-Kangaroo and Mammal Group, a local conservation organization, and other interested parties. Both Australian species of tree kangaroo are listed at Near Threatened by the Queensland government, consistent with evaluations by the IUCN (www.iucnredlist.org). This chapter presents an overview of the Australian tree kangaroo species, threats to their conservation and how those threats might be mitigated.

DISTRIBUTION AND POPULATION SIZE

Distribution

Both Australian tree kangaroos inhabit rainforest vegetation in north Queensland. The total extent of rainforest within the range of Lumholtz's tree kangaroo is approximately 5500 km^2 (Flannery et al., 1996; Newell, 1999a), however, the area occupied by the species is considerably smaller, around 1750 km^2. This includes 750 km^2 of higher elevation rainforests (>800 m a.s.l. {above sea level}), where the species is most abundant (Kanowski, 1999; Kanowski et al., 2001b), and approximately another 1000 km^2 of mid-elevation rainforest (400–800 m a.s.l.) across its range. Lumholtz's tree kangaroo is occasionally recorded in lowland rainforests as well as in eucalypt forests to the west of the Atherton Tablelands, but these records appear to be for transient (presumably dispersing) individuals (Kanowski et al., 2001a; Lumholtz, 1889). Bennett's tree kangaroo occupies an area approximately 1500–2500 km^2 in extent (Martin, 2005; Woinarski et al., 2014).

Population Size

Based on population densities calculated from systematic surveys at multiple sites (Kanowski et al., 2001b), the population size of Lumholtz's tree kangaroo in higher elevation rainforests of the Atherton Tablelands has been estimated to be 17,000 individuals (±50%) (Kanowski, 1999; Kanowski et al., 2001a). Assuming similar densities, the population in upland rainforests on the Carbine Tableland (the other main area of higher elevation forests within its range) would be in the order of 2500 individuals; another 5000 individuals may occur in mid-elevation rainforests across the range of the species. On this basis, the total population size of Lumholtz's tree kangaroo is estimated at around 25,000, noting the low precision of the estimate. The population of Bennett's tree kangaroo (Fig. 3.1) is estimated at

FIGURE 3.1 An adult Bennett's tree kangaroo (*Dendrolagus bennettianus*) photographed in the wild near Shipton's Flat, northern Queensland. *Source: Martin Willis.*

approximately 15,000, although the evidence base for this estimate is poor, as there are no systematic surveys of population density across its range (Woinarski et al., 2014).

Ecological Determinants of Distribution

In the rainforests of the Atherton Tablelands, Lumholtz's tree kangaroo (Fig. 3.2) is generally much more abundant in high elevation forests (>800 m a.s.l.) than at lower elevations (Kanowski et al., 2001b). Within a particular elevation, it tends to be more abundant in rainforests developed on fertile basalt soils than on poorer soils. The highest density populations are in remnant and regrowth rainforests developed on recent basalts in the drier, western edge of the Atherton Tableland (so-called "mabi" forests), where densities may exceed 1.5 individual per ha (Newell, 1999b). The relative abundance of tree kangaroos in forests on fertile soils is plausibly explained as a response to foliar chemistry, in particular the availability of nitrogen in foliage; the same patterns are observed for leaf-eating rainforest possums on the Atherton Tablelands (Kanowski, 2004), for leaf-eating possums elsewhere in Australia (Braithwaite et al., 1984), and for arboreal folivores worldwide (Oates et al., 1990; Ganzhorn, 1992). The relative abundance of Lumholtz's tree kangaroo in regrowth forests on the Atherton Tablelands (Coombes, 2005) is also explained by foliar chemistry, as regrowth forests are dominated by fast-growing pioneer trees and vines, which support relatively nutritious foliage (Kanowski, 1999).

FIGURE 3.2 An adult Lumholtz's tree kangaroo (*Dendrolagus lumholtzi*) feeding in the rainforest on the Atherton Tablelands, Queensland. *Source: Martin Willis.*

Variation in the abundance of Lumholtz's tree kangaroo (and leaf-eating possums) with elevation on the Atherton Tablelands does not appear to be a response to foliar chemistry (Kanowski, 2004) or to other hypothesized environmental drivers, such as floristic composition (Kanowski et al., 2003a). Instead, this pattern is most plausibly explained as a response to climate, particularly temperature (Kanowski, 1999; Krockenberger et al., 2012).

There are no data on how the population density of Bennett's tree kangaroo varies with elevation, geology or with other ecological drivers. Martin (2005) cites a number of early European collectors, all of whom reported that Bennett's tree kangaroo was primarily an inhabitant of higher elevation forests. Martin attributes this distribution to hunting pressure in lowland rainforests. At present, Bennett's tree kangaroo is relatively common in lowland gallery forests and in rainforest developed on basalt, such as the Shipton's Flat site studied by Martin (1996, 2005). These forest types support relatively high quality foliage (Martin, 2005). The Shipton's Flat site supports many of the same tree species as the "mabi" forests of the Atherton Tablelands (Tracey, 1982), where Lumholtz's tree kangaroo is particularly abundant (Newell, 1999a,b).

HISTORICAL THREATS

Hunting

Tree kangaroos are a prized target for traditional hunters (Flannery et al., 1996; Martin, 2005). Lumholtz (1889), Cairn and Grant (1890), and Waite (1894) describe Aboriginal people tracking tree kangaroos using dogs. Lumholtz (1889) remarked that Aborigines may have eliminated populations of Lumholtz's tree kangaroo from locations where hunting was relatively easy. Martin (1996) proposed that Aboriginal hunting eliminated Bennett's tree kangaroo from lowland rainforest (or at least the more accessible forests). Martin (2005) cited credible evidence from the Roberts brothers, who reside at Shipton's Flat, that the species has expanded its distribution in lowland rainforest following the cessation of traditional hunting. In contrast, there is no evidence that Lumholtz's tree kangaroo has expanded into lowland forests. Hunting has not been a significant threat to the Australian species of tree kangaroos for many decades.

Clearing for Agriculture

Only a small part of the habitat of Bennett's tree kangaroo has been cleared for agriculture. In contrast, a large area of high quality habitat for Lumholtz's tree kangaroo has been cleared for agriculture since European settlement in the late 19th century. Kanowski et al. (2003b) estimated that two-thirds of rainforest on basalt soil on the Atherton Tablelands (preferred habitat of Lumholtz's tree kangaroo) had been cleared, while nearly all (96%) of "mabi" rainforests—forests on recent basalt soils known to support the highest density populations of Lumholtz's tree kangaroo—had been cleared. Nevertheless, tree kangaroos persist in many of the remaining rainforest patches and areas of regrowth and replanted rainforest on the Atherton Tableland, and are capable of dispersing between these patches. However, in agricultural areas, tree kangaroos are vulnerable to road-kill and dog attacks (Newell, 1999a; Schmidt et al., 2000).

Clearing of rainforest on the Atherton Tablelands has largely ceased. Much of the remaining areas of rainforest that provides habitat of Lumholtz's tree kangaroo are protected, either as part of the Wet Tropics World Heritage Area, or by conservation covenants. In addition, efforts by the community group Trees for the Atherton and Evelyn Tablelands (TREAT) have led to the re-establishment of rainforest cover on some areas of degraded land, while regrowth rainforest has re-established on some areas of marginal farmland (Catterall et al., 2004). Changes in fire regimes following the dispossession of Aboriginal people have resulted in considerable expansion of rainforest into eucalypt forests on the margins of rainforest across the region (Stanton et al., 2014). For all these reasons, the extent of habitat for Lumholtz's tree kangaroo has likely expanded in recent decades, although it would still be well below pre-European levels.

Logging

The rainforests of north-east Queensland supported a major timber industry from the late 19th century until 1988, when declaration of the Wet Tropics World Heritage Area resulted in closure of the rainforest logging industry on public land. Logging was conducted under a selective silvicultural regime, rather than by clear-falling; consequently, it did not result in the wholesale loss of habitat for rainforest fauna. Indeed, logging probably improved habitat quality for tree kangaroos, as canopy disturbance promotes the growth of rainforest pioneers (trees and vines), which tend to have relatively nutritious foliage, and many of which are favored foods of tree kangaroos. The social conditions prevailing in Australia meant that the network of tracks established by the logging industry were not exploited by hunters.

CURRENT THREATS

As noted, the habitat of Australian tree kangaroos is well protected in public and private nature reserves, and hunting is negligible. Current threats include vehicle strikes and attacks by dogs, as well as habitat degradation in remnant forests.

Roadkill

In a survey of the community eliciting information on Lumholtz's tree kangaroo (Schmidt et al., 2000), 250 instances of roadkills were reported, the majority from the preceding decade (1990–99); another 31 tree kangaroo deaths were reported from dog attacks. Further records collected by wildlife carers M. Cianelli and K. Coombes (Cianelli and Coombes, personal communication) document an additional 149 roadkills and 25 records of dog-killed animals in the period 2002–08 (Chapter 6). The numbers of dog kills recorded are presumably underestimated, given that many dog kills are likely to go undetected or unreported. Many of the road-killed animals are young males (Newell, 1999a; Schmidt et al., 2000), presumably dispersing individuals attempting to establish a territory; for this reason, losses to roadkills were considered by Newell (1999a) to have little population-level impact. However, in more recently recorded roadkills, 43% of animals whose sex was recorded ($n = 83$) were females. The loss of females is more consequential for population dynamics. Hence, sub-populations vulnerable to roadkill—mainly on the Atherton Tablelands—may be more materially threatened by this factor than previously thought. Nonetheless, reported kills are <1% of the estimated population of Lumholtz's tree kangaroo, and do not pose a serious threat to the overall persistence of the species.

Dogs

Anecdotally, numbers of feral dogs have increased in remnant forests on the Atherton Tablelands, including a number of locations considered prime habitat for Lumholtz's tree kangaroo (M. Cianelli, personal communication; K. Coombes, personal communication; N. Preece, personal communication). Tree kangaroos are vulnerable to dogs when traversing open country, however the relative importance of this threat to tree kangaroos at the population level is unknown.

Habitat Degradation

As noted, extensive areas of habitat for Lumholtz's tree kangaroo on the Atherton Tablelands have been cleared, leaving small remnant patches that are vulnerable to weed invasion. In "mabi" forest, several weed species threaten the integrity of the ecosystem (Latch, 2008) including the canopy-invading introduced vines *Turbina corymbosa*, *Anredera cordifolia*, *Aristolochia elegans* and *Macfadyena unguis-cati*. Edges of these remnant forests are vulnerable to invasion by exotic grasses such as *Panicum maximum* which increase the risk of fire. Weed invasion increases in rainforest remnants following cyclone disturbance (Catterall et al., 2008).

FUTURE THREATS

Climate Change

The primary threat to Lumholtz's tree kangaroo in the long-term is climate change. Given the restricted altitudinal distribution of the species, and the limited extent of higher altitude rainforest in its range, increased temperatures are predicted to reduce the extent of suitable habitat and hence overall population size. Presuming that temperature is the primary driver of distribution, it has been estimated that an increase in temperature of 2 °C would reduce population sizes of Lumholtz's tree kangaroo by 75% (Kanowski, 1999). The impacts of climate change on Bennett's tree kangaroo are unknown, given we currently lack robust information on trends in its abundance with environmental drivers.

CONCLUSIONS

In the global context, the two species of Australian tree kangaroos are relatively secure. Most of their remaining habitat is protected, and anthropogenic impacts are relatively minor at the level of the entire populations. There is increasing community recognition of these unusual and distinctive animals, with positive consequences for conservation initiatives.

That said, knowledge of the ecology and conservation requirements of tree kangaroos lags well behind that iconic Australian arboreal folivore, the koala (*Phascolarctos cinereus*). Below is a summary of a couple of key areas where better knowledge of tree kangaroos would contribute to their conservation.

Ecological Determinants of Distribution

As noted, there has been no systematic study of the response of Bennett's tree kangaroo to environmental drivers. Previous studies (e.g., Martin, 1996) have been conducted on the margins of the species range, for logistical reasons, as the core part of its habitat is mountainous and difficult to access. However, without systematic study, we have no way of accurately evaluating the conservation status of Bennett's tree kangaroo or understanding its response to current and projected anthropogenic impacts.

In contrast, a number of studies have been conducted of Lumholtz's tree kangaroo, a consequence of its occurrence on the Atherton Tablelands, a closely settled area. The ecological determinants of its distribution have been assessed through systematic surveys and correlative studies (Kanowski, 1999; Kanowski et al., 2001a,b, 2003a), and we have a good understanding of its response to habitat clearing (Newell, 1999a,b) and fragmentation (Pahl et al., 1988; Laurance, 1990). However, the cited studies are now two to three decades old. Given changes in land-use in the region, particularly the contraction of the dairy industry, and anecdotal reports that climate change may already be affecting arboreal mammals in the wet tropics (S. Williams personal communication), it may be timely to repeat key studies.

Fortunately, new technologies may improve our capacity to survey tree kangaroos. Previous studies were conducted by hand-held spotlight and may underestimate abundance due to difficulties in locating tree kangaroos in the dense rainforest canopy. In recent years, ecologists have begun to use thermal cameras—handheld, or mounted on vehicles or drones—to survey a range of species including arboreal mammals. If these methods can be applied in rainforest, they have the potential to obtain relatively accurate estimates of population density of tree kangaroos and might for the first time provide an understanding of environmental factors driving the distribution of Bennett's tree kangaroo.

Response to Climate Change

There are currently serious concerns that increasing temperatures will threaten many species inhabiting the rainforests of north Queensland, particularly those species restricted to higher elevation forests (Shoo et al., 2011). These concerns are largely based on the correlation of species distributions with climate variables. However, a robust understanding of the consequences of climate change for rainforest species would require study of their thermoregulatory responses to elevated temperatures. Amongst fauna restricted to the wet tropics rainforests of north Queensland, such a study has only been conducted to date for the green ringtail possum (*Pseudochirops archeri*). Consistent with predictions from climate modeling, this species was shown to be intolerant of a sustained period of hot weather, and hence at risk from global warming (Krockenberger et al., 2012). However, it is possible that climate is confounded with other environmental determinants, such as floristic composition. It would be informative to extend thermoregulatory studies to the Australian tree kangaroos. Ideally, both species would be assessed, given that Lumholtz's tree kangaroo is predominantly restricted to upland forests, whereas Bennett's tree kangaroo extends to lower elevations (Martin, 1996).

Conserving Tree Kangaroos in the Anthropocene

If, as predicted, the endemic rainforest fauna of the wet tropics of north Queensland are vulnerable to climate change (Williams et al., 2003), then long-term conservation of that fauna may require the translocation of species to cooler, subtropical refuges (Krockenberger et al., 2004). On the face of it, translocation of tree kangaroos to subtropical rainforests is likely to be successful, given (i) subtropical rainforests are thought to have supported a diverse assemblage of arboreal marsupials including tree kangaroos in their evolutionary history (Flannery, 1994; Winter, 1997); and (ii), subtropical rainforests maintain strong floristic similarities with the upland rainforests of north Queensland (Webb, 1968).

However, translocations of species outside their current ranges can be contentious, because of potential impacts of translocated species on resident biota. At present, the folivorous arboreal mammal fauna of Australian subtropical rainforests comprises just two species: the common ringtail possum (*Pseudocheirus peregrinus*) and mountain brushtail possum (*Trichosurus caninus*), both relative generalists. Given the evolutionary loss of specialist folivores from subtropical rainforests, it may be predicted those rainforests will have a number of "empty niches" that were once occupied by rainforests specialists such as tree kangaroos. If so, there may be limited competition between resident and translocated species. Alternatively, with the passage of time, the possums inhabiting subtropical rainforests may have expanded their use of the available resource to take advantage of niches long-vacated by the rainforest specialists. In this case, there may be strong competition between translocated tropical folivores and resident subtropical species.

For these reasons, it would be useful to conduct research into the potential overlap in resource use between tree kangaroos and subtropical rainforest possums, ahead of any proposal to translocate tree kangaroos to the subtropics to save them from climate change. A comparison of host-plant selection (at the generic or family level) by tree kangaroos and subtropical possums, as well as a study of host-plant selection in relation to foliar chemistry (in particular, nitrogen availability) should provide relevant information. It should be noted that such a study would require additional surveys of the diets of tree kangaroos across their ranges in north Queensland, as current information (Proctor-Gray, 1984; Martin, 1996)—obtained from a couple of sites—is not comprehensive.

ACKNOWLEDGMENTS

I have a debt of gratitude to John Winter, *eminence grise* of north Queensland mammalogy, for introducing me to the rainforest arboreal mammals of the wet tropics, guiding my research on these species, and (along with Margit Cianelli and Graeme Newell) for involving me in the early activities of the Tree-kangaroo and Mammal Group. Thanks to Peter Valentine and Lisa Dabek for the invitation to contribute to this book.

REFERENCES

Braithwaite, L.W., Turner, J., Kelly, J., 1984. Studies on the arboreal marsupial fauna of eucalypt forests being harvested for wood pulp at Eden, N.S.W. III. Relationship between faunal densities, eucalypt occurrence and foliage nutrients, and soil parent materials. Aust. Wildlife Res. 11, 41–48.

Cairn, E.J., Grant, R., 1890. Report of a collecting trip to North-Eastern Queensland during April to September, 1889. Rec. Aust. Mus. 1, 27–31.

Catterall, C.P., Kanowski, J., Wardell-Johnson, G.W., Proctor, H., Reis, T., Harrison, D., Tucker, N.I.J., 2004. Quantifying the biodiversity values of reforestation: perspectives, design issues and outcomes in Australian rainforest landscapes. In: Lunney, D. (Ed.), Conservation of Australia's Forest Fauna. In: vol 2. Royal Zoological Society of New South Wales, Sydney, pp. 359–393.

Catterall, C.P., McKenna, S., Kanowski, J., Piper, S.D., 2008. Do cyclones and forest fragmentation have synergistic effects? A before-after study of rainforest structure at multiple sites. Aust. Ecol. 33, 471–484.

Coombes, K., 2005. The Ecology and Habitat Utilisation of Lumholtz's Tree-kangaroos, *Dendrolagus lumholtzi* (Marsupialia: Macropodidae) on the Atherton Tablelands, Far North Queensland (Ph.D. thesis). James Cook University, Cairns.

Flannery, T.F., 1994. Possums of the World: A Monograph of the Phalangeroidea. Geo Productions/Australian Museum, Sydney.

Flannery, T.F., Martin, R., Szalay, A., 1996. Tree Kangaroos: A Curious Natural History. Reed Books, Melbourne.

Ganzhorn, J.U., 1992. Leaf chemistry and the biomass of folivorous primates in tropical forests: test of a hypothesis. Oecologia 91, 540–547.

Kanowski, J.J., 1999. Ecological Determinants of the Distribution and Abundance of the Folivorous Marsupials Endemic to the Rainforest of the Atherton Uplands, North Queensland (Ph.D. thesis). James Cook University, Townsville.

Kanowski, J., 2004. Ecological determinants of the distribution and abundance of folivorous possums inhabiting rainforests of the Atherton tablelands, north-East Queensland. In: Goldingay, R., Jackson, S. (Eds.), The Biology of Australian Possums and Gliders. Surrey Beatty and Sons, Chipping Norton, NSW, pp. 539–548.

Kanowski, J., Felderhof, L., Newell, G., Parker, T., Schmidt, C., Stirn, B., Wilson, R., Winter, J.W., 2001a. Community survey of the distribution of Lumholtz's tree-kangaroo on the Atherton Tablelands, North-East Queensland. Pac. Conserv. Biol. 7, 79–86.

Kanowski, J.J., Hopkins, M.S., Marsh, H., Winter, J.W., 2001b. Ecological correlates of folivore abundance in North Queensland rainforests. Wildl. Res. 28, 1–8.

Kanowski, J., Irvine, A.K., Winter, J.W., 2003a. The relationship between the floristic composition of rain forests and the abundance of folivorous marsupials in north-East Queensland. J. Anim. Ecol. 72, 627–632.

Kanowski, J., Winter, J.W., Simmons, T., Tucker, N.I.J., 2003b. Conservation strategy for Lumholtz's tree-kangaroo on the Atherton Tablelands. Ecol. Restor. Manag. 4, 220–221.

Krockenberger, A.K., Kitching, R.L., Turton, S.M., 2004. Environmental Crisis: Climate Change and Terrestrial Biodiversity in Queensland. Rainforest CRC, Cairns.

Krockenberger, A.K., Edwards, W., Kanowski, J., 2012. The limit to distribution of a rainforest marsupial folivore is consistent with thermal intolerance hypothesis. Oecologia 168, 889–899.

Latch, P., 2008. Recovery Plan for Mabi Forest. Report to Department of the Environment and Water Resources, Canberra, Queensland Parks and Wildlife Service, Brisbane.

Laurance, W.F., 1990. Comparative responses of five arboreal marsupials to tropical forest fragmentation. J. Mammal. 71, 641–653.

Lumholtz, C., 1889. Among Cannibals. J. Murray, London (facsimile edition, ANU Press, Canberra, 1980).

Martin, R.W., 1996. Tcharibeena. Field studies of Bennett's tree-kangaroo. In: Flannery, T.F., Martin, R., Szalay, A. (Eds.), Tree Kangaroos: A Curious Natural History. Reed Books, Melbourne, pp. 36–65.

Martin, R.W., 2005. Tree-Kangaroos of Australia and New Guinea. CSIRO Publishing, Collingwood.

Newell, G.R., 1999a. Australia's tree-kangaroos: current issues in their conservation. Biol. Conserv. 87, 1–12.

Newell, G.R., 1999b. Home range and habitat use of Lumholtz's tree-kangaroo (*Dendrolagus lumholtzi*) within a rainforest fragment in North Queensland. Wildl. Res. 26, 129–145.

Nix, H.A., Switzer, M.A., 1991. Rainforest Animals: Atlas of Vertebrates Endemic to the Wet Tropics. Australian National Parks and Wildlife Service, Canberra.

Oates, J.F., Whitesides, G.H., Davies, A.G., Waterman, P.G., Green, S.M., Dasilva, G.L., Mole, S., 1990. Determinants of variation in tropical forest primate biomass: new evidence from West Africa. Ecology 71, 328–343.

Pahl, L.I., Winter, J.W., Heinsohn, G., 1988. Variation in responses of arboreal marsupials to fragmentation of tropical rainforest in north eastern Australia. Biol. Conserv. 46, 71–82.

Proctor-Gray, E., 1984. Dietary ecology of the coppery brushtail possum, green ringtail possum and Lumholtz's tree-kangaroo in North Queensland. In: Smith, A., Hume, I. (Eds.), Possums and Gliders. Surrey Beatty and Sons, Chipping Norton NSW, pp. 129–135.

Schmidt, C., Stirn, B., Kanowski, J., Winter, J.W., Felderhof, L., Wilson, R., 2000. Tree-Kangaroos on the Atherton Tablelands: Rainforest Remnants as Wildlife Habitat. Tree kangaroo and mammal group Inc., Atherton, Australia. Available at: https://www.tree-kangaroo.net/application/files/5514/4012/5019/reportMapFree.PDF. (28 October 2019).

Shoo, L.P., Storlie, C., VanDerWal, J., Little, J., Williams, S.E., 2011. Targeted protection and restoration to conserve tropical biodiversity in a warming world. Glob. Chang. Biol. 17, 186–193.

Stanton, P., Stanton, D., Stott, M., Parsons, M., 2014. Fire exclusion and the changing landscape of Queensland's wet tropics bioregion 1. The extent and pattern of transition. Aust. For. 77, 51–57.

Tracey, J.G., 1982. The Vegetation of the Humid Tropical Region of North Queensland. CSIRO, Melbourne.

Waite, E.R., 1894. Observations on *Dendrolagus bennettianus* De Vis. In: Proceedings of the Linnean Society of New South Wales IX, 571–582.

Webb, L.J., 1968. Environmental relationships of the structural types of Australian rain forest vegetation. Ecology 49, 296–311.

Williams, S.E., Bolitho, E., Fox, S., 2003. Climate change in Australian tropical rainforests: an impending environmental catastrophe. Proc. R. Soc. Lond. 270, 1887–1892.

Winter, J.W., 1997. Responses of non-volant mammals to late quaternary climatic changes in the wet tropics region of northeastern Australia. Wildl. Res. 24, 493–511.

Woinarski, J.C.Z., Burbidge, A.A., Harrison, P.H., 2014. Action Plan for Australian Mammals 2012. CSIRO, Melbourne.

CHAPTER

4

Threats to New Guinea's Tree Kangaroos

Bruce M. Beehler[a], *Neville Kemp*[b], *and Phil L. Shearman*[c]

[a]Smithsonian Institution, Washington, DC, United States

[b]US AID Lestari-Indonesia, Jakarta, Indonesia

[c]ANU College of Science, Australian National University, Canberra, ACT, Australia

OUTLINE

INTRODUCTION

The great island of New Guinea—the largest and the highest tropical island, is home to the most extensive tract of tropical humid forest in the Asia-Pacific region (Beehler, 1993). It is thus an important target for conservation planning and action. Identified by Conservation International as one of the Earth's major wilderness areas (Mittermeier et al., 1990), New Guinea supports untold stores of animal and plant biodiversity. Witness that botanists estimate from 15,000 to 25,000 species of plants inhabit this island, an indication of just how little we know about the details of the natural history of New Guinea (Beehler et al., 2002). Field naturalists, over the decades, have focused mainly on the most popular taxa for study and enumeration—birds, mammals, orchids, rhododendrons, butterflies, and beetles. Other groups, such as frogs and trees, have received inadequate attention, because of difficulty of access (e.g., trees) or less interest from the scientific establishment (e.g., frogs), although much progress has been made on these two groups in the last three decades (e.g., Allison and Tallowin, 2015; Johns et al., 2006). While scientists have dallied, certain mammal species have gone extinct (Plane, 1967) and others are today headed towards extinction, a process that is apparently running rampant in adjacent Australia.

https://doi.org/10.1016/B978-0-12-814675-0.00008-7

New Guinea's 14 tree kangaroo species (Chapter 1) have attracted the interest of naturalists since Salomon Müller in 1840 described the first species—the Vogelkop Tree Kangaroo (*Dendrolagus ursinus*). That said, for many decades after their discovery and naming, New Guinea's tree kangaroos were rarely encountered by western naturalists and scientists and remained very little-known (Beehler, 1991). During the 19th and first half of the 20th Century, the threats facing these creatures came mainly from traditional subsistence hunting. Today, tree kangaroos in New Guinea remain inadequately studied and little known, and yet the range of threats to them has now increased dramatically. The threats are reviewed in this chapter, and in Chapter 9 the opportunities for conserving the tree kangaroos inhabiting New Guinea are discussed. The taxonomic treatment of Groves (2005) for species-level treatment of tree kangaroos is followed in both this chapter and Chapter 9, mainly because this is the source for treatment by the IUCN Red List, a globally authoritative list of threatened species (IUCN Red List, 2018).

THREATS TO TREE KANGAROOS

Because nearly all of New Guinea's tree kangaroos are naturally uncommon or rare, with a low reproductive capacity (Flannery et al., 1996), their populations are inherently vulnerable. With growing economic development, habitat conversion, and expanding human populations, threats to tree kangaroos increase, decade by decade. These threats include loss of habitat, trade, subsistence hunting for food, traditional collection for pelts and other parts, and predation by feral dogs (Flannery, 1996). The various direct threats that impact the tree kangaroos inhabiting New Guinea are discussed followed by a discussion of the impact of habitat destruction, an important and broad-scale threat that perhaps overshadows all other indirect impacts.

Direct Threats

Subsistence hunting. Subsistence hunting is the most serious direct threat to all species of tree kangaroos inhabiting New Guinea. Even though many rural communities now are being connected by road to towns and urban centers, allowing for purchase of store-bought foods, most village families lack a regular cash income and hence cannot afford to subsist exclusively on store-bought staples. Therefore, they continue to depend on the land, gardening year-round to produce food crops of sweet potato, yam, taro, cassava, corn, pitpit, and also raising pigs and chickens. Rural families supplement their diet by fishing, collecting of forest products, and hunting wild game. In most rural areas, village men still make periodic hunting forays into the forest primarily in search of feral pigs, but also taking cassowaries, brush-turkeys, tree kangaroos, wallabies, possums, and other edible wild species when opportunities arise (Mack and West, 2008).

Among the many wild birds and mammals harvested by rural village hunters, tree kangaroos are considered a prime game animal because of their size, the quality and quantity of their flesh, and the value of their pelts. Although hunting parties are generally opportunistic, taking whatever is encountered in the forest, hunters often seek out tree kangaroos by traveling out to rarely-visited forest tracts where populations still persist. Hunting parties then deploy village hunting dogs to detect and chase down tree kangaroos, whose only recourse is to hide up in trees, both small and large. Such a retreat is only a temporary sanctuary, and local hunters use bows and arrows to kill the tree kangaroo or the tree kangaroo leaps down, and the dogs chase them down. Virtually all tree kangaroos located by hunting dogs end up being killed. The various ways hunters manage to capture tree kangaroos are well described in Flannery et al. (1996).

Over the years, hunting parties deploying dogs can devastate local tree kangaroo populations. As a result, tree kangaroos are, in most instances, extirpated from forests within a day's walk of a rural village. In watersheds where villages are abundant and human populations high, tree kangaroos are generally now absent, a product of chronic over-hunting. Populous rural regions today entirely lack tree kangaroos, so on the island of New Guinea tree kangaroos are only found in isolated areas far from villages and where hunting parties rarely tread. Hunting has produced a similarly patchy distribution for the three species of long-beaked echidna (*Zaglossus* spp.), another group of highly sought-after game animals (Helgen, 2007). In fact, because of hunting, remnant populations of echidnas and tree kangaroos appear to be principally confined to mountain redoubts far from human activity. With expanding human populations using ever-more of the landscape, the future of these game species is under dire threat.

Hunting for pelts. Tree kangaroos are also popular for their pelts. These glossy and beautiful fur-products are favored for body-adornment worn during traditional ceremonies. The pelt of the Huon tree kangaroo (*Dendrolagus matschiei*) is especially important to the people of the YUS ecosystem (an area encompassed by the Yopno, Uruwa, and Som rivers) on the Huon Peninsula in Papua New Guinea (PNG) for traditional adornment—with head-bands made from the long tail, and capes from the body pelage. The demand for pelts, therefore, adds to the hunting pressure discussed in the preceding section. Other tree kangaroo parts are treasured and displayed—teeth, bone, paws, and claws (Szalay, 1996).

Trade of live animals. As with most attractive species of New Guinea's bird and mammal fauna, tree kangaroos are occasionally trapped live and kept as pets in the village. In most cases, a female parent is killed during a hunt and the dependant joey is hand-captured, kept alive, and hand-reared. These are sometimes subsequently sold in the market or traded. The opportunity to capture and keep live individual tree kangaroos as pets is probably incidental to game hunting. The authors have seen pet tree kangaroos in villages in the Foja Mountains of Indonesian New Guinea, the Bewani Mountains of northwestern PNG, the Star Mountains of western PNG, and the Southern Highlands of PNG, so the practice is apparently widespread and thus of concern.

Wild dog predation. The New Guinea Singing Dog (*Canis familiaris hallstromi*) is nothing more than a population of the dingo living in New Guinea. More recently, these ancient lineages that came from Australia with human immigrants have inter-bred with modern dogs brought to New Guinea from all over the world by working western residents and missionaries. Today, packs of "wild dogs" in the high alpine zones prey upon a wide range of native mammals in the forest and alpine grassland. There is evidence that Doria's tree kangaroo (*Dendrolagus dorianus*) populations living at the verge of the alpine grasslands may fall prey to the packs of feral dogs (observations from the English Peaks area of south-eastern PNG). It is not known how important dog predation on tree kangaroos is, but it may be widespread and substantial, and merits study. These dog packs are probably also important predators on the species of long-beaked echidna (the senior author found dog-killed echidna remnants at the edge of alpine grassland in the English Peaks).

Indirect Threats

Tree kangaroos are mainly forest-dwellers, though species such as Doria's and the Huon tree kangaroos (and apparently the Dingiso,

Dendrolagus mbaiso) also forage at the verge of alpine grasslands and in rocky clearings at high elevation (Flannery et al., 1996). Presumably prior to the arrival of humans in New Guinea, tree kangaroos inhabited all forested habitats. Today, tree kangaroos probably occupy considerably less than half of the land area of New Guinea, mainly in circumscribed forest patches in the high mountains, isolated hill forests, and sparsely populated lowlands. As tree kangaroo populations become ever-more fragmented, effective population size of these species declines, and the pace of regional extirpation of isolated populations rises. Impacts on patches of habitat that support isolated tree kangaroo populations can have disproportionate effects on the long-term survival of each species. Thus, indirect threats include alteration of landscapes for human use as well as effects of climate change on the habitat.

Logging. Industrial logging has been carried out in Papua New Guinea since the 1950s and in Indonesian New Guinea since the 1960s. Logging operations greatly increased in PNG in the 1990s and in Indonesian New Guinea after the financial crisis of 1997 (Bryan et al., 2015; Angraenni, 2007). The largest logging operations carried out by several Asian-owned companies, mainly export of timber to China, primarily, to meet the growing demand there (Angraenni, 2007). Most logging on the island is highly selective, because loggers target trees over 40 cm diameter and most trees in New Guinea's forests are smaller than this. Unlike in parts of Southeast Asia, New Guinea's forests do not support dense stands of high-value dipterocarp tree species, therefore most forest logging projects extract relative few stems per hectare, making subsequent conversion to pasture or other non-forest uses uncommon (especially because of strong local land tenure). This means that most logged-over areas end up as degraded forest, which, when left undisturbed, regenerates naturally back to mature forest if it is not re-logged. The big question is whether tree kangaroo populations can subsist in logged forest. Given the relative abundance of grizzled tree kangaroo (*Dendrolagus inustus*) populations in coastal regions, it seems at least this species can, indeed, survive selective logging operations (NK, personal observation). NK has observed this species feeding in roadside regenerating forest (in the Bird's Neck Region), and the species may potentially thrive in these disturbed habitats. Regarding large-scale logging operations, it would be very useful to study the impact that widespread logging in the alluvial lowlands of southwestern PNG has had upon the population of the poorly-known lowlands tree kangaroo (*Dendrolagus spadix*).

The main threat posed by large-scale logging is probably secondary—adventitious hunting with firearms by employees of the logging companies. The logging roads and skid tracks open up the forest to weekend game hunting by the (mainly non-local) logging crew, and this may quickly lead to depletion of tree kangaroos and other large game in the logged forest. Hunters from local communities may also hunt here, but they have limited access to firearms.

Monoculture plantations. Industrial agriculture, whether tree plantations of *Acacia mangium*, rubber, or oil palm plantations, leads to the removal of large contiguous tracts of original forest (Angraenni, 2007). Whole local biotas are wiped out to create bare-earth landscapes for planting these monoculture crops. In New Guinea, oil palm and tree monocultures have been planted in a number of coastal lowland areas in both PNG and Indonesian New Guinea. Such converted habitats are entirely unsuitable for the survival of tree kangaroo populations. Plantation development has proceeded slowly in New Guinea, due to limited infrastructure and because of the complexity of local land tenure issues. Still, plantation development continues to pose a threat in the coastal and alluvial lowlands where plantation agriculture is most cost-effective or zoned for, thus potentially impacting populations of grizzled and lowlands tree kangaroos.

Mining, oil, and natural gas development. The impacts on the welfare of tree kangaroos of industrial mining and oil and gas development in New Guinea is probably indirect and similar to that of logging. Habitat is converted, roads are constructed, and non-resident work forces deployed into areas where tree kangaroos live. Dogs are introduced to these areas and weekend hunting by the work crews can lead to decimation of local game species, including tree kangaroos. The long-term impacts are probably substantial. At least some of these large operations put in place rules prohibiting wildlife harvest by workers who are not local landowners, but once the operation is completed and the big company departed, then the access created by the remaining road networks in the forest will lead to increases in local hunter use once the protections and controls are gone. Opening up access to old growth forests ultimately leads to the end of tree kangaroo populations.

Subsistence gardening. In interior highland valleys where human populations are on the increase, the impact of subsistence gardening can be serious. Where populations are dense, swidden gardening leads to permanent loss of forest because of the ever-shortening fallow cycle. As a result, anthropogenic grassland creeps ever-higher up the mountainside, leading to permanent loss of forest for the local tree kangaroo populations (Boissiére and Purwanto 2007). Of course, while this is happening, the local village communities are also hunting tree kangaroos in the adjacent forests, creating a "one-two" punch that inevitably leads to local extirpation.

Dry season burning and ENSO wildfire. One final threat related to growing human populations in rural forested habitats in New Guinea is fire. Annual adventitious dry-season burning of gardens and grasslands by villagers leads to accumulating losses of forest at the grassland-forest interface. Moreover, during very dry El Niño years, adventitious fires set by villagers traveling cross-country and by hunters in the alpine regions leads to loss of original closed forest in areas far from villages—areas of prime habitat for tree kangaroos (Bryan et al., 2015).

Climate change impacts. There is little doubt that climate change will have a growing impact on the future of tree kangaroos in New Guinea. Longer dry seasons will lead to more serious fires that threaten more forest (though recent data for Indonesian New Guinea show increasing rainfall according to data seen by NK). Increasing temperatures will cause the shifting upslope of forest and alpine habitats that will force species of tree kangaroos to move upward or adapt. A 6-year dataset of maximum daytime temperatures recorded at 3000 m elevation on Mount Wilhelm, PNG, showed a 0.5 °C increase over this brief period (P. Shearman, unpublished). We suspect there will be negative impacts for tree kangaroos. For instance, uphill shifting of species' ranges will result in a reduction of available habitat and also greater habitat fragmentation (because of the nature of the mountainous habitat). It would be useful to model these impacts on some of the more vulnerable populations of tree kangaroos.

THE PROSPECT FOR TREE KANGAROOS IN THE 22ND CENTURY

The combined impacts of growing industrial-scale resource exploitation in New Guinea, along with the relentless growth of local human populations and climate change, are expected to seriously threaten the future of all species of tree kangaroos inhabiting the island of New Guinea.

ACKNOWLEDGMENTS

Steven Richards critically read and commented on this chapter. We thank Conservation International and the Committee for Research and Exploration of the National Geographic Society for support of field work in New Guinea. We also thank the governments of Indonesia and Papua New Guinea for permission to conduct research in western and eastern New Guinea.

REFERENCES

Allison, A., Tallowin, O., 2015. Occurrence and status of Papua New Guinea vertebrates. In: Bryan, J.E., Shearman, P.L. (Eds.), The State of the Forests of Papua New Guinea 2014: Measuring Change over the Period 2002–2014. University of Papua New Guinea, Papua New Guinea, pp. 87–110.

Angraenni, D., 2007. Patterns of commercial and industrial resource use in Papua. In: Marshall, A.J., Beehler, B.M. (Eds.), Ecology of Papua. Periplus, Singapore, pp. 1149–1166.

Beehler, B.M., 1991. Papua New Guinea's wildlife and environments—what we don't yet know. In: Pearl, M., Beehler, B.M., Allison, A., Taylor, M. (Eds.), Conservation and Environment in Papua New Guinea: Establishing Research Priorities. Wildlife Conservation Society, New York, pp. 1–10.

Beehler, B.M. (Ed.), 1993. A Biodiversity Analysis for Papua New Guinea. In: Papua New Guinea Conservation Needs Assessment, Part 2Biodiversity Support Program, Washington, DC.

Beehler, B.M., Kula, G., Supriatna, J., Mittermeier, R.A., Pilgrim, J., 2002. New Guinea. In: Mittermeier, R.A., Mittermeier, C.G., Robles Gil, P., Pilgrim, J., da Fonseca, G.A.B., Brooks, T., Konstant, W.R. (Eds.), Wilderness: Earth's Last Wild Places. CEMEX/Agrupación Sierra Madre, Mexico City, pp. 134–163.

Boissiére, M., Purwanto, Y., 2007. Agricultural system of Papua. In: Marshall, A.J., Beehler, B.M. (Eds.), Ecology of Papua. Periplus, Singapore, pp. 1125–1148.

Bryan, J.E., Shearman, P.L., Aoro, G., Wavine, F., Zerry, J., 2015. The current state of PNG's forests and changes between 2002 & 2014. In: Bryan, J.E., Shearman, P.L. (Eds.), The State of the Forests of Papua New Guinea 2014: Measuring Change Over the Period 2002-2014. University of Papua New Guinea, Papua New Guinea, pp. 7–42. ISBN: 978-9980-89-106-8.

Flannery, T.F., 1996. Conservation. In: Flannery, T.F., Martin, R., Szalay, A. (Eds.), Tree Kangaroos: A Curious Natural History. Reed Books Australia, Melbourne, Victoria, pp. 84–89.

Flannery, T.F., Martin, R., Szalay, A., 1996. Tree Kangaroos: A Curious Natural History. Reed Books Australia, Melbourne, Victoria.

Groves, C.P., 2005. Order diprotodontia. In: Wilson, D.E., Reeder, D.M. (Eds.), Mammal Species of the World—A Taxonomic and Geographic Reference. Johns Hopkins University Press, Baltimore, Maryland, pp. 43–70.

Helgen, K., 2007. A taxonomic and geographic overview of the mammals of Papua. In: Marshall, A.J., Beehler, B.M. (Eds.), Ecology of Papua. Periplus, Singapore, pp. 689–749.

IUCN Red List, 2018. The IUCN Red List of Threatened Species. Version 2017-3. Available at www.iucnredlist.org (22 February 2018).

Johns, R.J., Edwards, P.J., Utteridge, T.M.A., Hopkins, H.C.F., 2006. A Guide to the Alpine and Subalpine Flora of Mount Jaya. Kew Publishing, Royal Botanic Gardens, Kew, England.

Mack, A.L., West, P., 2008. Ten Thousand Tonnes of Small Animals: Wildlife Consumption in Papua New Guinea, a Vital Resource in Need of Management. RMAP Working Papers 61, pp. 1–21.

Mittermeier, R.A., Mittermeier, C.G., Robles Gil, P., Pilgrim, J., da Fonseca, G.A.B., Brooks, T., Konstant, W.R. (Eds.), 1990. Wilderness: Earth's Last Wild Places. CEMEX/Agrupación Sierra Madre, Mexico City.

Plane, M.D., 1967. The Stratigraphy and Vertebrate Fauna of the Otibanda Formation, New Guinea. Bureau of Mineral Resources, Geology, and Geophysics, Australia, Bulletin 86, pp. 1–64.

Szalay, A., 1996. Ornaments, trophies, and charms. In: Flannery, T.F., Martin, R., Szalay, A. (Eds.), Trees Kangaroos—A Curious Natural History. Reed Books Australia, Melbourne, Victoria, pp. 24–31.

SECTION III

CONSERVATION SOLUTIONS: IN THE FIELD – AUSTRALIA

CHAPTER

5

Community Conservation of Tree Kangaroos

Peter Valentine

College of Science and Engineering, James Cook University, Townsville, QLD, Australia

OUTLINE

INTRODUCTION

The Wet Tropics World Heritage Area of northern Queensland includes almost a million hectares of tropical rainforest within which occur two species of tree kangaroos (Flannery et al., 1996). In the north, the large Bennett's tree kangaroo (*Dendrolagus bennettianus*) occupies secure habitat with little human settlement and currently does not appear to be in decline. Yet, this species has been assessed as Near Threatened on the IUCN Red List due to probable decline in habitat extent and quality (Winter et al., 2019). Bennett's tree kangaroo has been less well studied than the second species, the Lumholtz's tree kangaroo (Martin, 2005; Chapters 6 and 7). The Lumholtz's tree kangaroo (*Dendrolagus lumholtzi*) is the smallest of all tree kangaroos (Fig. 5.1) and listed as Near Threatened due to probability of population declines (Woinarski and Burbidge, 2016). This species occupies an extensive area further south, but much of its former range has been cleared for agriculture and grazing and is now dotted with small towns and extensive human settlement.

Unfortunately, their preferred rainforest habitat type also coincides with the volcanic soils

Tree Kangaroos
https://doi.org/10.1016/B978-0-12-814675-0.00022-1

FIGURE 5.1 Lumholtz's tree kangaroo is the smallest of all tree kangaroos, here shown close up in a remnant patch of forest, near Malanda. *Source: Martin Willis.*

preferred by humans for agriculture, so much of the most densely populated habitat has now been lost with small patches and fragments remaining (Martin, 2005). This proximity to settlement has led to the species being well known for a long time and frequently encountered in the wild by residents. They seemed to survive in the smaller fragments of forest but are under serious threat from the increased numbers of dogs and from traffic on the network of roads that dissect the Atherton Tablelands.

Tree kangaroos were seen as an interesting curiosity by local people who often called them simply "tree-climbers" and many people were aware of them (Chapter 8). Aboriginal people in their local language used the term *mabi* for the Lumholtz's tree kangaroo and botanists subsequently named the preferred habitat type (Complex Notophyll Vine Forest; Wet Tropics Management Authority, 2020) as Mabi Forest, itself now a threatened ecosystem. Fig. 5.2 shows an example of Mabi forest with a tree kangaroo present. Over the past three decades a significant amount of formal research and informal observations have contributed to a great deal of knowledge about Lumholtz's tree kangaroo.

FIGURE 5.2 A Lumholtz's tree kangaroo in a patch of Mabi Forest near Malanda, Atherton Tablelands. Visible near the middle of the image. *Source: Martin Willis.*

Much of this has been associated with the establishment and operation of a community conservation group, the Tree Kangaroo and Mammal Group (TKMG).

SCIENCE AND COMMUNITY ON THE ATHERTON TABLELANDS

Early studies of the Lumholtz's tree kangaroo were undertaken by a number of scientists in the 1980s and 1990s; from the Atherton Commonwealth Scientific and Industrial Research Organization (CSIRO) Research Station and from universities (Proctor-Gray, 1985, 1990; Proctor-Gray and Ganslosser, 1986; Pahl et al., 1988; Laurance and Laurance, 1996; Newell, 1999b). One study being undertaken near Yungaburra by Dr. Graham Newell of CSIRO was focused on establishing home ranges and behavior of a population of Lumholtz's tree kangaroos on private property (Newell, 1999a). Part way through the study the landowner decided to clear half the area of Mabi forest for agriculture and proceeded to log the forest. This enabled novel observations to be made on the population. One surprising outcome was the fact that almost all the tree kangaroos refused to abandon their territory even after the logging. Many simply lived among the debris and piles of branches on the ground, maintaining their home range territory.

Newell's study and finding alerted a community of researchers, scientists, conservationists, and wildlife carers to the critical role of habitat for the survival of Lumholtz's tree kangaroos. Several people came together to consider how to create a more secure future for this tree kangaroo species given that its prime habitat was mostly cleared and the remainder was mostly fragmented with threats from the urbanizing impacts (especially dogs and roads). Beth Stirn (a recent immigrant from the United States) and Margit Cianelli (a wildlife carer) put forward the idea of forming a group to address the conservation status of local mammals to several other residents, including researchers and veterinarians, and persuaded wildlife ecologist Dr. John Winter to become the first President of this group. Together, they established the Tree Kangaroo and Mammal Group (TKMG) in 1997. The name is meant to indicate that the group was also concerned with other species, given that several other mammals were in difficulties from the same processes that threatened tree kangaroos.

One of the earliest projects by TKMG was to engage with the wider community about where tree kangaroos had been seen across the Atherton Tablelands (Kanowski et al., 2001). There were two motives in this approach; to take advantage of the many potential observers across the known range of the tree kangaroo to provide effective and efficient survey data but equally important, by engaging the community to raise awareness of the tree kangaroos and their conservation challenges. More than 10,000 questionnaires were distributed across the region and these elicited nearly 800 responses with the response rate higher in the prime habitat locations such as Malanda. These provided some 2368 sighting records (15% of which were dead animals) and the outcomes gave a far greater database on distribution than any previous work. The significance of the Atherton Tablelands and particularly the fragments of ideal basaltic soil rainforest tracts for the long-term survival of the Lumholtz's tree kangaroo was evident and prompted discussions of how the remaining fragments might be managed to improve their value for tree kangaroos. The patches of dark green seen in Fig. 5.3 are mostly Mabi forest and even the smaller fragments (generally from as little as 2 or 3 hectares in area) will support tree kangaroos.

One method of habitat management would be to improve connectivity between fragments and/or conduct revegetation projects (Kanowski et al., 2003). The significance of road kills of this species was evident and the deaths caused by dogs were also documented. Kanowski et al. (2001) also comment on the need to consider the potential effect of global warming on the distribution of Lumholtz's tree kangaroo. These outcomes provided the foundation for further TKMG projects over the next two decades. The success of this project also led to a strong local membership base that continues (around 130 members) and saw the close collaboration between the TKMG and other conservation and land management groups including Trees for the Evelyn and Atherton Tablelands, Inc. (TREAT) (with its focus on tree plantings); the Queensland Parks and Wildlife agency (Government investments in science and conservation); local government programs (especially Visitor Centres and Council environmental management programs); the regional natural resource management body (Terrain), the Wet Tropics World Heritage Management Authority (WTMA) and various university researchers.

TWO DECADES OF COMMUNITY LEADERSHIP IN LAND MANAGEMENT ACTIVITIES

The Committee and members of TKMG used every opportunity to seek grants from Government and non-Government sources to implement a program of increased conservation for the tree kangaroos and other mammals of the Atherton Tablelands. In the year 2000 the organization attracted grants from the Threatened Species Network to work on spotted-tailed quolls and to trial tree planting and a novel idea of shelter poles for Lumholtz's tree kangaroos. The idea of these poles was to provide closely spaced escape options for tree kangaroos that were crossing open pastures between rainforest fragments and risked attacks from dogs. It was obvious to all that improved conservation would depend very much on cooperation with

FIGURE 5.3 The fragmented landscape of the Atherton Tablelands that was previously all Lumholtz's tree kangaroo habitat. This scene of the agricultural landscapes and remnant forests is focused on the area around Malanda (the small rural town) but similar dissected and fragmented habitat occurs across all the former Mabi forests on the Atherton Tablelands. *Source: Google Map 2020, CNES/Airbus, Landsat/Copernicus, Maxar Technologies.*

private landowners and so TKMG won a grant that supported careful assessment of remnant vegetation on private property across the Atherton Tablelands including individual property assessments. A community workshop was held to further develop ideas and processes. These activities were supported financially by the National Heritage Trust (NHT), a Federal Government initiative. Local people responded enthusiastically with 80 people attending the workshop.

Further funding in 2004 from the NHT and from the local Council along with other groups including TREAT and the Queensland Parks and Wildlife Service enabled extensive tree planting to begin, with a focus on re-creating forest types suitable for both the endemic musky rat-kangaroo (*Hypsiprymnodon moschatus*), and the Lumholtz's tree kangaroo. The partnership with TREAT has proved highly valuable and enduring, the efforts of the two groups achieving extensive new areas of revegetated forests, especially on private lands. Some of the success stories were show-cased in 2005 at an international conference organized by TKMG in partnership with a research group (the Rainforest Cooperative Research Centre) and the regional natural resource management body (Terrain). The Conference was focused on the "Ecology and Conservation of Tree-kangaroos: Current issues and Future Directions" and was held on the Atherton Tablelands.

Subsequently many more revegetation projects have been initiated and completed involving partners across the region including the natural resource management body (Terrain), the World Heritage Management Authority (WTMA) and various other groups but especially TREAT. Strategic site selection saw additions to the growing connectivity between isolated fragments, including large areas of forest outliers such as Lake Eacham. Just as important has been efforts to secure much better connectivity along stream lines and between the smaller patches of forest occupied by tree kangaroos across the Atherton Tablelands (see Fig. 5.4).

DEVELOPING THE COMMUNITY ACTION PLAN

In July 2012 in partnership with Conservation Volunteers, and the regional Natural Resource Management (NRM) body Terrain, TKMG members organized a workshop to develop an action plan for the conservation of Lumholtz's tree kangaroo. Thirteen different organizations participated including two universities, CSIRO, and other research providers; local and state government agencies with a focus on land management; community groups with a history of conservation work including TREAT and TKMG, and the School for Field Studies where many projects that

FIGURE 5.4 An aerial view showing riparian remnant forest along creek lines (lower left quadrant) and patches of revegetation plantings by TREAT connecting with the larger area of preserved Mabi forest in Lake Eacham National Park (upper right quadrant) designed to re-establish connectivity for wildlife, including tree kangaroos. *Source: Google Map 2020, CNES/Airbus, Landsat/Copernicus, Maxar Technologies.*

focus on tree kangaroos have been undertaken by students. A total of 43 individuals participated in the workshop. Following the workshop, the Community Action Plan (CAP) was prepared and signed off by the participants (Burchill et al., 2014). Fig. 5.5 shows the cover of the report using artwork of the Lumholtz's tree kangaroo gifted to the TKMG by internationally famous wildlife artist William T. Cooper.

The plan established five goals to address the key concerns about the long-term conservation of Lumholtz's tree kangaroo on the Atherton Tablelands. The plan also identified a number of objectives for each goal, to be implemented over a 5-year period by the various organizations participating or with responsibilities (2014–19).

Goal 1: An aware and engaged community

This goal will assist in engaging the local, Australian and international communities.

Goal 2: Adequate LTK habitat in sound condition, protected and well connected

This goal will assist in identifying areas for protection and measures to make their protection more effective.

Goal 3: Direct human-related threats are mitigated

This goal will assist in mitigating threats to LTK by introduced predators and traffic.

Goal 4: Protocols based on sound knowledge and experience are applied in LTK husbandry, rehabilitation and release

FIGURE 5.5 The Community Action Plan was developed by intensive community participation processes, led by the TKMG community group and supported by regional conservation and resource management bodies. *Source: Peter Valentine.*

This goal will assist in the detection, care, and release of injured LTK, and the integration of rehabilitated animals in research.

Goal 5: Knowledge of the species is adequate to guide conservation actions

This goal will promote research into the distribution, the abundance, the current and future population viability, and threat-mitigation techniques.

Following the completion of the Community Action Plan, TKMG has used the plan to help drive its subsequent projects and has sought various funds to enable many actions to meet various objectives. The Community Action Plan also informs many other organizations on the kinds of projects or activities that would support the conservation of Lumholtz's tree kangaroo. Under Goal 1 for example, TKMG has provided talks to local schools about the Lumholtz's tree kangaroo and its conservation, and has contributed numerous articles in local media to enhance awareness among residents and landowners. Under Goal 2, TKMG has successfully sought funds to expand connectivity between fragments in partnership with TREAT and other bodies. TKMG has also been active in both raising awareness about road-kills and also contributed to awareness of the problems of domestic and wild dogs in killing tree-kangaroos as indicated under Goal 3 (Shima et al., 2019). TKMG designed a road sign (Fig. 5.6) and identified hot spots for signage to alert motorists about the risks of tree-kangaroo road kill.

FIGURE 5.6 Road sign designed by the community group TKMG and deployed at hot spots of Lumholtz's tree kangaroo road kills along highways and rural roads. This one in the upper Barron River area. Many of the road kills are young males seeking territory and because of fragmentation having to cross roads with significant fatality risks. *Source: Peter Valentine.*

TKMG also established, through Committee member and veterinarian Amy Shima, a database of road kills of Lumholtz's tree kangaroos and fatalities by dogs. This subsequently led to partnership with the local Council which is now working with TKMG to promote better driver behavior and also to address the dog issues. New state-wide legislation makes it an offense for any dog to be in a public place without a leash (exceptions are in designated off-leash fenced areas). The Council is working actively to promote more responsible dog ownership and TKMG has developed temporary signs to alert the community to both tree kangaroo presence (recent sightings) and to tree kangaroo deaths (from vehicles or dogs). This a part of a new partnership with the local government to both raise awareness and reduce mortality.

Several members of TKMG are involved in wildlife care activities with orphaned tree kangaroos, especially those rescued when the mother was killed by a vehicle or by a dog. Considerable experience now exists, and successful rearing and release has occurred on numerous occasions (Chapter 6). The final goal of the CAP is for better investment in science and research and TKMG offers financial support for some projects that focus on tree kangaroo conservation. Thanks to the efforts of many individuals and groups, much more information is now available about tree kangaroos in the wild and conservation issues and solutions. The local School for Field Studies (SFS) often has students undertake projects that focus on tree kangaroo behavior or ecology. Some of these have led to published outcomes that contribute to improved knowledge (Heise-Pavlov et al., 2011, 2014, 2018; Heise-Pavlov and Gillanders, 2016).

A WIDER AND MORE INNOVATIVE COMMUNITY PROGRAM

The Tree Kangaroo and Mammal Group members have developed a number of new initiatives that both promote the importance of the Lumholtz's tree kangaroo and also help people learn more about them. The first of these is by working with the Malanda Visitor Centre to establish an excellent display about tree kangaroos (Fig. 5.7). Funds were raised more recently to develop computer-based technology (software and hardware) to create, in association with computer scientists at James Cook University, a virtual reality tree kangaroo. Named Kimberley (after an actual rescued tree kangaroo, raised by Margit Cianelli, Chapter 6), the outcome is a system that enables visitors to use specialized helmets to engage with virtual tree kangaroos behaving naturally. The system has proved popular especially with younger visitors, and newer versions have incorporated conservation themes and ideas. Members also took the original version to the Queensland State Show in the capital, Brisbane, to reach out to a greater audience.

The continuous presence of a stall at the monthly markets of Yungaburra has been a feature of TKMG for many years. While some funds for conservation are raised through the sale of items such as t-shirts, reusable shopping bags, photographs and books, the principal benefit is to maintain the high visibility of tree kangaroos in the community consciousness. Another means to achieve this has been through regular talks held every second month in the Malanda Hotel, where experts make enjoyable presentations about some aspect of tree kangaroo or other mammal ecology and conservation. Many of these focus on the latest research results but are presented for a general audience and all are open to the public free of charge. This regular activity usually achieves good publicity in the local community newspapers, adding further to the reach. Members of TKMG have also given talks to other community groups and to classes in local schools, reaching out to young and old to achieve better recognition and conservation. There is an ongoing project to try and encourage the Tablelands Regional Council to adopt the Lumholtz's tree kangaroo as its faunal emblem. The tree kangaroos themselves have proved to

FIGURE 5.7 Part of the Lumholtz's tree kangaroo display at the Malanda Visitor Centre (A) including the world's first virtual reality tree kangaroo named Kimberley (B). This has been a joint project between the TKMG community group and the Tablelands Regional Council. *Source: Peter Valentine.*

be excellent ambassadors through their appeal to people when seen at various popular locations on the Atherton Tablelands. For example, a small group of four or five animals have occupied a few hectares of forest adjacent to the Nerada Tea Plantation visitor center and are reliably seen by visitors over the past 3 years (Fig. 5.8).

Another novel idea has been implemented to make available wildlife cameras free of charge to members to record the presence of various wildlife on their properties. A similar project was undertaken at the local school in Malanda and resulted in a significant increase in awareness of wildlife with people becoming aware of many species that are present but had previously been unseen. TKMG has many sets of these cameras now and they get frequent use by members and by researchers who are able to borrow them for fauna-related projects. An example of an important use has been to check whether animals are using fauna overpass bridges established for that purpose. A recent example showed possums starting to use the faunal rope bridge overpass even when the new plantings were quite young.

Most recently the TKMG has produced temporary signs indicating the recent sighting of a Lumholtz's tree kangaroo in the local area. These are deployed along the road side for a period of 2 weeks as a means to heighten community awareness and to encourage reduced speed by drivers (Fig. 5.9). Two other temporary signs are being used to alert people to recent road kills and to recent dog kills, an attempt to encourage more appropriate driving and pet management behavior. This project is part of an agreement with the local Tablelands Regional Council to improve survival of tree kangaroos across the Atherton Tablelands.

FIGURE 5.8 Two of the group of Lumholtz's tree kangaroos in a small patch of forest adjacent to the Nerada Tea Plantation. On the left is the male showing more color typical of males and on the right is a young animal. *Source: Peter Valentine.*

FIGURE 5.9 A recent TKMG program to further raise awareness is the short-term placement of temporary signs along roadsides indicating recent sightings or road kills or dog kills. These are intended to stay in place briefly to encourage better behavior by drivers and dog owners. *Source: Peter Valentine.*

TKMG has not been confined to the Atherton Tablelands in its wildlife conservation activities and has participated in various regional and national efforts to ensure better outcomes for wildlife. For example, TKMG made a lengthy submission to the National inquiry into Australia's faunal extinction crisis and drew attention to a range of issues of concern. The potential negative impacts from climate change has also been highlighted in submissions to the National Government and TKMG does pursue a role of advocacy as appropriate.

The Tree Kangaroo and Mammal Group instituted the William T. Cooper Award for Conservation of North Queensland Mammals in memory of the internationally acclaimed artist, naturalist, conservationist, and member of TKMG, Bill Cooper and this has been awarded annually for several years now. In 2019, the award went to David Hudson, a person who significantly increased the area of revegetated rainforest cross the region through project development and fund-raising. He thereby contributed to a legacy that has substantially improved the habitat availability for tree-kangaroos and other wildlife.

CONCLUSIONS

Community conservation has very significantly improved both awareness and on ground conservation for the Lumholtz's tree kangaroo. Much has been achieved and the community group Tree Kangaroo and Mammal Group, Inc. has been at the forefront of conservation efforts for 20 years. Established following the shocking realization of the loss of preferred habitat for these astonishing marsupials some 20 years ago, the group has rallied the wider community around better conservation outcomes by working with local, state, and national authorities to better understand and value these animals. Much has been achieved but there is recognition that the future holds even more concerns in the face of climate change and an uncertain appreciation of how that might create new challenges for tree kangaroos. It seems likely that this community conservation group will play a significant role in the years ahead to try and ensure all levels of government invest in a future for our tree kangaroos.

ACKNOWLEDGMENTS

I am particularly indebted to the members of TKMG and the past and present office-bearers who both shared their knowledge of various activities and welcomed me into the group as President a few years ago. While some information is available in the public record through various publications, much has been gleaned from conversations with individuals and through access to the various newsletters (Mammal Mail) available on the TKMG website. In particular, I thank Simon Burchill, Margit Cianelli, David Hudson, John Kanowski, Amy Shima and John Winter.

REFERENCES

Burchill, S., Cianelli, M., Edwards, C., Grace, R., Heise-Pavlov, S., Hudson, D., Moerman, I., Smith, K., 2014. Community Action Plan for the Conservation of the Lumholtz's tree-kangaroo (*Dendrolagus lumholtzi*) and its habitat 2014–2019. Malanda, Australia. Available from: https://www.tree-kangaroo.net/application/files/2114/7791/0024/TKMG_CAP_2014-2019.pdf (18 June 2020).

Flannery, T.F., Martin, R.W., Szalay, A., 1996. Tree-Kangaroos: A Curious Natural History. Reed Books, Melbourne 202 pp.

Heise-Pavlov, S., Gillanders, A., 2016. Exploring the use of a fragmented landscape by a large arboreal marsupial using incidental sighting records from community members. Pac. Conserv. Biol. 22, 386–398.

Heise-Pavlov, S., Jackrel, S.L., Meeks, S., 2011. Conservation of a rare arboreal mammal: habitat preferences of the Lumholtz's tree-kangaroo, *Dendrolagus lumholtzi*. Aust. Mammal. 33, 5–12.

Heise-Pavlov, S., Anderson, C., Moshier, A., 2014. Studying food preferences in captive cryptic folivores can assist in conservation planning: the case of the Lumholtz's tree-kangaroo (*Dendrolagus lumholtzi*). Aust. Mammal. 36, 200–211.

Heise-Pavlov, S., Rhinier, J., Burchill, S., 2018. The use of replanted riparian habitat by the Lumholtz's Tree-kangaroo (*Dendrolagus lumholtzi*). Ecol. Manage. Restor. 19, 76–80.

Kanowski, J., Felderhof, L., Newell, G., Parker, T., Schmidt, C., Stirn, B., Wilson, R., Winter, J.W., 2001. Community survey of the distribution of Lumholtz's Tree-kangaroo on the Atherton Tablelands, north-east Queensland. Pac. Conserv. Biol. 7, 79–86.

Kanowski, J., Winter, J.W., Simmons, T., Tucker, N.I.J., 2003. Conservation strategy for Lumholtz's tree-kangaroo on the Atherton Tablelands. Ecol. Manage. Restor. 4, 220–221.

Laurance, W.F., Laurance, S.G.W., 1996. Responses of five arboreal marsupials to recent selective logging in tropical Australia. Biotropica 28, 310–322.

Martin, R., 2005. Tree-Kangaroos of Australia and New Guinea. CSIRO Publishing, Melbourne 158 pp.

Newell, G.R., 1999a. Australia's tree kangaroos: current issues in their conservation. Biol. Conserv. 87, 1–12.

Newell, G.R., 1999b. Responses of Lumholtz's tree-kangaroo (*Dendrolagus lumholtzi*) to loss of habitat within a tropical rainforest fragment. Biol. Conserv. 87, 181–189.

Pahl, L.I., Winter, J.W., Heinsohn, G., 1988. Variation in response of arboreal marsupials to fragmentation of tropical rainforest in North Eastern Australia. Biol. Conserv. 46, 71–82.

Proctor-Gray, E., 1985. The Behavior and Ecology of Lumholtz's Tree-Kangaroo, *Dendrolagus lumholtzi* (Marsupialia: Macropodidae). PhD thesisHarvard University.

Proctor-Gray, E., Ganslosser, U., 1986. The individual behaviors of Lumholtz's tree-kangaroo: repertoire and taxonomic implications. J. Mammal. 67, 343–352.

Proctor-Gray, E., 1990. Kangaroos up a tree. Natural History 60–67.

Shima, A.L., Berger, L., Skerratt, L.F., 2019. Conservation and health of Lumholtz's tree-kangaroo (*Dendrolagus lumholtzi*). Aust. Mammol. 41, 57–64.

Wet Tropics Management Authority, 2020. Notophyll rainforests and thickets of the wet tropics bioregion. Available from:https://www.wettropics.gov.au/site/user-assets/docs/factsheets/wtmaVMWTB6a13d.pdf (1 March 2020).

Winter, J., Burnett, S., Martin, R., 2019. *Dendrolagus bennettianus*. The IUCN Red List of Threatened Species 2019: e.T6426A21957127. Available from:https://doi.org/10.2305/IUCN.UK.2019-1.RLTS.T6426A21957127.en (01 March 2020).

Woinarski, J., Burbidge, A.A., 2016. *Dendrolagus lumholtzi*. The IUCN Red List of Threatened Species 2016: e.T6432A21957815. Available from:https://doi.org/10.2305/IUCN.UK.2016-1.RLTS.T6432A21957815.en (01 March 2020).

CHAPTER

6

Rehabilitation of Lumholtz's Tree Kangaroo Joeys

Margit Cianelli[a,b] *and Katrin Schmidt*[b]

[a]Tree Kangaroo and Mammal Group & Tablelands Wildlife Rescue, Atherton, QLD, Australia

[b]Lumholtz Lodge, Upper Barron, Atherton, Far North Queensland, QLD, Australia

OUTLINE

Tree Kangaroos
https://doi.org/10.1016/B978-0-12-814675-0.00002-6

INTRODUCTION

Rehabilitation of tree kangaroos is the attempt to raise an orphaned joey (marsupial pouch young) in a way that prepares it for life in its natural environment to ensure its survival after eventual release. Rehabilitation of orphaned tree kangaroos should only be considered when a wildlife carer is sure that:

- The animal is viable to raise
- Health status can be monitored throughout
- Physiological and psychological development can be nurtured
- Survival skills can be taught and muscle development encouraged
- The joey can be raised to be independent and eventually be released with a good chance of survival.

If a joey fails to thrive or is physically disabled, transfer to an ex situ facility (zoological institution) might have to be considered.

Because all native wildlife in Australia is protected, every wildlife carer needs a permit to look after any wildlife individual. The permits can be obtained from the Queensland Department of Environment and Science by private persons for every individual animal that comes into care. Alternatively, if a wildlife carer is a member of a registered Wildlife Rescue Group, he/she will be covered by a group permit.

The Lumholtz's tree kangaroo (*Dendrolagus lumholtzi*) is locally abundant on the Atherton Tablelands, although there have been occasional sightings as far south as the Cardwell Range. It is unknown if their distribution extends further North than this, overlapping with the other Australian tree kangaroo species, the Bennett's tree kangaroo (*Dendrolagus bennettianus*). The conservation status of the Lumholtz's tree kangaroo was listed as 'Least Concern' by the Queensland (QLD) Department of Environment and Science in 2015, but the status has since been elevated to 'Near Threatened' following a submission to the QLD Threatened Species Committee, matching that of the International Union for Conservation of Nature (IUCN) Red List (Woinarski and Burbidge, 2016). Therefore, a carer does not need a special permit to look after a Lumholtz's tree kangaroo, as would be the case when looking after an endangered species.

The 'Tree Roo Rescue and Conservation Centre Ltd' (TRRACC) in Malanda on the Atherton Tablelands is an organization looking after mainly adult tree kangaroos that have been injured or affected by car strikes, dog attacks, or other disabling incidences. It is a non-profit organization and registered charity in Australia (Tree Roo Rescue and Conservation Centre (TRRACC), 2019). They liaise closely with Australian zoos and wildlife parks (via the Zoo and Aquarium Association [ZAA]), where some of the un-releasable animals are placed for education, display, and captive breeding programs (Chapter 22).

This chapter will provide information on hand-rearing Lumholtz's tree kangaroo joeys from when they first come into care to eventual release. Included is an overview of the reasons for bringing in joeys for care and then the step by step process for rehabilitation and release. As tree kangaroos have such different needs and requirements compared to other macropods, only experienced carers with access to appropriate facilities should look after them.

REASONS AND CIRCUMSTANCE FOR BRINGING TREE KANGAROO JOEYS INTO CARE

Unfortunately, certain circumstances make it necessary for tree kangaroo joeys to be raised by a wildlife carer. This is most often the case when the mother is injured or killed due to car strike or dog attack, or when joeys are found seemingly abandoned.

Car Strike

There are at least 20 tree kangaroo fatalities a year due to car strikes (Shima, 2018), although the actual number may approach 50, as some fatally injured animals may be able to move off the road and therefore remain undetected. More males than females are hit and killed on the road, probably because they are dispersing in search of their own territory (Shima, 2018). The majority of road deaths of females are adults of reproducing age and joeys rarely survive. Those joeys that do survive are taken to rescue facilities, where they go into managed care by experienced wildlife carers. Terrestrial kangaroo species have a deep pouch that appears to protect the joey from being expelled upon impact with a car. Tree kangaroos have a shallower pouch and consequently joeys are often cast out upon impact with a vehicle.

Dog Attack

Dog attacks might be much more frequent than reported, but most attacks occur away from roads, therefore going unnoticed. The growing human population on the Atherton Tablelands has resulted in an increase in the abundance and widespread distribution of domestic dogs. Dingos naturally prey on tree kangaroos, and this threat has increased due to a rise in the number of dingo-domestic dog hybrids. Like dogs, hybrid offspring have two breeding seasons a year, compared to only one in pure-race dingos. If the macropod mother drops the joey during a dog/dingo pursuit, it often remains unnoticed by the dog, but is likely to perish if not found and rescued. Dog attack victims rarely survive due to the extent of injuries.

Further contributing to dog attacks is the tree kangaroo's natural flight behavior. When startled by a dog or dingo, tree kangaroos descend from a safe position in a tree to attempt to escape on foot. However, they are much slower on the ground than other terrestrial macropods and are no match for a canine predator. Notably, dingos usually consume their prey; in contrast, dogs mostly just chase and kill. Victims of dog attacks appear to have little external damage and death generally occurs due to internal injuries (McKenzie, personal communication).

Abandoned Joeys

There could be many reasons for mother and joey to become separated. The mother might have been injured, incapacitated, or otherwise unable to care for her joey. In other scenarios, a joey may take flight and become separated from the mother. However, joeys may also be abandoned by their mother.

Because the maternal care provided by marsupial mothers is energetically expensive and time consuming, sick, unviable, or deformed offspring are generally abandoned. This allows the female to return to estrus and produce a healthy infant that has a greater chance of survival. However, it appears that a greater number of seemingly healthy tree kangaroo joeys are found 'abandoned' than joeys of other macropod species (Cianelli, personal observation).

Based on personal observations, it may be possible that tree kangaroo mothers purposely abandon an overly demanding joey, if maternal care becomes too strenuous. The challenge of living in a three-dimensional habitat poses a high demand on the species. Joeys need to be taught to climb, recognize dead branches that might break underweight, select appropriate food sources, and develop other essential life skills. A boisterous joey can become too demanding for a first-time or run-down mother and could potentially exhaust her reserves.

A hand-raised young female living in the rainforest with a large milk-dependent joey lost close to 30% of her body weight, despite a daily intake of highly nutritious foods, such as almonds, chickpeas, and sweet potatoes

(Cianelli, personal observation). This may be an indication of the high energy expenditure and resultant vulnerability of tree kangaroo mothers raising joeys. Despite the significant reproductive investment, self-preservation instincts might override the maternal bond with their offspring and sometimes cause a joey's abandonment.

REHABILITATION PROCEDURES

Assessment of the Joey

In general, it is quite rare for more than three viable tree kangaroo joeys to come into care in any given year (Cianelli, personal observation). Some years, carers do not receive any joeys despite the occurrence of female fatalities. Upon reception into care, the first decision to be made is whether the joey is viable, which depends on its stage of development and the level of injury sustained (if any). Table 6.1 illustrates an age estimation matrix based on foot and head lengths.

The smallest hand-reared joey in the last 30 years weighed 110 g with an estimated age of two and a half months according to the age matrix (Table 6.1). His skin was very thin, blood vessels were visible, ear pinnae attached, and eyes closed. The immune system of such a small joey is still developing and a carer must take into consideration that it is difficult to provide the immunity that the joey would obtain from its

TABLE 6.1 Lumholtz's tree kangaroo joey age estimation matrix, using head and pes lengths, with prediction intervals based on error estimates.

Head length (mm)	Estimated age (days)	Prediction interval (days)	Pes length (mm)	Estimated age (days)	Prediction interval (days)
8	1	0–2	4	10	8–13
9	4	2–5	5	15	12–18
10	6	4–8	6	20	16–23
11	9	7–11	7	24	20–28
12	11	9–14	8	28	24–32
13	14	11–17	10	37	32–42
14	17	14–20	12	45	39–50
16	22	19–26	14	53	47–59
18	28	24–32	17	64	57–70
20	34	29–38	20	74	67–81
23	42	37–47	24	87	79–95
26	51	46–56	28	99	91–107
29	60	55–65	33	113	104–121
32	69	64–74	38	125	117–133
35	78	73–84	44	139	131–147
38	87	82–93	50	152	144–160

TABLE 6.1 Lumholtz's tree kangaroo joey age estimation matrix, using head and pes lengths, with prediction intervals based on error estimates—cont'd

Head length (mm)	Estimated age (days)	Prediction interval (days)	Pes length (mm)	Estimated age (days)	Prediction interval (days)
41	97	91–102	56	165	156–173
44	106	100–112	62	177	168–186
47	115	110–121	68	190	180–200
50	125	119–131	76	208	195–221
54	138	132–144	84	228	211–245
58	151	144–158	92	252	229–275
62	165	156–173	102	287	253–320
66	178	169–188			
72	200	188–212			
80	231	213–249			
88	266	239–292			
96	305	267–342			

Reproduced from Johnson, P.M., Delean, S., 2003. Reproduction of Lumholtz's tree-kangaroo, Dendrolagus lumholtzi *(Marsupialia: Macropodidae) in captivity, with age estimation and development of the pouch young. Wildlife Res. 30, 505–512, with permission from CSIRO Publishing.*

mother's colostrum-rich milk. From previous experience, a joey weighing less than 100 g may not be viable, as, due to a lack of immunity, ensuing health issues generally lead to life-threatening complications or even death before maturity.

The joey initially needs to be checked for internal and external injuries, such as possible organ damage, fractures, and/or lesions. The joey might be cold and dehydrated and must be kept warm and stabilized. Depending on the stage of dehydration, electrolytes must be administered either by mouth, subcutaneously, or in severe cases, an intravenous drip may need to be administered by a veterinarian.

Required Temperature and Protection

An abandoned animal, specifically a furless joey, is likely to have a low body temperature, as it is unable to thermoregulate at this stage. In hypothermic animals the digestive system is disrupted, as nutrients are not adequately absorbed and metabolized. Milk formula or electrolytes should only be given once the joey has reached the required body temperature of around 35 degrees Celsius.

To warm a joey, it is best placed in a pouch liner within a warmer fleece pouch and placed inside a carer's shirt for the absorption of body heat; this also promotes bonding. Other heat sources include a heat pad, warm water bottle (the water must not be at boiling temperature and the bottle needs to be completely wrapped in a towel), or a gel pack. A humidity crib is an ideal environment for a furless joey; the temperature should be maintained at 30–32 degrees Celsius. For a furred animal, the temperature should be about 28–30 degrees. It is strongly advised to use a temperature gauge to appropriately control the temperature of any artificial heat source.

The Rearing Environment

A new joey must be kept quiet and stress-free at all times. It can react adversely to loud or unfamiliar noises and other impacts, often without showing obvious signs of stress. It is important that only one carer handles the young animal when it first comes into care, which also assists bonding. Generally, contact with pets should be avoided, as domestic animals can cause stress and transmit diseases such as toxoplasmosis.

Joeys need to be kept hygienic and clean, which requires sterilizing bottles, teats, and work areas. Pouches need to be changed several times a day and furless joeys should be bathed daily. Furless joeys need to be kept in a pouch, rather than be covered with a blanket; the restricted space provided by a cloth pouch best simulates a mother's pouch. Once older, it is imperative to eliminate physically dangerous situations in a household setting. Although the joey must be safeguarded at all times, it should be given the opportunity to explore (Fig. 6.1).

FIGURE 6.1 Even at six and a half months old, with no climbing skills, tree kangaroo joeys do not look uncomfortable in a tree. In Upper Barron, Far North Queensland. *Source: Karin Semmler.*

Equipment Required

The equipment needed is mostly the same as for terrestrial macropod joeys and baby possums. Wildlife carers usually have standard equipment at hand, as would most zoos or wildlife parks. Equipment can be kept for long periods until needed, but will have to be sterilized when a joey comes in. Formula and medication have an expiry date and have to be checked and possibly replaced from time to time. Keeping formula and medication refrigerated usually extends shelf life.

Recommended Equipment

- Glass bottles and silicone teats, as well as baby bottles to make up formula
- Syringes and possibly cannulas for feeding very small joeys or administering medication
- Cotton pouch liners and pouches of a warmer material, such as flannelette, wool, or fleece (synthetic materials may lead to overheating)
- Electronic scales
- Mild antibacterial liquid soap for bathing the joey and for keeping personal hygiene before and after handling an animal
- Moisturizer such as Alpha Keri® oil or any baby lotion for joey's skin after a bath (baby oil is unsuitable, as it blocks pores)
- Tissues and cotton wool to toilet the joey
- Calipers and a tape measure to measure and record growth for age estimation with available charts (Table 6.1; Johnson and Delean, 2003)

Diet and Nutrition

Macropod milk composition changes throughout joey development, but there are few data on tree kangaroo milk specifically. Studies have shown that macropod milk will contain increased levels of sulfur-containing amino acids when the joey begins to grow fur and, toward late lactation, carbohydrates decrease, whereas lipid and protein levels increase as the joey develops (Jackson, 2003; Rich, 2012). Macropods must therefore be raised on specialized milk formulas; cow's milk containing lactose can lead to severe diarrhea, cause cataracts, and even result in death (Stephens, 1975).

Milk Formulas

There are a variety of milk formulas available; some brands offer specific formulas for different stages of joey development to imitate the change in macropod milk.

The three main macropod formulas are as follows:

1. Wombaroo® milk replacers (Wombaroo® Food Products)

Wombaroo® offers four different varieties of macropod formula to mimic the changes of macropod milk during lactation, with the following product names:

- **Kangaroo Milk Replacer <0.4**

For joeys with less than 40% pouch life completed; furless, pink skin, eyes closed, ears down.

- **Kangaroo Milk Replacer 0.4**

For joeys with 40% of pouch life completed; furless, darkening skin, eyes just opened, ears nearly erect.

- **Kangaroo Milk Replacer 0.6**

For joeys with 60% of pouch life completed; fine to short fur, ears erect.

- **Kangaroo Milk Replacer 0.7**

For joeys with 70% of pouch life completed; fur short to dense, spends time out of pouch.

Wombaroo® is the preferred formula of many carers for macropod species, such as agile wallabies (*Notamacropus agilis*, syn. *Macropus agilis*), red-legged pademelons (*Thylogale stigmatica*), and swamp wallabies (*Wallabia bicolor*). However, in Lumholtz's tree kangaroos, this formula has only been used once by M. Cianelli.

2. Biolac® (Biolac, 2019)

Biolac® is another formula designed to imitate macropod milk changes during lactation. There are three types for various developmental stages, with the following product names:

- **M100**—for furless joeys
- **M150**—for furred joeys
- **M200**—for pouch emerging joeys

Biolac® has not been tried on Lumholtz's tree kangaroos to date.

3. Divetelact® (Sharpe Laboratories)

Divetelact® is a generic lactose-free formula designed not only for macropods, but for various mammal species intolerant to lactose. It has been available for many years, over which the composition of the formula has been changed several times, not always to the satisfaction of macropod carers. Lumholtz's tree kangaroo joeys generally do well on this formula (Cianelli, personal observation).

A carer might use additives to Divetelact® formula, such as 'Impact,' an artificial colostrum made by Wombaroo® Food Products. Egg yolk or whey protein powder may be added for extra protein (however, some joeys can develop an allergy to egg yolk). Due to Divetelact®'s low energy content, it has been suggested to add mono- or polyunsaturated fats such as canola oil. The addition of saturated fats, such as cream, can lead to malabsorption of calcium (Rich, 2014).

Joeys should be fed 10–20% of their body weight over a 24-hour period. (As Wombaroo® is more concentrated, the amount should be at the lower scale of 10–15%, whereas for Divetelact® it should be 15–20%.) For example, if a joey weighs 500 g, it should be fed about 75–100 ml of Divetelact® over 24 hours, spread over several feeds (about 12–16 ml every 4 hours). A furless joey should be fed every 3 hours. A 200 g animal should be given 30–40 ml spread over eight feeds per day (5 ml every 3 hours). Older, furred joeys can be fed five times a day, especially once they show interest in solids. Formula should not be microwaved, as this can denature proteins. It should be warmed up to about 32 degrees Celsius in a container of hot (not boiling) water. Heated leftover formula must be discarded. Made-up formula is kept in the refrigerator to last for 24 hours, and only the amount needed per feed is warmed up.

Carers might have preferences or use different formulas for different macropod species. For Lumholtz's tree kangaroo joeys, Divetelact® is the preferred formula (Cianelli, personal observation). When Wombaroo® was tried on a female joey, the animal developed well, but the formula was considered inferior to Divetelact® (Cianelli, personal observation). As Wombaroo® is more concentrated than other formulas, less volume needs to be fed and the joey often appeared still hungry after a feed. To prevent dehydration, water was offered between feeds, which meant the animal had to be woken up more frequently. This often resulted in lack of sleep and weight gain while the joey was exclusively on milk formula.

Inoculating the Joey with Gut Content

The first tree kangaroo joey raised by M. Cianelli weighed only 110 g when he came into care and was less than 3 months old, with a low chance of survival. Despite all odds, he developed well, after overcoming some initial setbacks. He started to show interest in solids at about 6 months old and was introduced to some rainforest leaves. He learned to climb in a suitable enclosure and at around 15 months of age he was exposed to the real tree kangaroo world, the rainforest, for several hours of every day. Unexpectedly, he died less than a month after spending most of the day in the forest. A pathology report stated that he died from calcification of the heart muscle. His stomach was inflamed and swollen.

A second joey that came in soon after survived, another died, followed by two survivors and another fatality. It took a while to realize that all joeys that had died after partial release had weighed below 500 g when they came into care. Also, most of these joeys, at some stage, tried to intensely lick the carers mouth. This observation triggered the assumption that tree kangaroo mothers orally pass something essential to the joey for its survival. As 500 g is about the weight when joeys show interest in solids, this might be the time when special bacteria need to be introduced into their gut flora. Many of the tree species that tree kangaroos feed on are highly toxic. It is likely that the mother passes on bacteria to the joey that facilitate the digestion of foliage containing potentially harmful secondary plant compounds, such as toxins and digestion inhibitors. Such transfer was recently observed by Heise-Pavlov (Heise-Pavlov, personal communication).

Professor Rick Speare (formerly at James Cook University School of Public Health and Tropical Medicine), a leading parasitologist, confirmed this theory, stating that protozoa and large bacteria aid to break down the toxins, thus neutralizing them (Speare, personal communication). Professor Speare also speculated that nematodes in the tree kangaroos' stomach may live in symbiosis with the protozoa. He suggested administering stomach contents from fresh tree kangaroo road kills to the joeys. Preferably, this should be done as frequently as possible. Furthermore, it is important to source stomach content from the area where the esophagus enters the stomach, as this would be where the mother's regurgitate originates from. Stomach content needs to be transferred as fresh as possible, as exposure to air results in oxidization and darkening, rendering this part of the sample unsuitable. A sufficiently large sample size ensures that unexposed matter can be transferred. About 1–2 ml seems sufficient per dose. For a beneficial effect on the gut flora of the joey, six to eight inoculations during the critical 4-month period between the age of 5–9 months are desirable (Fig. 6.2 shows example data for age and intervals at which inoculations were provided). After inoculation, a large variety of leaf species has to be offered for the protozoa to thrive. Since giving stomach content to young joeys, survival rates have increased, most likely due to reduced adverse reactions to browse toxicity.

Introduction to Solid Food

Growing up with the biological mother, a joey is weaned at about 13–16 months of age (Johnson and Delean, 2003). From about 6 months old, the joey shows increased interest in solid foods. For hand-raised joeys, leafy greens such as rocket, water cress, kale, and Asian greens are a good start. These initial solids offer opportunities to explore textures and flavors, rather than to supplement milk feeds. Within a couple of weeks, the joey becomes more adventurous and a greater variety of solids can be introduced. Green beans, capsicums, sweet corn kernels, and thin slices of apple and sweet potato are favorites, as are cooked chickpeas. The joey should then slowly be introduced to rainforest leaves. It is important to only offer benign species at this stage, such as brown bollywood (*Litsea leefeana*) and hairy-leafed bollygum (*Neolitsea dealbata*), before inoculation with stomach content occurs and a tolerance to the leaf toxins can be established. Leaves should be offered in bunches of long branches so that the joey can choose between new shoots and older browse. Tree kangaroos will also nibble bark, blossoms, fruit, mosses, and epiphytes, but are predominantly browsers, and large quantities have to be ingested, as leaves are nutrient poor. They are foregut fermenters and

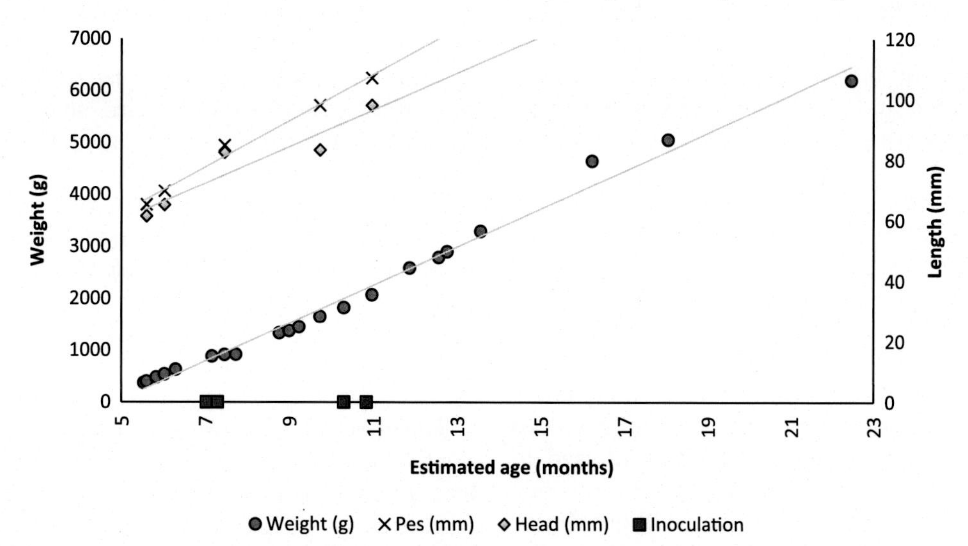

FIGURE 6.2 Weight and length measurements for a hand-raised, male Lumholtz's tree kangaroo joey against the estimated age in months. Trend lines were included for weight, pes, and head lengths. Inoculation with stomach content occurred on four occasions, marked on the *x*-axis according to the approximate age of the joey.

breakdown of cellulose takes place in the large sacculated stomach (Van Dyck and Strahan, 2008).

HEALTH, PHYSICAL, AND PHYSIOLOGICAL NEEDS

Hygiene, Toileting and Skin Care

Very young joeys cannot urinate or defecate on their own. Their mother stimulates excretion by licking the cloaca of the joey, which also helps to keep her pouch clean. Young hand-raised joeys must be cleaned to stimulate toileting, either before, or after feeding. This is best done with a tissue or cotton pad, moistened with warm tap water. Often a couple of gentle wipes over the cloaca are enough to start the process. It is important not to rub excessively, as this might damage the skin. Irritated skin can make the joey hold back urine or feces and, in extreme cases, even lead to a prolapsed rectum.

Furless joeys should be bathed at least once a day. The bath water should be about 35 degrees Celsius. Liquid soap is recommended and a moisturizer such as Alpha Keri® lotion can be applied after the joey is patted dry. There is no need to bathe furred joeys. If a furred animal has diarrhea and has soiled itself, a spot bath with soapy water is usually sufficient to adequately clean the animal. The cause of the diarrhea should be investigated, preventative measures taken, and appropriate treatment or medication given.

A high level of hygiene must be kept at all times. Before handling the joey, hands should be washed. The joey's bedding should be kept meticulously clean and pouches and liners should be changed regularly. When joeys are kept in a room unsupervised, it is advised to keep a few clean towels on the floor for the joey to urinate on. This way, the joey will not slip or jump around in the puddle, spreading it over the floor or bedding. Tree kangaroo joeys love to play; soft toys should be washed regularly in the washing machine and any enrichment equipment, such as logs, branches, and ropes, should be washed down or wiped with a diluted disinfectant solution.

Dehydration

When joeys first come into care, they often suffer from dehydration, particularly those that have been abandoned. A dehydrated joey looks skinny and listless and its eyes might be sunken. A good method to check for dehydration is pinching the skin at the back of the neck and pulling it up. In a dehydrated animal, the skin will stay raised, rather than gliding and flattening back into its natural position. If the animal is hypothermic, it has to be warmed to its normal body temperature of about 35 degrees Celsius before giving electrolytes. Vytrate® or other oral electrolytes are given by mouth. In severe cases, either sodium chloride solution or Hartmann®'s solution (compound sodium lactate) must be injected under the skin around the back of the neck (Fowler, 2007), at a rate of about 10% of the joey's body weight (e.g., if the joey weighs 400 g, it should be injected with 40 ml solution over the next 24–36 hours). Once the solution is absorbed, another dose can be administered. Full rehydration may require several days. Feeding milk formula should only be attempted once the joey shows signs of recovery, as a dehydrated animal cannot absorb food properly. Severely dehydrated animals may need intravenous electrolytes, which should only be administered by a veterinarian.

Diarrhea

Diarrhea can have different causes—it can be related to excess food intake, caused by poor care, or by fungal or bacterial infections. Joeys' feces are of a soft, toothpasty consistency while it exclusively feeds on milk in the pouch. A joey's milk intake must be monitored closely, as too much milk given in a single feed could result in diarrhea. Poor care due to lack of hygiene or unclean equipment can very quickly make a joey dangerously ill with bacterial or fungal diarrhea.

Nonbacterial diarrhea due to over-feeding or poor care can often be treated successfully with charcoal (Speare, 1988) or Bentonite. Charcoal is porous, which increases its surface area, enabling toxic molecules to be absorbed and transported out of the gut. Bentonite is a benign soil type that thickens the feces and supports better absorption during digestion. Both are effective and should be considered the first choice of treatment, as young tree kangaroo joeys are extremely susceptible to chemicals in medication.

Fungal and bacterial diarrhea should only be treated under a veterinarian's supervision. Nilstat® has been used for fungal or yeast-induced diarrhea (Sadler, 2009/2010). It should be given about 30 minutes before feeding and according to veterinary instructions (Campbell [Veterinarian Atherton], personal communication). Bacterial diarrhea is difficult to treat in tree kangaroos, as antibiotics also destroy beneficial bacteria that are essential for vital bodily functions. Even given as an injection, rather than orally, antibiotics can lead to more severe diarrhea. The reaction to the drug is often worse than the initial infection and joeys have died from treatment with antibiotics. If medications need to be used, veterinarians may consider Peptocyl® or Scourban®, as they contain only a mild dose of antibiotics.

Housing

As joeys grow and progress, they become increasingly active, climbing and swinging from curtain rods, and jumping on benches and furniture. It needs to be noted that the house environment is insufficient for the joeys' muscle and skill development. To stimulate mobility, they should spend several hours of every day in a well-equipped enclosure.

Physical Needs

As joeys grow older, food, exercise and housing requirements change. Their three-dimensional movements make looking after them a special challenge. Some animals are literally hyperactive and can wear a carer out. It is important that their agility is never underestimated. Joeys should be strictly supervised in any room with a variety of objects, glass, and heavy items that could expose them to accidents and injury. Too rapid an introduction of joeys from a domestic environment to the forest can be problematic, as their climbing ability would not have fully developed yet. An enclosure with challenging 'furniture' (climbing structures and ropes) is the best way to prepare a youngster for life in the forest.

The 'Jungle Gym'

A chain-link fenced enclosure covering an area of approximately 10 by 8 meters is an excellent 'jungle gym' for tree kangaroo joeys (Fig. 6.3). The fence is curved inwards at the top and wooden poles and large climbing logs hold up the shade cloth roof that spans over the entire structure and reaching 3.2 meters at its highest point. There is also 1-meter-high shade cloth covering the fence from the ground up to visually block the outside. Ropes, both tight and slack, are attached between the logs, and rope ladders can also be incorporated. Nets are a versatile enrichment material that can be installed, both horizontally and vertically, catering effectively toward developing climbing skills.

The ground should be covered with small grain gravel to avoid the risk of joeys contracting melioidosis, a soil-borne disease caused by the bacterium *Burkholderia pseudomallei*, that can be fatal (Shima et al., 2019). Vertical pipes attached to some of the support structures can serve to hold cut browse. Suitable ground cover legumes (e.g., pinto peanut) can be planted as an additional food source. Accessible from within the

FIGURE 6.3 A tree kangaroo joey climbing upside down in a 'jungle gym' enclosure in Upper Barron, Far North Queensland. Playing in the jungle gym encourages muscle development and teaches joeys the climbing skills they need for life in the forest. *Source: Katrin Schmidt.*

enclosure, a covered space (e.g., a modified tin garden shed), partially enclosed on two sides and with an elevated shelf that can be accessed via logs, is essential. The shelf serves as a resting space and feeding platform.

First Experiences in the 'Jungle Gym'

Lumholtz's tree kangaroo's permanent pouch emergence is at about 10 months of age (Johnson and Delean, 2003; Animal Diversity Web, 2020). This is roughly the age when they should get exposed to the enclosure for the first time. The first outings in the enclosure are only for short periods and a carer must stay for supervision and reassurance at all times. Tree kangaroo joeys have different body proportions to adults and feel very comfortable just hanging on a rope freely by their arms or upside down by their arms with feet in contact with the rope. Interestingly, in previous observations, when they first hang from a rope by their hands and feet, they do not seem to know in which direction to move. In such a position, anatomical restrictions do not allow them to move in the direction of their feet, which needs to be learned by repeated experience. It is observations like these that manifest that tree kangaroos are not 'born' climbers and some of their skills are learned rather than innate. This is also supported by the fact that tree kangaroos have the longest mother-joey relationship of all macropods; up to 3 years. When joeys first come out of the pouch to be exposed to a life in the rainforest, their mothers select 'safe' climbing areas, such as regrowth or forest with a lot of vines or undergrowth, so that joeys might not be prone to injury, should they fall from some height. Mothers have also been observed offering their shoulders as support for their joeys descending a tree trunk (Stirn, personal communication).

As gym outings are extended, joeys do not need to be supervised at all times. In preparation for the first experience in the rainforest, large live branches are moved into the gym. Joeys now also need to be exposed to bark and dead limbs and learn that a dead limb might break off under their weight. To fall from a dead limb from great height in the forest could prove dangerous or even fatal.

Making the 'Jungle Gym' Safe

In the past, a sub-adult male was kept in the jungle gym for a few hours most days. Incidentally, a wild adult male from the area reacted territorially to his presence and approached the enclosure. Both animals grew increasingly agitated and consequently the adult male climbed the fence of the enclosure and climbed onto the shade cloth roof. If the adult had broken through the shade cloth, neither animal could have escaped, which would likely have resulted in the sub-adult being seriously hurt or killed, and the adult becoming very stressed from being trapped in the enclosure. To prevent such dangerous escalations from occurring in the future, the enclosure was entirely covered with robust netting in addition to the shade cloth, creating a safe environment for joeys.

PSYCHOLOGICAL NEEDS AND BEHAVIOR

Psychological Needs

Psychological needs of a joey must include being nurtured from a young age. Physical closeness, stimulation, encouragement, and praise are extremely important for his/her mental health and behavioral development. Stimulation promotes the joeys' interest in what is going on around them and motivates them to be active. They react to vocal encouragement and praise, and the tone of voice, together with some physical closeness, are important for fostering their confidence and stamina. A carer can provide a lot for a hand-raised joey to successfully mature mentally, so that it is not disadvantaged later in adult life. Tree kangaroos appear to be more behaviorally sophisticated than other macropods (Cianelli, personal observation). This theory appears to be supported by the fact that they live mostly a solitary life, live in a three-dimensional environment of tree canopies, joeys evidently enjoy playing, and some tree kangaroo behavior suggests processes of heightened mental awareness.

Behavioral Observations

A hand-raised female tree kangaroo living in the rainforest displayed some strange behavior toward a joey that was brought into care. The joey was rescued on the highway situated alongside the carer's 160 acre property. The adult female had been exposed to several orphaned joeys previously and had never shown any aggression. In this case however, she behaved restlessly and was agitated for several days. She then proceeded to fiercely and repeatedly attack the carer, ripping clothes and causing bleeding wounds and lacerations. As soon as the joey was removed from the premises and passed onto another carer, the female changed back to her normal behavior. Due to the female's acceptance of previous orphaned joeys, the carer ruled out jealousy for the attacks, but rather assumed that the mother of this particular joey might have been known to the female and, living on the same property, that they might have had territorial or other issues. Possibly, the female recognized the connection and therefore rejected the joey.

The same hand-raised female raised two joeys within 3 years, both of which were fathered by wild-born males. When the joeys were about 9 months old, the female displayed the urge to distance herself from them for some hours of the day. She came into the house most evenings, after spending the day in the

FIGURE 6.4 Reuniting a tree kangaroo joey with its mother in Upper Barron, Far North Queensland, whilst an orphaned joey looks on. *Source: Amanda French.*

rainforest and enjoyed having a rest in the carer's room. At around 2:00 a.m. she wanted to go back to the forest, but would push her joey away, refusing to let it get near the pouch, and left without it. Initially, the joey was insecure about this rejection, but it soon seemed to enjoy the added hours of rest. In the morning, the carer would find the mother via radio tracking and reunite the joey with her (Fig. 6.4). The mother greeted her young vocally, climbed down from the tree and groomed and licked her offspring before they hopped off together. This behavior also shows the trust both the hand-raised mother and the wild-born joey displayed toward the carer.

This was a unique situation in which the relationship between a mother and her joeys could be observed (Box 6.1). Research on the behavioral traits of mother and joeys was done in collaboration with the Centre for Rainforest Studies at The School for Field Studies. The findings, summarized in Chapter 7 of this book, reinforce knowledge based on experience for the preparation and release of rehabilitated orphaned tree kangaroos into the rainforest. Unfortunately, the bond between mother and joeys, as well as the bond they shared with the carer, could not be further explored. The first joey was taken by an eagle, and when the female was 7 years old, carrying her third advanced pouch young, she was killed by feral dogs. The joey could not be saved.

RELEASE

The first visit to the forest is more unnerving to the carer than to the 15–18 month old juvenile, as it can get out of reach to worrying heights. A lower regrowth area is preferred to mature forest for the safety of the juvenile. Importantly, the carer must be confident that skill development has sufficiently progressed in the 'jungle gym' for the young animal to be introduced to its natural environment.

BOX 6.1

Kimberley's story (by Margit Cianelli)

Of all the tree kangaroos that I reared, Kimberley had, without a doubt, the most personality.

Kimberley came to me as a healthy and bright joey at about 8 months old, after falling from a tree into a swimming hole. Kimberley quickly adjusted to her new home. She was very active, practiced her climbing skills on furniture and curtains around the house, and in the jungle gym. Introducing her to the forest was nerve-racking; being boisterous and confident, she didn't stick to the safer regrowth area and got herself up to some daring stints in established rainforest. At 2 years old, she was spending most of the day in the forest, but still came home for the night. When she saw me, she would climb down from the tree, hop on my shoulder, and enjoy the ride back to the house. Kimberley loved her quiet time at home. She usually came into my bed early in the morning, wanting cuddles and a long scratch behind her ears. If I was too sleepy to respond, she would try to wake me by putting her head under mine and lifting it up. After our cuddle time, she was eager to go out for another day in the forest.

At close to 3 years old, Kimberley showed interest in a male tree kangaroo close to the house. Kimberley and her new friend, Gregory, spent many weeks together. Six months later, Kimberley held her pouch open to reveal a tiny joey. She became very moody over the next few weeks, cuddling up one moment and biting the next. But she was a loving mum and when her little one, Monty, started spending time out of the pouch, she looked after him caringly, grooming him in the evenings (Fig. 6.5) and teaching him skills during the day. Even though Kimberley was well established in the forest by now, she came home most evenings. When she felt like coming in, she would come down from the tree, followed by Monty; Kimberley hopped after me and Monty hopped after her, back to the house. In the forest, Monty was very wary, always in flight mode, and I could not get close to him. But once inside the house, he was calm and never showed signs of being frightened.

FIGURE 6.5 Kimberley is nursing her joey, Monty, while gently fending off the orphaned Dobby. In Upper Barron, Far North Queensland. *Source: Margit Cianelli.*

When Monty was older, Kimberley often refused to take him with her when she went back outside. She seemed to enjoy her outings alone and Monty did not mind sleeping in with me. In the mornings, I took Monty outside and Kimberley would descend from where she was and pick up her son for another day in the forest together. I cannot put into words these precious times I shared with Kimberley and Monty. She trusted me to watch Monty, often letting me take over in the evenings, and Monty enjoyed being in my shirt while his mum had a break. Monty also had a special relationship with Gregory—they often displayed affectionate behavior, cupping each other's faces in their hands, licking and grooming each other. Monty had a snowy white claw on his right little finger. This was quite unusual, and a photo later revealed that Gregory had a white claw on the same finger. There was no doubt they were father and son!

At 15 months old, Monty was tragically grabbed by a wedge-tailed eagle and died when the eagle dropped him. Kimberley was very

Continued

BOX 6.1 (*cont'd*)

confused and appeared distressed and listless the following weeks. I often found her sitting by the creek, looking at the spot where I found Monty. It took her months to get over the loss of Monty. Eventually, Kimberley appeared to be her old self again, and cared for her second joey, Holly, as she had done for Monty. When Kimberley's third joey, a boy, was about 8 months old, I saw her for the last time. That night, they were both killed by feral dogs. A week later Kimberley would have been in my life for 6 years.

Kimberley left such a void and it took a long time before the sadness could be replaced by happy memories. As able as Kimberley was to live in the rainforest, she chose not to give up the world she grew up in. We had such a strong bond and she was full of love and affection. To know this special, gorgeous animal was a privilege and I will cherish our relationship forever.

Presence of some thick, established vines is beneficial for the juvenile to grab and hold on to, should it fall. Sometimes the joey ascends to a place where it becomes too scared to climb down; it might remain in the same spot for hours, even overnight. Eventually, most juveniles find their way back down. Fortunately, their climbing skills and confidence improve rapidly. The outings should initially be relatively short and always under supervision.

Once the juvenile spends many hours in the forest and does not need constant supervision anymore, it is fitted with a light-weight radio-collar (less than 30 g). While most juveniles come back to the house on their own accord, they might get stuck somewhere, get lost, chased, or be otherwise unable to return. A radio-collar gives peace of mind to the carer, ensuring that the juvenile can be found and brought back to safety at any time. Tree kangaroo juveniles accompany their mothers longer than other macropods, up to 3 years. The survival chances of a young, inexperienced sub-adult would be limited without the opportunity to return to the carer when they feel the need to.

Once the joey spends most of the day in the forest, it mainly eats rainforest leaves and occasionally fruit, bark, blossoms, mosses, and epiphytes. It has been observed that tree kangaroos supplement their diet with large insects like cicadas (Schuerler, 2019), or even an occasional kill, which makes them opportunistic meat eaters. Some carers thus offer older joeys cooked chicken about twice a week, which they eagerly devour. As potential predation on animals such as frogs, snakes and birds has been observed in captivity (Johnson et al., 2002), they could take advantage of the opportunity to grab an egg or young bird from a nest to supplement their nutrient-poor browse diet. This is supported by the behavior of a 15 month old male juvenile; when his carer picked up a small native bird egg from the forest floor, he eagerly snatched it from her hand and put the egg into his mouth, tilting his head backwards as he devoured the liquid contents. He instinctively knew, not only to recognize and eat eggs, but also how to eat them most efficiently.

Lumholtz's tree kangaroos are considered to be cathemeral (i.e., active periods during day and night) rather than nocturnal (active only at night), as previously described in older publications (Strahan, 1983). Most hand-raised juveniles will come into the house at night for many months, possibly because they feel insecure in the forest due to increased nocturnal activity. When they come in, the carer offers a bowl of food, like sweet potato, kale, almonds, and chickpeas, in addition to some browse. They are also offered a small bowl of rainforest soil, which they eagerly eat from. This is collected from a location where released adult tree kangaroos were observed eating soil, which aids in detoxification of browse and provides them with minerals and trace elements.

Initially, juveniles like to stay in the house overnight, but this changes as they grow older and reach sexual maturity. An adolescent of between 2 and 3 years of age will look for his own territory in the wild. Some of the hand-reared juveniles may be able to establish themselves close to the carer's house if the area is vacant, which would enable access to a familiar environment and special treats. Others may search for a territory well away from the house. The carer follows them around (by radio-tracking) as they move to different areas in search of a vacant habitat until they appear to have established their own territory. At this time, their collar is removed, and the carer fares them goodbye, hoping they will do well in their newly chosen habitat.

REHABILITATION DATA

Importance of Record Keeping

It is vital to keep records of animal data in order to later reference valuable details to facilitate comparisons and observe patterns in an individual's health, development, and progress. As little is known about raising tree kangaroo species, records are invaluable for exchanging information with other tree kangaroo carers and zookeepers. It was only through such records, that it became apparent that once hand-raised joeys lived partially in the forest and were exposed to leaf toxins, they died. Since carers offer stomach contents of fresh tree kangaroo road kills to joeys, the survival rate has increased dramatically.

Important Types of Data to Record

Weight records and developmental milestones are the most important data to collect. Records must be kept throughout the joey's care. Weights should be recorded consistently, preferably on the same weekday (Fig. 6.6). Any relevant behavioral observations and milestones should be noted (e.g., very young joeys grabbing objects for the first time, older joeys climbing up a free-hanging rope or climbing upside down, sniffing or nibbling solid food items for the first time, etc.).

Head and pes measurements recorded from several individuals raised by their natural

FIGURE 6.6 Recording a tree kangaroo joey's weight, in Upper Barron, Far North Queensland. *Source: Margit Cianelli.*

mothers in a captive environment at Queensland Parks and Wildlife Service facilities at Pallarenda, Townsville provide guidance for estimating age of pouch young (Johnson and Delean, 2003). Measurements of hand-raised joeys proved to be very similar to these records (Cianelli, personal observation).

When Animals Cannot Be Released

There are a few health issues in incoming tree kangaroo joeys that would render them un-releasable. Most un-releasable tree kangaroos are adults suffering from debilitating, often permanent, injury after a car strike or dog attack. There was an instance when a juvenile became blind after suffering stress following a dog attack. Also, an unusually high number of displaced tree kangaroos seem to be vision-impaired and some of these might also be un-releasable. However, a large proportion of animals deemed un-releasable can still become valuable breeding individuals and serve educational purposes in zoos or wildlife parks, after going through the legalities with the Queensland Department of Environment and Science (Chapter 22).

SUMMARY OF LESSONS LEARNED

There has been much progress in the care and rearing of orphaned tree kangaroo joeys over the past three decades up to 2020. The most significant lesson learned is possibly the administration of stomach content from tree kangaroo carcasses to introduce appropriate gut bacteria in young joeys that are deemed releasable. This is hardly relevant for captive zoo animals, as they are not usually exposed to toxic browse. However, for joeys that are being released back into the wild, giving stomach content is vital.

Physical preparation for release has also seen improvement. As climbing does not come completely natural to tree kangaroos and skills need to be learned by repeat experience, exposure to a 'jungle gym' with rigid climbing facilities, as well as ropes and nets simulating moving branches, improves preparation for release into the rainforest. Joeys that are reared in tree kangaroo habitats have to be kept in enclosures that are completely safe from aggression of wild tree kangaroos.

The knowledge of tree kangaroo food plant species has significantly increased. Carers work with botanists and rangers to learn more about the distribution and location of food plants. An added bonus is that food tree lists are now available to landholders wanting to grow tree kangaroo browse on their properties (Tree Kangaroo and Mammal Group (TKMG), 2020).

Apart from practical improvements, significant knowledge has been gained in the medical treatment of tree kangaroos (Chapters 20 and 25). Some veterinarians on the Atherton Tablelands now have extensive experience in treating tree kangaroos. Cautious administration of drugs, especially antibiotics and wormers, is of vital importance.

Rearing *Dendrolagus* species is in many ways very different from raising terrestrial kangaroos. Current knowledge of raising tree kangaroo joeys should be shared and published, enabling more carers to engage in looking after these demanding species. Currently, a husbandry manual for Lumholtz's tree kangaroos in Australian zoos and wildlife parks is being compiled by a group of wildlife carers and scientists with relevant experience. It will contain a section on the hand-raising and rehabilitation of orphaned tree kangaroo joeys.

The rehabilitation of orphaned Lumholtz's tree kangaroo joeys by a human carer contributes to the conservation of this species primarily through release of animals back into the wild that would otherwise have been lost to the natural gene pool. Secondarily, un-releasable animals are provided to zoos and other facilities for educational purposes, fostering public awareness and supporting managed breeding programs (Box 6.2).

BOX 6.2

Kimberley, a hand-reared tree kangaroo as ambassador for her species (by Margit Cianelli)

Kimberley, a healthy and boisterous tree kangaroo joey, came into care at about 8 months old. Her curious nature enabled me to take her for talks at high schools, Kindergartens and other local venues, where she never ceased to captivate her audience. I run a Bed & Breakfast and my guests had often already heard of Kimberley through word-of-mouth. When Kimberley started venturing into the forest, the sight of her returning 'home' at the end of the day, fascinated both local and overseas visitors. It was usually their first encounter with this little-known species. Even when Kimberley was already well established in the forest, she dropped in for visits, often followed by a wild-born joey. Many times, I was told that this was one of the most inspirational wildlife experiences guests had ever had. Although wild in the forest, Kimberley was trusting and confident at home, even among strangers. Film crews from all over the world have come to film her and have obtained wonderful footage of her and her joeys. Through these documentaries, Kimberley has raised the awareness of literally millions of people for her species.

Kimberley was a true ambassador to all tree kangaroos.

ACKNOWLEDGMENTS

The authors would like to thank Lars Kazmeier for help with editing.

REFERENCES

Animal Diversity Web, 2020. *Dendrolagus lumholtzi*: Lumholtz's Tree Kangaroo. Available from: https://animaldiversity.org/accounts/Dendrolagus_lumholtzi/#9ED9ECA6-9396-11E1-8C44-002500F14F28. (19 April 2020).

Biolac, 2019. Marsupials: User Guide for Milk Formulas. Available from: https://www.biolac.com.au/marsupials (23 February 2020).

Fowler, A., 2007. Fluid Therapy in Wildlife. National Wildlife Rehabilitation Conference Proceedings. Available from: http://www.awrc.org.au/uploads/5/8/6/6/5866843/fowler_anne_fluid_therapy.pdf (23 February 2020).

Jackson, S., 2003. Australian Mammals – Biology and Captive Management. CSIRO Publishing, Collingwood, VIC, Australia.

Johnson, P.M., Delean, S., 2003. Reproduction of Lumholtz's tree-kangaroo, *Dendrolagus lumholtzi* (Marsupialia: Macropodidae) in captivity, with age estimation and development of the pouch young. Wildl. Res. 30, 505–512.

Johnson, P.M., Hawkes, M., Sullivan, S., 2002. Predation by Lumholtz's Tree Kangaroos *Dendrolagus lumholtzi* in captivity. Thylacinus 26, 6–7.

Rich, G., 2012. Nutritional Requirements of Juvenile Marsupials. Available from: http://www.awrc.org.au/uploads/5/8/6/6/5866843/rich_gordon_nutritional_requirements_juv_mars-2012.pdf (8 March 2020).

Rich, G., 2014. Which formula? Adding supplements to Wombaroo milk replacers. In: Staker, L. (Ed.), Macropod Husbandry, Healthcare and Medicinals. In: vols. 1 & 2. Lynda A. Staker, Mackay, Queensland, Australia.

Sadler, S., 2009/2010. Husbandry Guidelines for Agile Wallaby (*Macropus agilis*), Mammalia: Macropodidae. Course Compendium for Certificate 3 Captive Animals. Western Sydney Institute of TAFE, Richmond.

Schuerler, U., 2019. Zur Geschichte der Baumkaenguru-Haltung in europaeischen und einigen anderen Zoos. Bulette Berlin 7, 7–45.

Shima, A.L., 2018. Factors affecting the mortality of Lumholtz's tree kangaroo (*Dendrolagus lumholtzi*) by vehicle strike. Wildl. Res. 45, 559–569.

Shima, A.L., Berger, L., Skerrat, L.F., 2019. Conservation and health of Lumholtz's tree kangaroo (*Dendrolagus lumholtzi*). Aust. Mammal 41, 57–64.

Speare, R., 1988. Clinical assessment, disease and management of the orphaned macropod joey. In: Australian Wildlife. The John Keep Refresher Course for Veterinarians. Proceedings 104: 15–19. University of Sydney, pp. 211–296.

Stephens, T., 1975. Cataracts and kangaroos: nutrition of orphan hand-reared macropod marsupials. Aust. Vet. J. 5, 453–458.

Strahan, R. (Ed.), 1983. The Australian Museum Complete Book of Australian Mammals/The National Photographic Index of Australian Wildlife. Angus & Robertson, London; Sydney.

Tree Kangaroo and Mammal Group (TKMG), 2020. Lumholtz's Tree Kangaroo Food Plant List. Available from: https://www.tree-kangaroo.net/download_file/view/61/157 (19 April 2020).

Tree Roo Rescue and Conservation Centre (TRRACC), 2019. Tree Roo Rescue and Conservation Centre Ltd. Available from: https://www.treeroorescue.org.au/about_us (25 June 2019).

Van Dyck, S., Strahan, R., 2008. The Mammals of Australia, third ed. New Holland Publishers, Sydney.

Woinarski, J., Burbidge, A., 2016. *Dendrolagus lumholtzi*. The IUCN Red List of Threatened Species. Available from: https://www.iucnredlist.org/species/6432/21957815 (23 February 2020).

CHAPTER

7

How an Understanding of Lumholtz's Tree Kangaroo Behavioral Ecology Can Assist Conservation

Sigrid Heise-Pavlov[a] *and Elizabeth Procter-Gray*[b]

[a]Centre for Rainforest Studies at The School for Field Studies, Yungaburra, QLD, Australia

[b]Department of Medicine, University of Massachusetts Medical School, Worcester, MA, United States

OUTLINE

INTRODUCTION

In the late 20th and early 21st centuries, the science of animal behavior became more and more integrated into attempts to explain ecological patterns and dynamics (Sutherland, 1996). Studies demonstrated how behavioral traits and their variations influence population ecology phenomena such as a species' distribution, abundance, dispersal patterns, impact of predators, and responses to human-induced environmental changes (Sih et al., 2012). The importance of the inclusion of behavioral traits into research of animal population dynamics became more apparent with increasing demands for efficient species management and conservation (Blumstein and

Tree Kangaroos
https://doi.org/10.1016/B978-0-12-814675-0.00028-2

Fernández-Juricic, 2004; Buchholz, 2007). Cases in which behavioral aspects were integrated into species conservation gave evidence of how behavioral studies of species can inform conservation planning (Beissinger, 1997; Berger-Tal et al., 2015; Campbell-Palmer and Rosell, 2011; Knight, 2001; Martin, 1998).

Despite these encouraging examples, it took considerable effort and time before the sciences of ethology and conservation started to connect (Buchholz, 2007; Curio, 1996) and a new discipline, "Conservation behavior," emerged (Caro, 2007; Nelson, 2014). Lack of funding for behavioral studies, the perception that wildlife conservation deals with practical problems while the field of ethology is a "soft" science, the ignorance of relevance of a species' evolution for conservation, and differences in methodological approaches are a few obstacles that challenge interdisciplinary work in species conservation (Caro and Sherman, 2013). Furthermore, behavioral studies of cryptic species are problematic and thus often result in neglecting the inclusion of behavioral traits in planning conservation strategies for these species. However, recent advances in non-invasive and indirect methods of animal observations allow the detection of behavioral patterns, their links to ecological factors and processes (e.g. Crofoot et al., 2010; Kelly, 2008; Treves, 2000), and facilitate the assessment of the vulnerability of a species to environmental changes (Chapter 26; Cattarino et al., 2016; Van Dyck and Baguette, 2005).

This chapter summarizes the current knowledge of behavioral traits of the Lumholtz's tree kangaroo (*Dendrolagus lumholtzi*) (LTK), an endemic species in rainforests of the Wet Tropics in far north Queensland, Australia. It focuses on those traits that explain the species' habitat requirements, its intra-specific interactions, the responses to native and introduced predators, and to human-induced landscape modifications. The chapter then concludes with the relevance of this knowledge for the conservation of this species.

Despite the emphasis of research on behavioral aspects of this species for conservation in the "Community Action Plan for the Conservation of Lumholtz's tree-kangaroo (*Dendrolagus lumholtzi*) and its Habitat 2013 to 2018" (Burchill et al., 2014), studies on behavioral aspects of LTKs' ecology have been largely neglected during the last decades. Because of the cryptic nature of the species' life, our knowledge of its behavior and associated ecological pattern and processes is only rudimentary and often relies on incidental observation of wild, captive, and rehabilitated animals (Heise-Pavlov, 2017). However, the description of incidental observations can assist in the extraction of general behavioral patterns that occur under certain conditions and the underlying ecological factors that drive them. Further studies will allow conclusions on population and metapopulation dynamics. So far, only a few aspects of LTKs' behavior have been sufficiently investigated and resulted in publications in scientific journals. Many aspects need more long-term studies and first results are documented only in research reports. By summarizing published and unpublished studies, this chapter identifies areas in which more research is needed.

BEHAVIORAL TRAITS THAT DEFINE MICROHABITAT REQUIREMENTS

Microhabitat features such as vegetation structure, presence of various growth forms, tree branching, canopy density and undergrowth density are likely to determine the degree of habitat quality for this species that is one of the largest arboreal marsupials of Australia. This follows from the species' needs to move within a habitat that consists mainly of the canopy of rainforest trees where LTKs spend most of their time (Procter-Gray and Ganslosser, 1986).

LTK's movements within the canopy appear to be clumsy although they possess a repertoire of locomotion patterns that enables them to use different structural features encountered in the rainforest canopy. Individuals can apply a hopping motion along larger (>10 cm in diameter) horizontal branches and slightly upward-sloping branches or a slow quadrupedal walking along small branches and twigs (<2 cm in diameter) (Procter-Gray and Ganslosser, 1986). Springing or jumping from branch to branch was rarely observed in adult LTKs, which may explain why canopy connectivity could not be shown to be an important structural feature of rainforest for LTKs (Toumpas, 2016).

Resting usually occurs high in the forest canopy while feeding takes place in the canopy or middle zone between ground and canopy, especially on the forest edge where vines and subcanopy leaves are abundant in the middle zone (Procter-Gray, 1985). LTKs prefer resting spots in dense canopies or vine thickets (Martin, 2005) and have been seen resting on relatively small branches (4.5 cm in diameter) (Procter-Gray and Ganslosser, 1986). Rehabilitated released LTKs were observed resting at a height between 4 and 24 m which accounted for between 18% and 97% of the available maximum canopy height (Schuh, 2014). These observations support conclusions by Coombes (2005) that there is no relationship between canopy height and the intensity of its use by LTKs. However, at sites with uneven canopies, LTKs were also frequently seen in emerging trees (Vernier, 2011) which supported the hypothesis that the architecture of trees, such as their branching and canopy shape, may result in preferences of LTKs for certain tree species as resting spots. This was confirmed in a study of LTKs in a restored riparian habitat. The study showed that LTKs preferably rested in Blue Quandong (*Elaeocarpus grandis* or *E. angustifolius*) and a species of Eucalyptus (specifically tallowwood [*Eucalyptus microcorys*]) (Heise-Pavlov et al., 2018). These tree species develop an extensive and wide crown while possessing branches which protrude from the trunk at nearly 90 degrees (Australian Tropical Rainforest Plants, 2016; Lamb and Borschmann, 1998). The importance of horizontal branches for resting LTKs was also reported by Procter-Gray and Ganslosser (1986). However, whether LTKs have favorite trees for resting seems to depend on their home range size and habitat quality (Coombes, 2005; Newell, 1999a).

One reason for the low importance of canopy connectivity for LTK habitat use is the species' habit of moving on the forest floor between trees. Their relatively large average body weight between 7.1 kg for females and 8.6 kg for males (Martin, 2005) ultimately restricts their movements between trees within the canopy and rather forces them to descend to the ground when intending to move over longer distances within the forest or in response to threats. Repeated observations of female LTKs with their young revealed regular descending during dusk after extensive resting during the day (Heise-Pavlov personal observations, 2014). Descent from a tree is usually either by quadrupedal walking or sliding backwards (Procter-Gray and Ganslosser, 1986). This mode of descending led to the hypothesis that trees of certain sizes and with clear trunks (less obstructions in the form of low amounts of protruding branches, touching branches from neighboring trees, attached epiphytes and vines) may be preferred by LTKs for a quick descent. This hypothesis was proved correct in a study by Heise-Pavlov et al. (2011) who showed an association between trees with signs of LTK activity in the form of scratch marks and their size and amount of obstruction along their trunks.

While LTKs usually freeze when threatened while in the canopy, if they are in the subcanopy, they may leap from the tree to the ground with a great crash and immediately bound away. LTKs were never observed to be stunned or injured in any way by these leaps, even from heights of 15 m (Johnson and Newell, 2008; Procter-Gray, 1985; Procter-Gray and Ganslosser, 1986). When moving on the ground, LTKs use either bipedal leaping (Procter-Gray and Ganslosser, 1986) or quadrupedal walking (Heise-Pavlov, personal observations, 2015). A strong affiliation to the ground when moving within a forest has repeatedly been observed during the early release stage of orphaned LTKs in which the unfamiliar environment caused a threat response (Heise-Pavlov, personal observations).

Apart from using the forest floor to move and to retreat from threats, individual LTKs have also been observed performing geophagy, the consumption of soil, which has been described for a range of species including other mammals (Klaus and Schmid, 1998). It is likely to provide LTKs with mineral supplementation (Robbins, 1993) and assists in the detoxification of toxic food compounds (Hladik, 1978). Geophagy in LTKs has been described for LTKs under managed care by Johnson and Delean (2003) and also by various caretakers, rehabilitators, and zookeepers (Anderson, Cianelli, personal observations; Chapter 6). It has also been observed in juvenile orphaned LTKs that were released into the forest (Heise-Pavlov and Cianelli, personal observations). The use of the forest floor may also be of relevance for copulations, but it is currently unknown whether copulations take place only on the ground, or also on large branches in the subcanopy. Heise-Pavlov observed one copulation incident on the ground and Johnson and Delean (2003) report that most mating of captive LTKs took place on the ground.

The utilization of the forest floor by LTKs for movements, geophagy and potentially for copulations may assign an important role to the density of the undergrowth vegetation. Thicker undergrowth vegetation within a habitat may restrict access and movements on the ground. However, LTKs have been observed using cyclone-affected habitat with dense understory (Kanowski et al., 2008; Lufty, 2017), forests subjected to logging (Laurance and Laurance, 1996; Newell, 1999b), and riparian rainforest remnants (Laurance and Laurance, 1999). On the other hand, studies on habitat use of a mother LTK with her juvenile suggest that areas with denser understory may play a role in providing a certain degree of safety for the juvenile while facilitating its explorative behavior (Blomberg, 2016). When young first come out of the pouch at about 300 days of age, they appear to be very clumsy, springing from branch to branch but staying within about 2 m of the mother, who shows very little response or even awareness of the joey's precarious position high in the trees. Juveniles falling while learning to move within the canopy have been reported (Procter-Gray, 1985; Heise-Pavlov, personal observations) and their landing on dense understorey may be a safety precaution of the otherwise unsuspecting mother (Blomberg, 2016; Procter-Gray, 1985).

When ascending from the ground into the canopy, LTKs generally chose trees less than 40 cm DBH (diameter at breast height) (Procter-Gray and Ganslosser, 1986). Also, Heise-Pavlov et al. (2011) noted that most of the trees with scratch marks (which originated from LTKs) on their trunks were between 9 and 29 cm of DBH. The use of trees with smaller trunks may allow LTKs a better grip around the trunk when climbing (Martin, 2005). When ascending into the canopy, LTKs apply a bipedal hopping motion (Procter-Gray and Ganslosser, 1986), very similar to Doria's tree kangaroo (*D. dorianus*) and Goodfellow's tree kangaroo (*D. goodfellowi*) as illustrated by Ganslosser (1980) (Fig. 7.1). This is probably attributable to their evolutionary links to rock-wallabies (Flannery et al., 1996; Chapter 1).

LTKs spent the majority of their time resting or alert (Procter-Gray and Ganslosser, 1986). During a typical 3-h observation period, individual LTKs were observed resting or alert for more than 2.5 h with occasional comfort movements. The remaining time generally included one feeding bout during which the animal moved within its tree. Procter-Gray (1985) reported that feeding occurred at all hours of the day and night, approximately every 4 h in bouts lasting 2–20 min while the majority of longer-distance movements took place at night. In a study by Newell (1999a), animals were mostly active at night. Coombes (2005), on the

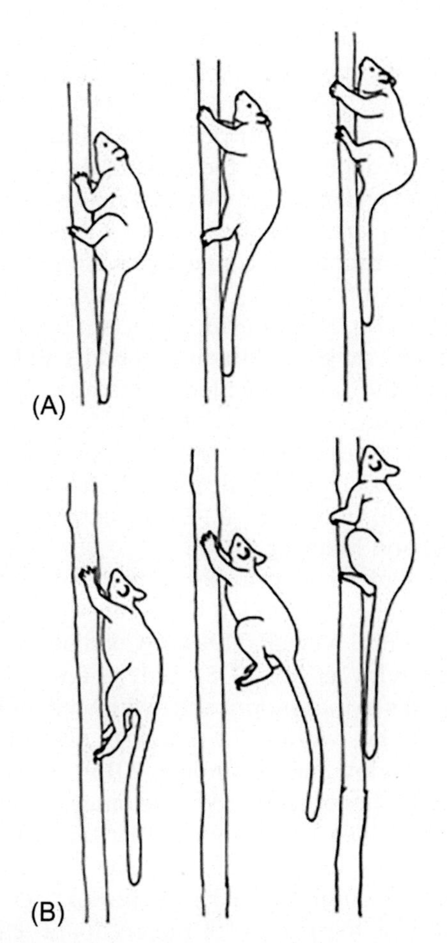

FIGURE 7.1 Vertical locomotion of *Dendrolagus* along trees (drawing from Ganslosser, 1980): (A) *D. dorianus;* (B) *D. goodfellowi.*

other hand, observed her radio-collared animals being active during the day, engaging in moving throughout their home range and feeding. Observations of rehabilitated and released individuals by Heise-Pavlov showed that individuals engaged in feeding after their daily release into the forest in the morning, which was followed by an extensive resting period from about 11 AM to 3 PM when feeding resumed. Based on these reports, the species has been categorized as cathemeral.

LTKs are considered generalist folivores as they have been observed most frequently consuming leaves. Foliage from a wide variety of rainforest and non-rainforest plants has been reported to be consumed by the species (Coombes, 2005; Martin, 2005; Newell, 1999a; Procter-Gray, 1984; Tree-Kangaroo and Mammal Group, 2014). New plant species are continuously added to the repertoire of food plants (Anton, 2014) suggesting that the species may consume only small amounts of foliage from selected plant species. This is likely a mechanism to avoid the accumulation of plant toxins in the consumed diet (according to the detoxification limitation hypothesis by Marsh et al., 2006). Although most of the food consists of leaves, LTKs have also been observed to consume mosses, lichens, ferns (*Asplenium* spp.), flowers (Procter-Gray, 1984; Heise-Pavlov, personal observations, 2016) and fruit (Cianelli, personal communication). Furthermore, a study by Heise-Pavlov et al. (2014) revealed that LTKs show seasonal preferences in their food choice which may reduce the consumption of plant toxins whose concentrations vary between seasons. In conclusion, LTKs should be regarded as selective folivores as suggested by Coombes (2005) and Newell (1999b).

BEHAVIORAL TRAITS ASSOCIATED WITH THE USE OF FRAGMENTED LANDSCAPES

Previously described as obligate rainforest species by Williams (2006), analyses of sighting data sets revealed that LTKs utilize a wide range of habitats (Hauser and Heise-Pavlov, 2017). Studies of LTKs in secondary forests, riparian areas, and wildlife corridors have proved that the dependence of this species on rainforest habitat is not as strong as originally thought (Laurance and Laurance, 1996, 1999) which is likely to be attributable to their lower degree of arboreality (in comparison with other endemic arboreal species of this area), and their predominantly generalist diet (Martin, 2005).

Laurance and Laurance (1996) conclude that LTKs may possess a preference for regrowth and secondary rainforest over old growth mature rainforest. This preference could have been attributable to the higher abundance of pioneer plant species which LTKs have been observed consuming in the habitats studied by Laurance and Laurance (1996). Also Newell (1999a) observed LTKs in young rainforest regrowth.

Kanowski et al. (2001a,b) emphasize that LTKs are most abundant in forests on fertile basalt soils which may be caused by a higher nutritional quality of foliage and the need of soil consumption by this species. However, observations of LTKs in riparian restoration sites on loamy soils (Heise-Pavlov, 2017) suggest that LTKs are not bound to habitat on fertile soils as long as the habitat provides sufficient food resources and appropriate climbing and resting structures.

The species is also found in wet and dry sclerophyll forests (Hauser and Heise-Pavlov, 2017). Sclerophyll forests are characterized by trees with hard, leathery, evergreen foliage specially adapted for prevention of moisture loss. In dry sclerophyll forest, LTK sightings are often in proximity to rainforest patches or patches with rainforest undergrowth. Sclerophyll forests may therefore pose a challenge to this species and should be regarded as marginal habitat since they offer a lower variety of available food, and also restrict the movement of individuals within the canopy of trees as well as on the ground (the ground is usually covered by a thick undergrowth of grasses, ferns and seedlings in these forests) (Ashton and Attiwill, 1994).

The preferred rainforest habitats as well as most of the habitats in sclerophyll forests are highly fragmented, particularly on the Atherton Tablelands where the majority of LTK populations occurs (Johnson and Newell, 2008; Hauser and Heise-Pavlov, 2017; Kanowski et al., 2001a; Winter et al., 1987). Fragmentation, or the breaking apart of habitat into smaller fragments and their separation by a matrix of unsuitable habitat, can lead to dysfunctionality of a landscape for a species (Pascual-Hortal and Saura, 2006; Villard and Metzger, 2014). Whether a species can persist in a fragmented landscape depends on the structural connectivity of a given landscape (which is made up of the amount of area of remaining suitable habitat, its configuration and connectivity through various types of matrices [Villard and Metzger, 2014]), and species-specific traits (such as a species' home range requirements and its ability to move within a fragmented landscape, specifically within the matrix of habitat unsuitable for the species [often referred to as a species' "matrix resistance"]) (Bentley, 2008; Cattarino et al., 2016; Cooney et al., 2015; Debinski, 2006; Villard and Metzger, 2014).

Although several studies were devoted to the assessment of the species' home ranges (Procter-Gray, 1985; Newell, 1999a; Coombes, 2005; Heise-Pavlov et al., 2018), only a few related the results to the amount and availability of suitable habitat within the species' distribution (Heise-Pavlov and Gillanders, 2016). However, the variability of reported home range sizes suggests that they are influenced by the size and connectivity of available habitat, the habitat quality (Coombes, 2005; Kanowski et al., 2001b) and the gender of the occupant. Home ranges between 0.58 and 2.6 ha with up to 15 ha have been reported for female LTKs, while values between 1.01 and 4.4 ha were reported for males (Coombes, 2005; Heise-Pavlov et al., 2018; Newell, 1999a; Procter-Gray, 1985, 1990). In general, male home ranges are larger than those of females, although this seems to depend on the available habitat (Coombes, 2005). Larger home ranges may exist on less fertile soils or in continuous forests (Kanowski et al., 2001b).

While reports on home ranges of male LTKs show that their home ranges are exclusive and can include home ranges of several females, most reports note exclusive female home ranges (Coombes, 2005; Procter-Gray, 1985) or minor overlaps between them (Newell, 1999a).

However, habitat quality may play a role in this respect since extensive overlaps of female home ranges are reported from a restored riparian habitat that has been colonized by LTKs (Heise-Pavlov et al., 2018). On this site, core areas of female home ranges were exclusive, except when females were related. Similarly, a report from the Tree-Kangaroo and Mammal Group (2000) noted three females sharing a 1 ha fragment. Home ranges of males seem to change more often than those of females which may be due to fights between males over home ranges and access to females (Procter-Gray, 1985).

The observed variability/plasticity in home range sizes seems to enable LTKs to occupy habitat fragments of varying sizes, and their presence in a fragment may not depend on its size (Laurance, 1989, 1990; Pahl et al., 1988). For example, within the 200 ha patch of complex notophyll forest at the Curtain Fig National Park on the Atherton Tablelands, nearly all of the mapped study area was being used by LTKs (Procter-Gray, 1985).

An analysis of 1481 sightings of LTKs revealed that most sightings were associated with landscapes that contain habitat patches which are either less than two hectares or between 20 and 50 ha (Fielding, 2016). For instance, Coombes (2005) reported a case of a 0.1 ha home range of a female LTK using a small patch of rainforest around a spring. In contrast, a home range of 332 ha was reported by Newell (1999a) for a juvenile male that used several forest fragments and other habitats, presumably during its post-natal dispersal. Also, Procter-Gray (1985) reported a radio-collared subadult male that was relocated for 10 days after which he completely disappeared despite extensive searches which could indicate male dispersal from forest patch to forest patch across open ground.

The above-mentioned spatial analysis of LTK sightings showed that sightings were more abundant in areas in which habitat patches featured an intermediate degree of clustering and had large edge-to-area ratios (Fielding, 2016). LTKs may be more detectable in areas with more densely clustered habitat fragments and along edges of fragments which are often dominated by pioneer plant species for which LTKs seem to have a preference (Pahl et al., 1988).

Since on the Atherton Tablelands the majority of suitable fragments are up to 2 ha in size (which is the average size of an adult LTK's home range) (Heise-Pavlov and Gillanders, 2016), it can be concluded that suitable habitat is available for this species, specifically when it includes several habitat patches into its home range. Small habitat patches could also be beneficial as steppingstones for long-distance dispersal (Saura et al., 2014).

Various studies point to a low matrix resistance of LTKs. Although the species hasn't been much observed within the matrix (between forests), sightings of LTKs from unforested areas have been reported from unusual places such as domestic gardens, orchards, and maize fields (Table 7.1) (Heise-Pavlov and Gillanders, 2016; Kanowski et al., 2001a).

The majority of LTK sightings within the matrix was within 150 m from the nearest suitable habitat patch while unconnected suitable habitat patches were on average 77 m apart. From this it can be concluded that LTKs are able to reach most of the currently available unconnected patches with suitable habitat (Heise-Pavlov and Gillanders, 2016).

Despite the species' ability to move through a range of matrices, this ability can be detrimental for individuals since the crossing of open, non-forested habitat within a matrix is associated with the risk of falling victim to encounters with introduced predators and large animals such as cattle in pastures (Newell, 1999b; White and Ward, 2010), or to collisions with vehicles. Sixty eight percent of dead LTKs reported in a community survey between 1998 and 1999 were road-kill animals (Kanowski et al., 2001a). Road-kill "hot-spots" and "cold-spots" were identified in a pilot study by Cohen (2013) who also identified the degree of fragmentation to be a major

TABLE 7.1 Distribution of LTK sighting records within the matrix ($n = 257$) across different types of land use of the matrix (Heise-Pavlov and Gillanders, 2016).

Types of land use	Number of sighting records associated with a certain type of land use
Airports/aerodromes	1
Beverage and spice crops	2
Commercial services	2
Grazing modified pastures	45
Grazing native vegetation	100
Irrigated cereals	1
Irrigated cropping	3
Irrigated perennial horticulture	4
Irrigated tree fruits	3
Managed resource protection	1
Manufacturing and industrial	4
Conserved areas	5
Other minimal use	6
Plantation forestry	3
Production forestry	7
Public services	1
Recreation and culture	4
Research facilities	1
Residential	9
Residual native cover	24
Rural living	14
Rural residential	17

determinant of roadkill "hot-spots" for this species (Heise-Pavlov, 2017). The majority of roadkill victims consists of males who are probably seeking to establish their own home range or trying to access females in estrus (Shima et al., 2018; Tree-Kangaroo and Mammal Group, 2000).

INTRA-SPECIFIC BEHAVIOR

Interactions between individual LTKs are rarely observable in the wild and are rare even on occasions when two adults are seen in close proximity (Procter-Gray and Ganslosser, 1986). Of 221 sightings of adult LTKs in Procter-Gray's (1985) study, 206 were of single individuals or mother with young. Animals in close proximity seem to ignore each other. Newell (1999a) gives an account of 6% and 2.7% of telemetry fixes for females and males, respectively, that include the presence of another animal in the same or adjacent tree at the time of location. For most of the time in which adults were observed in the same tree by Procter-Gray (1990), they showed no awareness of each other. Occasionally one adult would approach to within about 1 m of the other but appeared "uninterested." LTKs are therefore regarded as solitary animals.

Aggressive male-male interactions have been reported and attributed to fights over access to females in estrus. These interactions can result in severe injuries (Fig. 7.2) and usually include biting and scratching (Johnson and Newell, 2008; Newell, 1999a; Cianelli, personal communication). Martin (2005) describes the presence

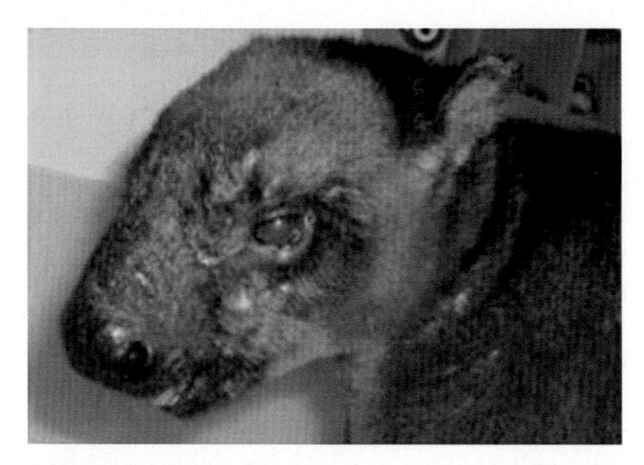

FIGURE 7.2 Facial injuries of a male *D. lumholtzi*, potentially due to aggressive interactions with another male. *Source: S. Heise-Pavlov.*

of numerous scars from fighting in male Bennett's tree kangaroos. Newell (1999a) reports antagonistic male-male encounters along the boundaries of male home ranges. However, Procter-Gray (1985) did not observe active territorial defense in LTKs.

Females seem to avoid contact with each other. Procter-Gray (1985) has never seen a female within another female's home range, although some were seen outside or on the borders of those ranges. In cases of aggressive interactions between females, females have adjacent home ranges (Newell, 1999a). This suggests a certain level of defensiveness for home ranges in females and may explain the observation by Newell (1999b) in which individual LTKs remained in their original rainforest habitat even though it was cleared and continued to live among the woody debris.

Despite aggressive encounters between adult LTKs, Procter-Gray (1985, 1990) reported groups of two adults in 14 out of 221 LTK observations and three adults in 1 out of 221 LTK observations. While most groups of two adults consisted of one male and a female and were attributable to mating (Procter-Gray, 1990; Cianelli, personal communication), groups of females are likely to consist of related females such as a mother, her daughter and one or both of their young (Heise-Pavlov and Burchill, personal observations).

Male-female interactions due to mating were observed at different times of the year which indicates the absence of a defined breeding season in this species (Procter-Gray, 1985, 1990). Johnson and Delean (2003) observed births and pouch joeys in captive LTK throughout the year (except February and June). The births of seven young LTKs in a wild population of LTKs, recorded by Procter-Gray (1985), were estimated to be in: January (one young), July (two), October (three), and December (one) from which the author concluded the absence of a specific breeding season. However, other observations of wild LTKs suggest that more pregnancies may be established in the late dry season/early wet season (Gillanders, personal communication, 2016). For female LTKs which were observed regularly within a restored riparian habitat (Heise-Pavlov et al., 2018), conception dates for their joeys were calculated by applying an average age of 195 days of first facial emergence from the pouch and a gestation time of 45 days (Johnson and Delean, 2003). The applied age of 195 days of first facial emergence of a joey from the pouch was an average that takes variations of reported ages for this event into account, but also allows variations due to restrictions in observations. Most of the conception dates (with one exception) fell within the first half of a year (Table 7.2). Despite the low sample size of eight joeys, the results may indicate the existence of a breeding season in this species which coincides with the wet/late wet season and the beginning of the tropical winter season. More observations of reproducing wild LTKs are needed. Bennett's tree kangaroos (*D. bennettianus*) in lowland rainforests seem to have seasonal breeding (Martin, 2005).

LTKs possess the longest mother–young relationship among macropods. Joeys remain in the pouch for 246–275 days (Johnson and Delean, 2003) and have been observed staying with their

TABLE 7.2 First facial emergence from the pouch and calculated potential conception dates for eight individual joey LTKs within a restored riparian habitat.

Individual	First facial emergence	Potential conception date
1	9/01/2014	14/05/2013
2	9/01/2014	14/05/2013
3	28/08/2014	1/01/2014
4	10/11/2014	15/03/2014
5	23/11/2014	28/03/2014
6	12/07/2015	14/11/2014
7	31/10/2015	5/03/2015
8	30/11/2015	4/04/2015

mother for up to two years (Martin, 2005; Procter-Gray, 1985). Procter-Gray (1985) reported frequent body contact between mothers and young that were mostly initiated by the young, a behavior that was confirmed by Finn (2016). Pouch young also playfully reach out to touch the mother's nose, neck, or arms. The mother usually shows no response except to turn her head away (Procter-Gray, 1985). Speculation has arisen about the reasons for this long mother-young relationship in LTKs (Olds and Collins, 1973, Chapter 2), likely associated with the arboreal life of this species and its diet consisting of partly toxic foliage (see details below).

Training with regard to food selection starts when the joey begins to look out of the pouch (approx. at 250 days of age) and takes food from its mother (Procter-Gray, 1985). After leaving the pouch completely at an age of approximately 300 days, the young feeds on its own while staying in close contact with the mother (often not moving further away than 2 m) (Procter-Gray, 1985). From observations of captive LTKs, Johnson and Delean (2003) conclude that the mother plays an important role in the juvenile's acquisition of the ability to move within the canopy of a structurally complex rainforest. Because young LTKs move rather "clumsily" through rainforest structures, but are highly "explorative," their mothers appear to facilitate this exploration ("forced exploration") by promoting a strong bonding between them and the juveniles through intensive grooming and body contacts (Finn, 2016; Procter-Gray, 1985). Prevention of pouch entries and the selection of densely branched canopies for resting (that reduce the risk of the juvenile falling to the ground [Blomberg, 2016]) by the mother may promote explorative behaviors of the juvenile (Finn, 2016). Cases of infanticide of non-related young have been recorded in captive Matschie's tree kangaroos (*D. matschiei*) (Hutchins et al., 1991) and cannot be ruled out in LTKs.

While Procter-Gray (1985) referred to an age of 330–350 days when the joey enters the pouch less frequently, a study by Finn (2016) found that pouch entries were frequently prevented by a mother LTK when the joey was 210 days of age. Both authors noted that joeys follow close behind their mothers (within 1 m), sometimes butting them with their heads. Often mother and juvenile were observed in a suckling posture: the young with his head in the pouch, the mother curving her upper body over him and resting her hand or chin on his back (Procter-Gray, 1985; Finn, 2016).

Not much information is available on vocalizations of LTKs. Fights between male LTKs involve harsh grunts (Flannery et al., 1996) during which individuals also have been heard emitting loud shrieks and coughing sounds (Flannery et al., 1996). Procter-Gray (1985) describes a "squeak-huff" emitted by the female when she was head-butted by the male, and a clicking "tsk tsk tsk" sound with approximately two clicks per second, also when a female was approached by a male. Vocalizations during interactions between individual LTKs have also been described for communication between females and their young in which cases females have been heard emitting soft clicking sounds (Flannery et al., 1996), "hissing noises", "ekking" sounds and "huffing" (produced by strong exhaling) (Finn, 2016). Clicking sounds were also recorded by Procter-Gray (1985) from a female LTK or her young (age 340 days) when they were in especially dense vegetation, possibly to maintain contact.

Clicking sounds have also been noted during male-female interactions in captive LTKs (Heise-Pavlov, personal observations). Similar sounds have been reported for male-female interactions in captive Goodfellow's tree kangaroo (Schreiner et al., 2015). Loud grunts and coughing sounds were heard during male-female LTK interactions in the wild (Heise-Pavlov, personal observations, 2015).

While vocalization is part of interactions between LTK individuals, communication via odor cues is likely to play a much greater role in social communications. Given the arboreal life of LTKs within the dense foliage of rainforest canopies, it can be assumed that odor cues are more reliable than visual and auditory cues. Furthermore, odor molecules are more persistent and easier to transmit in humid environments such as rainforests (Müller-Schwarze, 2006). The role of communication via odor cues between LTKs has already been hypothesized by Procter-Gray (1985) who expected that LTKs maintain territorial boundaries by scent-marking. Newell (1999a) came to the same conclusion and based it on the presence of various well-developed glands in LTKs. Paracloacal glands are present in both sexes and exude a thick, creamy colored material with a pungent and acrid smell (Newell, 1999a). Procter-Gray (1985) and Newell (1999a) also mention the presence of sternal glands which might be used to scent mark branches and boles. The use of glands was investigated in more detail in a preliminary study on marking behaviors of captive LTKs (Egelkamp, 2013). Marking behaviors were initiated by swapping provided browse material between individually kept LTKs after the material had been used by them for one day. Five different marking behaviors were observed which potentially involve several different glands:

- "Browse handling" was observed in males and females when freshly cut browse was introduced into the enclosures and when browse was swapped between individuals. This behavior consisted of the individual sitting on its hind feet, reaching browse material with its front paws and manipulating it. Eccrine glands positioned on the palms of the front paws may be involved.
- "Browse licking" was only observed in males when browse was introduced and browse was swapped. When performing this behavior, the individual extends its tongue and moves it across the leaves. Individuals are thus likely to spread saliva on the material suggesting that salivary glands may be involved.
- "Browse hugging" was observed only in males when browse was introduced while in the presence of a female. The individual would grab browse material with its forearms and push it towards the lower ventral body. Sometimes the individual would drag its body over the browse material once it had pushed the material under its body. Most likely secretions from paracloacal, circumanal, and sternal glands are distributed onto the material. Chin and sternal marking have also been reported for *D. dorianus* (Brown and McDonald, 1985).
- "Cloacal rubbing" was observed only in females when receiving browse from another individual. It was also observed independently of browse swap. During this behavior, the individual would perform sidewards and circular movements with its anal region while pushing this region downwards towards the surface. This behavior most certainly involves paracloacal glands and may serve in territorial marking.
- "Feet dragging" was observed only in males when new browse was introduced. When on a horizontal surface the individual would push its weight onto the hind feet and move backwards using a modified quadrupedal gait in which it drags rather than sets its hind feet. This behavior most likely involves eccrine glands although the presence of these glands has not been reported in tree kangaroos. If these glands do exist in tree kangaroos, it can be expected that individuals also leave gland secretions on branches and tree trunks when moving through the forest. It can therefore be hypothesized that eccrine glands on front and hind paws may assist in advertising the presence of an individual in its home range and therefore may result in the avoidance of this area by potential intruders.

Some of these behaviors have been observed in forest settings. Osterman (2018) observed "Browse licking," "Browse handling" and "Browse hugging" performed by a released rehabilitated orphaned male LTK in the forest. In this study "Browse handling" was described as moving the front paws back and forth past stems of seedlings. On another occasion a released juvenile female LTK was observed dragging its cloaca close to the surface of a branch while moving in a quadrupedal gait along the nearly horizontal branch (Heise-Pavlov, personal observations, 2014). A yellow secretion was left on the surface of the branch.

The results suggest that LTKs use a range of marking behaviors, with some of them already described for *D. dorianus* (Brown and McDonald, 1985). Studies on the morphology of involved glands are urgently needed. At present we can only speculate about the role of these marking behaviors within a social context of LTKs. Parallel observations of wild LTKs are needed.

FIGURE 7.3 LTK performing "sniffing" ("air tasting") using its vomeronasal organ. *Source: S. Heise-Pavlov.*

That chemical communication plays a role in LTKs can be inferred from observations in which strong pungent musky smells are noted from the bark of trees at certain locations in a forest where no animal was detected (Heise-Pavlov, personal observations). This suggests either recent interactions between individuals or the scent marking of exclusive resources by an individual. The described observations also indicate the existence of a highly developed olfactory sense in this species (Goldberg, 2011). In this respect, one of the most frequently observed behaviors was the "intense nasal sniffing" which consists of a profound movement of the nostrils while the head is stretched forward with an occasional head motion towards the odor source. Occasionally, studied captive LTKs were seen to exhale intensively during which they produced a distinct "whuff" sound ("huffing"). This behavior is likely to be attributable to a cleaning of the nasal epithelium from odor molecules to allow an undiminished perception of new odor molecules. Most interestingly, individuals were seen exposing the tip of their tongue repeatedly when a novel odor was present in their enclosures, a behavior described as "air tasting" (Goldberg, 2011) (Fig. 7.3). This suggests the presence of a well-developed Jacobson's organ (or vomeronasal organ), which has been described in a range of marsupials and is linked with processing odor molecules of lower molecular weight by the accessory olfactory bulb (Müller-Schwarze, 2006).

Indication for the high degree of sensitivity at which LTKs can detect odor cues comes from observations of caretakers. On one occasion a released juvenile orphaned female LTK, which wore a radio-collar, was approached by an adult male LTK from a distance via the canopy and then attacked vigorously (Cianelli, personal communication, 2009; Chapter 6). After the female's head had slipped through her radio-collar allowing her to fall to the ground, the male was observed holding the radio-collar in its front paws and shredding it to pieces with his teeth. The radio-collar had been fitted previously with another LTK male and the remaining odor from this male had potentially agitated the attacking male from a considerable distance and stimulated it to move to the potential intruder to attack him. Similarly, radio-collared juvenile male LTKs were frequently chased by residential males after their release into the forest. During one chase, a radio-collar was lost and after its recovery, it released a profound fear response in the juvenile rehabilitated male, presumably due to odors remaining on the radio-collar from residential males. These observations represent a strong case for the role odor cues play in social interactions of LTKs but studies have only begun to gain insight into the odorsphere of this species (Müller-Schwarze, 2006). More research is needed to understand the role of odors as communication means of LTKs.

BEHAVIORAL TRAITS ASSOCIATED WITH PREDATOR DETECTION AND AVOIDANCE

LTKs have arboreal as well as terrestrial predators. Within the canopy, predators are most likely to be pythons, as in a case reported by Martin (1995) where a Bennett's tree kangaroo was consumed by a python. However, it cannot be excluded that juvenile LTKs may become victim to raptors when resting high in the canopy (Chapter 6; Cianelli, personal communication). While on the ground, LTKs are killed by dingoes and dogs. Although introduced predators have been identified as major reason for the decline and extinction of native Australian mammals (Woinarski et al., 2015), tree kangaroos may possess anti-predatory strategies against these predators since they have co-evolved with terrestrial predators due to their relatively recent divergence from terrestrial rock-wallabies (Flannery et al., 1996). Furthermore, their ability to climb into the canopy of forests provides opportunities to avoid encounters with terrestrial predators. It is therefore surprising to note that fatalities of LTKs by dogs and dingoes occur and were estimated to account for 10% of 300 dead individuals reported between 1985 and 2000 (Tree-Kangaroo and Mammal Group, 2000) or 31 records out of 367 reported dead tree kangaroos during a community survey (Kanowski et al., 2001a). Shima (2014) attributed one of every 12 reported deaths of wild LTKs to predation by dogs.

Given the well-developed olfactory sense and the role of olfactory cues in intra-specific interactions of this species, studies were conducted to test whether LTKs could recognize odors from predators and associate them with a threat. The detrimental effect of introduced predators on Australian native marsupials was claimed to be linked with a general lack of the ability of Australian marsupials to recognize odors originating from novel predators (Mella et al., 2010; Russell and Banks, 2007). Banks and Dickman (2007) termed this failure as "naiveté level 1", which may be caused by differences in olfactory cues emitted by native marsupial predators and novel introduced predators (Cox and Lima, 2006). To test whether this applies to LTKs, captive LTKs were exposed to fecal material from dingoes (*Canis dingo*), domestic dogs (*Canis lupus familiaris*), amethystine pythons (*Morelia amethystina*) and Tasmanian devils (*Sarcophilus harrisii*). Due to phylogenetical relationships between dingoes and dogs (Anson and Dickman, 2013; Carthey and Banks, 2012) and between the extant Tasmanian devil and extinct thylacines such as *Thylacoleo carnifex* (a large, leopard-sized carnivore with the ability to climb, that lived in the early to late Pleistocene) and *Thylacinus cynocephalus* (rendered extinct around 1933) (Burbidge and Woinarski, 2016; Flannery et al., 1996; Jones and Stoddart, 1998; Wroe et al., 2007) with which LTKs have co-evolved (Black et al., 2012; Calaby and White, 1967; Dawson, 1982; Flannery and Szalay, 1982; Horton, 1977; Muirhead, 1992), it was hypothesized that LTKs would recognize odor cues from these terrestrial predators as well as the arboreal predator as threat. The analyses of the subjects' comfort and vigilance behaviors in response to the presence of odors from these predators revealed that they indeed associate all the presented cues with a threat. A case for naiveté level 1 can therefore be rejected as explanation for the observed fatalities of LTKs by dingoes and domestic dogs (Heise-Pavlov, 2016). Similarities in chemical composition of fecal odor cues from predators they evolved with (python, Tasmanian devil and dingo [LTKs had spent at least 3000 years with dingoes], Smith and Savolainen, 2015) and those from novel predators (domestic dog) may be responsible for LTKs' recognition of these odor cues as originating from predators (Nolte et al., 1994).

A further hypothesis for the impact of novel predators on native Australian mammals is the application of an inappropriate antipredator strategy after the predator has been recognized (naiveté level 2; Banks and Dickman, 2007). Because LTKs are known to descend from the canopy when threatened and flee on the ground (Martin, 2005; Procter-Gray and Ganslosser, 1986), the possibility of the presence of naiveté level 2 in this species' predator response was also tested in the study of captive LTKs by analyzing the durations and frequencies of their displayed directional movements and "freeze" behaviors in response to the presented odor cues from different predators. The co-evolution of LTKs with predators of different archetypes (arboreal and terrestrial predators) in a multi-predator environment (Cruz et al., 2013) should have been a selective force for the evolution of archetype-specific antipredator responses (Blumstein, 2006). It was therefore expected that LTKs would descend to the ground and flee when threatened in the canopy by an arboreal predator (expressed in an increase in their movements), but would climb up and adopt a "sit and wait" strategy (expressed in higher amounts of "climbing" and "freeze" behaviors) when threatened from the ground by terrestrial predators. However, results of the study suggest the absence of clear archetype-specific antipredatory behaviors in LTKs since tested individuals displayed flight responses and "freeze" behavior in an inconsistent pattern.

Both responses are likely to be attributable to anti-predatory strategies of LTKs' terrestrial ancestors, the rock-wallabies (Flannery et al., 1996). Rock-wallabies first adopt the "freeze" strategy as "sit-and-wait-strategy" to avoid being detected by a predator (Hagenaars et al., 2014) before applying a "flight" mode during which they develop high speeds on the ground supported by their long hind legs and long tail that propel them across rocky outcrops and boulders in their dry, open habitat (Sharman et al., 1995).

The relatively recent divergence of LTKs from ancestral rock-wallabies and their adaptation to a new habitat, the canopy of forests, has resulted in morphological changes of their tibio-fibular articulations (Barnett and Napier, 1953; Warburton and Prideaux, 2010), making climbing feasible. This apparently was not accompanied by changes in behavioral traits in the way of developing predator-archetype specific anti-predatory behaviors that utilize the safety of the canopy. This is in contrast to the general assumption that behavioral adaptations precede morphological changes (Foster, 2013) in the course of adaptations to a new environment. Since morphological changes of their tibio-fibular articulations (Barnett and Napier, 1953; Warburton and Prideaux, 2010) also resulted in a restriction of tree kangaroos' ability to develop high speeds when moving on the ground, LTKs often succumb to terrestrial predators, particularly when these predators hunt in groups.

The presence of both arboreal and terrestrial predators during tree kangaroos' divergence from rock-wallabies may have resulted in the persistence of ancestral antipredator responses (according to the multi-predator hypothesis from Blumstein, 2006) which likely has been genetically manifested as a hard-wired behavioral trait (Blumstein et al., 2004, 2010; Reimers et al., 2012).

LEARNED OR INHERITED BEHAVIOR—THE RELEVANCE OF BEHAVIORAL STUDIES IN REHABILITATION AND RELEASE OF ORPHANED LUMHOLTZ'S TREE KANGAROOS

Behavioral traits of a species are of particular importance when it comes to raising and releasing orphaned tree kangaroos since this species has an exceptional long mother-joey relationship that indicates the potential role of the mother in preparing the young for a life in the canopy of rainforests. Observations, experiences, and the knowledge of wildlife caretakers in rehabilitating and releasing tree kangaroos (Chapter 6) have inspired many research projects that have been conducted by students at the Centre for Rainforest Studies at The School for Field Studies in Yungaburra, Queensland.

This section summarizes the results of research projects that focused on comparing behavioral traits between orphaned LTKs and young LTKs with a natural mother. The research was done in collaboration with two wildlife caretakers, Margit Cianelli and Karin Semmler, and complements their experiences in releasing rehabilitated orphaned LTKs.

Encounters of LTKs with predators and collisions with vehicles result in injured individuals and orphaned juveniles whose mothers were killed. Rehabilitation can be a time-consuming process which is particularly the case when juvenile orphaned individuals come into care. Raising juvenile orphaned tree kangaroos poses several challenges to caretakers since the exceptionally long mother-offspring relationship in this macropod points to an important role the mother plays in the development of the juvenile. This role must be taken over by the caretaker and necessitates profound knowledge of behavioral traits the juvenile has inherited and traits the juvenile has to learn to achieve a successful release.

During the last decade, observations of orphaned juvenile rehabilitated LTKs have increased our knowledge of learned and inherited behavioral traits that can now be integrated in their rehabilitation, training and release. Juvenile LTKs possess inherited abilities for the selection of food items and climbing, but observations suggest that the expression of these inherited abilities needs to be enforced and established by training. Additionally, self-learning by trial and error seems to play an important role in a juvenile's adjustment to a life in the forest.

Food items selected by juvenile orphaned tree kangaroos after their release into the forest were identified and compared to those the animals had previously received from their caretaker, and to those listed as known food plants for LTKs (Anton, 2014; Cook, 2017). It was found that juvenile tree kangaroos consumed leaves from species which they had not received previously from the caretaker. The number of previously unknown consumed plant species gradually increased with progressing time the animals spent in the forest. These results point to the presence of an inherited ability of LTKs to select appropriate food plant species and may also support the hypothesis that the wide range of plant species being consumed by LTKs assists in avoiding the accumulation of toxic foliar compounds in LTKs' diet (Marsh et al., 2006). Besides limiting the intake of toxic compounds by feeding from a range of available food plants, the presence of certain microbiota within the digestive system of LTKs may enable them to detoxify consumed toxic foliar compounds. Observations of an experienced LTK caretaker suggest that the close mother-joey relationship may contribute to the transmission and establishment of a microflora in the joeys' digestive system that allows the break-down of toxic plant compounds (Chapter 6; Cianelli, personal communication). The caretaker repeatedly noticed a strong intention of orphaned joey LTKs at a certain age to lick the mother's mouth region (in this case the mouth of the caretaker). This behavior may indicate that natural LTK mothers provide essential microbiota to their young via regurgitation resulting in an inoculation of the joeys with essential microbiota to allow the digestion of rainforest foliage. Associated behaviors such as licking of the mother's face initiated by young LTKs were reported in wild LTKs by Procter-Gray (1985). Intensive licking of a juvenile's head and specifically its nose and mouth region by its natural mother was also repeatedly observed in LTKs at a restoration site (Anton, 2014; Heise-Pavlov, personal observations, 2014). On one occasion, slime was seen transferred between the mouth of the licking mother and the nose-mouth region of the juvenile, which may support the "transmission of microflora via regurgitation" hypothesis. Geophagy was observed in juvenile orphaned tree kangaroos suggesting that this behavior is inherited.

Juvenile orphaned LTKs show a strong urge to climb and to explore the stability and suitability of all kinds of structures for climbing (Levin, 2014; Cianelli, personal communication; Heise-Pavlov, personal observations) suggesting an inherited component of arboreality. Observations of juvenile orphaned LTKs after their release into the forest revealed that a natural mother plays an important role in the developing juvenile's familiarity with the forest structures although mother LTKs often appear indifferent to the explorative behavior of their young (Procter-Gray, 1985; Finn, 2016; Blomberg, 2016). Immediately after their release, orphaned juvenile LTKs often spent considerable time on the forest floor and were rather reluctant to ascend into the canopy (Schuh, 2014; Cianelli and Heise-Pavlov, personal observations). This may be a reflection of the animals' association of the forest floor with a state of safety and attributable to the species' linkage to terrestrial ancestors. Preference for moving on the forest floor may also be indicative of the existence of an inherited element of LTKs' usual antipredator response of fleeing on the ground (Jenike, 2014). Movement of a young LTK (which was about 14 months of age) on the forest floor was repeatedly observed in the wild after its mother descended at dusk from their daily resting spot to move on the forest floor (Heise-Pavlov, personal observations).

A released orphaned juvenile LTK also frequently attempted to climb dead trees (Schuh, 2014). It appears that the avoidance of unstable (dead) climbing structures may require the training by a natural mother. However, the subject seemed to learn the necessary skills for tree selection by trial and error (Schuh, 2014). The majority of trees the subject selected for climbing and resting was within the range of heights and DBHs (diameter at breast height) of the climbing structures that were provided to the subject in its pre-release training enclosure (Schuh, 2014, Chapter 6). However, the subject quickly became more comfortable in using higher trees with larger circumference.

Similar to juvenile LTKs with their natural mothers, released orphaned LTKs showed a high degree of explorative behavior which was expressed by active climbing on a variety of forest structures. Procter-Gray (1985) reported a young tree kangaroo springing from branch to branch in small leaps and landing unsteadily while the mother seemed to be unresponsive even in a situation of a near fall of the young from a height of 20 m. During observations of a mother and her young at a restoration site, the juvenile (approximately 14 months of age) was frequently seen venturing away from the mother to feed (Heise-Pavlov, personal observations, 2016).

When descending from the canopy, a juvenile orphaned LTK frequently selected trees with DBHs which seemed to be too large for clinging to the trunk with its forearms to allow a slow descend of the animal to the ground (Schuh, 2014). Incidents were recorded in which the subject was unable to leave large trees as it could not find suitable connections from large trees to trees of appropriate sizes for descending. These observations support the hypothesis that training of a young LTK by its natural mother contributes greatly to the ability of the young to use available forest structures appropriately. However, the orphaned juvenile showed the ability to learn the use of various structures of the forest by trial and error and adapted to its natural environment gradually even without the presence of a natural mother (Schuh, 2014).

When descending, LTKs use the posterior descending method (bottom first) (PDM). Studies by Chavez (2014) and Klein (2017) investigated the degree to which juvenile orphaned LTKs apply the correct descent method in relation to incorrect methods (such as the anterior descending method [head first] [ADM] and a combination of anterior and posterior [APDM]). The results revealed that the age at which orphaned LTKs had the opportunity to practice in a natural forested environment has an impact on the selection and performance (number of slips) of the correct descending method. In comparison to an LTK joey with a natural mother, it appeared that the preference for the PDM is inherited, depends on the angle of the selected structure for descent (Chavez, 2014), and that the familiarity of an inhabited forested area may contribute to the ability to perform the PDM.

Release of orphaned juvenile LTKs into a forest may result in conflicts with residential individuals due to the species' strong affiliation with established home ranges. Released animals need to establish their own home range. Aggressive encounters initiated by residential LTKs are more likely to take place when juvenile male LTKs are released, and their occurrence has been concluded from the behavior of frightened released male juveniles (Cianelli and Heise-Pavlov, personal observations). In some cases, rehabilitated male juvenile LTKs were followed by agitated male residential LTKs when they retreated into the caretaker's home (Cianelli and Semmler, personal observations).

CONCLUSIONS—THE RELEVANCE OF CURRENT KNOWLEDGE ON BEHAVIORAL TRAITS OF LTKs FOR THE CONSERVATION OF THE SPECIES

Studying behavioral traits of LTKs has relevance for its conservation in many ways. It not only allows targeted habitat conservation and restoration; it also enables conservationists to assess how the species responds to a highly fragmented landscape in order to make appropriate decisions on landscape-scale conservation actions. Our knowledge of social interactions between LTKs and their antipredatory responses assists in rehabilitation and the control of introduced predators.

Specifically, the analyses of the species' multidimensional use of its habitat and the use of various structural features within its habitats directs conservation to the protection and restoration of rainforest habitat that is characterized by a high complexity of growth forms such as vines and trees of different branching types. Habitat should contain a wide variety of plants to minimize the accumulation of consumed plant toxins. However, habitats with canopy gaps or/and dense undergrowth are not unacceptable to LTKs, as this species seems to be able to utilize areas with diverse microhabitat features. Therefore, habitat protection for LTKs should encompass a range of habitats.

That LTKs can persist in small patches of suitable habitat, can use wet and even some dry sclerophyll forests, and can move through non-forested matrices between forests suggests that this species can cope with the current degree of fragmentation within its distributional range in far north Queensland. Landscapes within its core distribution still provide functional connectivity for this species whose response to fragmentation was described as negative but intermediate by Laurance (1990). Pahl et al. (1988) base their conclusion that LTKs can persist in a fragmented landscape on the species' generalist diet and its dispersal abilities. Because we are lacking information on the threshold of fragmentation at which a landscape becomes unsuitable for LTKs, we should apply the precautionary principle and prevent further clearing and fragmentation of LTKs habitat. Preservation of remnant fragments emerged as essential for this species as the species seems to have the capacity to incorporate several fragments into its home range or/and use them as steppingstones when crossing non-forested matrices. Appropriate matrix management and restoration of previously lost habitat are valuable conservation mechanisms for this species allowing it to reach and colonize restored habitat.

If suitable habitat becomes less available, declines in local populations are inevitable due to the predominantly solitary life of LTKs, their strong affinity to their home range, and aggressive male-male fights over access to females. Habitat loss will result in more human-LTK interactions which will increase the number of animals coming into wildlife care. For their rehabilitation and release our knowledge of LTKs' intra-specific interactions and the role of the species' long mother-joey relationship is of great importance. Despite inherited traits such as food selection, climbing, and a strong explorative behavior, orphaned juvenile LTKs acquire many of their abilities to move in a structurally diverse forest through learning or trial and error. Prerelease training of orphaned LTKs in structurally diverse enclosures using their inherited strong explorative behaviors appears to be an essential part of the rehabilitation process in which orphans become familiar with structural features of the forest such as vines, pole-like trunks and epiphytes at an early age. Our knowledge of the role of olfaction in the species' communication and predator recognition may provide rehabilitators with the opportunity to implement antipredator response training of rehabilitated LTKs before their release to increase their survival when encountering introduced predators. Antipredator response training has been proved to increase the survival rate of released marsupials (McLean et al., 2000; Blumstein and Daniel, 2002), although the control of introduced predators remains one of the major conservation measures for the protection of LTKs (Burchill et al., 2014).

Climate change will ultimately have an impact on LTK populations. Habitat loss and modifications through more intense fires are likely to happen. The species' behavioral traits may be affected since the effectiveness in which LTKs use odor cues for communication between individuals, for advertising home ranges, and to detect predators may be altered. In such a world, higher temperatures and changed humidity will prevail, affecting the persistence and distribution of volatile and non-volatile odor cues.

The outlined examples demonstrate how merging our knowledge of the distribution and abundances of LTK populations with information on behavioral traits of this species will increase and strengthen our capacity to be more effective in planning and implementing conservation actions for this species.

ACKNOWLEDGMENTS

Studies on behavioral traits of LTKs would not have been possible without the collaboration with wildlife caretakers on the Atherton Tablelands, specifically Margit Cianelli and Karin Semmler, and zookeepers of the Habitat in Port Douglas, specifically Clare Anderson. They not only granted access to their animals, but also contributed with many ideas and observations to the research. Driving forces to further explore behaviors of these remarkable animals were, however, many inquisitive students of the Centre for Rainforest Studies at The School for Field Studies who spent many hours observing and filming these animals, often under uncomfortable conditions.

The Centre for Rainforest Studies at The School for Field Studies and its staff provided logistic support for undertaking many of the research projects described in this chapter.

Studies on the response of Lumholtz's tree kangaroos to various predators could not have been done without the provision of fecal material that came from a Leanne Kruss in Malanda, Queensland, Lyn Watson at the Dingo Discover Sanctuary and Research Centre in Toolern Vale, Victoria,

Peter Krauss in Biboohra, Queensland, and Tim Faulkner from the Australian Reptile Park and Devil Ark, NSW.

Field observations of radio-tracked tree kangaroos were made possible by many individuals who assisted in their capture: John Winter, Rupert Russell, Wayne Gray, Daryn Storch, and Liz Thomas.

We are also grateful to the Queensland Government in granting ethics and scientific permits to conduct research on this species. Research done by the Centre for Rainforest Studies at The School for Field Studies was conducted under the ethics permits CA2011/02/487, CA2014/04/755, CA2014/08/804, and CA2018/02/1160 and under the scientific permits WIP15108614, WISP1760916, and WISP08961011 granted to S. Heise-Pavlov.

REFERENCES

Anson, J., Dickman, C., 2013. Behavioral responses of native prey to disparate predators: naiveté and predator recognition. Oecologia 171, 367–377.

Anton, T., 2014. To eat or not to eat? A question of diet detoxification mechanisms and survival of the Lumholtz's tree-kangaroo (*Dendrolagus lumholtzi*). Unpublished Directed Research Report, Centre for Rainforest Studies at The School for Field Studies.

Ashton, D.H., Attiwill, P.M., 1994. Tall open forests. In: Groves, R.H. (Ed.), Australian Vegetation. Cambridge University Press, Cambridge, pp. 157–192.

Australian Tropical Rainforest Plants, 2016. *Elaeocarpus angustifolius*. Available from: <http://keys.trin.org.au/key-server/data/0e0f0504-0103-430d-8004-60d07080d04/media/Html/taxon/Elaeocarpusangustifolius>. [15th May 2016].

Banks, P.B., Dickman, C.R., 2007. Alien predation and the effects of multiple levels of prey naiveté. Trends Ecol. Evol. 22, 229–230.

Barnett, C.H., Napier, J.R., 1953. The form and mobility of the fibula in metatherian mammals. J. Anat. 87, 207–213.

Beissinger, S.R., 1997. Integrating behavior into conservation biology: Potentials and limitations. In: Clemmons, J.R., Buchholz, R. (Eds.), Behavioral Approaches to Conservation in the Wild. Cambridge University Press, Cambridge, pp. 23–47.

Bentley, J.M., 2008. Role of movement, interremnant dispersal and edge effects in determining sensitivity to habitat fragmentation in two forest-dependent rodents. Austral Ecol. 33, 184–196.

Berger-Tal, O., Blumstein, D.T., Carroll, S., Fisher, R.N., Mesnick, S.L., Owen, M.A., Saltz, D., Claire, C.C., Swaisgood, R. R., 2015. A systematic survey of the integration of animal behavior into conservation. Conserv. Biol. 30, 744–753.

Black, K.H., Archer, M., Hand, S.J., Godthelp, H., 2012. The rise of Australian marsupials: A synopsis of biostratigraphic, phylogenetic, palaeoecologic and palaeobiogeographic understanding. In: Talent, J.A. (Ed.), Earth and Life: Global Biodiversity, Extinction Intervals and Biogeographic Perturbations Through Time. Springer, Netherlands, Dordrecht, pp. 983–1078.

Blomberg, J., 2016. Influence of a Lumholtz's Tree-Kangaroo Pouch-Young on Food Selection and Habitat Use of its Mother. Unpublished Directed Research Report, Centre for Rainforest Studies at The School for Field Studies.

Blumstein, D.T., 2006. The multipredator hypothesis and the evolutionary persistence of antipredator behavior. Ethology 112, 209–217.

Blumstein, D.T., Daniel, J.C., 2002. Isolation from mammalian predators differentially affects two congeners. Behav. Ecol. 13, 657–663.

Blumstein, D.T., Fernández-Juricic, E., 2004. The emergence of conservation behavior. Conserv. Biol. 18, 1175–1177.

Blumstein, D.T., Daniel, J.C., Springett, B.P., 2004. A test of the multi-predator hypothesis: rapid loss of antipredator after 130 years of isolation. Ethology 110, 919–934.

Blumstein, D.T., Lea, A.J., Olson, L.E., Martin, J.G.A., 2010. Heritability of anti-predatory traits: vigilance and locomotor performance in marmots. J. Evol. Biol. 23, 879–887.

Brown, D., McDonald, D., 1985. Social Odours in Mammals. Clarendon Press, Oxford, England.

Buchholz, R., 2007. Behavioural biology: an effective and relevant conservation tool. Trends Ecol. Evol. 22, 401–407.

Burbidge, A.A., Woinarski, J., 2016. *Thylacinus cynocephalus*. The IUCN red list of threatened species 2016: e.T21866A21949291. Available from: https://doi.org/10.2305/IUCN.UK.2016-2.RLTS.T21866A21949291.en [21 October 2019].

Burchill, S., Cianelli, M., Edwards, C., Grace, M., Heise-Pavlov, S., Hudson, D., Moerman, I., Smith, K., 2014. Community Action Plan for the Conservation of the Lumholtz's Tree-Kangaroo (*Dendrolagus lumholtzi*) and its Habitat 2014–2019. Tree-kangaroo and Mammal Group, Malanda.

Calaby, J.H., White, C., 1967. The Tasmanian devil (*Sarcophilus harrisii*) in northern Australia in recent times. Aust. J. Sci. 29, 473–475.

Campbell-Palmer, R., Rosell, F., 2011. The importance of chemical communication studies to mammalian conservation biology: a review. Biol. Conserv. 144, 1919–1930.

Caro, T., 2007. Behavior and conservation: a bridge too far? Trends Ecol. Evol. 22, 394–400.

Caro, T., Sherman, P.W., 2013. Eighteen reasons animal behaviourists avoid involvement in conservation. Anim. Behav. 85, 305–312.

Carthey, A.J.R., Banks, P.B., 2012. When does an alien become a native species? A vulnerable native mammal recognizes and responds to its long-term alien predator. PLoS One 7, e31804.

Cattarino, L., McAlpine, C.A., Rhodes, J.R., 2016. Spatial scale and movement behaviour traits control the impacts of habitat fragmentation on individual fitness. J. Anim. Ecol. 85, 168–177.

Chavez, H., 2014. Learned or inherited?: Investigating descending and threat response behaviours in juvenile Lumholtz's tree-kangaroo (*Dendrolagus lumholtzi*). Unpublished Directed Research Report, Centre for Rainforest Studies at The School for Field Studies.

Cohen, J., 2013. A Spatial Study of Factors Influencing the Distribution of Roadkill of Lumholtz's Tree-Kangaroo (*Dendrolagus lumholtzi*). Unpublished Directed Research Report, Centre for Rainforest Studies at The School for Field Studies.

Cook, S.M., 2017. Foraging behavior and food selection learning in juvenile Lumholtz's tree kangaroos (*Dendrolagus lumholtzi*). Unpublished Directed Research Report, Centre for Rainforest Studies at The School for Field Studies.

Coombes, K., 2005. The ecology and habitat utilisation of the Lumholtz's tree-kangaroo, *Dendrolagus lumholtzi* (Marsupialia: Macropodidae), on the Atherton Tablelands, far north Queensland. PhD Thesis, James Cook University, Cairns.

Cooney, S.A., Schauber, E.M., Hellgren, E.C., 2015. Comparing permeability of matrix cover types for the marsh rice rat (*Oryzomys palustris*). Landsc. Ecol. 30, 1307–1320.

Cox, J.G., Lima, S.L., 2006. Naiveté and an aquatic–terrestrial dichotomy in the effects of introduced predators. Trends Ecol. Evol. 21, 674–680.

Crofoot, M.C., Lambert, T.D., Kays, R., Wikelski, M.C., 2010. Does watching a monkey change its behaviour? Quantifying observer effects in habituated wild primates using automated radiotelemetry. Anim. Behav. 80, 475–480.

Cruz, J., Sutherland, D., Anderson, D., Glen, A., Tores, P., Leung, L.P., 2013. Antipredator responses of koomal (*Trichosurus vulpecula hypoleucus*) against introduced and native predators. Behav. Ecol. Sociobiol. 67, 1329–1338.

Curio, E., 1996. Conservation needs ethology. Trends Ecol. Evol. 11, 260–263.

Dawson, L., 1982. Taxonomic status of fossil devils (Sarcophilus, Dasyuridae, Marsupialia) from late Quaternary eastern Australian localities. In: Archer, M. (Ed.), Carnivorous Marsupials. Royal Zoological Society of New South Wales, Sydney, pp. 517–525.

Debinski, D.M., 2006. Forest fragmentation and matrix effects: the matrix does matter. J. Biogeogr. 33 (10), 1791–1792.

Egelkamp, C., 2013. A preliminary study on the marking behaviors of the Lumholtz's tree-kangaroo (*Dendrolagus lumholtzi*). Unpublished Directed Research Report, Centre for Rainforest Studies at The School for Field Studies.

Fielding, I., 2016. Making connections: Measuring the functional connectivity of the Atherton tablelands for *Dendrolagus lumholtzi*. Unpublished Directed Research Report, Centre for Rainforest Studies at The School for Field Studies.

Finn, S., 2016. Examining Mother–Joey Relationships and the Development of Learning in Lumholtz's Tree Kangaroo (*Dendrolagus lumholtzi*, Marsupialia: Macropodidae). Unpublished Directed Research Report, Centre for Rainforest Studies at The School for Field Studies.

Flannery, T., Szalay, F., 1982. Bohra paulae: a new giant fossil tree kangaroo (Marsupialia: Macropodidae) from New South Wales, Australia. Aust. Mammal. 5, 83–94.

Flannery, T.F., Martin, R.W., Szaley, F., 1996. Tree-Kangaroos: A Curious Natural History. Reed Books, Melbourne.

Foster, S.A., 2013. Evolution of behavioural phenotypes: influences of ancestry and expression. Anim. Behav. 85, 1061–1075.

Ganslosser, U., 1980. Vergleichende Untersuchungen zur Kletterfähigkeit einiger Baumkänguruharten (*Dendrolagus*, Marsupialia). Zool. Anz. 205, 43–66.

Goldberg, J., 2011. The evolution of antipredator responses in Lumholtz's tree-kangaroos (*Dendrolagus lumholtzi*). Unpublished Directed Research Report, Centre for Rainforest Studies at The School for Field Studies.

Hagenaars, M.A., Oitzl, M., Roelofs, K., 2014. Updating freeze: aligning animal and human research. Neurosci. Biobehav. Rev. 47, 165–176.

Hauser, W., Heise-Pavlov, S., 2017. Can incidental sighting data be used to elucidate habitat preferences and areas of suitable habitat for cryptic species? Integrat. Zoology 12 (3), 186–197.

Heise-Pavlov, S., 2016. Evolutionary aspects of the use of predator odors in antipredator behaviors of Lumholtz's tree-kangaroos (*Dendrolagus lumholtzi*). In: Schulte, B., Goodwin, T. (Eds.), Chemical Signals in Vertebrates 13. Springer, New York, pp. 261–280.

Heise-Pavlov, S., 2017. Current knowledge of the behavioural ecology of Lumholtz's tree-kangaroo (*Dendrolagus lumholtzi*). Pac. Conserv. Biol. 23, 231–239.

Heise-Pavlov, S., Gillanders, A., 2016. Exploring the use of a fragmented landscape by a large arboreal marsupial using incidental sighting records from community members. Pac. Conserv. Biol. 22, 386–398.

Heise-Pavlov, S.R., Jackrel, S.L., Meeks, S., 2011. Conservation of a rare arboreal mammal: habitat preferences of the Lumholtz's tree-kangaroo, *Dendrolagus lumholtzi*. Aust. Mammal. 33, 5–12.

Heise-Pavlov, S., Anderson, C., Moshier, A., 2014. Studying food preferences in captive cryptic folivores can assist in conservation planning: the case of the Lumholtz's tree-kangaroo (*Dendrolagus lumholtzi*). Aust. Mammal. 36 (2), 200–211.

Heise-Pavlov, S., Rhinier, J., Burchill, S., 2018. The use of a replanted riparian habitat by the Lumholtz's tree-kangaroo (*Dendrolagus lumholtzi*). Ecol. Manage. Restor. 19, 76–80.

Hladik, C.M., 1978. Adaptive strategies of primates in relation to leaf-eating. In: Montgomery, G.G. (Ed.), The Ecology of Arboreal Folivores. Smithsonian Institution Press, Washington, DC.

Horton, D.R., 1977. A 10,000-year-old Sarcophilus from Cape York. Search 10, 374–375.

Hutchins, M., Smith, G., Mead, D., Elbin, S., Steenberg, J., 1991. Social behavior of Matschie's tree kangaroos (*Dendrolagus matschiei*) and its implications for captive management. Zoo Biol. 10, 147–164.

Jenike, K., 2014. Implications for Genetic Inheritance of Predator Odor Detection and Avoidance in the Lumholtz's Tree-Kangaroo (*Dendrolagus lumholtzi*). Unpublished Directed Research Report, Centre for Rainforest Studies at The School for Field Studies.

Johnson, P.M., Delean, S., 2003. Reproduction of Lumholtz's tree-kangaroo, *Dendrolagus lumholtzi* (Marsupialia: Macropodidae) in captivity, with age estimation and development of pouch young. Wildl. Res. 30, 505–512.

Johnson, P.M., Newell, G., 2008. Lumholtz's tree-kangaroo (*Dendrolagus lumholtzi*). In: Van Dyck, S., Strahan, R. (Eds.), The Mammals of Australia. Reed New Holland, Sydney, Auckland, London, Cape Town, pp. 310–311.

Jones, M.E., Stoddart, M.D., 1998. Reconstruction of the predatory behaviour of the extinct marsupial thylacine (*Thylacinus cynocephalus*). J. Zool. 246, 239–246.

Kanowski, J., Felderhof, L., Newell, G., Parker, T., Schmidt, C., Stirn, B., Wilson, R., Winter, J.W., 2001a. Community survey of the distribution of Lumholtz's Tree-kangaroo on the Atherton Tablelands, north-East Queensland. Pac. Conserv. Biol. 7, 79–86.

Kanowski, J., Hopkins, M.S., Marsh, H., Winter, J.W., 2001b. Ecological correlates of folivore abundance in North Queensland rainforests. Wildl. Res. 28, 1–8.

Kanowski, J., Catterall, C.P., Winter, J.W., 2008. Impacts of cyclone Larry on arboreal folivorous marsupials endemic to upland rainforests of the Atherton Tablelands, Australia. Austral Ecol. 33, 541–548.

Kelly, M.J., 2008. Design, evaluate, refine: camera trap studies for elusive species. Anim. Conserv. 11, 182–184.

Klaus, G., Schmid, B., 1998. Geophagy at natural licks and mammal ecology: a review. Mammalia 62, 481–497.

Klein, J., 2017. Investigating the Relationship between Climbing Training Methods and Descending Abilities in Lumholtz's Tree-Kangaroo Joeys (*Dendrolagus lumholtzi*). Unpublished Directed Research Report, Centre for Rainforest Studies at The School for Field Studies.

Knight, J., 2001. If they could talk to the animals…. Nature 414, 246–247.

Lamb, D., Borschmann, G., 1998. Agroforestry with high value trees. Publication no. 98/142; project no. UQ-18A.

Laurance, W.F., 1989. Ecological impacts of tropical forest fragmentation on non-flying mammals and their habitats. PhD Thesis, University of California, Berkeley.

Laurance, W.F., 1990. Comparative responses of five arboreal marsupials to tropical forest fragmentation. J. Mammal. 7, 641–653.

Laurance, W.F., Laurance, S.G.W., 1996. Responses of five arboreal marsupials to recent selective logging in tropical Australia. Biotropica 28, 310–322.

Laurance, S.G., Laurance, W.F., 1999. Tropical wildlife corridors: use of linear rainforest remnants by arboreal mammals. Biol. Conserv. 91, 231–239.

Levin, M., 2014. Dropbears or tree-kangaroos? Unpublished Directed Research Report, Centre for Rainforest Studies at The School for Field Studies.

Lufty, G., 2017. Through Thick and Thin: The Effects of Understory Density on the Lumholtz's Tree-Kangaroo's (*Dendrolagus lumholtzi*) Habitat Use in the Atherton Tablelands. Unpublished Directed Research Report, Centre for Rainforest Studies at The School for Field Studies.

Marsh, K.J., Wallis, I.R., McLean, S., Sorensen, J.S., Foley, W. J., 2006. Conflicting demands on detoxification pathways influence how common brushtail possums choose their diets. Ecology 87, 2103–2112.

Martin, R., 1995. Field observations of predation on Bennett's tree-kangaroo (*Dendrolagus bennettianus*) by an Amethystine python (*Morelia amethistina*). Herpetol. Rev. 26, 74–76.

Martin, K., 1998. The role of animal behavior studies in wildlife science and management. Wildl. Soc. Bull. 26, 911–920.

Martin, R., 2005. Tree-Kangaroos of Australia and New Guinea. The University of Chicago Press, Chicago.

McLean, I.G., Schmitt, N.T., Jarman, P.J., Duncan, C., Wynne, C.D.L., 2000. Learning for life: training marsupials to recognise introduced predators. Behaviour 137, 1361–1376.

Mella, V.S.A., Cooper, C.E., Davies, S.J.J.F., 2010. Predator odour does not influence trappability of southern brown bandicoots (*Isoodon obesulus*) and common brushtail possums (*Trichosurus vulpecula*). Aust. J. Zool. 58, 267–272.

Muirhead, J., 1992. A specialised thylacinid, *Thylacinus macknessi* (Marsupialia: Thylacinidae) from Miocene deposits of Riversleigh, northwestern Queensland. Aust. Mammal. 15, 67–76.

Müller-Schwarze, D., 2006. Chemical Ecology of Vertebrates. Cambridge University Press, Cambridge.

Nelson, X.J., 2014. Animal behavior can inform conservation policy, we just need to get on with the job - or can it? Curr. Zool. 60, 479–485.

Newell, G.R., 1999a. Home range and habitat use by Lumholtz's tree-kangaroo (*Dendrolagus lumholtzi*) within a rainforest fragment in North Queensland. Wildl. Res. 26, 129–145.

Newell, G., 1999b. Responses of Lumholtz's tree-kangaroo (*Dendrolagus lumholtzi*) to loss of habitat within a tropical rainforest fragment. Biol. Conserv. 91, 181–189.

Nolte, D.L., Mason, J.R., Epple, G., Aronov, E., Campbell, D. L., 1994. Why are predator urines aversive to prey? J. Chem. Ecol. 20, 1505–1516.

Olds, T.J., Collins, L.R., 1973. Breeding Matschie's tree kangaroo (*Dendrolagus matschiei*) in captivity. Int. Zoo Yearbook 13, 123–125.

Osterman, A., 2018. Scent marking behavior in a young rehabilitated male Lumholtz's tree-kangaroo (*Dendrolagus lumholtzi*). Unpublished Directed Research Report, Centre for Rainforest Studies at The School for Field Studies.

Pahl, L.I., Winter, J.W., Heinsohn, G., 1988. Variation in responses of arboreal marsupials to fragmentation of tropical rainforest in north eastern Australia. Biol. Conserv. 46, 71–82.

Pascual-Hortal, L., Saura, S., 2006. Comparison and development of new graph-based landscape connectivity indices: towards the priorization of habitat patches and corridors for conservation. Landsc. Ecol. 21, 959–967.

Procter-Gray, E., 1984. Dietary ecology of the coppery Brushtail possum, green ringtail possum and Lumholtz's tree-kangaroo in North Queensland. In: Smith, A.P., Hume, I. D. (Eds.), Possums and Gliders. Australian Mammal Society, Sydney, pp. 129–135.

Procter-Gray, E., 1985. The Behavior and Ecology of Lumholtz's Tree-Kangaroo, (*Dendrolagus lumholtzi*, Marsupialia: Macropodidae). Dissertation. Harvard University, Cambridge, MA.

Procter-Gray, E., 1990. Kangaroos up a tree. Nat. Hist. 1 (90), 60–67.

Procter-Gray, E., Ganslosser, U., 1986. The individual behaviors of Lumholtz's tree-kangaroo: repertoire and taxonomic implications. J. Mammal. 67, 343–352.

Reimers, E., Røed, K., Colman, J., 2012. Persistence of vigilance and flight response behaviour in wild reindeer with varying domestic ancestry. J. Evol. Biol. 25, 1543–1554.

Robbins, C.T., 1993. Wildlife Feeding and Nutrition. Academic Press, New York.

Russell, B.G., Banks, P.B., 2007. Do Australian small mammals respond to native and introduced predator odours? Austral Ecol. 32, 277–286.

Saura, S., Bodin, Ö., Fortin, M.J., 2014. EDITOR'S CHOICE: stepping stones are crucial for species' long-distance dispersal and range expansion through habitat networks. J. Appl. Ecol. 51, 171–182.

Schreiner, C., Schwarzenberger, F., Kirchner, W.H., Dreßen, W., 2015. Hormonphysiologische und ethologische Untersuchung am Goodfellow-Baumkänguruh (*Dendrolagus goodfellowi* Thomas, 1908). Der Zoologische Garten 84 (1), 45–60.

Schuh, B., 2014. How the rehabilitation process affects the use of structural components of the rainforest by a released orphaned Lumholtz's tree-kangaroo (*Dendrolagus lumholtzi*). Unpublished Directed Research Report, Centre for Rainforest Studies at The School for Field Studies.

Sharman, G.B., Maynes, G.M., Eldridge, M.D.B., 1995. Yellow-footed rock-wallaby. In: Strahan, R. (Ed.), The Mammals of Australia. Reed Books, Sydney, pp. 391–393.

Shima, A., 2014. Mortality report: calendar year 2013. Mammal Mail (Newsletter of the Tree Kangaroo and Mammal Group), 14(3), 6–7. Available from: <http://www.tree-kangaroo.net/documents> [16 June 2016].

Shima, A.L., Gillieson, D.S., Crowley, G.M., Dwyer, R.G., Berger, L., 2018. Factors affecting the mortality of Lumholtz's tree-kangaroo (*Dendrolagus lumholtzi*) by vehicle strike. Wildl. Res. 45, 559–569.

Sih, A., Cote, J., Evans, M., Fogarty, S., Pruitt, J., 2012. Ecological implications of behavioural syndromes. Ecol. Lett. 15, 278–289.

Smith, B., Savolainen, P., 2015. The origin and ancestry of the dingo. In: Smith, B. (Ed.), The Dingo Debate—Origins, Behaviour and Conservation. CSIRO, Clayton, South Victoria, pp. 55–80.

Sutherland, W.J., 1996. From Individual Behaviour to Population Ecology. Oxford University Press, Oxford.

Toumpas, A., 2016. Assessing factors that affect the use of rainforest restoration sites by Lumholtz's tree kangaroo (*Dendrolagus lumholtzi*). Unpublished Directed Research Report, Centre for Rainforest Studies at The School for Field Studies.

Tree-Kangaroo and Mammal Group, 2000. Tree-Kangaroos on the Atherton Tablelands: Rainforest Fragments as Wildlife Habitat. Information for shire councils, land managers and the local community, Available from: http://www.tree-kangaroo.net/documents/tkinfo/report MapFree.pdf. [23 August 2013].

Tree-kangaroo and Mammal Group, 2014. Habitat plantings—food plantings. (Ed. K. Freebody.) May 2010. Available at: http://FoodPlantsList_10May2010.pdf [26 November 2014].

Treves, A., 2000. Theory and method in studies of vigilance and aggregation. Anim. Behav. 60, 711–722.

Van Dyck, H., Baguette, M., 2005. Dispersal behaviour in fragmented landscapes: routine or special movements? Basic Appl. Ecol. 6, 535–545.

Vernier, A., 2011. Factors affecting colonization of riparian zones by Lumholtz's tree-kangaroos (*Dendrolagus lumholtzi*) in the Atherton tablelands, Northeast Queensland, Australia. Unpublished Directed Research Report, Centre for Rainforest Studies at The School for Field Studies.

Villard, M.A., Metzger, J.P., 2014. Review: beyond the fragmentation debate: a conceptual model to predict when habitat configuration really matters. J. Appl. Ecol. 51, 309–318.

Warburton, N.M., Prideaux, G.J., 2010. Functional pedal morphology of the extinct tree-kangaroo Bohra (Diprotodontia: Macroodidae). In: Coulson, G., Eldridge, M.D.B. (Eds.), Macropods—the Biology of Kangaroos, Wallabies and Rat-Kangaroos. CSIRO Publishing, Collingwood, Victoria, Australia, pp. 137–151.

White, P.C.L., Ward, A.I., 2010. Interdisciplinary approaches for the management of existing and emerging human–wildlife conflicts. Wildl. Res. 37, 623–629.

Williams, S.E., 2006. Vertebrates of the wet tropics rainforests of Australia—Species distribution and biodiversity. Cooperative Research Centre for Tropical Rainforest Ecology and Management, Rainforest CRC, Cairns, Australia. 282 pages.

Winter, J.W., Bell, F.C., Pahl, L.I., Atherton, R.G., 1987. Rainforest clearing in northeastern Australia. Proc. Roy. Soc. Queensland 98, 41–57.

Woinarski, J., Burbidge, A.A., Harrison, P.L., 2015. Ongoing unraveling of a continental fauna: decline and extinction of Australian mammals since European settlement. Proc. Natl. Acad. Sci. 112, 4531–4540.

Wroe, S., Clausen, P., McHenry, C., Moreno, K., Cunningham, E., 2007. Computer simulation of feeding behaviour in the thylacine and dingo as a novel test for convergence and niche overlap. Proc. Biol. Sci/Royal Soc. 274, 2819–2828.

CHAPTER

8

Tree Kangaroo Tourism as a Conservation Catalyst in Australia

Alan Gillanders[a] *and Clevo Wilson*[b]

[a]Alan's Wildlife Tours, Yungaburra, QLD, Australia

[b]QUT Business School, School of Economics and Finance, Queensland University of Technology, Brisbane, QLD, Australia

OUTLINE

INTRODUCTION

"Are there really tree kangaroos in this forest?" asked the middle-aged foreign tourist as the group prepared for a nocturnal tour.

"Yes, more than a hundred in the national park and adjoining private lands."

"That's wonderful. I don't really need to see one. Just knowing they are there is enough."

Such a well-developed conservation ethic in an ecotourist is not rare but is rarely expressed. It does not take the pressure off the guide to find the elusive rare species which the guest has come to see; guides suffer from their own

Tree Kangaroos
https://doi.org/10.1016/B978-0-12-814675-0.00012-9

expectations and knowledge of human behavior. While many visitors will be satisfied to have spent time in the same forest as the animals, for most people the experience is enhanced manyfold by actually encountering their target and spending time observing and learning. The cathemeral Lumholtz's tree kangaroo (*Dendrolagus lumholtzi*) is not a reliable target. This chapter will outline the development of tourism around tree kangaroos in north Queensland, Australia and elucidate the implications for the conservation of the species. Some conclusions about the efficacy of tree kangaroo tourism as a conservation measure will be drawn from anecdotes of the experiences of north Queensland operators.

FIGURE 8.1 Lumholtz's tree kangaroo is the smallest but most easily accessible species of tree kangaroo for tourists to see. *Source: Sandy Carroll Photography.*

HISTORY OF TREE KANGAROO TOURISM IN NORTH QUEENSLAND, AUSTRALIA

James English of Malanda, in northern Queensland, may have been guiding tourists to see tree kangaroos as early as 1913 but it is known that by 1928, he had established a banqueting hall in the "Jungle" between the village and nearby Malanda Falls (P. English, personal communication, 2018). Guests arrived by train from the regional coastal city of Cairns and overnighted in the Malanda Hotel. Tours were conducted on wide well-maintained paths through the forest. These paths were lined with bird's nest ferns sustained by a system of sprinklers. Local indigenous people were employed to demonstrate their remarkable ability to "run" up trees using a loop of lawyer vine (*Calamus* spp.) and to find tree kangaroos (Fig. 8.1).

In addition to the banqueting hall and its kitchen, James kept an aviary in which he had a brolga (crane) (*Antigone rubicunda*) called Joe and a Victoria's riflebird (*Ptiloris victoria*), a bird-of-paradise, which would take grubs from James' lips. The "Jungle" continues as forest held by the English family with few signs of the old infrastructure remaining. Along with the conservation reserve across the North Johnstone River, "The Jungle" supports a small but continuing population of Lumholtz's tree kangaroo while the surrounding area is either urban or pasture. The edge of the conservation reserve has been reinforced by tree planting by a local community group, Trees for the Evelyn and

Atherton Tablelands (TREAT). The Malanda Falls Visitor Centre is on land adjoining the reserve and provides information on all the attractions of the area, but their displays have an emphasis on the natural history including a taxidermied Lumholtz's tree kangaroo on loan from the Tree-Kangaroo and Mammal Group (TKMG). Audio-visuals on tree kangaroos, their lives and their rehabilitation are featured at the visitor center.

Businesses in north Queensland which are involved in tree kangaroo tourism are small, often single operator firms having the ideals of ecotourism in mind in their procedures. Much of the tourism which occurs in natural areas is not ecotourism in that it has little educational value, does not support local conservation efforts and engages little, if at all, with local communities. However, most operators with a focus on tree kangaroos are involved in local conservation organizations, participate in education sessions and give of their free time to further the aims of conservation. They also call on their guests to consider what they can be doing in their local communities to progress conservation ideals and practice there. While this group of travelers tends to self-select from those already involved in conservation to some degree at home, reflection and moral support can strengthen their resolve. As part of a guide training program, it is impressed upon guides and trainees that it is up to them to bring home the educational and conservation message to visitors. Guides have the advantage of location, time, and timing. When people are on holidays, freed from normal stressors, they are in a good position to evaluate the impacts they have on the world.

As most tourists visit in the winter and dry season, it is not easily possible to involve them in habitat restoration plantings, but some do visit the TREAT nursery and display center at Lake Eacham where they can become involved in preparations for this work. Some guests become long term supporters of local community restoration projects. Most tree kangaroo-focused restoration works occur in and around Mabi Forest, but others have taken place at higher altitudes (Fig. 8.2).

FIGURE 8.2 A young Lumholtz's tree kangaroo utilizes one of the species' favorite food, umbrella tree, *Schefflera actinophylla* in a revegetation site. *Source: Sandy Carroll Photography.*

LUMHOLTZ'S TREE KANGAROO HABITAT

The core distribution of Lumholtz's tree kangaroo extends along the uplands of the Wet Tropics World Heritage areas of north Queensland,

Australia from west of Ingham in the south to west of Mossman in the north. Most sightings occur in the forests growing on rich basal-derived soils (Burchill et al., 2014). This forest is important to tree kangaroos because of the rich soil leading to nutritional levels in the leaves which may be double that in nearby forest on other soil types. Lumholtz's tree kangaroo naturally reaches its highest density in this forest type as does the green ringtail possum (*Pseudochirops archeri*), although they occur widely within the wet tropics bioregion.

This Mabi Forest was the first regional ecosystem to be listed as threatened (EPBC, Act) under the Commonwealth Environment Protection and Biodiversity Conservation Act 1999 (EPBC Act) (Australian Government Department of Agriculture, Water and the Environment (DAWE), 2002). At the time of listing of Queensland Regional Ecosystem 7.8.3, also known as Complex Notophyll Vine Forest 5b (Tracey, 1982), there were about one thousand hectares of the forest remaining and it was under threat from invasive weeds, livestock, and clearing. The forest is characterized by growing on rich young basalt soils in the drier zone capable of supporting rainforest. Canopy height may exceed 45 m, but the canopy is uneven and deciduous, semievergreen trees are common. This leads to high light levels in the understory which is dense compared to other rainforests and a multitude of indistinct layers in the forest. Epiphytes are not common but those which grow there often attain great size.

Muppee and Mabi are Aboriginal names given to the Lumholtz's tree kangaroo in its range across the Atherton Tablelands. Tony Irvine, an ecologist with the Commonwealth Scientific, Industrial, and Research Organization, in his role as a committee member of TREAT, suggested that Mabi be adopted as the name for the forest type which is the preferred home of this species. "Forest type 5b" was not very exciting when trying to enlist the endeavors of local conservationists, let alone the general public.

A recovery plan (Latch, 2008) for Mabi Forest was developed by local community leaders in association with Queensland National Parks officers, but despite being signed off by both state and federal governments, no funding was forthcoming to directly implement this plan. This state of affairs is not unique to this habitat or species. TKMG, a community conservation and education group (Chapter 5) which has been involved in tree plantings and bimonthly public meetings with guest speakers, developed a five goal Community Action Plan to facilitate the conservation of this species in 2014 (Burchill et al., 2014). This plan aims to raise awareness, protect and connect habitat, mitigate threats, develop knowledge, and institute protocols for rehabilitation of injured and orphaned animals.

Lumholtz's tree kangaroo has been the major impetus for conservation in Mabi Forest, although not the only one. Among the 13 plant species and 12 animal species listed as threatened are trees such as pink silky oak (*Alloxylon flammeum*) and coorangaloo quandong (*Elaeocarpus coorangooloo*), the shrubs Atherton sauropus (*Sauropus macranthus*) and Atherton turkey bush (*Hodgkinsonia frutescens*), a bird, Macleay's double-eyed fig-parrot (*Cyclopsitta diophthalma macleayana*), the yellow-blotched forest-skink (*Eulamprus tigrinus*), and a mammal, the green ringtail possum already mentioned (Australian Government Department of Agriculture, Water and the Environment (DAWE), 2002).

Community conservation plantings have resulted in the creation of new habitat at Freeman's Forest and links between the forest reserves of Lake Eacham and Lake Barrine that form the Crater Lakes National Park, and the Curtain Fig National Park. Work to expand and secure these corridors is continuing and tree kangaroos are utilizing the plantings.

COMMUNITY AWARENESS AND CONSERVATION

As clearing of vegetation for farming progressed across the Atherton Tablelands, settlers came into contact with Lumholtz's tree kangaroo which were accidentally or purposefully killed. Some of these were females with advanced pouch young which were taken to be raised as pets. Although there was no organized tourism around such animals, they were displayed to tourists at least in Kuranda, Malanda, and in Yungaburra. It seems that this did not raise significant local pride in the species as they were considered ubiquitous and having one was as normal as was raising a cassowary chick for Christmas dinner. A photographic collection contains shots of girls in their finery with their pet tree kangaroos (Fig. 8.3). While viewing these pictures, the author heard two older ladies comment, "well, we all did that, and it was so common we never thought to take a picture".

As the rate of clearing vegetation greatly reduced, interactions with wildlife occurred less frequently and people adapted to the new paradigm of rarely seeing tree kangaroos. When tourism around tree kangaroos increased and community awareness was raised by conservation efforts, awareness of the species rebounded. Pride in hosting such special creatures is now high in the community.

FIGURE 8.3 Hislop sisters-Ann, Julie and Janetta-Whyalla Plains at Bloomfield—circa 1905—with their pet Bennett's tree kangaroos. *Source: Cairns Historical Society.*

The establishment of TREAT in 1982 brought together a number of people concerned for the future of tree kangaroos and other wildlife to act by restoring connectivity and strengthening resilience of remnants of rainforest on the Atherton Tablelands (TREAT, 2020). This group has continued in strength and plants as many as 15 thousand trees annually in community plantings on public and private lands, some of which have been purchased by benefactors for the purpose of conservation. While tree kangaroos are not the focus of many of these plantings, they are able to benefit from them all. A list of tree kangaroo food species has been developed and published on the TKMG website (TKMG, 2020). One land holder started with a forestry plot which has morphed into a restoration project with a population of at least seven tree kangaroos on what was once 3.5 ha of grazing land and a swamp. This planting also forms an important node in the middle of a corridor between two national parks. Tourists visit the TREAT display centre where they learn about the work of the group and its impact in increasing tree kangaroo numbers. Text is supported by before and after pictures of the landscape, of plantings in progress, and of their development. Display panels describe the tropical rainforests of the area, outline Aboriginal use, and the impacts of settlement. Human stories and case studies add to the interest of the restoration work. There is the opportunity for visitors to join the organization and or to make donations.

With the listing of Mabi forest as a threatened ecological community by the state and commonwealth governments in 2008, more local people realized that tree kangaroos were limited in their distribution. A few years earlier, Graham Newell was conducting research on tree kangaroos in an area of forest regrowth with a substantial population. When the farmer cleared this land and was chastised in the local community, his response was, "But there are tree-climbers everywhere." And so there were, in his limited experience. Community reaction to this was a catalyst to the formation of the Tree Kangaroo and Mammal Group (TKMG) (Chapter 5). Community surveys, restoration plantings, and public meetings conducted by TKMG from 1997 also raised awareness and gave the public an opportunity to be involved in data gathering and conservation efforts.

The sale of Queensland state-owned timber assets to Hancock Timber Resource Group in 2008 raised significant concerns in the community for the large stand of Mabi Forest in Wongabel State Forest. This forest area, known to be rich in tree kangaroos and utilized by locals for recreation and by tourism enterprises, was mapped as "Red Cedar Plantation." As a plantation, it could be logged but if it had been accurately mapped as mature Mabi regrowth, this wouldn't have been the case. In the past, foresters new to the rainforests of north Queensland were trained in the recognition of local tree timber species: not a simple task given its diversity. The historic, recreational, and environmental values of the forest could be clearly demonstrated. After eighteen months of lobbying, a clause to exclude this section of Wongabel State Forest from timber harvesting was inserted into the agreement between Hancock and the government of Queensland.

TOURISM HISTORY AND EXPERIENCE

Tourism involving tree kangaroos on the Atherton Tablelands was established before the great depression but has grown significantly in the past three decades. The establishment of natural history tours with professional guides occurred from the 1980s. One company, Wait-a-While Tours, developed personalized guiding with a focus on mammals and birds also involving nocturnal spotlighting and was a forerunner for several more guided tours that included the tree kangaroo, an appealing species. Unfortunately, tree kangaroos are rarely predictable and tour operators needed to focus on a range of species to satisfy their guests.

Up until the year 2000, there was only one tour company offering tours to see tree kangaroos. By the end of the next decade, there were three and now there are at least six. In the 1990s, Fur and Feathers at Tarzali instituted night walks to see tree kangaroos and Lumholtz Lodge at Upper Barron often had them in managed care as well as in the forest. Mt. Quincan Crater Retreat, Yungaburra had opened by 2000, and tree kangaroos were being seen occasionally in 2005. Only five years later they were a feature of the place.

LUMHOLTZ'S TREE KANGAROO ECOTOURISM ACCOMMODATIONS AND TOURS

The Quincan Crater Retreat Experience

Mt. Quincan Crater Retreat, a nature-based tourist accommodation facility, was started by Barbara and Kerry Keough who at the time adopted the Rainbow Bee-eater as their logo but said that they would today have chosen

Lumholtz's tree kangaroo (B. Keough, personal communication, 2017). Building of holiday cabins commenced in 1998 and consists of separate high-quality pole-house units. Some guests came purposely for the tree kangaroos which are highlighted on the website and in the brochure. However, all guests on check-in were given the challenge of finding their own tree kangaroo. Mt. Quincan's Facebook page features a video of Lumholtz's tree kangaroos, and this video also runs in reception. The compendium in the rooms has information about Lumholtz's tree kangaroo. Also, in each room, is a copy of Jane Hopkinson's book, "Tails from the Treetops" (Hopkinson, 2010). This delightful picture book also provides basic information on tree kangaroo biology and threats to them.

The owners purchased trees to plant for privacy around the cabins. Through talking to their neighbor about his plantings, they became aware of TREAT and increased their planting efforts. With the Barron River Catchment Authority, they planted along their creek and in 2002, planted a tree kangaroo specific corridor. There has been an interpretive trail established in this planting with the help of Sue Mathams of TKMG. The plaque was unveiled and the trail opened by the Queensland Governor. With knowledgeable community groups, a supportive landholder, and volunteer labor from the broader community, the appeal of the resort has been increased by successfully providing for the needs of a population of Lumholtz's tree kangaroo.

Tours were conducted by buggy around the crater of the volcano for guests with opportunities for viewing Lumholtz's tree kangaroos. Seeing three or four was not unusual but one day they saw a record eight individuals. One couple reported that while they were enjoying the spa bath, they witnessed tree kangaroos mating in the tree near their treehouse.

The population of Lumholtz's tree kangaroo at Mt. Quincan was not just used for the entertainment and enlightenment of tourists (Figs. 8.4 and 8.5). Scientists Lars Katzmyer and John Kanowski conducted radio tracking surveillance of the animals in 2004 and found that the resident population consisted of six adult females, two adult males plus their young (Katzmyer, personal communication, 2004).

The Keoughs said of their time at Mt. Quincan, "There were no tree kangaroo there when we arrived in 1998 but there are now!" This demonstrates how effective regeneration plantings have been in providing food and shelter to support a permanent wild population. The new owners adopted Lumholtz's tree kangaroo as their emblem (Fig. 8.6).

Lumholtz Lodge, Where Tree Kangaroos Come to Visit

Lumholtz Lodge is a bed-and-breakfast on the southern Atherton Tablelands where guests are often lucky enough to interact with wild tree kangaroos. It provides comfortable accommodation in a rainforest setting with of the potential for seeing many wild animals and also some in managed care and various stages of rehabilitation (Chapter 6).

To some extent Lumholtz Lodge "just happened" according to the owner, Margit Cianelli. Travelers made contact because Margit published articles regarding the husbandry of sick and injured tree kangaroos, and they wished to see her work firsthand (Fig. 8.7). Her early career had involved tree kangaroos under managed care, and she took on rehabilitating this species in the early 1980s. In the mid-1990s, Margit developed her home into a bed-and-breakfast in a more formal manner by decking it out with art works and books of natural history beyond those she already possessed (Fig. 8.8).

Guests are fascinated by the tree kangaroos, but their interests are not limited to mammals with some being serious bird watchers. However, given a chance to interact with a wild tree kangaroo returning "home" for food or rest is

FIGURE 8.4 Mt. Quincan Crater Retreat sits on the rim of an explosive volcano, giving wonderful views over the rich farming lands of the Atherton Tablelands, once covered in the preferred rainforest habitat of Lumholtz's tree kangaroos. From here a revegetation corridor traversing private lands and linking two National Parks is visible. *Source: Mt. Quincan Crater Retreat.*

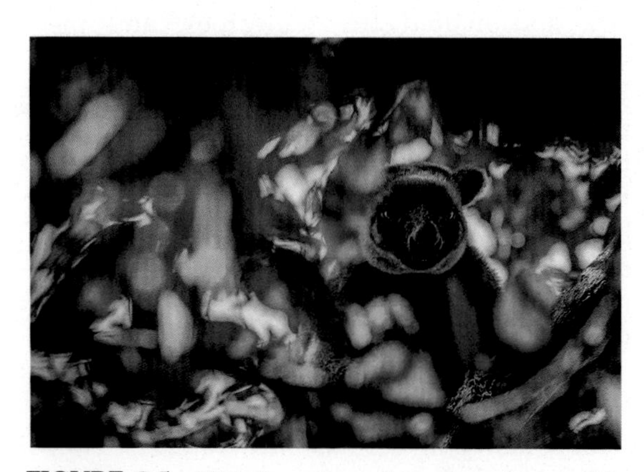

FIGURE 8.5 With the return of tree kangaroos to Mt. Quincan, it is not uncommon for guests to find they are being watched. *Source: Brendan MacRae.*

FIGURE 8.6 Mt. Quincan's logo used to feature a bee-eater, but now it is a tree kangaroo. *Source: Lumholtz Lodge.*

FIGURE 8.7 Orphaned tree kangaroos are taught climbing skills before their soft release at Lumholtz Lodge. *Source: Lumholtz Lodge.*

such a special experience, few have ever refused it and most find the experience spellbinding. Fifty per cent of the lodge's guests are international, and they are usually very well prepared, having studied the birds and mammals of the area before arriving.

"We feel absolutely certain, Margit, that your incredible lodge is one of the most memorable and unique places that we have EVER stayed in our three years of adventures. You are such a DIFFERENCE MAKER … Thank you for sharing your piece of heaven with us," Tom and Janet of Ohio wrote in the Lumholtz Lodge guest book, before going on to list the names of the animals with which they had interacted. Another wrote in the guest book, "There is a much-used word in Australia, "awesome." It is how I would describe Lumholtz's Lodge. … All I can say is that I am planning my return visit before I leave." These and the majority of comments show the appreciation guests have for the opportunity to interact with a rare species, to support their rehabilitation and aid in their conservation.

Guests have, "… decided to come to where they can see them (Lumholtz's tree kangaroos) up close and personal," said Margit. Film crews avail themselves of the opportunity to film wild but habituated animals behaving naturally, without fear of the humans. The reputation of this lodge for wildlife and hospitality is such that one must book well in advance. The guests who stay here also spend considerable sums elsewhere on their holidays.

Nerada Tea Plantation

Nerada Tea Plantation on the outskirts of Malanda, Atherton Tablelands, north Queensland is one of the most reliable places for tour parties or independent travelers to see Lumholtz's tree kangaroo during the day (Fig. 8.9). Many visitors to the plantation are there just for that purpose and the manager, along with the operators of the Tea House are supportive. Effort is made to ensure that both the visitors and the tree kangaroos are safe. Some income in the form of increased sales of souvenirs and meals occurs, but the greatest benefit to Nerada is the generation of goodwill.

On the establishment of the plantation, Nerada Tea Estates sought advice on plants for wildlife with a particular emphasis on tree kangaroos. These animals were sometimes seen in the remnant vegetation along the creeks of the plantation. Amenity plantings between the tea house and the factory and along the roadside have been particularly successful in attracting Lumholtz's tree kangaroos, with an extremely high-density present.

FIGURE 8.8 Set in a large private rainforest at nearly 1000 m above sea level, Lumholtz Lodge is known not only for its hospitality, tree kangaroos, and other animals in care but the abundant and diverse wildlife. *Source: Lumholtz Lodge.*

FIGURE 8.9 The Tea House at Nerada Tea Plantation, Malanda, is a wonderful spot to rest and refresh after the excitement of watching tree kangaroos or while waiting for the sleepy heads to wake and move. *Source: Nerada Tea Estates.*

Of note is also the reduced territoriality observed in this population with young females remaining to breed in their natal territory (Fig. 8.10). This is despite the presence of their dam. Elsewhere observations showed that the last female offspring can remain in the territory of their mother, but do not exhibit signs of estrus until the older female ceases to breed. A subadult male returning to his natal territory was tolerated by the dominant male here too. Indeed, playful interactions between the young of this population and the series of dominant males have been observed.

Nerada Tea Estates are committed to not only maintaining the amenity planting but plan to extend the revegetation. This provides evidence of the perceived value of tree kangaroos in attracting people to the site and in private businesses investing in their habitat.

TREE KANGAROO TOURISM BEST PRACTICES

The Alan's Wildlife Tours Experience with Lumholtz's Tree Kangaroo

Alan Gillanders established Alan's Wildlife Tours, a nature tour business in 2003 and adopted the Lumholtz's tree kangaroo as the business emblem and logo (Fig. 8.11). It has proved to be one of the important attractions for guests. Nocturnal tours are conducted on a dairy farm with a World Heritage rainforest boundary under an exclusive access arrangement (Fig. 8.12). Care is taken to ensure the safety of the tree kangaroos and minimize their stress. Tree kangaroos are known to avoid disturbance threats and will even jump from heights over 15 m to the ground if menaced. Over the first seven years of operation a detailed diary of sightings was maintained and showed that in an area not visited by other guides or the general public, the rate of sighting tree kangaroos by guests increased from just over 34% to nearly 86% of tour nights, until the advent of Cyclone Larry which destroyed much of the canopy cover, reduced habitat, and led to the disruption of the population. While the operator may have become more skilled at finding the animals, the conclusions drawn from this experience are that they will become habituated if not stressed and the maintenance of quality habitat is continued. This is borne out by the experience of Nerada Tea Plantation as shown. Guests were supportive of avoiding stress to the animals though at times it took considerable self control from the guide not to push for a better view of an animal which was too close to the party for the animal to feel safe.

Specific animals were able to be recognized, not just by their territories but by appearance. This allowed the guide to adjust to the individual personal space required by each animal and to become responsive to the subtle changes in behavior which precede an animal evacuating the roost or feeding site. Tree kangaroos are not the easiest animals to "read" so a precautionary approach is best and familiarity with individuals a great advantage.

In areas of high visitation, tree kangaroos and indeed other mammals can become stressed through the irresponsible use of high-powered lights for lengthy periods. These animals then become conditioned to avoiding the areas or at least avoiding the lights. Light avoidance can include turning away from the lights and hiding in dense vegetation as well as fleeing. The use of red filtered lights and an infrared monocular can negate the second effect, but if an excursion follows the use of numerous lights or very bright lights by others at that site then the sightings are reduced. It is suspected that this response will continue for some nights and if the human interference continues at a high level, will produce an avoidance area.

FIGURE 8.10 Mating tree kangaroos at Nerada Tea Plantation don't mind an audience of humans, but junior sleeps on. *Source: Sandy Carroll Photography.*

FIGURE 8.11 Alan's Wildlife Tours Logo. *Source: Alan's Wildlife Tours.*

Appropriate practices for tree kangaroo viewing include:

(a) Never touch the tree or vine a wild tree kangaroo is sitting in - this is a threat to the animal, and it may jump from the tree and injure itself. Tree kangaroos avoid mild threats by climbing higher but will exit a tree if under significant pressure.

(b) Maintain a horizontal distance of ten meters from a wild tree kangaroo unless the tree kangaroo is more than 15 m from the ground. This avoids intruding excessively into its space and gives more in the group a chance for a good view.

(c) Keep your group together on one side of the tree as a slightly nervous animal may try hiding behind a branch or leaves and to prevent this multiplies stress.

(d) Limit movement and noise around tree kangaroos because they may feel threatened, move to places more difficult to see or even jump from the tree. Tree kangaroos are not very demonstrative animals so can be hard to read. If an animal is twitching its head a lot and there are no flies bothering it, if it is making a "woofing" sound like a hollow bark, and particularly if it starts sliding down a tree at speed, back off quietly.

(e) Do not try to attract the animal's attention with noises. This is the most tempting of all. We all want to see the face of an animal, but they need their sleep and to feel secure. Yours might not be the only group to encounter that

FIGURE 8.12 Nocturnal tours to view tree kangaroos and possums have become popular on the Atherton Tablelands. *Source: Alan Gillanders.*

animal. If you make a small noise and it does not work do you or others make louder disturbances or even perturb the forest?

(f) If spotlighting, limit the number and brightness of lights. The use of red filters or lights is highly recommended as they have less impact on night vision. Use just enough light to see the animals. Search with the light as close to your eye as feasible to increase the chance of seeing their eye reflection. Once the animal is found, switch other lights off. This enables the entire group to know which light is in play and reduces disturbance.

(g) If photographing the animals, do it ethically. Lumholtz's tree kangaroos are active during the day as well (Lumholtz's tree kangaroos have a cathemeral activity pattern). Try to photograph them during the day, and keep other recommendations (see above) in mind including distance, noise, etc. You will get better photos of the animal in daylight. Photographing Lumholtz's tree kangaroos at night is not advised, as the animals very rarely continue their normal behavior when disturbed by light or flashes—even red light.

(h) Limit the size of groups. This both reduces the potential to scare the animals and increases the value of the experience for guests. It is easier for them to hear the guide and to find a position from which to view the tree kangaroo which may be visible through only a narrow window. If the animal is barely visible through a small space, keep one light on it and have a member of the group move around, without lighting the trees, to see if they can find a better angle.

The use of a professional guide will enhance the experience of the group and the target animals by maximizing ethical behavior.

ECONOMIC IMPLICATIONS OF TOURISM INVOLVING LESSER KNOWN SPECIES WITH SPECIAL EMPHASIS ON TREE KANGAROOS

Not much published literature exists on the use of lesser known species in tourism and the implications for their conservation (Tisdell and Wilson, 2012). Lesser known species do have tourism values, but not all species have the potential to attract sufficient tourism dollars to justify the protection of their habitat outside nature refuges (e.g. national parks). There are alternative uses for the private land on which they live (e.g. clearing their habitat for agriculture, grazing) and it is this fact in which the problem lies. This section discusses some of the issues related to lesser known species involved in tourism, the extent and type of values generated, the effectiveness of tourism dollars generated, the need to consider nonuse values stemming from such tourism and their conservation implications.

Despite their iconic species' status, tree kangaroos are not well known to the public. This lack of knowledge also extends to some insect species such as glow-worms. An experiment conducted by Tisdell and Wilson (2004a) in 2002 on the knowledge of Australian tropical wildlife among 204 residents of Brisbane, Australia, revealed that only 36% of the respondents knew of the existence of tree kangaroos. Most of those who stated they knew about tree kangaroos said that their knowledge was poor. Another study conducted by the same authors during the period 2002–2004 (Tisdell and Wilson, 2004b), showed that 80% of the visitors to a glow-worm site in South-East Queensland, Australia knew about the existence of glow-worms, but did not have a good knowledge of them.

Despite tourists' poor knowledge of lesser known species, they (especially nature-based tourists) are keen to see such animals and insects in the wild (or in captivity) when the opportunity provides. Nature-based tourism focused on tree kangaroos has existed in Australia from the early part of the 20th century. However, tourist numbers have been gradually increasing since the 1980s, more so after the 2000s although the numbers are subject to tourism's booms and busts.

As is the case with wildlife species which are the object of tourism, lesser known species such as tree kangaroos have tourism value, but not all of it is captured adequately. Lesser known species do not generate large tourism revenues to an individual or to a property owner, but instead, most of the revenue is spread-out. That is, the use of some of these resources (sightings) is free (e.g., seeing tree kangaroos in a national park in the Atherton Tablelands where there is no entry charge) or under-priced. Furthermore, tourist numbers visiting private properties are low or the tourist numbers are not evenly spread through the year. In short, the extent of tourism revenue that can be generated by a private property from tree kangaroo-based tourism is low. On the other hand, if tourism numbers are steady throughout the year and if places of such viewings are restricted, then it is likely that private landholders will derive a financial gain in having tree kangaroos on their properties. In the case of tree kangaroos, there is a limit to the size of party involved per excursion. Furthermore, sightings may not be guaranteed.

A survey conducted by Tisdell and Wilson (2004a) on tree kangaroo-based tourism revealed more than 90% of tourists found it was worth the time and effort spent in looking for Lumholtz's tree kangaroo. A third of those expressed that the experience was worth more than their costs, demonstrating an economic surplus. Of the surveyed visitors, 87% incurred expenditure at the nearby village or within 60 km of it with the average amount being three times what they spent on the tree kangaroo tour itself. It must be pointed out here that it is difficult to show how much of this expenditure can be attributed to the desire to see tree kangaroos, but it is indicative of the desire to see them.

In addition to their marketed value, tree kangaroos also have nonmarketed values (nonuse values) which account for the major part of their total economic and tourism use values. Nonuse economic values involve intangible benefits to society from the existence of wildlife as referred to in a quote from a visitor in the introduction to this chapter: "That's wonderful. I don't really need to see one. Just knowing they are there is enough". The reasons people are prepared to pay for the continuance of species include its existence value, its option value (they may seek it in the future) and its bequest value (leaving it for future generations). Tisdell and Wilson (2004a) found that nonuse values accounted for more than 80% of the total economic value of tree kangaroos for more than half of a sample of over 200 respondents in an experiment conducted in Brisbane, Australia.

While private landholders gain little from tree kangaroo tourism, monetary benefits accrue to areas proximate to where they exist. Governments have an incentive to conserve wildlife because the overall benefits to an area or region from wildlife tourism are positive and enhanced by favorable wildlife experiences. State intervention to expand the protected estate and compensating landholders based on the total economic value of the species can benefit other species and have wider ecological gains.

There is room for the development of strategies to further increase the involvement of tourists in conservation efforts. School and university groups are often tasked with tree planting, weeding, or monitoring activities. Echidna Walkabout Tours in Victoria include between 5 and 30 minutes of conservation activities "ranging from light weeding to bagging up discarded fishing net, to helping a guide record birds for atlas submission" in their tours. There is no reason why that cannot become a component of more tour and accommodation experiences.

IMPLICATIONS AND FUTURE

Ecotourism around seeing rare, little known but iconic species in remote places will not by itself save those species, their environments, or the communities of people who live with them. However, it may be one important element in the process. Just as bird tourism has had positive impacts on some places, it is likely with growing affluence and demand for further new experiences that mammal tourism will increase outside the well-known African experience centered on large animals. These tourists are likely to be well resourced, mostly of mature age and prepared to pay for distinctive experiences with a degree of comfort. This has the potential to provide development and employment opportunities.

By converting all ecotourism effects—positive and negative—to ecological parameters and including known data for nine species of threatened animals, researchers from Griffith University used population viability modeling to quantify the impact of ecotourism on threatened species (Buckley et al., 2016). Population viability models estimate cumulative population changes by simulating births and deaths repeatedly. Their research showed that while there was a possibility for negative effects, the impact of ecotourism would in most cases make the difference between survival and extinction. For the success of the tourism enterprises, it is important that there is a reasonable chance of successfully showing the animals to their guests.

One possible strategy is to have only one guide or group of guides using each area so that the animals are known to the guide and the guide to the animals. This might reduce the risk of stressing the animals for the one-off view as it is in the guide's interests to be able to show these animals many times in the future. Guests are usually amenable to poor views or even no views when such a decision is couched in terms of the animal's welfare (Fig. 8.13).

It is important that local communities retain some control of these experiences as when the benefit is going elsewhere, they will not act to preserve the resource. The call for "authentic" experience is also in favor of the provider being local (Star et al., 2020). Ecotourism requires the maintenance of high-quality landscapes and environments and while this is to the long term advantage of communities, they may not see it as the best short term means to satisfy their needs (Balmford, 2012). It is in this regard that governments, other agencies, and tourist operators can lead with provision of infrastructure (Rolfe and Flint, 2018), tools, plans, and even things as simple as appropriate language for this development to occur and be sustained by first engaging with the local people involved.

People are excited to see this little known and little understood species. They can be moved to tears or bouts of laughter by their antics. Some become regular supporters of local conservation efforts or have stepped up their involvement at home through the challenge set by their guide. The tourism around tree kangaroos is of low impact on the environment and has some benefit to the community beyond the income to guides, accommodation providers, and other service employees. The value tourism adds is demonstrated by the increased conservation activity it encourages.

FIGURE 8.13 Female Lumholtz's tree kangaroo with advanced pouch young and yearling at foot. This female shared the territory of her dam and had three young while her mother had two. Now she shares that territory with her youngest sister and her offspring. *Source: Sandy Carroll Photography.*

ACKNOWLEDGMENTS

Professor Emeritus Clem Tisdell, School of Economics, University of Queensland, Brisbane for insights on the economics of ecotourism; Kerry & Barb Keough founders of Mt. Quincan Crater Retreat, Yungaburra; Tony & Bev Poyner of Nerada Tea Plantation, Malanda for access and insights; Margit Cianelli of Lumholtz Lodge for photos and historical information; Paul English of Malanda for family history; Pauline O'Keeffe of Cairns Historical Society for historical photos; Sandy Carroll of Sandy Carroll Photography for photographs and encouragement; Ms. Jodie Eden Community Education Ranger, Department of Environment and Science for restoring to me an old spotlighting picture; Patrick De Geest of Eyes on Wildlife for discussions on best practice; Maria Gillanders of Alan's Wildlife Tours for a reflective ear; and editors Peter Valentine and Karin Schwartz for incisive suggestions and sympathetic treatment of the text and authors.

REFERENCES

Australian Government Department of Agriculture, Water and the Environment (DAWE), 2002. Environment Protection and Biodiversity Conservation Act 1999. https://www.environment.gov.au/epbc [5 Mar 2020]

Balmford, A., 2012. Wild Hope. On the Front Lines of Conservation Success. Chicago University Press, Chicago, IL.

Buckley R.C., Morrison C., Castley, J.G., 2016. Net effects of ecotourism on threatened species survival. PLoS One 11(2), e0147988. https://doi.org/10.1371/journal.pone.0147988 [2 Jan 2019]

Burchill, S., Cianelli, M., Edwards, C., Grace, R., Heise-Pavlov, S., Hudson, D., Moerman, I., Smith, K., 2014. Community Action Plan for the Conservation of the Lumholtz's tree-kangaroo (*Dendrolagus lumholtzi*) and its habitat 2014–2019, Malanda, Australia. https://www.tree-kangaroo.net/download_file/view_inline/186 [3 Feb 2019]

EPBC, 2020. https://www.environment.gov.au/epbc. Accessed February 2020.

Hopkinson, J, 2010. Tails from the Treetops. Jane Hopkinson Photography.

Latch, P., 2008. Recovery plan for Mabi Forest. Report to Department of the Environment, Water, Heritage and the Arts, Canberra. Environmental Protection Agency, Brisbane. https://www.environment.gov.au/biodiversity/threatened/recovery-plans/recovery-plan-mabi-forest [2-Jan 2019]

Rolfe, J., Flint, N., 2018. Assessing the economic benefits of a tourist access road: a case study in regional coastal Australia. Econ. Anal. Policy 58, 167–178. https://doi.org/10.1016/j.eap.2017.09.003.

Star, M., Rolfe, J., Brown, J., 2020. From farm to fork: is food tourism a sustainable form of economic development? Econ. Anal. Policy 66, 325–334. https://doi.org/10.1016/j.eap.2020.04.009.

Tisdell, C., Wilson, C., 2004a. The public's knowledge of and support for conservation of Australia's tree-kangaroos and other animals. Biodivers. Conserv. 13 (12), 2339–2359.

Tisdell, C., Wilson, C., 2004b. Economics of wildlife conservation. In: Higginbottom, K. (Ed.), Wildlife Tourism—Impacts, Management and Planning. Sustainable Tourism CRC, Common Ground.

Tisdell, C., Wilson, C., 2012. Nature-Based Tourism and Conservation: New Economic Insights and Case Studies. Edward Elgar, Cheltenham, UK.

Tracey, J.G., 1982. The Vegetation of the Humid Tropical Region of North Queensland. CSIRO, Melbourne.

Tree-Kangaroo and Mammal Group (TKMG), 2020. Habitat Plants—Food Plants for tree kangaroos. From: Coombes, K., Simmons, T. and Jensen, R., 2006. Tree use by Lumholtz's tree-kangaroo, *Dendrolagus lumholtzi*, on the Atherton Tablelands. https://www.tree-kangaroo.net/download_file/view/68/157 [22 February 2020]

Trees for the Evelyn and Atherton Tablelands Inc. (TREAT), 2020. Mabi Forest. http://www.treat.net.au/mabi/index.html [22 Feb 2020].

SECTION IV

CONSERVATION SOLUTIONS: IN THE FIELD – NEW GUINEA

CHAPTER

9

Opportunities for Tree Kangaroo Conservation on the Island of New Guinea

Bruce M. Beehler[a], *Neville Kemp*[b], *and Phil L. Shearman*[c]

[a]Smithsonian Institution, Washington, DC, United States

[b]US AID Lestari-Indonesia, Jakarta, Indonesia

[c]ANU College of Science, Australian National University, Canberra, ACT, Australia

OUTLINE

INTRODUCTION

This chapter discusses the opportunities for conservation of New Guinea's tree kangaroos in relation to the threats to these species (Chapter 4). Conservation for these beleaguered creatures can take the form of protecting habitat, creation of local constituencies actively protecting local tree kangaroo populations, and involvement in an international program of captive breeding and management.

CONSERVING TREE KANGAROO HABITAT

On the island of New Guinea, places that support one or more viable populations of tree kangaroos all tend to have similar characteristics. These are large tracts of old growth forest isolated from human populations and free of hunting. Despite the considerable informal and formal economic development of New Guinea over the last three decades, there are still pockets

https://doi.org/10.1016/B978-0-12-814675-0.00018-X

of mature forest habitat that are rarely visited by humans. These places tend to be physically isolated by their ruggedness, high rainfall, elevation, or distance from roads and economic development (Saulei and Beehler, 1993; Supriatna, 1999). In Indonesian New Guinea, most of these undeveloped areas are currently identified as national protected areas, though active management and protection is limited or absent. Because of land tenure issues, Papua New Guinea (PNG) has fewer national level protected areas.

A recipe is offered for the selection of a critical network of tree kangaroo conservation areas across the New Guinea mainland. For each New Guinean tree kangaroo species, the authors suggest that a contiguous block of more than 100,000 hectares of forest be allocated from near the heart of the species' range, further delineated by the focal species' elevational requirements. The selected conservation area should be as isolated from roads, areas of economic development, and villages as possible. Ideally, these forest tracts should be gazetted as strict conservation areas with full protection—no economic development, no forest clearance, no agriculture, and no hunting. In addition, the management of each area needs to engage local landowning populations in ways that make them proactive stakeholders in the conservation process (Beehler and Kirkman, 2013, Chapter 10). In this section the authors discuss candidate sites and in the following section, the necessary community engagement for each site. Selection of sites is driven by the distribution of the various species of tree kangaroos.

Doria's Tree Kangaroo (*Dendrolagus dorianus*) (IUCN status Vulnerable—VU). The undisturbed forest block in southeastern PNG above 600 m elevation between Mount Victoria and Mount Scratchley that drains northeastward into the Mambare River catchment is ideal as a reserve for this species. The Biagge people of Kanga village are traditional landowners of the eastern scarp of this large upland area and have demonstrated an interest in conserving the area for its ecotourism potential (Beehler, unpublished data).

Goodfellow's Tree Kangaroo (*Dendrolagus goodfellowi*) (IUCN status Endangered—EN). The hill forest block north and northeast of Kokoda (the western sector of the Ajule Kajale Range) in Oro Province of PNG is locally reported to support a good population of this species (Beehler, unpublished data). This uninhabited old growth lower montane forest is unhunted and rarely visited, and thus a good site to conserve this species.

Grizzled Tree Kangaroo (*Dendrolagus inustus*) (IUCN status VU). The isolated southern foothills of the Foja Mountains (100–1400 m) in the north-central sector of Indonesian New Guinea is ideal as a reserve for this species. This pristine forest tract is already designated by the government of Indonesia as the Mamberamo-Foja Wildlife Reserve. Additional monitoring, patrol, management, and local community engagement are required to establish a credible conservation regime here.

Huon Tree Kangaroo (*Dendrolagus matschiei*) (IUCN status EN). The 87,000-hectare YUS Conservation Area, a PNG national conservation area, was established to conserve a population of this species and has just expanded the dimensions of this current protected area. This is the gold-standard in tree kangaroo reserves being managed by local communities with assistance from local and international NGOs (Chapters 10 and 13).

Wondiwoi Tree Kangaroo (*Dendrolagus mayri*) (IUCN status Critically Endangered–CR). This hyper-endemic is restricted to the uplands of Mount Wondiwoi, on the Wandammen Peninsula, on the eastern Bird's Neck of Indonesian New Guinea. The tiny patch of upland forest of this narrow peninsula is already designated as a national nature reserve named Pegunungan Wondiboy and it is adjacent to Cenderawasih Bay Marine National Park. The entire range of this species is probably considerably less than 50,000 hectares, so the final size of this action area will probably be small because of the limits of montane forest present on the peninsula.

Dingiso (*Dendrolagus mbaiso*) (IUCN status EN). A substantial portion of this species' high elevation range lies within Lorentz National Park, the largest protected area in Indonesia. Active management of Dingiso within this protected area is now required, as indicated by the most recent Red List assessment, indicating a drastic population decline over the last three decades.

Ifola (*Dendrolagus notatus*) (IUCN status EN). The Kaijende Uplands above Porgera, PNG, are in the process of community-designation for conservation status under the Forest Stewards program, a partnership of Monclair State University (Monclair, New Jersey), the Porgera Mine, and conservation authorities at the provincial and national level. The upland forest of this large montane area can serve as a reserve for this species.

Golden-mantled Tree Kangaroo (*Dendrolagus pulcherrimus*) (IUCN status CR). The unhunted and unvisited foothills and uplands of the Foja Mountains are ideal as a conservation reserve for this species and are already designated within the Mamberamo-Foja Wildlife Reserve.

Tenkile (*Dendrolagus scottae*) (IUCN status CR). Conservation of upland forest sections of the PNG North Coastal Ranges is being pursued through the Tenkile Conservation Alliance (Chapters 16 and 22). In addition, all of the uplands of Mount Menawa further to the west merit protection to encompass the range of additional populations of this critically endangered species. Because of the fragmented nature of the species' population, several distinct forest blocks need to be conserved.

Lowland Tree Kangaroo (*Dendrolagus spadix*) (IUCN status VU). A large lowland and foothill area in southern Papua New Guinea, between the Wawoi River in the west and Purari River in the east should be delineated and protected for this species, though ideally, before that happens, additional surveys documenting the current distribution of this mysterious species should guide the geography of the habitat delineation. The forest tract should range from sea level to 800 m (or however high the species is shown to range once more field data are available).

Seri's Tree Kangaroo (*Dendrolagus stellarum*) (IUCN status VU). Lorentz National Park encompasses a large block of this species' montane range and should serve as an adequate preserve for this particular species.

Vogelkop Tree Kangaroo (*Dendrolagus ursinus*) (IUCN status VU). The entire uplands of the Tamrau Mountains of the Bird's Head of Indonesian New Guinea can serve as a reserve for this species. This is currently encompassed in the Tamrau Utara nature reserve, a national protected area. The single concern is the presence of a population of the exotic Sunda Sambar deer (*Rusa timoriensis*) in this forest to an elevation of ca. 1000 m (BB, unpublished data). In places, the exotic deer have over-browsed the understory vegetation. Does this impact the welfare of the native tree kangaroo at present? Perhaps not, but it is certainly having an impact on the forest ecology of palatable species and sapling regrowth, which may impact tree kangaroos in the long-term. Secondarily, the lower portions of this reserve can serve as a back-up conservation area for the western subspecies of grizzled tree kangaroo, which is common here (Beehler, unpublished data).

The value of establishing species-focused protected areas is that it provides a focus for action by the communities and government authorities. Each reserve has a measurable purpose—to safeguard a substantial population of a threatened species of tree kangaroo. It also provides a focus for any management plan, and also a means of measuring the effectiveness of management. So long as the focal species is stable or increasing, then the conservation interventions are adequate. Having a named focal species (or two) provides valuable focus for the conservation practitioners at the provincial, national, and international levels. The local communities will of course do their part (see next section) but should also hold the provincial and national authorities responsible for properly managing the protected area (e.g., to not give out logging, agricultural, or mining permits within these areas). The YUS management model is worth exporting to all of these possible tree kangaroo protected areas (Beehler and Kirkman, 2013).

CREATING LOCAL CONSTITUENCIES FOR TREE KANGAROOS

Because of the importance of local communities to each of the above proposed protected areas, full engagement of all communities living adjacent to these areas is mandated. All of these areas are the traditional forest lands of one or more local community. Without the strong community support of tree kangaroo conservation and of protection of these areas, success cannot not be guaranteed. Luckily, there are two models upon which to base this program—those of the YUS Conservation Area (Chapters 10 and 13) and the Tenkile Conservation Alliance (Chapters 16 and 22). In the complex social setting of rural New Guinea, a multi-decade stakeholder engagement process must be carried out, with substantial local leadership and wide citizen participation. Selection of appropriate management incentives is a major challenge in this process. In most instances, initially these community programs will need to be instigated and managed by outside institutions (national or international), with recurrent annual funding (as is happening with Tenkile and YUS). In addition, the Pride programs pioneered in many places around the tropical world by RARE, a US-based conservation NGO, might be worth considering deploying to these new areas (RARE, 2019). Rare's Pride campaigns are programs that provide community-based solutions for conservation of threatened species in the world's areas of highest biodiversity—from Latin America and the Caribbean to Africa and India to Asia and the Pacific Islands.

Most traditional societies that hunt tree kangaroos have certain taboos regarding these creatures. These traditional taboos are excellent starting points for re-establishing community-led sensibilities about protecting treasured local resources for future generations. The YUS teams use the concept of a "wildlife bank," wherein the protected forest (the "bank") generates "interest"—new populations of tree kangaroos that disperse into the adjacent sustainable-use zones where hunters can harvest them. So long as the hunters respect the no-hunting zones, such a model may be sustainable. This is in the testing phase in YUS today.

Building on the YUS and Tenkile models, local engagement needs to create strong alliances in support of each local conservation area and each local species of tree kangaroo. Regular patrolling of the conservation area and regular monitoring of the tree kangaroo population need to be carried out by the local community. Moreover, economic or other incentives need to be provided until these initiatives achieve self-sustainability. Such incentives must create a true sense of local community ownership of the conservation area and of the flagship species. It will be local communities that ensure the long term survival of wild populations of tree kangaroos into the next century.

One final point. While the local communities are critical, the projects cannot succeed without the sincere and strong support of provincial and national conservation and forest authorities. Creating the environment that fosters the growth and responsibility of these institutions remains a serious challenge. In developing nations, true site-based long-term conservation success requires layers of commitment and support at all levels of government, from ward to the national capital. This has proven elusive in Indonesia and Papua New Guinea over the past decades.

CAPTIVE BREEDING AND MANAGEMENT

Several species of tree kangaroos have been managed in zoological institutions over the decades, and breeding under managed care has been successful in at least two species (Chapter 18). Thus, it is likely that some threatened species from New Guinea could be established in an array of international zoo-based breeding facilities to establish a self-sustaining collection of tree kangaroos across the zoo world. These species would be deployed to multiple accredited facilities that are involved in managed breeding programs to ensure long-term genetic and demographic sustainability (Chapters 18–21). Suffice it to say, some of these species should be established in globally-recognized captive breeding facilities as soon as possible. Every effort should be made to keep these managed populations healthy and expanding as an insurance population in the event that a re-introduction to New Guinean localities becomes necessary to restore the species to areas where they have been extirpated.

ACKNOWLEDGMENTS

Steven Richards critically read and commented on this chapter. We thank Conservation International and the Committee for Research and Exploration of the National Geographic Society for support of field work in New Guinea. We also thank the governments of Indonesia and Papua New Guinea for permission to conduct research in western and eastern New Guinea.

REFERENCES

Beehler, B.M., Kirkman, A.J. (Eds.), 2013. Lessons Learned From the Field: Achieving Conservation Success in Papua New Guinea. Conservation International, Arlington, VA.

RARE, 2019. A RARE Approach. http://www.rare.org/our-approach/ [30 July 2019].

Saulei, S., Beehler, B.M., 1993. Biodiversity and conservation of humid forest environments in Papua New Guinea. In: Beehler, B.M. (Ed.), A Biodiversity Assessment for Papua New Guinea. Conservation Needs Assessment, Biodiversity Support Program, Washington, DC, pp. 423–431.

Supriatna, J. (Ed.), 1999. The Irian Jaya Biodiversity Conservation Priority-Setting Workshop, Biak, Indonesia, 7–12 January 1997. Conservation International, Jakarta, Indonesia.

CHAPTER

10

Creating the First Conservation Area in Papua New Guinea to Protect Tree Kangaroos

Lisa Dabek[a] *and Zachary Wells*[b]

[a]Tree Kangaroo Conservation Program, Woodland Park Zoo, Seattle, WA, United States

[b]Conservation International, Crystal City, VA, United States

OUTLINE

INTRODUCTION

Papua New Guinea (PNG) occupies the eastern half of the equatorial island of New Guinea. The forests of New Guinea are still largely intact, and the island is considered one of the three high biodiversity wilderness areas remaining on earth (Beehler, 1993; Mittermeier et al., 1997; Lipsett-Moore et al., 2010).

The mountainous Huon Peninsula in Morobe Province contains the Finisterre, Sarawaged, Cromwell, and Rawlinson Ranges. It is isolated from the central mountain ranges of PNG by the Markham Valley, making it one of the most distinct biogeographic areas in PNG (Fig. 10.1). More endemic bird and mammal species are found here, including the endangered Matschie's tree kangaroo (*Dendrolagus matschiei*) (Ziembicki and Porolak, 2016), than in any other like-sized area in the nation (Wikramanayake et al., 2001) (Fig. 10.2). In addition to endemic species, several other mammal and bird species

https://doi.org/10.1016/B978-0-12-814675-0.00032-4

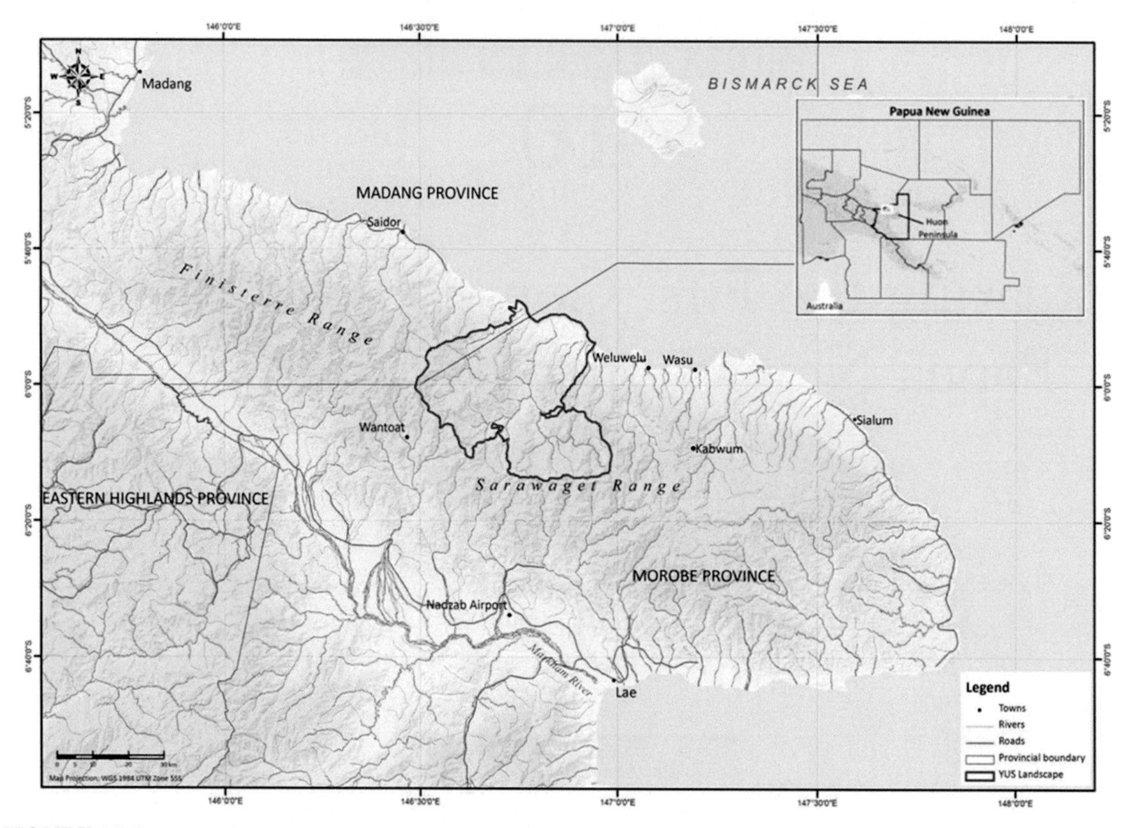

FIGURE 10.1 Map of the Huon Peninsula and the YUS area (outlined). *Source: TKCP.*

of particular conservation concern occur on the Huon Peninsula, such as the long-beaked echidna (*Zaglossus bruijni*), New Guinea Vulturine Parrot (*Psittrichas fulgidus*), and New Guinea Harpy Eagle (*Harpyopsis novaeguineae*).

While the Huon Peninsula is a storehouse of PNG's rich biodiversity, the nation itself is globally exceptional when it comes to human diversity. The geographic isolation of the island and rugged topography that have produced New Guinea's biological diversity have also contributed to the development of an incredible variety of human cultures and languages. PNG is the most linguistically diverse area in the world, with more than 820 distinct indigenous languages spoken (World Atlas, 2020). The people of Papua New Guinea, who own the land in a system of customary land tenure, therefore play a great role in the fate of their nation's environmental management. Customary land ownership in Papua New Guinea is enshrined in the National Constitution. It is commonly held that 97% of land is owned by Papua New Guinea's citizens who retain traditional clan-based structures. Such a system of customary ownership provides a rich opportunity for engaging local people in the creation of protected areas to conserve nature and empower stewardship of their land and sea for cultural purposes (Fig. 10.3). Worldwide, successful experiences in Papua New Guinea could build momentum towards greater community participation in conservation, particularly in countries with state land

FIGURE 10.2 Landscape of YUS. *Source: Bruce Beehler.*

FIGURE 10.3 Young woman with tree kangaroo tail and Bird of Paradise headdress showing the balance of conservation and cultural use of animals in YUS. *Source: Lisa Dabek.*

ownership and a more centralized system of protected areas.

The Yopno-Uruwa-Som (YUS) area is situated on the Huon Peninsula, with the majority of its extent located in Kabwum District, Morobe Province (Fig. 10.1). It roughly follows the jurisdictional boundaries of the YUS Local Level Government (LLG) with a few exceptions based on cultural relationships, i.e., clans, families, and language groups, and includes some portions of the Tewai-Siassi District of Morobe Province. These boundaries closely reflect the natural physiographic boundaries of three watersheds and are thus named for the three main river systems in the area-Yopno, Uruwa, and Som (YUS). The YUS area is a mixed landscape of forests, villages, grasslands, and agricultural areas (Fig. 10.4). All YUS land is owned by local clans. Based on data from the 2010 census, the YUS LLG population was approximately 12,000 living in approximately 50 villages and hamlets. The area is remote, rugged and roadless. Access is either by plane, foot, or in coastal areas by boat.

FIGURE 10.4 Multiple land-use in YUS—gardens and forest in background. *Source: TKCP.*

Many steep foot trails link villages across the landscape. Six airstrips are maintained in YUS, and weekly flights service specific airstrips for delivery of goods and transport of people. Scheduled flights are few, thus time-bound travel schedules and deliveries require expensive aircraft charters. Sea access is possible by hiring private local boats from Madang to the west and Wasu to the east. As a result, there is no large-scale commercial development in the area. Large blocks of forest remain unlogged and many can be classified as pristine.

The YUS landscape is a priority for conservation for many reasons, including high biodiversity value, intact habitat, unique species, and a high degree of endemism. There are many plant and animal species of global conservation significance present in YUS requiring habitat protection (Table 10.1). Of these species, those identified by IUCN as critically endangered, endangered, vulnerable, or near-threatened include the flagship species for the YUS Conservation Area, the Matschie's tree kangaroo (Tree Kangaroo Conservation Program, 2012). The area has remained of high value for nature conservation because of the subsistence lifestyle and sustainability of the land-use practices of the people of the YUS area. Threats to biodiversity in YUS have so far been prevented largely due to the rugged terrain which deters large-scale development. This geography has also greatly constrained economic development in the villages by restricting access to employment, markets and government services.

The history of the YUS Conservation Area, starting with the introduction of a protected area idea, to the primary gazettal and finally to the re-gazettal, has provided fertile ground for identifying the challenges and opportunities that will improve the process leading to long term sustainable conservation of tree kangaroos and other wildlife (Wells et al., 2013).

TABLE 10.1 Fauna in YUS of conservation concern on the IUCN Red List (International Union for Conservation of Nature, 2020).

Latin/Scientific Name	English Name	IUCN Status
Dendrolagus matschiei	Matschie's (Huon) tree kangaroo	Endangered
Zaglossus bruijni	Eastern long-beaked echidna	Critically Endangered
Dasyurus albopunctatus	New Guinea quoll	Near Threatened
Dorcopsulus vanheurni	Small Dorcopsis	Near Threatened
Casuarius bennetti	Dwarf Cassowary	Near Threatened
Harpyopsis novaeguineae	New Guinea Harpy Eagle	Vulnerable
Psittrichas fulgidus	Vulturine Parrot	Vulnerable
Parotia wahnesi	Wahne's Parotia	Vulnerable
Paradisaea guilielmi	Emperor Bird of Paradise	Near Threatened

THE CONSERVATION AREAS ACT

In PNG there are currently three existing legal structures for protected areas in PNG: National Parks, Wildlife Management Areas (WMAs), and Conservation Areas (CAs). National Parks are a rare occurrence since most land is privately owned. A significant aspect of protected areas in PNG is that in both WMAs and CAs, local landowners maintain ownership of their land. WMAs have been by far the most prevalent classification of protected area in PNG. They have been widely criticized because the legislation in which they are enshrined, the Fauna (Protection and Control) Act of 1966 (PACLII, 1966), protects the wildlife, but does not preclude destructive economic development on the lands encompassed by the WMA. This can (and does) occur when government priorities for economic development override environmental conservation concerns.

Consultation on the setup of a protected area must properly engage the necessary government line agencies in discussions about alternative pathways to development prior to establishment of a protected area. Moreover, increasing natural resource demands by villagers in growing rural populations may put dangerous pressures on WMAs, which depend entirely on self-regulation by local communities. In other cases, landowners simply change their minds when the benefits of the protected area do not meet their aspirations. This is a particularly relevant threat when communities expect, often to their ultimate disappointment, that protected areas will be a source of substantial income generation and development. A conclusion may be that the Fauna Act is not strong enough to provide long-term protection through changes in public and private priorities. The Conservation Areas Act of 1978 (PACLII, 1978) places a greater degree of authority with the Environment Minister who must authorize any development activities within a gazetted Conservation Area. The governance of Conservation Areas is clearer (and more restrictive) than that defined by the other protective acts. The Conservation Areas Act gives the mandate of management and monitoring to a multi-stakeholder management committee which reports to the Minister through the Conservation and Environment Protection Authority (CEPA), formally known as the Department of Conservation.

Whimp (1998) states that "Conservation Areas provide the only conservation mechanism that clearly allows management of the area to extend to controls over development. Existing land uses are not allowed to be changed unless either (a) the management plan [explicitly] allows it or (b) the Minister has authorized the development." These provisions regulating development activities in CA's are more rigorous than those regulating WMA's. However, they are counterbalanced by an onerous gazettal and management process for which the onus of executing often falls to partners in conservation in order to meet the aspirations of landowners.

The Conservation Areas Act, CA gazettal may begin in two ways. The first option is that the Minister may seek the recommendation that an area be declared a Conservation Area. The second option is that a person, group or authority makes a written request to the Minister.

THE PROCESS TO GAZETTAL OF THE YUS CONSERVATION AREA

The YUS Conservation Area on the Huon Peninsula evolved organically as many conservation initiatives do. A 1991 PNG Conservation Needs Assessment (Beehler, 1993) considered the Finisterre Mountain range of the Huon Peninsula to be a "scientific unknown" and in need of scientific and conservation attention. Dr. Lisa Dabek and her team first came from the United States to the YUS area in 1996 to determine the conservation status of the endemic Matschie's tree kangaroo. This species is listed as endangered under the IUCN Red List (2020) and is also exhibited in zoos in North America. Dabek had previously completed research on the reproductive biology and behavior of this species and wanted to connect research in zoos with accomplishing conservation efforts in the wild (Dabek, 1994; Chapter 22). No long-term biological research on the Matschie's tree kangaroo had been conducted previously. This initial work helped to determine basic population density estimates (Betz, 2001) as well as collect information on the key threats to tree kangaroos such as hunting and forest clearing.

The idea of creating a protected area developed early on through discussions between local YUS landowners and Dabek's team. Local YUS hunters had perceived a decreasing trend in local wildlife populations, particularly of the Matschie's tree kangaroo. One solution that resonated with local hunters was setting aside portions of hunting land to allow wildlife to reproduce and ensure the sustainability of hunted species. The concept of setting aside a portion of one's hunting land for a protected area was described as a "wildlife bank." The protected area would serve as a safe place for tree kangaroos and other wildlife to reproduce, and when the young dispersed from the protected lands into the buffer areas the hunters would be able to harvest them sustainably. Through these discussions it became clear that there had been a similar practice of culturally-based tambu (taboo) areas in the past. Communities in YUS continued to express interest in conservation over successive years of working with Dabek and team. In response, Dabek's team formalized a partnership with YUS landowners and developed a community-based conservation program called the Tree Kangaroo Conservation Program (TKCP) which is now based at the Woodland Park Zoo, Seattle.

Initially one clan in the Yopno region of the YUS area, led by a local well-known tree kangaroo hunter, Mr. Mambawe Manuno, set aside forest for conservation and research (Fig. 10.5). TKCP's small international research team, employing local assistants, began to facilitate informal community meetings with landowners in surrounding villages, and ultimately in the other YUS regions, to discuss the decline of tree kangaroo populations and the desire by the hunters to create a sustainable resource for subsistence hunting. Dabek also shared information about tree kangaroo reproduction from zoo

FIGURE 10.5 Yopno tree kangaroo hunter and first TKCP Conservationist Mr. Mambawe Manauno. *Source: TKCP.*

research (e.g., that the species is slow to reproduce, producing only one offspring every one to two years) to emphasize the vulnerability of tree kangaroos as a game species and the need to manage the rate of harvest.

In all community meetings TKCP emphasized that there would be no compensation for setting land aside, highlighting that the purpose of the protected area was for the long-term sustainability of YUS natural resources for the local communities. A second benefit presented was the diverse employment opportunities offered by the TKCP program in YUS. TKCP subsequently initiated several community-wide initiatives that addressed local needs regarding education, health, and livelihoods.

From the early stages of partnership with the YUS community, TKCP took the approach of "community-based conservation," focusing on local ownership, and prioritizing community development needs as well as environmental needs (Chapter 16). TKCP researchers took guidance from YUS leaders, and the local Member of Parliament from Kabwum District, Mr. Ginson Saonu, who came from YUS. The services that the YUS communities were most concerned about were education and health. Many schools in the villages were closed because of a lack of teachers. As a response, TKCP sponsored a community education project in 1998 which provided teacher training scholarships for YUS students, as well as teacher training workshops for existing local teachers facilitated by international educators. This approach was appreciated by the YUS communities and helped garner support for conservation efforts. Additionally, training of local assistants to increase scientific research capacity became another benefit of TKCP's presence in YUS. It was evident that YUS communities could benefit from a protected area in direct economic terms, even without paid compensation for pledging land.

TKCP's community approach sought to address services that were not being met by the government. The goal has been to initially fill gaps and then strengthen the direct link between YUS communities and the provincial government agencies rather than replace government services. It was essential to build working relationships with the provincial government as well as with local community leaders.

In 2005 TKCP initiated a community health project in collaboration with the Morobe Provincial Government and YUS communities (Chapter 15). Two US volunteer physicians (Drs. Nancy Philips and Blair Brooks) did a health patrol in YUS with TKCP staff and community health workers to determine priority health issues in YUS. The initial health project supported training workshops for village midwives and an immunization project by establishing solar refrigerators to store vaccines in the village aid posts and health centers. Both of these needs were identified as top priorities by the community and the Provincial Health Department. It was significant that TKCP focused on community needs in addition to its conservation and research efforts. This approach helped build the long- term relationship with the broader community and worked to improve access to government services for YUS. As the development of a protected area agenda advanced, it became evident that community development is a necessary complementary strategy in YUS.

Many aspects of the protected area approach in the early years were modeled after the Crater Mountain WMA, a product of the pivotal work of the Research and Conservation Foundation of PNG and Wildlife Conservation Society (WCS) out of Goroka, Eastern Highland Province. TKCP was able to utilize lessons learned from Crater WMA. In 2001 and 2003, collaborative biodiversity surveys (or Rapid Biodiversity Assessments) were conducted in YUS with the WCS, Binatang Research Centre, and other scientists to document the flora and fauna to be conserved in a proposed protected area. These were important biodiversity data sets that underpinned the proposal developed for a protected area in YUS.

From 1996 to 2003, local clans collaborated with TKCP in drawing a proposed protected area boundary based on their pledges to contribute land, sea, and nearshore coral reefs (TKCP, 2013). The Conservation Area creation process became much more formalized when, in 2005, a team from TKCP, Conservation International (CI), and CEPA facilitated a workshop with representatives from 26 local YUS clans to develop a plan for creating a protected area. TKCP collaborated with Dr. Mac Chapin of Support for Native Lands (United States), the Cartography Department of the University of Technology in Lae, and YUS landowners on an indigenous mapping project which resulted in local language community resource maps for Yopno, Uruwa, and Som regions. This was another first for PNG and had the benefit of visualizing the proposed areas for conservation as well as for other land uses (Fig. 10.6).

TKCP staff collaborated with Dr. Robert Horwich of Community Conservation Inc. to create a PNG Landowner Environmental Law Handbook which outlined all the options for landowners to protect their resources (Horwich, 2005). Subsequently YUS landowners decided to attempt to create PNG's first Conservation Area. The principle consideration in that decision was the

FIGURE 10.6 Dono Ogate, YUS landowner, looking at one of the YUS Indigenous Maps. *Source: Lisa Dabek.*

significantly stronger levels of protection from threat of large-scale resource extraction afforded by the Conservation Areas Act in comparison with the Fauna (Protection and Control) Act.

In addition, in 2005 the first YUS clan members used GPS units to map the land and sea they were protecting, marking a shift from earlier estimates based on hand-drawn maps, and the beginning of a clan-based GIS mapping program which continues to the present.

Compilation of the CA proposal for CEPA included comprehensive information about YUS. This primary formation of the Conservation Area was slowed by the unfamiliarity with the process. TKCP staff worked closely with, and provided support to, CEPA throughout all steps of the Conservation Area gazettal process. Conservation International staff also served in an advisory role. The requirements of the Act strengthen the foundations of the proposed Conservation Area by obliging due diligence in relationship-building and understanding local customs. The process was resource-intensive and in the future could be a main hurdle toward establishment of other CAs.

After convening all involved stakeholders, the local landowners requested that the proposal for creation of the YUS Conservation Area (YUS CA) be taken to the Provincial, and then the National Government by TKCP. It is important to understand the roles and responsibilities of the agencies and partners in the CA process.

One year after the formal landowner workshop, in 2006, the resulting proposal gained approval from the Morobe Provincial Executive Council with strong leadership by the Governor (Morobe PEC Approval Direction No. 03/2006 and Meeting No. 01/2006). This step is not a requirement of the CA's Act, yet project proponents were advised that Provincial buy-in and support was both necessary for the long-term sustainability of the YUS CA and for advancing the application. The Governor of Morobe Province, Mr. Luther Wenge, was a crucial supporter of the YUS CA and was instrumental in delivering the submission to then Minister for Environment and Conservation, Mr. Benny Allen. Moreover, from the viewpoint of implementing the Organic Law and the Government Planning Framework, which establishes the multiple levels of PNG Government, this step is very important to gain access to resources at the Provincial, District and Local Level Governments (LLG).

Following a meeting in which the Governor presented the approved proposal to the Minister for Environment and Conservation, a letter signed by representatives from YUS wards was sent to the Minister stating their approval of the proposal. The Minister subsequently published the proposal in the National newspaper for a 90-day comment period. The public notice requirement simultaneously functions to notify landowners within and outside of the proposed protected area of the pending submission, necessary because engaging every individual in rural areas like YUS would be extremely difficult. The approval process for YUS served to highlight the locations where boundary disputes might arise while indicating which landowners were most strongly in support of the Conservation Area. Formal inclusion of the resolution of these disputes in the legislation is important to avoid conflicts over land use at the National and Provincial planning levels. In effect, it aids the mainstreaming of the Conservation Area agenda across other sectors in Papua New Guinean planning, government, and civil society.

During the public comment period, several concerns were raised, indicating that the landowners and partners had not been able to generate universal support among the more than 12,000 individuals living in the YUS area. Individual landowners in one of the YUS villages protested the inclusion of their clan lands within the proposed Conservation Area, because surface oil had been found and they wished to explore options for developing that resource. These clans chose to remove their land from the pledged protected area. To date, the exempted village has not rejoined the YUS Conservation Area although individual landowners from that village have expressed interest in re-pledging land and no oil resources have been pursued.

Following the public comment period, the next step in the gazettal process was submission of the proposal to the National Executive Council (NEC). The process towards completion of the NEC proposal is an arduous one, requiring substantial time and financial resources. The collection of data itself can be a major undertaking, including biodiversity studies that require teams of scientists (Conservation International publishes these data sets as Rapid Biodiversity Assessments) to document key species in the proposed area, demographic data for the LLG, and boundary estimations with local landowners. The YUS CA proposal sent to NEC also included supporting letters from key government agencies, namely Forestry and Mining and Petroleum, and letters from outside organizations (e.g., TKCP, CI) stating ongoing financial and technical support for the creation of the YUS CA.

From Ministerial agreement to gazettal a number of steps remained. Following the public notice period, the submission to the National Conservation Council (NCC) is to obtain advice for the Minister on a Conservation Area submission. However, at the time of the YUS CA submission, the NCC did not exist, so the YUS CA proponents sought and achieved approval from the Attorney General to proceed along an alternate route. With the Attorney General approval, proponents obtained letters of support from key National Ministries: Mineral Resources Authority, Department of Community Development, PNG Forest Authority, and the Department of Provincial and Local Government Affairs. Approval was also obtained from the State Solicitor of Landowner Land Pledges. This process followed the NEC submission format and is intended primarily to obtain comments on the validity of the landowner pledges and seek comment from agencies for which gazettal may have development and/or policy implications.

A revised proposal was then submitted to additional authorities for approval, beginning with the Department Heads Economic Sector Committee and proceeding to the Ministerial Economic Sector Committee. These committees are important in the approval of any NEC submission, leading to the recommendation to the NEC for approval. Again, this stage required addressing any concerns brought up. A final proposal was then submitted to the National Executive Council for decision and recommendation for the Governor General endorsement. With the arrival of that endorsement the CA was published in the National Gazette, and gazettal was formally achieved. The gazetted reef-to-ridge YUS Conservation Area included 76,000 hectares of pledged core areas, plus additional buffer, and multiple-use zones (Fig. 10.7).

A huge celebration of the YUS Conservation Area took place in April 2009 in Teptep Village in YUS involving the PNG National Government, Provincial Governments, District Governments, Local Level Governments, local YUS communities, TKCP, Woodland Park Zoo, and Conservation International (Figs. 10.8–10.10).

IMPLEMENTATION OF THE CONSERVATION AREAS ACT

Operationalizing the YUS CA included activities aimed at land-use planning and management, as well as capacity-building for future ownership of the management process by local landowners and a local Non-Governmental Organization (NGO) (Fig. 10.11). A major component of the Conservation Areas Act for the YUS CA was the creation of PNG's first Conservation Area Management Committee (CAMC) for the YUS CA. Under the Conservation Areas Act of 1978, each Conservation Area must have a Management Committee (CAMC) to

1

National Gazette

PUBLISHED BY AUTHORITY

(Registered at the General Post Office, Port Moresby, for transmission by post as a Qualified Publication)

No. G5] PORT MORESBY, FRIDAY, 9th JANUARY [2009

CONSTITUTION

Broadcasting Corporation Act (Chapter 149)

APPOINTMENT OF CHAIRMAN, DEPUTY CHAIRMAN AND MEMBERS OF THE NATIONAL BROADCASTING CORPORATION BOARD

I, Sir Salamo Injia, Kt., Acting Governor-General, by virtue of the powers conferred by Section 193(2) of the Constitution and Sections 12 and 16 of the *Broadcasting Corporation Act* (Chapter 149) and all other powers me enabling, acting with, and in accordance with, the advise of the National Executive Council, given after consultation with the Public Sevices Commission and the Permanent Parliamentary Committee on appointments, hereby—

(*a*) appoint Paul Raptario, Timothy Maisu, Moale Rivu, Theresea Jainton and Henao Iduhu to be members of the National Broadcasting Corporation Board for a period of three years; and

(*b*) appoint Paul Raptario to be the Chairman and Timothy Maisu to be the Deputy Chairman of the National Broadcasting Corporation Board for a period of three years,

with effect on and from the date of publication of this instrument in the *National Gazette*.

Dated this 9th day of January, 2009.

SALAMO INJIA.
Acting Governor-General.

Conservation Areas Act 1978

DECLARATION OF CONSERVATION AREA

I, Sir Salamo Injia, Kt., Acting Governor-General, by virtue of the powers conferred by Section 17 of the *Conservation Areas Act* 1978, and all other powers me enabling, acting with, and in accordance with, the advice of the National Executive Council, hereby declare Yus Conservation Area in the Yus Local-Level Government (LLG), in Kabwum District of the Morobe Province to be a conservation area.

Dated this 9th day of January, 2009.

SALAMO INJIA.
Acting Governor-General.

FIGURE 10.7 Gazettal of YUS Conservation Area 2009. *Source: Lisa Dabek.*

"(a) manage the Conservation Area; and (b) to make recommendations to the Minister on the making of rules applicable within the Conservation Area; and (c) to advise the Minister in respect of co-ordination of development within the Conservation Area; and (d) to prepare a management plan for the Conservation Area outlining the manner in which land use will be managed and features of special significance conserved; and (e) to direct the work of rangers; and (f) such other functions as are determined by the Minister."

FIGURE 10.8 YUS Conservation Area Celebration in Teptep Village, 2009. *Source: Lisa Dabek.*

FIGURE 10.9 Woman dancing at the YUS Conservation Area celebration 2009. *Source: Ryan Hawk.*

The Act stipulates that the CAMC should reflect the interests of the local landowners and the provincial government. The reality of the YUS CAMC was that a committee was needed to engage multiple stakeholders in making decisions over the present and future of the YUS CA. Therefore, the structure is made up of key positions within different levels of government, NGO and local representatives (Chapter 13).

While the YUS CAMC is tasked under the Act with the management of the YUS Conservation Area, the reality of the geographic distribution of its members, combined with the recognized need for local ownership of the management process, means that day-to-day management lies with the Tree Kangaroo Conservation Program—PNG (TKCP-PNG), a PNG technical NGO legally formed in 2013, and is focused on management and support of the YUS CA and the local YUS community. It is currently staffed 100% by Papua New Guineans, with all field staff directly from the YUS area. The development of a site-based technical NGO is an

FIGURE 10.10 Dancers at the YUS Conservation Area celebration, 2009. *Source: Ryan Hawk.*

important and unique strategy for the management of the YUS CA, and one which may provide a model for other sites where CAs are currently being developed. Technical work has included the creation of a landscape-level Management Plan for the area, The YUS Landscape Plan (Chapter 11). It includes a management plan for the protected area as well as land-use plans and guidance on a wide range of activities across the YUS landscape. Its development is achieved through a nested approach, aligning activities in YUS with goals set out in the PNG Vision 2050, and Provincial and District Five-Year Development Plans (Chapter 13). Within this YUS Landscape Plan lies the core management strategy for YUS CA, the development and implementation of a community-based Ranger and Ecological Monitoring Program, as dictated by the Act (Chapter 12).

TKCP also benefits from a partnership with an advisory community-based organization made up of local landowner representatives called the YUS Conservation Organization (YUS CO) which was registered in 2009. In its early stages, project partners understood that each clan pledging land to the CA needed to have a voice in landscape-level management and those representatives would need to make communal decisions as in the development of the YUS CA bylaws (Fig. 10.11).

The YUS Conservation Organization's mandate is "to foster wildlife and habitat conservation while also improving livelihoods for local communities within the YUS Conservation Area of Papua New Guinea."

While TKCP-PNG and advisory YUS CO ensure that local landowners create the management agenda for YUS CA, questions about the purpose and benefits from the Conservation Area are common occurrences. From the earliest stages, project proponents stressed that there would be no compensation for land pledges and that the major benefits of the protected area would be sustainable natural resources, opportunities for employment as staff, and opportunistic responses to communal development

FIGURE 10.11 TKCP-PNG's staff member Karau Kuna discussing conservation planning with YUS landowners. *Source: TKCP.*

priorities in partnership with the government. The project has sought to harness opportunities to promote local development and improve livelihoods where possible. Examples of development projects in YUS include support to coffee and cocoa farmers through market partnerships and technical trainings as part of a conservation coffee initiative and development of a conservation coffee and cocoa cooperative (Chapter 14); a teacher scholarship program and Junior Ranger Program (Chapter 16); and health trainings and clinical support (Chapter 15).

Priorities for these projects are locally driven and are focused on economic capacity-building as often as possible, yet questions of compensation and motivation still arise and will likely be a continual discussion. In the YUS Conservation Area, the model has been to view the CA as a wildlife bank providing sustainable hunting for the local landowners, a point that is emphasized in community meetings and in talks among landowners and local TKCP and YUS CO staff. However, misunderstandings, misinformation, and sometimes conflicting priorities among villages do arise. The necessity of clear, honest, and constant communication is paramount. Clarity among thousands of locals, outsiders, and government, particularly on the benefits of conservation and the roles and responsibilities of each stakeholder, is crucial.

The roles of the organizations involved in the YUS CA have evolved over time as have their staffing and structure. The local NGO partner, TKCP-PNG, carries out day-to-day management, provides technical support, does some fundraising, and facilitates strategic partnerships with government and civil society. TKCP-PNG is staffed to provide the expertise needed to support the YUS CO in managing YUS CA and to ensure that local ownership is enabled throughout the initiative. TKCP-PNG has a Board of Directors which provide guidance, governance, and fiscal oversight. The YUS CO plays a crucial role in giving each clan a representative voice, facilitating landscape-level decision making, and advising TKCP-PNG. Woodland Park Zoo's Tree Kangaroo Conservation Program is the third partner and serves as a fiscal sponsor, supports the work of TKCP-PNG, and facilitates funding and financial oversight for the management of the YUS CA.

A significant constraint in the creation of CAs is the logistical and financial ability of the National Government to fund long-term management of protected areas. CA proposals with indications of long-term, third party, financial support may be more likely to reach gazettal.

The most significant sources of support for the YUS Conservation Area were grants from Conservation International Global Conservation Fund and a major five-year grant from the German Ministry of the Environment (BMU) through the German Development Bank (KfW) to Conservation International and Woodland Park Zoo's TKCP. This funding allowed the project partners to build the infrastructure and organizational structure necessary for supporting a Conservation Area. The increased amount of funding was also critical in allowing TKCP to expand and address some of the communities' pressing development priorities, conduct land-use planning and management workshops, develop a landscape-level management plan, and establish a conservation ranger program.

Woodland Park Zoo and Conservation International (CI) collaborated to create a YUS Conservation Endowment to help fund the long-term costs of managing the YUS CA through the NGO. The endowment is housed at Woodland Park Zoo and funds are dispersed to the NGO annually. After the CI and BMU/KfW grants were completed, TKCP was successful in obtaining a five-year grant from the Global Environment Facility (GEF) facilitated through the United Nations Development Program (UNDP). This additional substantial funding has allowed TKCP to continue to strengthen the programs and infrastructure of TKCP-PNG and YUS CO. Starting in 2019, TKCP is part of another five-year grant through the USAID Biodiversity program and Cardno Corporation.

RE-GAZETTAL OF THE YUS CONSERVATION AREA ON A LANDSCAPE LEVEL

In 2014 PNG created a new Policy on Protected Areas that further defines the range of protected areas in PNG (Independent State of Papua New Guinea, 2014). Along with this policy, a Protected Areas Bill is currently being introduced in Parliament. As part of this policy with a focus on landscape level protection, TKCP-PNG and the YUS landowners in partnership with CEPA once again led the country in working towards re-gazettal of the YUS Conservation Area as a landscape-level protected area. In May 2020 the YUS Conservation Area was gazetted as a landscape level Conservation Area for tree kangaroos and other wildlife. This newly re-gazetted YUS CA supports the land-use plans and future management planning by the YUS landowners and stewards of the YUS (Chapter 11). Having a landscape approach to conservation that focuses on tree kangaroos and other wildlife as well as the local communities is yet another way in which the YUS CA can be a model program for PNG (Fig. 10.12).

FIGURE 10.12 Tree Kangaroo in YUS Conservation Area Forest. *Source: Bruce Beehler.*

CONCLUSIONS

Key lessons learned from establishing the YUS Conservation Area:

- Have clear reasons for choosing a proposed protected area site based on factors including biodiversity, scientific knowledge, and community support and interest.
- Realize that creating a Conservation Area is a long term process; for YUS CA it took 13 years to reach gazettal.
- Make sure that the intentions and expectations of landowners and outside conservation organization are clear, transparent, and realistic. Building trusting and respectful long-term relationships is essential.
- Recognize that community needs must be incorporated **with** conservation goals.
- Build relationships with all levels of PNG government including LLG, District, Province, and National.
- Collaborate with local and international organizations and universities to strengthen the approach to conservation and obtain community support.
- Plan for long-term sustainable financing.
- Build local capacity to ensure long term success. Goals should include hiring local staff members and building leadership.
- Share knowledge and challenges with other community and conservation organizations in order to support protected area programs throughout the nation and beyond.

ACKNOWLEDGMENTS

The authors thank the numerous partners who have been a part of the YUS CA initiative over the years, including Conservation International and James Cook University. Special thanks go to the YUS LLG, the Morobe Provincial Government and the PNG Government's CEPA for their ongoing support. Thanks to Conservation International Global Conservation Fund (CI-GCF), German Ministry of the Environment (BMU) through the German Development Bank (KfW), GEF/UNDP, Rainforest Trust, Zoos Victoria, and USAID Biodiversity as well as the multitude of other donors over the years for their generous financial support. We acknowledge the ongoing support of the Woodland Park Zoo and the Tree Kangaroo Conservation Program staff. Thanks go to the Association of Zoos and Aquariums' Tree Kangaroo Species Survival Plan®, and many AZA institutions that have supported TKCP's work. Thank you to Mr. Gai Kula who was instrumental in supporting the YUS CA process and in co-writing the original article on creating the YUS Conservation Area. Thank you to current Morobe Provincial Governor Ginson Saonu. Most importantly we wish to extend sincere appreciation to the people of YUS for their dedication, partnership, and stewardship.

REFERENCES

Beehler, B.M. (Ed.), 1993. Papua New Guinea Conservation Needs Assessment. In: vol. 2. Biodiversity Support Program, Washington, DC.

Betz, W., 2001. Matschie's Tree Kangaroo (Marsupialia: Macropodidae, *Dendrolagus matschiei*) in Papua New Guinea: Estimates of Population Density and Landowner Accounts of Food Plants and Natural History. Master's thesisUniversity of Southampton, Southampton, England.

Dabek, L., 1994. The reproductive biology and behavior of captive female Matschie's tree kangaroos *(Dendrolagus matschiei)*. Ph.D. dissertationUniversity of Washington, Seattle, Washington Abstract in Dissertation Abstracts International 55(11B), 4748.

Horwich, R.H., 2005. A Landowner's Handbook to Relevant Environmental Law in Papua New Guinea. Community Conservation, Gay Mills, WI. Available from: https://www.zoo.org/file/visit-pdf-bin/LawBooklet-FinalVersion-July2009.pdf (17 July 2020).

Independent State of Papua New Guinea, 2014. Papua New Guinea Policy on Protected Areas. Conservation & Environment Protection Authority, Waigani, National Capital District, Papua New Guinea. Available from: https://info.undp.org/docs/pdc/Documents/PNG/PNG%20Protected%20Areas%20Policy-NEC%20Approved_Signed.pdf (17 July 2020).

International Union for Conservation of Nature, 2020. Available from:www.iucnredlist.org (27 July 2020).

Lipsett-Moore, G., Hamilton, R., Peterson, N., Game, E., Atu, W., Kereseka, J., Pita, J., Ramohia, P., Catherine Siota, C., 2010. Ridges to Reefs Conservation Plan for Choiseul Province, Solomon Islands. The Nature Conservancy Pacific Islands Countries Report No. 2.

Mittermeier, R.A., Robles-Gil, P., Mittermeier, C.G. (Eds.), 1997. Megadiversity. Earth's Biologically Wealthiest Nations. CEMEX/Agrupaciaon Sierra Madre, Mexico City.

Pacific Island Legal Information Institute (PACLII), 1966. Papua New Guinea Consolidated Legislation: Fauna (Protection and Control) Act 1966. Available from: http://www.paclii.org/pg/legis/consol_act/faca1966290/ (17 July 2020).

Pacific Island Legal Information Institute (PACLII), 1978. Papua New Guinea Consolidated Legislation: Conservation Areas Act 1978. Available from: http://www.paclii.org/pg/legis/consol_act/caa1978203/ (17 July 2020).

Tree Kangaroo Conservation Program, 2012. YUS Landscape Plan 2013–2015. Tree Kangaroo Conservation Program, Lae, Papua New Guinea. Available from: https://www.zoo.org/document.doc?id=904 (17 July 2020).

Tree Kangaroo Conservation Program (TKCP), 2013. Annual Report. Woodland Park Zoo, Seattle, Washington. Available from: https://www.zoo.org/document.doc?id=1727 (17 July 2020).

Wells, Z., Dabek, L., Kula, G., 2013. Establishing a Conservation Area in Papua New Guinea – lessons learned from the YUS Conservation Area. In: Beehler, B., Kirkman, A. (Eds.), Lessons Learned From the Field: Achieving Conservation Success in Papua New Guinea. Conservation International, Arlington, Virginia.

Whimp, K., 1998. Some issues of law and policy relating to landowner organization and representation mechanisms. In: Paper given at the 'Conference on Incorporated Land Groups', Organised by the Department of Petroleum and Energy and Chevron, Granville Motel, Port Moresby, 8–9 September 1998.

Wikramanayake, E., Dinerstein, E., Loucks, C., Olson, D., Morrison, J., Lamoreux, J., McKnight, M., Hedao, P., 2001. Terrestrial Ecoregions of the Indo-Pacific: A Conservation Assessment. Island Press, Washington, DC.

World Atlas, 2020. What Languages Are Spoken in Papua New Guinea? Available from: https://www.worldatlas.com/articles/what-languages-are-spoken-in-papua-new-guinea.html (20 July 2020).

Ziembicki, M., Porolak, G., 2016. *Dendrolagus matschiei*. The IUCN Red List of Threatened Species. Available from: https://doi.org/10.2305/IUCN.UK.2016-2.RLTS.T6433A21956650.en (26 July 2020).

CHAPTER

11

Land-Use Planning for a Sustainable Future in Papua New Guinea

Danny Nane[a], *Modi Pontio*[a], *and Trevor Holbrook*[b]

[a]Tree Kangaroo Conservation Program, Lae, Papua New Guinea

[b]Tree Kangaroo Conservation Program, Woodland Park Zoo, Seattle, WA, United States

OUTLINE

INTRODUCTION

With vast tracts of pristine forest containing unparalleled biodiversity, Papua New Guinea (PNG) is considered to be one of three major Tropical Wilderness Areas left on earth. The country is home to approximately 20,000 plant species, more than 700 bird species and more than 230 mammal species, many of which are only found on the island of New Guinea (TKCP, 2012). PNG's immense biodiversity is rivalled only by its cultural diversity. With great pride and respect for its traditions and belief systems, customary ownership and control of land in the country is protected by the Constitution—securing more than 95% of PNG's land under customary land tenure. As such, the bond between the land and its people is unbreakable, and heavily influences each community's livelihood and identity. With the great majority of the country's predominantly rural population following a traditional lifestyle based on subsistence agriculture, hunting, and gathering, the long-term health and sustainability of the local ecosystems, livelihoods, and culture require the effective management of land and resources.

The YUS Conservation Area, on the north side of the Huon Peninsula in PNG's Morobe Province, is home to approximately 15,000 people living among 50 villages throughout the 162,000-hectare landscape. Similar to other parts of the country, the people of YUS maintain

Tree Kangaroos

https://doi.org/10.1016/B978-0-12-814675-0.00021-X

close connections with their land, which forms a fundamental component of the area's socio-economic structure and the way society operates. Having collectively pledged portions of their customary land for protection by establishing the country's first Conservation Area in 2009 in collaboration with the PNG government and the Tree Kangaroo Conservation Program (TKCP), the YUS communities sought to strengthen their coordinated management of land and resources under their control in order to maintain a socially and environmentally compatible, desirable, and economically sound landscape (TKCP, 2009). Home to the endemic and endangered Matschie's tree kangaroo (*Dendrolagus matschiei*), the YUS Conservation Area was established to protect critical habitat and promote sustainable local hunting practices for the Matschie's tree kangaroo and other native species.

With the support of TKCP, the leaders and landowners of YUS adapted the internationally-recognized land use management planning method to accommodate the local context, with particular emphasis on the integration of cultural perspectives and traditional ecological knowledge. Serving to complement government planning processes for conservation and development, the land-use plans are wholly owned by the local communities to support decision-making and coordinated resource management across the landscape.

COMMUNITY-BASED LAND-USE PLANNING PROCESS IN THE YUS CONSERVATION AREA

To support the communities of YUS to sustainably meet their local demand for natural resources and strengthen local management of the YUS Conservation Area, the Tree Kangaroo Conservation Program collaborated with the local landowners and clan leaders who had pledged portions of their customary land for conservation in order to integrate traditional resource management practices with scientific information and PNG governance structures. As many wildlife species within the YUS Conservation Area are endemic and/or listed under the International Union for the Conservation of Nature (IUCN) Red List, the land-use plans provide a platform for enhancing local conservation action. Developed at the Ward level—Papua New Guinea's most localized level of government administration—the Land-Use Plans are comprised of eight different use classifications—each with agreed rules and restrictions to guide the community's activities within the areas:

- Conservation Area: Primary protected area serving as a "wildlife bank" supporting sustainable wildlife populations, including Riparian Zones which serve as wildlife corridors. The Conservation Area includes critical habitat for endangered and endemic species including the Matschie's tree kangaroo.
- Buffer Zone: Acts as boundary between the Conservation Area and the Reforestation Zone. Local communities may hunt or extract limited resources in a sustainable manner.
- Reforestation Zone: Specific areas designated for the cultivation of native trees to meet local resource needs, as well as for forest regeneration.
- Livelihood and Agroforestry Zone: Areas allocated for subsistence gardening and income-generating agricultural activities.
- Grassland Zone: Reserved for the protection and management of alpine grasslands and savannah resources.
- Village Zone: Main areas of development and human settlements.
- Reef Zone: Recreation and fishing zones.
- Tambu Reef Zone: Primary marine conservation zone, in which all fishing and other marine resources extraction are prohibited.

Under this zoning system, each zone is defined by the types of activities that are permitted and those that are prohibited. For example, coffee farming and other agricultural activities are permitted only in the Livelihood Zones, while hunting and fishing activities are prohibited within the Conservation and Tambu Reef Zones. The Community Land-Use Planning process helps to create community consensus for resource use, and helps neighboring clans to discuss and agree on how to improve the use of their existing land and sea, which forests and marine habitat to use, and which to protect and declare off-limits. Decisions are reached based on consensus through dialogue among the local landowners and stakeholders, resulting in mutually-agreed needs and priorities regarding conservation and local development (Fig. 11.1).

TKCP utilizes a suite of tools to facilitate each ward's development and refinement of land use plans, relying on a series of consultative workshops supported by technical expertise to inform the process. This process is designed to be highly consultative and participatory, serving to engage a wide range of perspectives and priorities from various groups within each community. The Land-Use Planning process involves three rounds of consultative workshops:

- Round 1: Community entry
  This introductory workshop involves participatory exercises to establish the ward's common vision for the future, an analysis of their current situation, and initial identification of key needs and priorities regarding the area's environmental, social, cultural, and economic well-being.
- Round 2: Detailed planning
  Participants conduct a detailed review with consideration for the ward's conservation and development priorities,

FIGURE 11.1 Members of the YUS community participate in a Land-Use Planning workshop, reviewing satellite imagery for comparison with clan boundaries and land-use zones. *Source: Tree Kangaroo Conservation Program.*

habitat restoration, as well as identification of key community actions and activities. Where relevant, further consultation with other stakeholders and interest groups is incorporated into the ward's land-use plans. This round includes consultations among land-owning clans to encourage habitat connectivity and alignment of land-use designations between adjacent clan lands.

- Round 3: Confirmation

  Following collation of the ward's agreed needs, priorities, and land-use zoning assignments, participants conduct a final review of the plan prior to its endorsement by the ward's community leaders and landowners. The plan is then printed and disseminated throughout all villages within the ward to support local awareness and adherence to agreed resource use.

The final Land-Use Plans (Fig. 11.2) are also shared with government representatives at the local, district, provincial, and national levels to further support bottom-up planning, budget allocations, and policy development (Fig. 11.3).

LAND-USE PLANNING APPLICATIONS

The participatory local approach to Land-Use Planning as developed for the YUS Conservation Area aligns closely with PNG's concept of rural development, providing a "bottom-up" process based on the community's self-sufficiency and self-responsibility. Beginning with the review and consideration of the community's existing resources and capacities, and leading to the identification of specific priorities for environmental conservation and local development, the Land-Use Plans are designed to inform Local Level Government and District plans and budgets to complement local efforts. The process also serves to operationalize PNG's National Development Goal of Ecologically Sustainable Development, which aims to meet the needs of the current generation without compromising the ability of future generations to meet their own needs. This includes the conservation of natural life-supporting systems, protection of biological diversity, and ensuring the sustainable management of natural resources within the planet's carrying capacity (Mowbray, 2017). The community-level Land-Use Plans in YUS serve as a local-level foundation, paving the way for communities within the YUS landscape to maintain a healthy environment, community cohesion and cultural vitality, secure customary land tenure, as well as sustainable livelihoods and well-being (TKCP, 2012). The Land-use Planning process helps to ensure that communities continue to benefit from local natural resources, and guides landowners in coordinating resource management among stakeholders. In support of these aspirations, the Tree Kangaroo Conservation Program developed a range of complementary programs and initiatives using a holistic approach to wildlife conservation. Both driven by and complementary to the community's Land-Use Plans, TKCP facilitates protected area management throughout the landscape (Chapters 10 and 12), builds local leadership capacity (Chapters 13 and 16), strengthens community livelihoods (Chapter 14), and promotes the health of wildlife, people, and the environment (Chapter 15) (Fig. 11.4).

LAND-USE PLANNING BENEFITS AND IMPACTS

The YUS communities' adherence to their ward-level Land-Use Plans has resulted in numerous benefits, particularly with regard to the management and effectiveness of the core protected areas within the YUS Conservation Area. As the Land-Use Plans play a vital role in raising local awareness of the protected area's boundaries and prohibited activities, as well as

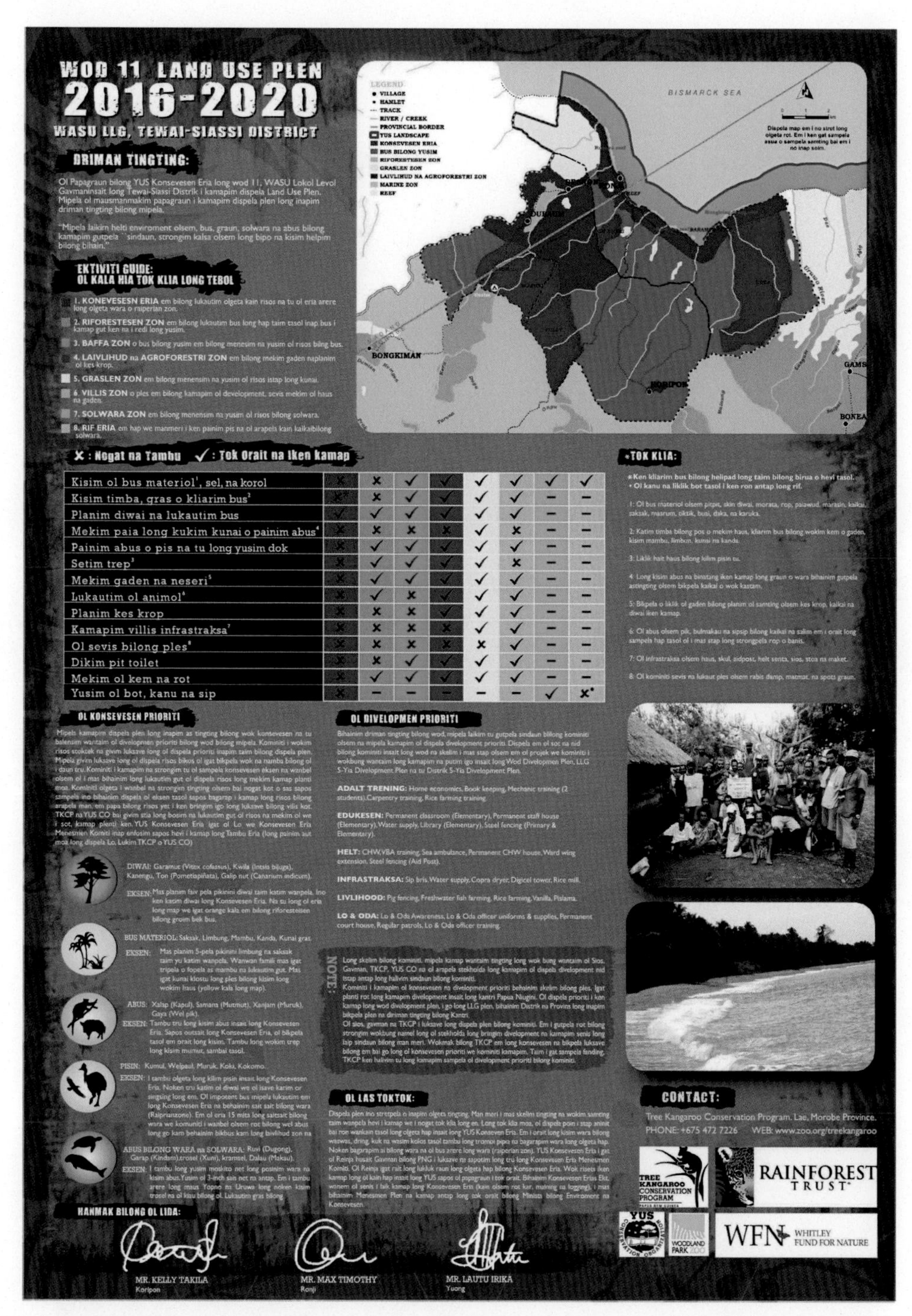

FIGURE 11.2 Each of the YUS Conservation Area's ward-level Land-Use Plans are designed and distributed in poster form, and displayed prominently throughout each of the 50 villages across the YUS landscape. The posters promote local awareness regarding the communities' agreed land-use plans and priorities for conservation and development, and support accountability. *Source: Karau Kuna, Tree Kangaroo Conservation Program.*

FIGURE 11.3 Morobe Provincial Governor Kelly Naru (right) and MP from Kabwum District, Bob Dadae (center) accepting YUS land-use plans from YUS Conservation Organization President Timmy Sowang (left) in 2014. *Source: Tree Kangaroo Conservation Program.*

FIGURE 11.4 Members of the YUS community review and discuss progress toward conservation and development priorities as stated in their ward's Land-Use Plan. *Source: Tree Kangaroo Conservation Program.*

encouraging habitat connectivity and the establishment of riparian zones, the reduced human disturbances within the designated conservation zones is resulting in notable ecological improvements. Increasingly, key wildlife species are being observed both within and beyond the core protected areas, suggesting the effectiveness of the YUS Conservation Area as a "wildlife bank" enabling reproduction and territorial expansion into buffer and livelihood zones (Fig. 11.5). In recent years, community members have reported increased sightings of species near Village and Livelihood Zones, including Matschie's tree kangaroos, bowerbirds (Family: *Ptilonorhynchidae*), eclectus parrots (*Eclectus roratus*), orange-footed scrub fowl (*Megapodius reinwardt*), New Guinea Vulturine Parrots (*Psittrichas fulgidus*; also known as Pesquet's Parrot), and Niugini pademelons (*Thylogale* spp.).

By incorporating focus and consideration for local-level priority species of flora and fauna requiring conservation action, the Land-Use Plans have also created new opportunities for discussion, learning, and comparison between scientific/biological information and traditional knowledge relating to the ecology, behavior, reproduction, and diet of various species including the endangered Matschie's tree kangaroo. Such information has proven valuable in guiding community members and TKCP in determining wildlife conservation needs and priorities, thereby supporting the adaptive management of the YUS Conservation Area (Fig. 11.6).

The Land-Use Plans also play an important socioeconomic role throughout YUS, as they comprise one of the only avenues for soliciting broad input and consensus relating to local resource use, livelihoods, and community development. The collaborative planning process facilitates increased dialogue and coordination among clans and village groups, supporting the clarification and resolution of potential conflicts relating to clan land boundaries or use agreements. Encouraging collective decision making and group dialogue both within and among clans, the process enables consideration for both current and future responsibilities, needs, and priorities.

LOOKING FORWARD: COMMUNITY-LED LANDSCAPE MANAGEMENT

To support the engagement of a broader range of stakeholders and perspectives in both the Land-Use Planning process and the ongoing implementation and monitoring of the agreed plans, TKCP has partnered with PNG-based Partners with Melanesians to facilitate a series of Participatory Three-Dimensional Modelling (P3DM) workshops throughout the YUS landscape (TKCP, 2017). Sharing many similarities in terms of its approach and objectives, the P3DM workshops solicit various inputs and perspectives with regard to an area's land and resources, and how they are utilized and managed by the local community. The P3DM approach integrates indigenous spatial knowledge with satellite elevation to produce standalone, scaled, and geo-referenced 3D relief models representing each zone (Rambaldi, 2010).

Increasingly, TKCP views P3DM as an integral part of its community-based Land-Use Planning program. The approach serves as a key component in building relationships with communities, spatially analyzing conservation priorities, and translating local commitment into conservation action. The model has also proved advantageous in mapping clan and village land boundaries, and facilitates the engagement of a broad range of community groups in discussions relating to local resource needs. The modeling exercise has offered a platform for maximizing inclusion and empowerment of the local people, including youth and women, to plan for sustainable resource use from their own village to ward and then to the entire zone as a whole.

The workshops result in the construction of a three-dimensional scale model of the area,

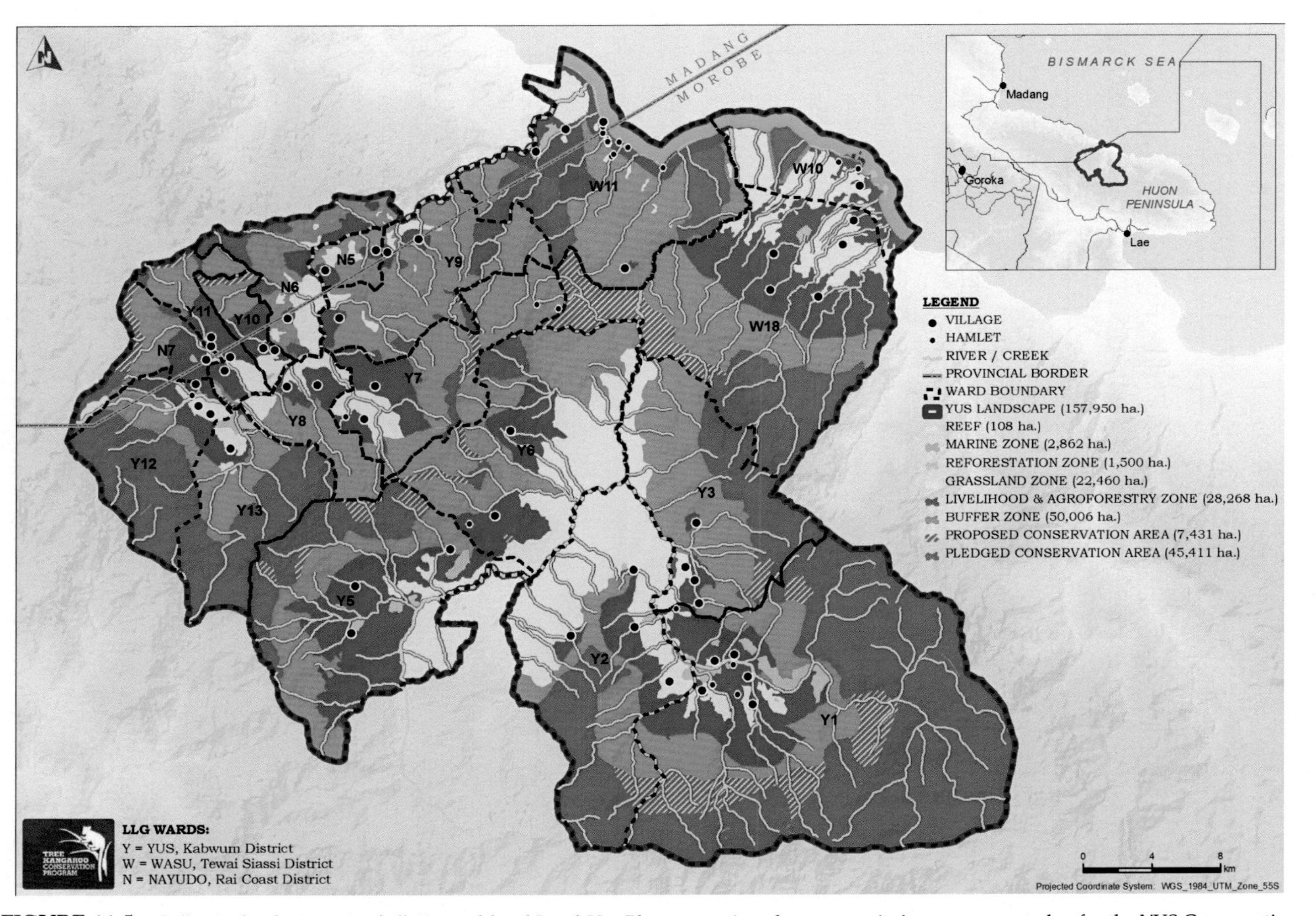

FIGURE 11.5 Collectively, the mosaic of all 18 ward-level Land-Use Plans comprises the community's management plan for the YUS Conservation Area. *Source: Karau Kuna, Tree Kangaroo Conservation Program.*

FIGURE 11.6 The Tree Kangaroo Conservation Program's Timmy Sowang facilitates a community discussion of resource use and conservation priorities in Yawan village. *Source: Tree Kangaroo Conservation Program.*

which remains in the community as a valuable resource and reference point for ongoing discussions around land-use management (see Fig. 11.7). As TKCP approaches the YUS community's review and updating of its 3rd-cycle ward-level Land-Use Plans, for the period 2021–25, the insights and perspectives raised through the P3DM process will be further incorporated into the Land-Use Planning process. Likewise, the process will increasingly utilize the three-dimensional models located in six central locations throughout YUS to encourage the continued participation of various groups within the community (Fig. 11.8).

Additionally, as the YUS communities increasingly take ownership in implementing their respective ward-level Land-Use Plans, TKCP will assist in developing complementary methods for monitoring progress as well as adherence to agreed land-use zoning and permitted activities. The monitoring methods will involve both community-based methods, supported by TKCP's YUS Conservation Area Rangers, Mapping Officers, and Conservation Officers, as well as Geographic Information Systems (GIS) based methods to monitor broader, landscape-level land-use changes over time.

Particular focus will be given to TKCP's continued efforts in building local leadership capacity in YUS, including the community-based YUS Conservation Organization. Recognizing the value and importance of the ward-level Land-Use Planning process for both the coordinated community-level management activities for the YUS Conservation Area as well as the implementation of key development efforts (including

FIGURE 11.7 Members of the YUS community participate in the construction of a three-dimensional model of their landscape, painted in accordance with the ward-level Land-Use Plan zoning designations. *Source: Tree Kangaroo Conservation Program.*

economic development, sustainable livelihoods, education, and health), there is great opportunity to further leverage the Land-Use Plans among all levels of PNG government to increasingly incorporate the community's defined needs and priorities into the government's development plans and budgets. TKCP will support the YUS Conservation Organization in its capacity to represent and advocate on behalf of the YUS community, utilizing the ward-level Land-Use Plans to pursue financial, technical, and infrastructural assistance from its Local Level, District, and Provincial Governments. To complement these capacity-building efforts, TKCP will also continue to expand and build upon its relationship with such government administrations and representatives, including through the YUS Conservation Area Management Committee.

CONCLUSIONS

The community-based Land-Use Planning process, developed and contextualized for Papua New Guinea's unique customary land tenure system, has proven to be a highly effective approach to engaging all members of the local communities in the management of the YUS Conservation Area and the practice of sustainable resource management. By integrating globally-recognized land-use planning methods with traditional knowledge—and emphasizing the incredible value of a healthy YUS landscape

FIGURE 11.8 The fully completed three-dimensional model for the Yopno 1 zone; a land owner of Wungon village, and the first person who pledged his land for conservation, Mabawe Beka (middle) viewing the fully completed 3D model, together with the workshop participants at Tapmange village. *Source: Danny Nane (TKCP-PNG).*

in securing the cultures, lifestyles, and livelihoods of the YUS people for generations to come—the Land-Use Planning process serves to empower the people of YUS for both the protection of their environment as well as the well-being of their communities while maintaining its cultural and aesthetic value. Collectively, the mosaic of the 18 ward-level Land-Use Plans comprising the YUS landscape represents the cohesive, coordinated management plan for the YUS Conservation Area as a whole. Moving forward, TKCP and community leaders in YUS aim to further strengthen the local management and implementation of Ward-level Land-use Plans, including the collaborative response to community-defined development aspirations and conservation priorities.

REFERENCES

Mowbray, D., 2017. Tools for Environmental Management & Decision Making. University of Papua New Guinea Printery, pp. 2–3.

Rambaldi, G., 2010. Participatory Three-dimensional Modelling: Guiding Principles and Applications. CTA, Wageningen, Netherlands.

Tree Kangaroo Conservation Program (TKCP), 2009. 2009 Annual Report. TKCP, Lae, Papua New Guinea. Available from: https://www.zoo.org/document.doc?id=1723 (28 July 2020).

Tree Kangaroo Conservation Program (TKCP), 2012. YUS Landscape Plan 2013–2015, TKCP, Lae, Papua New Guinea. Available from: https://www.zoo.org/document.doc?id=904 (28 July 2020).

Tree Kangaroo Conservation Program (TKCP), 2017. 2017 Annual Report. TKCP, Lae, Papua New Guinea. Available from: https://www.zoo.org/document.doc?id=2407 (31 July 2020).

CHAPTER

12

A Model Tree Kangaroo Conservation Ranger Program in Papua New Guinea

Daniel Solomon Okena[a] *and Paul van Nimwegen*[b]

[a]Tree Kangaroo Conservation Program - Papua New Guinea, Lae, Morobe Province, Papua New Guinea

[b]International Union for Conservation of Nature, Oceania Regional Office, Fiji

OUTLINE

PROGRAM HISTORY

The land tenure system of Huon Peninsula, where Yopno-Uruwa-Som (YUS) Conservation Area is located, is passed down from generation to generation through clan and descendant inheritance, which is common across Papua New Guinea (PNG). Lands are traditionally owned by indigenous communities and are handed down to clan and children. Patrilineal societies exist in YUS, where males own the land and dominate decision-making. Working in this context, it is important to respect traditional owners' rights and customary practices.

In 1996, Dr. Lisa Dabek traveled to YUS to study the Matschie's (Huon) tree kangaroo (*Dendrolagus matschiei*) in the wild (Fig. 12.1). This research examined population, status,

Tree Kangaroos
https://doi.org/10.1016/B978-0-12-814675-0.00014-2

FIGURE 12.1 Matschie's Tree kangaroo in YUS CA—the species that brought about the YUS Ranger Program. *Source: Daniel Solomon Okena.*

ecology, and threats, which confirmed that the tree kangaroo was endangered and that for its long-term survival, hunting pressure needed to be reduced and habitat sufficiently protected. In order to do so, Dr. Dabek initiated a collaborative process with the landowners and all levels of PNG governments (Chapter 10). This led to the establishment and gazettal of the YUS Conservation Area (YUS CA) in 2009, which was the first of its kind in Papua New Guinea under the 1978 Conservation Areas Act (PacLII, 1978). Importantly, the gazettal had the full consent and involvement of the traditional owners. A total of about 78,000 ha was set aside as a protected area to conserve not only the tree kangaroos but also many other significant species.

Despite this official status, there were violations of the agreed rules by people hunting tree kangaroos and other species. In response, the ranger program was established to monitor and report infringements to relevant authorities.

In 2011, local conservation champions were selected across the landscape to be community rangers. After several months of training, the actual start of operation of the ranger program began in 2012 with a total of 12 terrestrial rangers (Fig. 12.2). Since then, the rangers' numbers increased to cover protection of the entire landscape and extending down to the marine protected area. Currently there are 18 rangers including 16 terrestrial rangers and 2 marine rangers. After extensive community consultations, bylaws to govern the protected area were drafted. In 2014, the YUS Conservation Area Bylaws were put into effect (Chapters 10 and Chapter 13).

PROGRAM OVERVIEW

Overall Partnership

The Tree Kangaroo Conservation Program PNG (TKCP-PNG) partners with the Woodland Park Zoo to manage the ranger program, which is collectively known as TKCP. This partnership also collaborates with other local and international organizations, YUS Community Based Organization (YUS CBO), and government (local, provincial, and national) to run the ranger program. The main body that these stakeholders represent is called the YUS Conservation Area Management Committee (CAMC). The YUS CAMC was established as a requirement of gazettal under the Conservation Areas Act. The nine-member management committee reflects the interests of the landowners and the authorities that govern the conservation area. Below is the composition of the committee:

1. Three Executive Members from the YUS Community Based Organization

FIGURE 12.2 Sapmanga Station, the birthplace of YUS CA rangers. This is where the initial recruitment and training of the first rangers took place. *Source: Daniel Solomon Okena.*

2. District Administrator of the Kabwum District or their nominee
3. Program Advisor of the Division of Mining, Natural Resources & Environment, Morobe Provincial Administration, or their nominee
4. Head, Terrestrial Environment Programs, PNG Conservation and Environment Protection Authority (CEPA) or their nominee
5. President, YUS Local Level Government (LLG)
6. President, Wasu LLG
7. Program Manager/Country Director, TKCP or their nominee

The purpose of the CAMC is to oversee the management of the YUS CA and support strategic planning. The CAMC serves both the Minister and CEPA for all national and international requirements, while concurrently serving the YUS landowners with strategic guidance, and organizational and policy support to protect and sustainably use natural resources. Thus, the CAMC acts across all vertical and horizontal levels of government relating to the landscape. The CAMC meets twice a year to collate reports and data, and to discuss and respond to any development or alteration of land use. Any development applications or actions that could be in breach of either the Landscape Plan or The Act are referred to the Minister of CEPA. During the fourth year of a five-year Landscape Plan, the CAMC also contributes to the formulation of the new plan and submits it to the Minister of CEPA for approval.

YUS Ranger Program Overview

The Rangers monitor the YUS Conservation Area and report back to their field supervisors (Conservation Officers) and TKCP-PNG through the Program's Research and Conservation Manager. All rangers are locals selected from across the YUS landscape. Table 12.1 has the full listing of rangers and their duty stations. Priority for ranger selection is given to communities who pledge their land for conservation. The main criteria that is used to select rangers are that they should be literate and have good numeracy skills. Other criteria used for the selection process include:

1. Well-developed interpersonal and communication skills; both written and oral
2. Commitment to the goals and philosophy of the TKCP-PNG and YUS Community Based Organization
3. Be available to travel at various times throughout the year
4. Be physically fit, able to spend time in difficult conditions on patrol in YUS CA

Based on these criteria, rangers are nominated and selected by their respective communities to represent and patrol the conservation area. Rangers are only replaced if they have breached their contract or resign. In September 2014, the first marine ranger was recruited to patrol coastal waters. Notably, the ranger on his first patrol sighted a critically endangered Leather-back Turtle (*Dermochelys coriacea*) nesting. Currently (as of

TABLE 12.1 YUS rangers and their local village within the YUS landscape (as of 2019).

	YUS Ranger	Village	Zone	Local Level Government & Ward
1	Obtuse James	Towet	Uruwa	YUS LLG Ward 1
2	Moses Nasing	Yawan	Uruwa	YUS LLG Ward 1
3	Geno Yuwoc	Worin	Uruwa	YUS LLG Ward 1
4	Kemo Robert	Kotet	Uruwa	YUS LLG Ward 1
5	Tommy Narete	Sugan	Uruwa	YUS LLG Ward 2
6	Robson Soseng	Gomdan	Uruwa	YUS LLG Ward 3
7	Danny Wande	Kalaset	Som	YUS LLG Ward 5
8	Tingke Sapenu	Gogiok	Som	YUS LLG Ward 5
9	Hermon Yangeng	Bungawat	Som	YUS LLG Ward 6
10	Tamina Findeng	Mek/Isan	Yopno	YUS LLG Ward 7/8
11	Dogem Mirande	Bonkiman	Yopno	YUS LLG Ward 9
12	Weo Bafinuc	Nian/Nokopo	Yopno	YUS LLG Ward 9
13	Nelson Teut	Weskokop	Yopno	YUS LLG Ward 11
14	Soya Werave	Wungon	Yopno	YUS LLG Ward 13
15	Mono Sam	Singirokai	Nambis	Wasu LLG Ward 10
16	Stanis Max	Ronji	Nambis	Wasu LLG Ward 11
17	Mike Barup	Yuong	Nambis	Wasu LLG Ward 11
18	Manrex Yausi	Bonea	Nambis	Wasu LLG Ward 18

December 2019), there are 2 marine and 16 terrestrial rangers patrolling YUS CA.

YUS Rangers Training

Most ranger training is conducted in-house but has been outsourced to external experts as needed (Ziembicki, 2012). This training is designed to ensure the rangers have necessary skills to effectively carry out their role. It includes the basics of using GPS, data collection, and sampling (Fig. 12.3). Recently, the program has been focusing on a relatively new conservation technology called the Spatial Monitoring and Reporting Tool (SMART) (described below).

Rangers Roles and Responsibilities

Rangers have four core functions. These functions are:

1. Participate in the YUS Ecological Monitoring Program (EMP)
2. Conduct monthly patrols, keep a diary/YUS Record Book
3. Serve as Community liaisons and provide advocacy, and feedback of monitoring and ranger work results
4. Assist visiting researchers in the field

YUS Ecological Monitoring Program

The YUS Ecological Monitoring Program (EMP) is a long-term biodiversity monitoring plan developed by TKCP and James Cook University in Queensland, Australia. This program was designed to assess the status of environmental features, wildlife population trends, and effectiveness of management practices. The information provides essential feedback to donors, funding agencies, scientists, managers, and policy makers as well as local communities. The study design is based on 12 transects distributed across the YUS region, which are monitored every five years.

Rangers provide general assistance for the EMP work. The most appropriate sampling method for the focal taxa is based on fecal standing crop counts. This method determines fecal pellet density at one point in time and relates this to fecal pellet decay rate and mean defecation rate (Campbell et al., 2004). The method is relatively simple to employ (therefore may be used by local community members and rangers without the need for specialized training and equipment) and may be used to survey several different taxa at one time (Fig. 12.4). Five years after its initial inception and implementation, the EMP was replicated in 2016 with an additional inclusion of sampling of the long-beaked echidna.

The field sampling activities were carried out seasonally from August to October. A team of 10 people sampled each transect. The team included the lead coordinator, 2–4 rangers, a research assistant and the rest were local landowners. It took approximately a week to sample each transect. Findings were reported back to the stakeholders through CAMC for further deliberation on the management of the YUS Conservation Area.

Conduct Monthly Patrols, Maintain a Diary/YUS Record Book

The ranger patrols are conducted for a period of one week every month. Duties include monitoring the area for the presence/absence of priority species (Fig. 12.5) and checking for signs of non-compliance with bylaws within the conservation areas. In 2017, the rangers were introduced to SMART patrol system (SMART, 2020). This system uses a GPS enabled Personal Digital Assistant (PDA) device through the CyberTracker application (CyberTracker, 2020) to record field observations. This information can then be transferred to the SMART computer database (see section on SMART below).

FIGURE 12.3 Paul van Nimwegen training YUS CA rangers in the field on SMART and use of a Personal Digital Assistant to collect data. *Source: Daniel Solomon Okena.*

FIGURE 12.4 YUS CA Rangers collaboratively sampling and sorting field data during YUSCA Ecological Monitoring Program, which occurs every five years. *Source: Daniel Solomon Okena.*

FIGURE 12.5 Daniel Okena and YUS CA Ranger Moses Nasing identifying and confirming a nose poke evidence of long-beaked echidna during a regular ranger monthly patrol. *Source: TKCP-PNG.*

Community Liaison, Advocacy and Feedback of Monitoring and Ranger Work Results

Rangers represent TKCP and YUS CBO at a range of community and other meetings in the YUS landscape. They also raise awareness and promote conservation efforts. For example, rangers are involved in Environment Day celebrations every year (5 June) and assist with tree planting activities in collaboration with the YUS CBO.

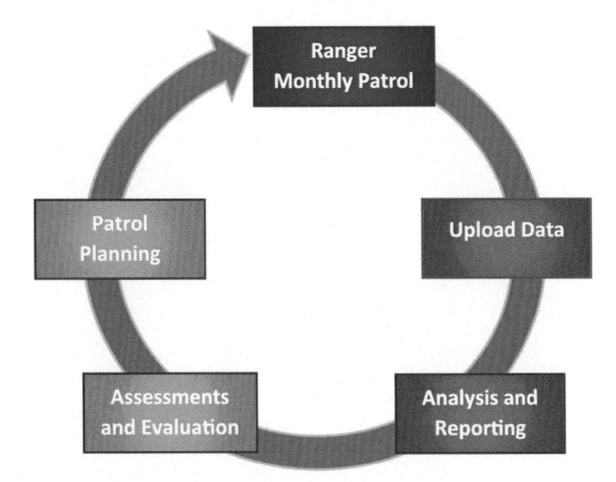

FIGURE 12.6 SMART is an adaptive management approach for protected area (SMART, 2017).

Field Assistants to Visiting Researchers

The rangers have excellent knowledge of the environment and biodiversity within YUS Conservation Area. This has been gained through growing up in the landscape and through their activities associated with the ranger program (patrolling, research, and conservation work). They are also trained in the use of field equipment such as GPS and possess the knowledge to work with a range of taxa groups. For this reason, the rangers are very good candidates for visiting researchers to recruit as field assistants. Importantly, being engaged as a field assistant provides the rangers with additional skills and income.

SPATIAL MONITORING AND REPORTING TOOL (SMART) IN YUS CONSERVATION AREA

The use of technology to support field activities and monitoring can result in improved management of protected areas. SMART is an open-source integrated geospatial database that allows field data to be systematically recorded, stored, and retrieved to support monitoring, compliance, planning, and adaptive management (Fig. 12.6). Rangers record field data with handheld devices using the CyberTracker application. This makes the process much easier and quicker than conventional methods. SMART also records patrol effort, which can improve staff management, internal accountability, and institutional governance. The technology offers the opportunity to measure protected area effectiveness and support reporting on national targets.

The system was developed and is maintained by the SMART Partnership, a broad consortium of global conservation organizations (SMART Partnership, 2018). It is now one of the most widely used conservation tools on the planet, being adopted by over 765 sites in more than 60 countries (SMART Partnership, 2018). Importantly, YUS CA is the first site in the Oceania region to use SMART.

The journey for YUS CA to adopt SMART started in March 2017 (TKCP, 2017). The TKCP Conservation Manager, who attended a regional workshop in Cambodia, adapted the database (or data model) in collaboration with field staff and conducted training. This was then field trialed in YUS. In January 2018, further training was carried out with the rangers and conservation officers. The information within the system was adapted based on feedback from the trial and translated to Tok Pidgin (an official language of PNG). Later that year, field staff commenced using the updated system. Fig. 12.7 is an output map from SMART showing patrol intensity for 2018. Further training was also conducted in January 2019.

Reporting

There are three standard reporting systems for rangers. Rangers report monthly, quarterly, and bi-annually to field conservation officers and TKCP supervisor.

Monthly Ranger Report

Each month, after patrolling and monitoring for one week, rangers report to the field conservation officers within their management zones. They report on tasks conducted, transfer data, present equipment for inventory, and review their monthly plan. Data from field devices are submitted to the conservation officer and checked. Any issues are addressed before the ranger's next patrol (e.g. broken equipment required, changes to workplans).

Quarterly Ranger Report

The Quarterly Ranger Report is done after every three months where the conservation officers report to the TKCP supervisor in person. The TKCP supervisor has the opportunity to meet with conservation officers (and rangers if practical) to address any issues that may have arisen in the last three months. It is also an opportunity for the TKCP supervisor to conduct checks of whether tasks are being conducted correctly, transfer SMART data to TKCP Lae office, and review work plans. The rationale behind addressing work plans during this time (in addition to during six monthly meetings) is to enable more flexibility in schedules and work activities, if required, due to unforeseen circumstances.

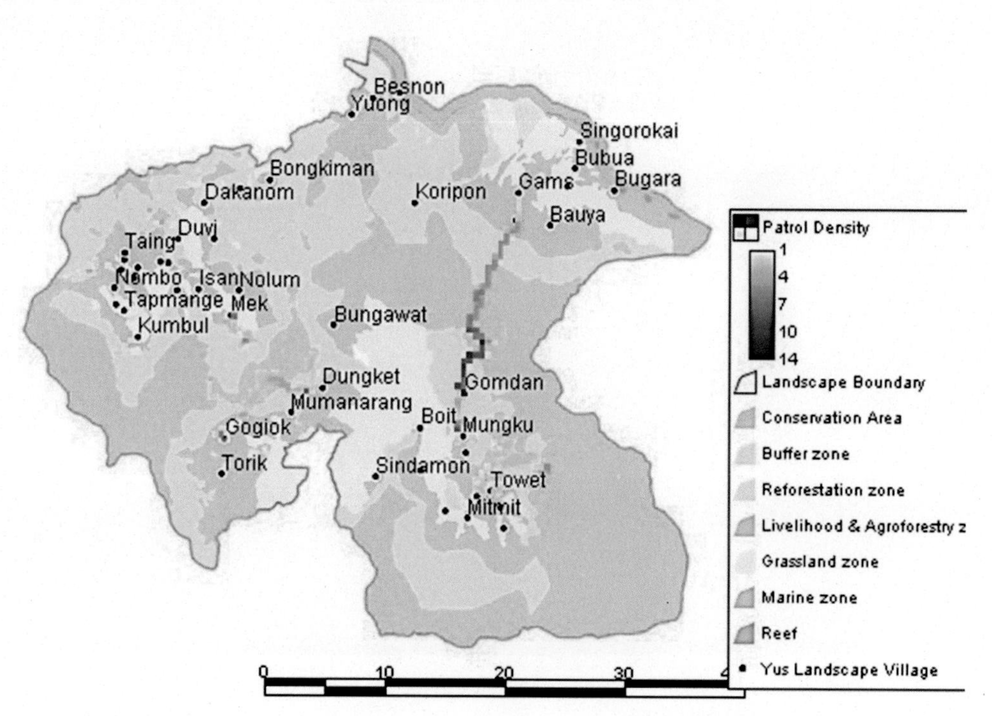

FIGURE 12.7 Patrol intensity of YUS Conservation Area Rangers automated from SMART for 2018.

Biannual Report

Biannual meetings are opportunities for rangers, conservation officers and TKCP staff to work through issues and develop work plans for the following six-month period. These gatherings are occasionally linked with YUS Community-based Organization meetings so that members can interact with rangers. Performance appraisals of rangers and conservation officers typically happen at this time. The meetings are also an opportunity for ranger training. Training requirements are assessed based on the activities planned. In most cases, training is conducted by outside consultants or organizations as appropriate, which can require additional funding.

IMPORTANCE OF HAVING COMMUNITY RANGERS IN THE YUS CONSERVATION AREA

The concept of conservation is not new to the indigenous landowners. Traditionally, village chiefs and elders were entrusted to be stewards of the land and coastal waters. Some of the lands outside of the villages are declared off-limits, or **tambu**. These areas are believed to be controlled by powerful unseen beings. Trespassers and their families will be cursed with bad fortune or death. These tambu areas then serve as reserves, regenerating and repopulating the entire forests, thus truly a system of sustainable resource management (Fig. 12.8).

FIGURE 12.8 Orokorokbo, the largest of the land parcels that was pledge as part of the YUS CA where most is covered by alpine vegetation. *Source: Daniel Solomon Okena.*

It is a great honor to continue the legacy of being stewards of the environment and resources. Although the challenge is greater than it once was, Woodland Park Zoo's Tree Kangaroo Conservation Program (TKCP) has trained the locals as rangers to monitor changes in their unique forest, with its incredible terrains flowing with breathtaking creeks and rivers. The YUS Conservation Area Rangers spend one week every month patrolling their forest. Equipped with a Global Positioning System (GPS) device, a pen and datasheet, safety boots, camping gear, and rations, Rangers look for the presence of some of the important fauna. They also observe and record any signs of illegal activity within the YUS Conservation Area, and report violators to the landowner to take the matter before the local courts. These responsibilities help them maintain and manage the forested lands even in the face of current threats. The Rangers also work to identify whether key faunas are present or not so that conservation action plans can be developed to address these challenges. It is a privilege indeed looking after the forest, which is also their livelihood.

FUTURE ASPIRATIONS OF THE PROGRAM

There are plans to do many things with the Ranger Program in the future. However, the priorities at the moment are:

1. Finalize effective usage of SMART in our program
2. Get the Marine Monitoring Program up and running
3. Continue capacity building of Rangers
4. Continue to be a role model conservation program for the country

FIGURE 12.9 Some of the YUS CA Rangers. *Source: Daniel Solomon Okena.*

At the moment, all of these plans look promising and achievable. However, resource and availability of skilled and determined people will help drive this forward.

CONCLUSIONS

The TKCP started as a research project on the endangered Matschie's tree kangaroo. The need to protect their wild population was identified as the priority. In doing so, the first conservation area in Papua New Guinea was established in YUS in 2009 to protect tree kangaroos. In creating the protected area, it also harbored protection of other important fauna and flora. Over the years it has grown into a holistic approach where education, livelihood, and health programs were introduced into the communities and people within the YUS landscape. Then started the most effective ranger program to monitor and report back on the status of the protected area. After years of noticing violations within the Conservation Area, the violations decreased significantly with the introduction of the ranger team.

The TKCP Ranger program all started with the conservation efforts for the endangered Matschie's tree kangaroo and is now the model Conservation Ranger program in the country, expanding new conservation boundaries and influencing protected area policies within Papua New Guinea (Fig. 12.9).

REFERENCES

Campbell, D., Swanson, G.M., Sales, J., 2004. Comparing the precision and cost-effectiveness of fecal pellet group count methods. J. Appl. Ecol. 41, 1185–1196.

CyberTracker, 2020. Available from:http://cybertracker.org/ (30 July 2020).

Pacific Islands Legal Information Institute (PacLII), 1978. Conservation Areas Act 1978. Available from:http://www.paclii.org/pg/legis/consol_act/caa1978203/ (27 June 2019).

SMART, 2017. Technical Training Manual SMART 5.0. Available from:https://smartconservationtools.org/wp-content/uploads/2020/06/SMART-Mobile-Data-Collection_June22nd.pdf.

SMART (Spatial Monitoring and Reporting Tool), 2020. https://smartconservationtools.org/ (March 2020).

SMART Partnership, 2018. 2018 Annual Report. Available from:https://smartconservationtools.org/wp-content/uploads/2019/07/SMART%202018%20Annual%20Report.pdf.

Tree Kangaroo Conservation Program (TKCP), 2017. Tree Kangaroo Conservation Program Annual Report 2017. Available from:https://www.zoo.org/document.doc?id=2407 (30 July 2020).

Ziembicki, M., 2012. Towards a YUS Conservation Ranger Program Operations and Training Manual. Tree Kangaroo Conservation Program, Lae, Papua New Guinea.

CHAPTER

13

Community-Based Conservation on the Huon Peninsula

Mikal Eversole Nolan[a], *Timmy Sowang*[b], *and Karau Kuna*[c]

[a]Lae, Papua New Guinea [b]Tree Kangaroo Conservation Program, Lae, Papua New Guinea [c]Kainantu, Eastern Highlands Province, Papua New Guinea

OUTLINE

INTRODUCTION

"As Papua New Guineans, our land is the most valuable thing we have. The landowners of the YUS Conservation Area pledge this land and the resources on top of it for indefinite protection. We are committed to maintaining our forests and wildlife across the YUS landscape to ensure their availability and existence for future generations. It is our people's job to preserve the environment for their grandchildren so they can see with their naked eye and not just learn from stories in newspapers and magazines." ***Timmy Sowang, President of the YUS Conservation Organization***

Papua New Guineans maintain a strong connection to their environment. For generations, the indigenous people have relied upon the local natural resources to meet their basic needs of food, clean water, shelter, and warmth. With over 85% of the national population continuing to live in rural areas today (The World Bank, 2018), it is little surprise to find that the culture, language, livelihoods, and traditions of each community are strongly influenced by their surrounding environment. This close relationship between humans and the environment

Tree Kangaroos
https://doi.org/10.1016/B978-0-12-814675-0.00025-7

propagates the renowned richness of Papua New Guinea's cultural diversity, which mirrors the natural biodiversity found from high in the cloud forests down to the network of coral reefs.

Extending out from the northeast corner of mainland Papua New Guinea (PNG), the Huon Peninsula hosts the largest and highest montane region outside of the Central Ranges (Chapter 4). Comprised of the Finisterre, Sarawaget, and Cromwell Mountain Ranges, the region rises to peak elevations of 4176 m in the Finisterre range. The unique topography on the Huon Peninsula has led to geographic isolation resulting in many species found nowhere else on Earth, including the Matschie's tree kangaroo (*Dendrolagus matschiei*). Bound by the lowlands of the Ramu-Markham Basin, a geographic barrier restricts the range of many of the area's montane fauna and flora. This isolation has promoted evolutionary divergence leading to the area being "one of the most distinct biogeographic provinces in PNG, with more endemic bird and mammal species than any other like-sized area in the nation" (Wikramanayake et al., 2001). This is a significant distinction in a country that harbors a vast array of fauna and flora, with estimates suggesting that within its borders 5–9% of the world's terrestrial biodiversity can be found on less than 1% of the land area (Adams et al., 2017). Yet it is this uniqueness and isolation that threatens its biodiversity with extinction including the endangered Matschie's tree kangaroo.

Culturally significant to the indigenous people, the tree kangaroo occupies a central role in local customs and ceremonies (Sowang, personal observations). Body decoration with tree kangaroo fur has traditionally served as a symbol of status. During traditional sing-sings, ceremonial displays of chanting, song, and dance, it is customary for people to adorn their bodies and heads with the fur, most commonly with the tail fur being used as a forehead decoration (Fig. 13.1). These customs carry through from generation to generation as elders and communities share stories and folklore where the tree kangaroo plays the central role. While dogs have lent assistance in more recent years for tracking, hunters continue to use traditional methods involving homemade bows and arrows to pursue the elusive tree kangaroos in the forest canopy. Yet, it is this very local cultural importance that has contributed to the overhunting of the tree kangaroo. Representing the largest native mammal on the peninsula, the tree kangaroo provides the indigenous people with an important source of protein often lacking in their traditional diets. Owing to its status of being both endemic and classified as Endangered by The International Union for Conservation of Nature (IUCN) (Ziembicki and Porolak, 2016), conservation efforts are imperative for local communities to maintain their culture and way of life.

Together with the customary landowners and the Papua New Guinea government, the Tree Kangaroo Conservation Program (TKCP) helped to establish the country's first nationally-recognized Conservation Area in 2009. Establishment of the Yopno, Uruwa, and Som (YUS) Conservation Area (CA), represented a paramount achievement taking 13 years to realize. YUS is named after the three main rivers in the area. The TKCP exists as an umbrella organization which characterizes the cooperation between the Tree Kangaroo Conservation Program of the Woodland Park Zoo (WPZ) and the Tree Kangaroo Conservation Program – Papua New Guinea (TKCP-PNG) (Fig. 13.2). To ensure projects maintain local relevance, the WPZ provided organizational guidance, capacity development, and financial support to facilitate the establishment of TKCP-PNG (Chapter 22). Today, TKCP-PNG operates as the local technical partner responsible for implementing the management plan of the YUS Conservation Area. The YUS Conservation Organization (YUS CO), a local community-based organization comprised of indigenous landowners, completes the collaboration by participating in governance functions.

The protected area aims to preserve the environment and the natural cultural inheritance

FIGURE 13.1 Tree kangaroo tail fur is used as a traditional forehead decoration during a welcome sing-sing to one of the villages of the YUS Conservation Area.

through conservation of important biological, topographical, geologic, historic, scientific, and social resources. Named for the three major rivers found within the landscape's borders, the Yopno, Uruwa, and Som rivers, the YUS Conservation Area Landscape, covers a total area of 158,271 ha encompassing both the core protected areas as well as buffer areas of mixed use and village agricultural zones (TKCP, 2012) (Fig. 13.3). The protected area provides ridge-to-reef ecosystem level protection extending across a narrow coastal plain to foothills, before rising over a distance of 40 km to altitudes in excess of 4000 m in the Sarawaget and Finisterre Mountain Ranges. In addition to the Matschie's tree kangaroo, the YUS Conservation Area protects 13 other species on the IUCN Red List including the critically endangered long-beaked echidna (*Zaglossus bartoni*), the dwarf cassowary (*Casuarius bennetti*), and several species of birds of paradise.

In Papua New Guinea, 97% of the land is owned and controlled by the customary landowners (AusAid, 2008). The Constitution of 1975 (DPLGA, 1975) guarantees protection of these property rights "as a special right of citizens", thus successful protected areas necessitate the continued commitment of local landowners. Papua New Guinea's 5th Report to the Convention on Biological Diversity (PNG Department of Conservation, 2014) identifies the two greatest threats to biodiversity as forest loss and indigenous hunting, notably both human induced. Unlike other parts of the world in which deforestation has taken a heavy toll,

Three Organizations, One Common Vision

The ***Tree Kangaroo Conservation Program (TKCP)*** is the umbrella name for the partnership between Woodland Park Zoo's TKCP and TKCP-PNG.

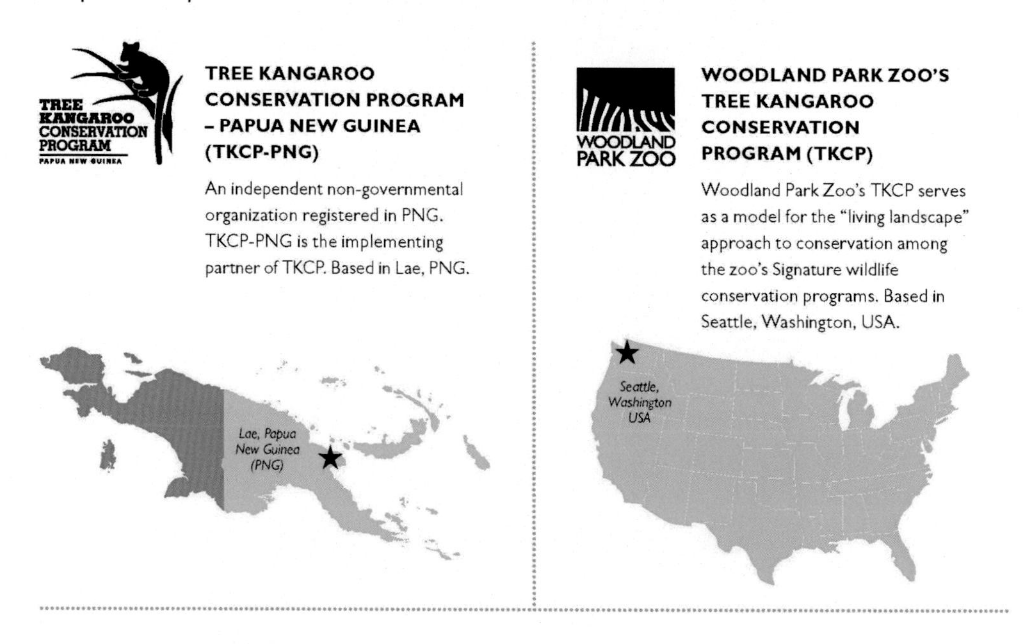

YUS CONSERVATION ORGANIZATION (YUS CO)

To ensure local ownership and continued community support of our work, TKCP partners with the community-based YUS Conservation Organization which represents the interests of local landowners and their communities. Based in YUS, PNG.

FIGURE 13.2 The Tree Kangaroo Conservation Program (TKCP) is the umbrella name for the partnership between Woodland Park Zoo's TKCP and TKCP-PNG. *Source: Tree Kangaroo Conservation Program (2017).*

about 70% of PNG rainforests remain intact. However, national spatial analysis reveals that during the period of 2002–2014, the rate of deforestation was 4.1%. Of the land area logged, 21% was cleared for the first time (Bryan and Shearman, 2015). Forest loss not only equates to habitat loss for endangered species, it also contributes to fragmentation which negatively impacts connectivity and species movement through wildlife corridors.

With all 50 villages across the YUS Landscape located within 500 m of rainforest, the local people rely heavily on forest products for food, fuel, and construction (Fig. 13.4). Subsistence hunting of wildlife plays an important part in traditional diets, constituting the primary source of protein and fats in many highland and isolated areas of Papua New Guinea. The increased hunting, like many other threats, may prove to be linked directly to increasing human population size,

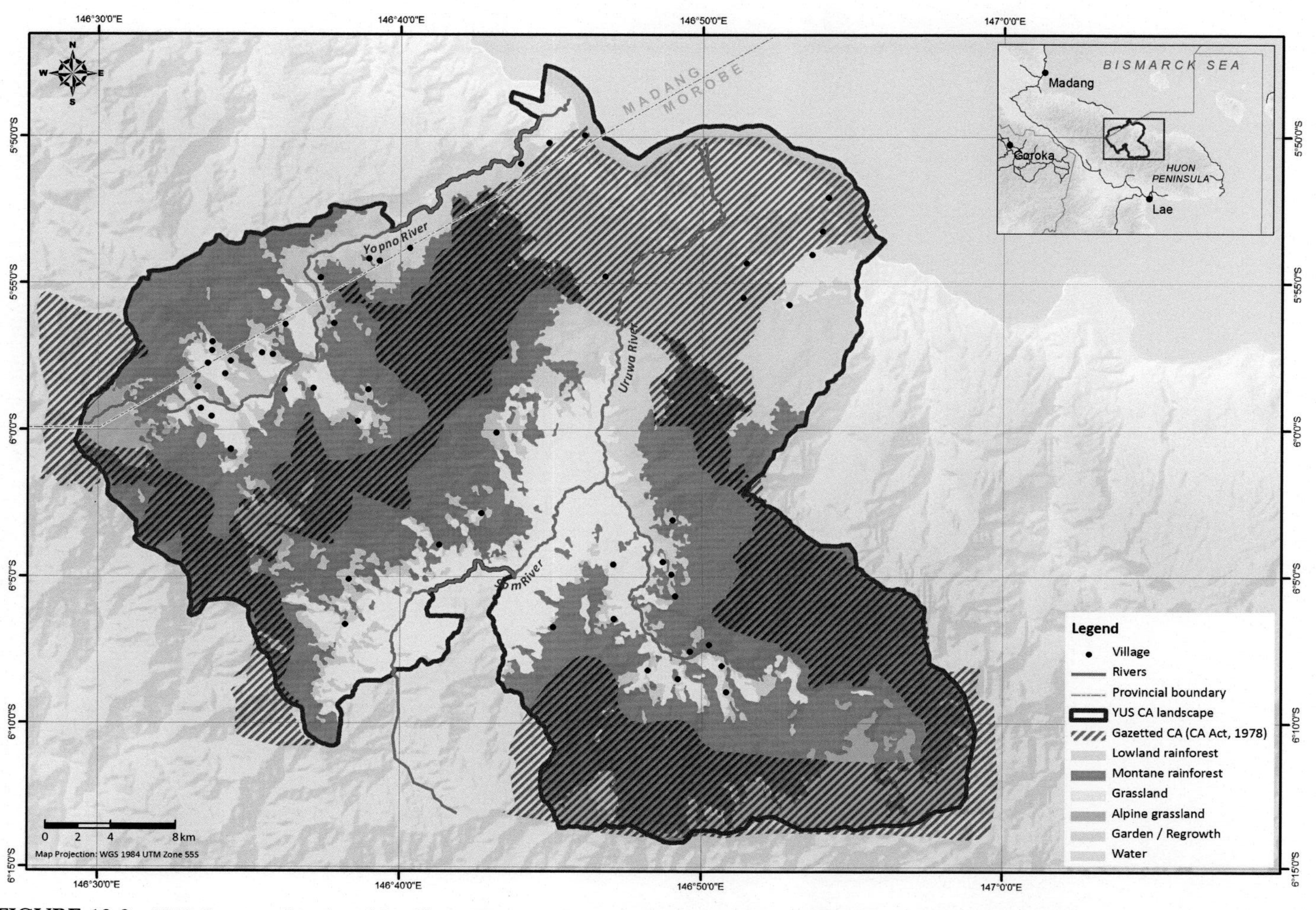

FIGURE 13.3 YUS Conservation Area Map illustrating proximity of villages to forest and legally gazetted conserved areas.

FIGURE 13.4 Villages are within close proximity to forests as people rely on forest products to meet their daily needs.

livelihood opportunities, and the ability of the people of YUS to sufficiently provide for themselves and their families. The YUS Conservation Area mitigates the threat of unsustainable resource use by serving as a "wildlife bank", providing safe refuge for wildlife within interconnected no-take zones. As wildlife populations grow within the no-take zone, offspring disperse to buffer zones where they can be sustainably locally hunted to fulfill local needs for protein and cultural use. This model, coupled with investments into community livelihoods, health, and education, has proven to be an appropriate model for community-based conservation in Papua New Guinea (Chapter 10).

The YUS community-based conservation model actively involves community members directly or through a community-based organization. The community engages in high level participation with the ultimate goals of wildlife protection, sustainable resource use, and effective management systems for the protected area. The YUS Conservation Area Landscape Management Plan identifies five "strategies" through which to foster community ownership of the conservation program (TKCP, 2012).

1. YUS Conservation Area Management
2. Research to inform resource and landscape management
3. Sustainable resource use and environmental services
4. Community services, livelihoods, and healthy families

5. Implementation and management

The aim of these strategies seeks to provide direct and indirect benefits to landowners and the community. The strategies above are not implemented in isolation but strive to link the various elements of the program for a holistic living landscape approach. The model leverages enhanced biodiversity protection and long-term conservation outcomes through the attention rendered by IUCN Red List Species, specifically for the Matchie's tree kangaroo. The TKCP employs this model to support its vision of a sustainable, healthy, and resilient Huon Peninsula landscape which supports the area's unique biodiversity, human communities, and culture.

YUS CONSERVATION AREA MANAGEMENT

Since independence, the government of Papua New Guinea has established various legal and formal institutions for the protection of fauna and flora. From the highest law of the Constitution (1975), concepts of conservation have been imbedded into the governing institutions. "We declare our fourth goal to be for Papua New Guinea's natural resources and environment to be conserved and used for the collective benefit of us all, and to replenished for the benefit of future generations" (DPLGA, 1975).

Globally, protected areas produce a positive benefit for habitat protection. The World Database on Protected Areas lists 57 protected areas across Papua New Guinea, affording protection to a total of 17,248 km^2 (UNEP-WCMC, 2020). According to the 2017 Assessment on Management Effectiveness report, the majority of protected areas in the country were found to be inadequately managed or without established management (Leverington et al., 2017). Of the area under formal protection, the vast majority of current protected areas across the country were gazetted under the Fauna (Protection and Control) Act (PACLII, 1966). This Act provides for the creation of Wildlife Management Areas (WMA), the most widely used protected area type nationally, which affords protection to specific species, yet neglects to recognize the importance of protecting habitat. The Act does little to explicitly prohibit mining, forestry, or other environmentally destructive economic activities within the protected area. This is clearly a concern given forest loss is the greatest threat to biodiversity. Conversely, the Conservation Areas Act of 1978 (PACLII, 1978) affords enhanced protection for both biodiversity and the habitat in which they live. The YUS Conservation Area was the first protected area to utilize this legislation to gazette a protected area in 2009, thus recognizing an opportunity to concurrently manage a species within the surrounding landscape. If biodiversity is to be protected, recognizing, and addressing the threats with holistic management is critical.

A history of failed conservation projects in Papua New Guinea has been linked to external interests dominating the planning and decision-making, negatively affecting the sense of ownership of the initiatives (Beehler and Kirkman, 2013). During the early stages of program development, the WPZ and TKCP sought to learn from these failures. The WPZ sought to first understand the traditional systems of natural resource management, then to incorporate these systems into the planning, design, and management model of the protected area, thus creating synergies between local and global knowledge. This hybrid of knowledge and experience formed the basis for the YUS Conservation Area Landscape Management Plan, the guiding management document which serves as a "functional mechanism for both the protection of endangered species and habits, and the preservation of cultural practices and sustainable resource use" (TKCP, 2012).

The theme of community-led conservation has continued to resonate throughout project implementation and management of the YUS Conservation Area. Today the TKCP not only partners with local and international organizations, but also employs the indigenous landowners of the protected area across all areas of the project including scientific research, land mapping, ecological monitoring, ranger patrols, community awareness and education, and conservation outreach activities.

The indigenous people of the YUS Conservation Area, as with many customs across Papua New Guinea, have a history of identifying land as *tambu* (taboo). This traditional custom prohibits access and activity within a defined area. The concept of a "wildlife bank" drew parallels with this cultural practice. Landowners identify and voluntarily pledge land as *tambu*, areas set aside for conservation and protection. Within the *tambu* areas of the YUS Landscape, hunting, cutting of trees and harvest of plants, introduction of dogs and pigs, and use of fire are among the prohibited activities under the YUS Conservation Area Bylaws. A set of nationally gazetted rules and penalties, the YUS Conservation Area Bylaws outline the governing principles of the protected area. The establishment of these rules and penalties involved a devolution of power and decision making to the clans who pledged land parcels for indefinite protection. This devolution of authority has empowered local ownership of the conservation program, recognizing the traditional customs of the area while providing the legal system and support to governance structures.

In 2017, the national government engaged a team of consultants to implement a Protected Area Management Effectiveness (PAME) assessment. The team utilized the Management Effectiveness and Tracking Tool (METT), the most widely used methodology globally, to evaluate the effectiveness of 57 gazetted and one proposed protected area across the country. Out of the 58 protected areas assessed using the METT only 4 showed "very good progress" toward management functions (Leverington et al., 2017). This assessment was based on a set of 30 questions which considered a number of factors such as protected area design, planning, enforcement, human resources, benefits to community, and condition of values. The YUS Conservation Area scored the highest nationally for protected area management recognizing strong governance through inclusive management arrangements, enforcement capacity, ongoing research, and meaningful improvements to community welfare (Leverington et al., 2017).

Papua New Guinea's Conservation Areas Act (PACLII, 1978) provides for collaborative management and governance of the protected area under the Conservation Area Management Committee (CAMC). However, the on-the-ground day-to-day management rests with the customary landowners of the protected area. The YUS CAMC draws together representatives from the indigenous landowning communities, the TKCP, represented local level governments, district administration, provincial government, and national government via the Conservation Environment Protection Authority (CEPA). Each member is entitled one voting representative except for the local landowners who comprise 1/3 of the total membership. This composition ensures the local voice remains strongest in decision making. Furthermore, the structure enables vertical two-way communication from grassroots through to the highest levels of government.

The indigenous people, TKCP, and government bodies represented on the YUS Conservation Area Management Committee (CAMC) collaborate to develop institutional support and oversee enforcement of the YUS CA bylaws. Initially developed by the YUS landowners, the rules and penalties are nationally gazetted laws endorsed by the Minister for Environment and Conservation. These relate to the core conservation areas or 'no-take' zones only, thus allowing the communities to independently manage the other established zones as per their land-use plans ensuring they maintain appropriate areas to meet their immediate needs. The YUS CA Rangers conduct monthly patrols of the core conservation areas to monitor and report infractions of the YUS CA Bylaws (Fig. 13.5). Empowered and supportive communities, village magistrates, and local court systems ensure perpetrators realize associated penalties.

FIGURE 13.5 YUS Conservation Area Rangers play a key role in monitoring and reporting. *Source: TKCP.*

In addition to the site specific rules, penalties, and fines (Fig. 13.6), the gazetted bylaws of the YUS Conservation Area (Government of Papua New Guinea, 2014) state that any person who enters the protected area must:

(i) Look after all the resources in the Conservation Area; and
(ii) Look after all of the different kinds of flora and fauna in the Conservation Area; and
(iii) Respect any traditional belief or sacred site of the traditional landowner of the area; and
(iv) Look after all the different kinds of non-living or abiotic resources; and
(v) Increase or extend or connect the YUS Conservation Area as and where necessary

The YUS Conservation Area Rangers also play an important function encouraging local ownership of the conservation program (Chapter 12). Rangers are appointed by clans pledging land to the core area and are endorsed by the TKCP once the neighboring clans reach consensus on the appointment. In the PNG

SCHEDULE 3

ENGLISH VERSION OF SITE SPECIFIC PENALTIES, FINES AND FEES

1. YUS Conservation Area is a subject of the content of the *Conservation Areas Act* (Chapter 362).
2. The YUS Conservation Area Committee may make regulations for changes to the use of land within the Conservation Area in accordance with Section 28 of the Act.
3. In Section 35(1), a person who develops or alters or permits the development or alteration of the existing use of land in a Conservation Area except —

   (*a*) in accordance with the terms of the Management Plan for that Conservation Area; or

   (*b*) in accorance with written approval from the Minister under Section 34(1), is guilty of an offence.

   Penalty: A fine not exceeding K40,000.00.
   Defaulty Penalty: A fine not exceeding K4,000.00.

FIGURE 13.6 The gazetted penalties, fines and fees as per Schedule 3 of the YUS Conservation Area Bylaws. *Source: Government of Papua New Guinea, 2014. National Gazette No. G338: Declaration of the Establishment and Appointment of Conservation Area Management Committee Members for YUS Conservation Area, Its Rules, Fees and Penalties. Port Moresby.*

(Continued)

No. G338—7th August, 2014 4 National Gazette

Declaration of the Establishment and Appointment of Conservation Area Management Committee Members for YUS Conservation Area, its Rules, Fees and Penalties—*continued*

Schedule 3—*continued*

English Version of Site Specific Penalties, Fines and Fees—*continued*

4. A person who develops or alters or permits the development or alteration of the existing use of land in an area in respect of which a notice of recommendation has been given under Section 12(1) of the Act except in accordance with written approval from the Minister under Section 34(1) of the Act, is guilty of an offence.

   Penalty: A fine not exceeding K40,000.00.
   Defaulty Penalty: A fine not exceeding K4,000.00.

5. It is shall not a defense to an actin for an offense committed under this Section that the development or alteration to the existing use of land did not adversely affect the environment.

6. In accordance with Section 28 of the *Conservation Areas Act* (Chapter 362), the YUS Conservation Area Management Committee, in consultation with the local landowners, has developed and approved the following site-specific rules and fines:

   (*a*) It is illegal to light a fire with the Conservation Area (does not include fires for cooking).

   Fine: = K200.00

   (*b*) It is illegal for a person to hunt or fish an animal or a fish which is an endangered species.

   Fine: = K100.00.

   (*c*) It is illegal for a person to hunt a Tree Kangaroo.

   Fine: = K500.00.

   (*d*) It is illegal for a person to hunt or fish within the Conservation Area.

   Fine: = K50.00.

   (*e*) It is illegal for a person to hunt for a pig with the Conservation Area.

   Fine: = K30.00.

   (*f*) It is illegal for a person to hunt for a dog within the Conservation Area.

   Fine: = K100.00.

   (*g*) It is illegal for a person to release a pig within the Conservation Area.

   Fine: = K30.00.

   (*h*) It is illegal for a person to release a dog within the Conservation Area.

   Fine: = K50.00.

   (*i*) It is illegal for a person to make a garden within the Conservation Area.

   Fine: = K150.00.

   (*j*) It is illegal for a person to pollute a water body or soil inside the Conservation Area.

   Fine: = K10.00

   (*k*) It is illegal to cut a tree or destroy a plant within the Conservation Area.

   Fine: = K50.00.

   (*l*) It is illegal for a person to go into a sacred site within the Conservation Area without permission.

   Fine: = K100.00.

7. A penalty fee or fine may be imposed by the relevant YUS Conservation Area Management Committee.

8. The penalty fee or fine shall be paid to the Committee.

9. The Committee shall manage and control the penalty fee or fine paid to the Committee in accordance with its Constitution.

FIGURE 13.6, cont'd

context, the ability to earn and keep community trust serves as an essential skill, especially when working with diverse stakeholders, cultures, and traditional clans. Rangers conduct regular patrols of the *tambu* areas and no-take zones recording observable infractions of the YUS CA Bylaws as well as undertake ecological monitoring of selected fauna. The TKCP provides training, support, and tools to undertake the work. Not only are Rangers required to have an understanding of specific scientific knowledge, they must be able to provide a basic analysis and synthesize conclusions from the information they collect to share with their local communities. The TKCP collects the data and undertakes a more thorough analysis which is shared with project partners, communities, and CAMC to support adaptive management.

Implementation and Management

Transparent institutions with clear common goals have proven advantageous in the pursuit of accomplishing the outcomes identified within the YUS Conservation Area Landscape Management Plan. Institutions have the power to shape the community's behavior, perception, and choices which ultimately sway performance and outcomes of the conservation goals. A lesson learned from partners in the YUS Conservation Area is that local institutions necessitate home-grown leadership and empowered communities to formulate preferences and assume decision making roles. Acknowledgment of indigenous cultural beliefs and norms, the role of local interest groups and political entities, and existing internal conflicts (such as land rights, clan disputes, etc.) cannot be underestimated as these will influence the mechanisms and entry into institutional development. The YUS Conservation Area Landscape Management Plan seeks to recognize and incorporate local institutions into the framework for effective management planning and implementation.

The TKCP and YUS Conservation Organization (YUS CO) represent two key grassroot organizations responsible for achieving objectives identified under the management planning instruments of the YUS Conservation Area. The role of the TKCP seeks to facilitate community engagement, provide technical assistance and training, secure funding for implementation of the landscape management plan, and perform the functions of monitoring and evaluation. The TKCP as an NGO partner plays a key role in strengthening community-based organizations, such as the YUS CO, who must assume leadership for driving and managing local projects which directly impact their communities. While the YUS CO is comprised entirely of indigenous landowners, both men and women, from across the landscape, the TKCP staff represent locally recruited members of the YUS communities (~80%) and non-local Papua New Guineans (~20%) from other parts of the country. International consultants are recruited to fulfill specialized roles and build local capacity as the current needs of the program necessitate. In Papua New Guinea, community politics, culture, and social responsibilities to one's clan often sway decisions. The TKCP exists as a self-directed NGO, and while it represents the interests and values of the customary landowners, it does not need to be beholden to local alliances within the communities. Local staff and YUS CO partners serve as an advisory body ensuring programs are aligned with the social and cultural norms found across the YUS Landscape. This has proven influential in cementing the relationship for achieving common goals.

The YUS CO, a community-based organization, provides a platform for landowners to self-organize and affect decisions relating to their mandate "to foster wildlife and habitat conservation while also improving livelihoods for local communities within the YUS Conservation Area of Papua New Guinea" (YUS CO Constitution) (Fig. 13.7). The YUS CO is a 22-member body with representatives elected from each ward (the foundation level of government)

FIGURE 13.7 Representatives of the YUS Conservation Organization.

across the program area. An additional four females or "mama reps" represent the Yopno, Som, Uruwa, and Nambis regional areas within the landscape, with the aim to improve representation and inclusion of their particular needs and priorities into discussions. The YUS CO holds an Annual General Meeting (AGM) where representatives gather to discuss management, program implementation, monitoring, and challenges of the protected area. Prior to the AGM, leaders hold ward-level and regional (Yopno, Som, Uruwa, and Nambis) meetings to collect input from various stakeholders who may not be able to travel the distances required to personally attend. The YUS CO also serves a role in disseminating information regarding conservation awareness, the status of biodiversity, and other information derived from meetings with partners such as the TKCP and government bodies. Sustaining a culture of respect, transparency, and trust, between and within the TKCP and YUS CO, guides constructive collaborative management and impl1ementation of the landscape management plan.

Recognizing that long-term management of the YUS Conservation Area requires continuous financial stability, the TKCP with the help of partners such as Conservation International and the Woodland Park Zoo in Seattle, Washington established the YUS Conservation Endowment Fund. This endowment fund, managed by the WPZ, provides for annual distributions to the project to support effective management as per the remit of the YUS Conservation Area Landscape Management Plan and assists with expenses associated with development of the local communities' sustainable livelihood activities. The YUS Conservation Endowment ensures continuity of basic management functions of the protected area, a comfort afforded to few conservation initiatives.

Research to Inform Resource and Landscape Management

A 1991 conservation needs assessment considered the area encompassing the YUS Conservation Area in the Finisterre Mountain Ranges on the Huon Peninsula to be a "scientific unknown" requiring scientific and conservation attention (Beehler and Kirkman, 2013). Filling the gaps for this "unknown", positively contributes to both the global scientific community as well as to local decision making and management effectiveness of the protected area. Home-grown researchers and research assistants provide a unique opportunity to engage local people for integrated indigenous knowledge. The indigenous landowners offer intrinsic knowledge of the vast biodiversity found within their landscape. Socio-economic and linguistic research has also positively supported appropriate planning and implementation of conservation activities. Understanding of the culture and society are inherently linked to the local environment, thus a broad research focus enables effective management and decision making of the YUS Conservation Area. This opportunity to enrich the scientific community proves advantageous to both outside researchers and indigenous people who gain specific skills and specialized knowledge.

Tree kangaroo research has been the keystone inquiry into local biodiversity since 1996. Long term studies have provided the global community with scientific information regarding tree kangaroo biology, feeding ecology and nutrition, genetics, behavior, habitat and home range, and population densities (Chapters 2, 7, 19, 20, 23, 25, 26). Traditional hunters have been converted into key members of the research team providing expert skills to find, capture, and safely release tree kangaroos.

An elevational transect established by Conservation International attracts an array of international researchers to the YUS area. The transect spans from 200 m to 3700 m above sea level. The elevational transect has been likened to "a giant earth thermometer" for monitoring shifts of fauna and flora with the impacts of climate change. This provides local communities with information and observations that may allow them to adapt effectively to their changing environment. It also encompasses an exceptional tool for adaptive management of the protected area and informs community response to changes for enhanced resilience.

Sustainable Resource Use and Environmental Services

The TKCP leads participatory land-use planning and monitoring projects with each community within the YUS Landscape (Chapter 11). An indigenous mapping project, facilitated by the Center for Native Lands (Washington D.C.) and PNG's University of Technology in 2005, produced local language community resource maps for the Yopno, Uruwa, and Som regions of the YUS Landscape (WPZ, 2019). This tool served as an important foundation for later developments to maintain local names, sacred sites, and traditional boundaries as identified by the customary landowners. It also proved advantageous in the early stages of the project to identify boundaries for inclusion in the government submission to establish the protected area.

From 2011 to 2013 the TKCP's land-use planning team facilitated a series of community workshops. These workshops aimed to help communities agree on a plan for local development, sustainable resource use, and the allocations of space for both people and wildlife to co-exist. Communities collectively identified conservation and development priorities as part of the project. These were coupled with visual representations of area maps generated using GIS modeling to agree on core conservation 'no-take' areas, buffer, livelihood, reforestation, marine/reef, and village-use zones. Ward level plans have driven small-scale, flexible, appropriate projects aimed at meeting the local needs and aspirations. The inclusivity of the project promotes local ownership and devolution of decision making. The culmination of the land-use planning project was a public endorsement of the final 5-year plans by the provincial Governor, Member for Parliament, local level government Presidents, and several ward Councillors.

Spanning across the entire YUS Conservation Area landscape, land-use planning and the establishment of the various zones have produced a number of benefits including:

- Increased connectivity of habitat and wildlife corridors;
- Increased land pledges from local landowners;
- A more effective "wildlife bank";
- Maintains traditional use and local value of wildlife and natural resources;
- Addresses needs and priorities of men, women, and young people;
- Adds clarity to land tenure and land boundaries; and
- Supports the engagement and empowerment of landowners in decision making.

During the 2017 METT Assessment for Papua New Guinea, the YUS Conservation Area was the only protected area to receive a "very good score" for resource management (Leverington et al., 2017). The TKCP's land-use planning model has been adopted as an example of best practice in PNG's national policy on protected areas.

Community Services, Livelihoods, and Healthy Families

For conservation models to thrive, indigenous people must be engaged as the local custodians of their lands. A continuous value must be derived from the project by local communities. This may be an economic, social, or cultural value. The customary landowners and communities of the YUS Conservation Area do not directly receive financial compensation for land pledges made to the protected area. However, indirect opportunities, employment, development input, and livelihood alternatives have been afforded as a result of the project.

A culture of enduring community consultation assists the TKCP to identify the areas of need that would produce the greatest impact for people and biodiversity. As the people of YUS rely solely on their locally available natural resources to meet their community and personal needs, recognizing and addressing the human and social dimensions associated with the local conservation efforts will alleviate certain pressures on the environment. Families across the YUS Conservation Area landscape must find ways to pay school fees, obtain health services, and purchase the necessities for life in local markets. Yet for the majority of families who maintain a subsistence lifestyle built around hunting, fishing, and farming, finding the cash to fulfill these necessities can be challenging. The extremely rugged terrain coupled with the obstacles of life in a remote location, such as lack of roads and electricity, hinder the ability to generate a cash income. The TKCP partners with local landowners to address some of these economic challenges through improved livelihoods and community managed resources for both present and future use. The program also strives to advance access to education and health services for community members. The TKCP attempts to consider the unintended consequences and not stimulate jealousy within the community by striving to work through the appropriate local institutions and by focusing on communal benefits.

Sustainable Livelihoods

The TKCP's sustainable livelihoods project aims to mitigate human threats to the conservation area by providing economic incentives based on existing local and conservation-friendly livelihoods (Chapter 14). The TKCP partners with farmers, local cooperatives, and socially-responsible businesses to establish market linkages and develop production systems for conservation-friendly commodities such as coffee and cocoa. The model supports the removal of obstacles for the quality production of coffee and cocoa in livelihood zoned areas, market integration, and handover of management to local producer organizations. The project engages the traditional Melanesian approach to integrated food production systems. These traditional gardens may be close to village centers or more than a day's walk. Primarily used for subsistence agriculture, the gardens fuse forest and food production with a wide range of crops being interspersed. Long-established cash crops, such as coffee and cocoa, are punctuated throughout the gardens to provide traditional farmers with economic opportunities. The TKCP and specialist partners facilitate training to optimize production and quality of these already existing cash crops. This model enhances the productivity of already cleared land under traditional gardening systems, and thereby mitigates the need to clear existing forests for agriculture.

The TKCP has assisted the local producer organizations to establish inclusive membership, appropriate structures, systems, and policies. Sound financial models assist leaders in making informed decisions and demonstrate transparency to stakeholders. Ensuring the efforts and benefits are community-focused, TKCP strives to build local leadership through development of a local cooperative to drive the initiative and connect farmers directly with international value-added markets. In 2011, the conservation coffee project began by exporting 1.7 tons of green bean coffee to an independent specialty roaster in the United States (US) (Fig. 13.8). By 2017, the project had grown to successfully export over 45 tons of conservation coffee collectively to the US and Australia, with 424 registered farmers all committing to uphold environmental standards. Under the project, farmers earn substantially higher profits while generating local awareness for wildlife conservation. Buyers of the coffee provide a premium price above the

FIGURE 13.8 A local coffee producer drying coffee beans for export.

international market and share the story of the conservation efforts with a broader audience.

A clear linkage between biodiversity and economic incentives encourages local leaders to assume responsibility for on-the-ground monitoring and sustainable management of natural resources. The project has encouraged additional landowners to voluntarily pledge their land to the core conservation area. More land pledges have positively translated into more effective connectivity of wildlife corridors and enhanced long-term protection for the biodiversity found within the YUS CA. Not only are farmers pledging more land, but the integrated coffee gardens have also proven to be attractive habitat and display sites for Birds of Paradise. The project has served as an additional mechanism to enforce the YUS Conservation Area Bylaws as the communities have decided that any person found in violation will not be eligible to sell their commodities under the conservation program. The success of removing obstacles to the development of coffee, market integration, and handover of management to local producers has proved an effective model, with the same approach being replicated to include conservation cocoa in communities at lower elevations (Chapter 14).

Healthy Village, Healthy Forest

Health services across Papua New Guinea fail to meet the needs of the country's growing population. Poor or non-existent infrastructure and services intensify the problem in rural areas. The communities of the YUS area are not immune to the problem of being critically underserved by health care professionals. Government-run aid posts often lack skilled workers to provide services and medicines to the community. This necessitates travel to urban centers in order to seek medical attention, a costly endeavor which may require a two to three day walk. With no doctors or hospitals, a handful of community health workers provide basic care to YUS communities. While careful not to replicate government services, the TKCP and a team of volunteer international health care professionals conduct site visits to identify the greatest needs impacting the health and wellbeing of communities. The TKCP has utilized this information to advocate for improved government services and provide recurring training to community health workers on the most pressing concerns. The TKCP hopes that by addressing health needs, the pressure to harvest local natural resources will be alleviated.

The TKCP's Healthy Village, Healthy Forest project launch an endeavor in 2011 aimed to reinforce the connections between human and environmental health and wellbeing (Chapter 15). The foundations of the Population-Health-Environment (PHE) project, funded by a subgrant through the BALANCED Project, USAID, have inspired a holistic EcoHealth approach for YUS. The concept has been supported by the findings reported in the 2017 Assessment of Management Effectiveness for Papua New Guinea as it builds upon people's traditional understanding of their link to the natural world (Leverington et al., 2017). The project has implemented a series of village level workshops touching on topics such as the importance of clean water and hygiene, the links between human and environmental health, and family planning and reproductive health topics. The workshops sought to train and develop a network of village-level peer educators serve as an important resource for communities underserved by health care professionals. Following the workshops, the TKCP has continued to partner with community health workers to facilitate the collection of information on family planning, births and deaths, and immunization rates for improved delivery of government services. The TKCP seeks to build an integrated initiative fostering beneficial outcomes for human, wildlife, and environmental health.

Education and Leadership Development

Papua New Guineans embrace a strong cultural history of environmental awareness, respect, and protection of their natural resources. Many rural

FIGURE 13.9 A local volunteer engages young Junior Rangers to learn about their environment. *Source: TKCP.*

people continue to value the traditions and customs passed down from generation to generation with the hope that the young people within their communities become empowered individuals who will lead through future conservation stewardship and sustainable community development. Young people express a clear interest in opportunities that foster the development of skills which will enable them to become positive members and leaders in their own communities. Yet educational institutions underperform in conveying these skills. This is validated as education levels across the YUS area are overall lower than the national and provincial averages with some exceptions. With limited access to formal education, young people are hungry for knowledge and skills to position them for future leadership opportunities.

The TKCP's Junior Ranger Program seeks to alleviate the educational gap through positive youth development opportunities. This project provides hands-on training to local teachers and volunteers to facilitate experiential environmental programing for children which reflects the local situation (Fig. 13.9). Junior Rangers are encouraged to realize their personal role, both present and future, as environmental stewards and conservation leaders (Chapters 16 and 22).

Since 2002, the TKCP has provided scholarships to 33 men and women from YUS communities to complete a three year teaching accreditation from the Balob Teacher's College in Lae (TKCP, 2017). As part of the scholarship, the newly qualified teachers agree to return to one of the 16 YUS primary schools to undertake teaching positions. This approach, coupled with the Junior Ranger Program, seeks to address educational deficiencies through both formal and informal learning models.

LESSONS LEARNED FOR COMMUNITY-BASED CONSERVATION

For conservation efforts to succeed in Papua New Guinea, programs should reflect the local values, customs, and institutions. As such, it is imperative that the indigenous people and communities fill leadership roles and assume ownership of the project and its components.

A number of lessons have been identified from work completed to establish and manage the YUS Conservation Area. These lessons have shown positive benefit to communities and biodiversity such as the Matchie's tree kangaroo.

- A holistic approach addressing environmental priorities, human needs, and capacity development, individual and institutional, has helped to mitigate negative impacts while supporting overall project goals. The approach should be local in nature and respond to the specific context shaped by the areas culture, traditions, needs, priorities, and both formal and informal institutions.
- Utilizing a hybrid of traditional and formal institutions has proven advantageous to encouraging local ownership. This cross-section recognizes the importance of integration into traditional and local institutions. It promotes the effective use of established local influencers, such as village elders, *bigman*, mama groups, and teachers, while seeking to mitigate "elite capture" of decision making and project benefits by defining formal rules and monitoring mechanisms. Strong home-grown leadership and investment into young people as the future stewards of the land further advances the effectiveness of a hybrid model.
- Community members have shown an increased willingness to adopt the program when they derive meaningful value from the project. Notably, various community members possess a diverse array of needs and priorities, as well as hope and expectations of the program's goals. Program managers should monitor the project to reduce the risk of conservation goals being over shadowed by development, therefore striving to link all projects directly to conservation outcomes and maximize the value of wildlife and biodiversity to local people. Appropriate and sustainable economic opportunities may provide individual and community incentives to conservation. Numerous non-economic benefits may also be derived from the project especially when integrated into local customs and traditions.
- Land-use plans provide a useful tool to appropriately make space for people and wildlife. These plans have shown to positively support improvements to wildlife corridors and add clarity to land boundaries. A participatory model of land-use planning, one which involves men, women, and young people, supports the engagement and empowerment of landowners in making decisions which impact their futures. Development of land-use plans may also provide a useful platform to identify and plan for future changes with potential impacts on the project, such as population growth, introduction of new infrastructure and development activities, substitution of values derived from external influence, and shifts in political institutions.
- The identification and creation of a sustainable funding mechanism seeks to mitigate negative impacts of funding fluctuations. It provides for long-term planning and on-going management of the conservation program. The YUS Conservation Area Endowment Fund is managed externally to reduce the risk of mismanagement and misappropriation.
- Establish institutional support for monitoring and evaluation that appropriately reflects the scale of the conservation program and actively engages the various actors and beneficiaries. Through the sharing of program results and challenges, a more cohesive and improved national system of protected areas can be supported.
- The involvement of the indigenous landowners and communities in developing appropriate and representative rules and penalties for the protected area fosters community ownership through the devolution of authority. By utilizing local institutions, such as village elders, community appointed Rangers, local courts and

government representatives, to enforce the rules and penalties, communities maintain ownership and influence over the protected area's success. The process of legally gazetting rules and penalties has given greater authority to these institutions.

- The establishment and promotion of strong partnerships with communities, civil society, government, and private sector has proven essential. The various stakeholders should be encouraged and proved with opportunities to meet, reflect and discuss the project, and build rapport. Partners must be prepared to invest time, patience, and resources as necessary to establish trust and respectful long-term relationships.

REFERENCES

Adams, V.M., Tulloch, V.J., Possingham, H.P., 2017. Land-Sea Conservation Assessment for Papua New Guinea. Available from: https://doi.org/10.13140/rg.2.2.26219.13606 (27 June 2019).

AusAid, 2008. Land tenure systems in the Pacific. In: Making Land Work. Reconciling Customary Land and Development in the Pacific, vol. 1. Australian Government, Canberra Available from: www.dfat.gov.au/sites/default/files/MLW_VolumeOne_Bookmarked.pdf>.

Beehler, B., Kirkman, A. (Eds.), 2013. Lessons Learned From the Field: Achieving Conservation Success in Papua New Guinea. Conservation International, Arlington, VA.

Bryan, J.E., Shearman, P.L. (Eds.), 2015. The State of the Forests of Papua New Guinea 2014: Measuring Change Over the Period 2002–2014. University of Papua New Guinea, Port Moresby.

Department of Provincial and Local Government Affairs (DPLGA), 1975. Papua New Guinea, 2018. Constitution of the Independent State of Papua New Guinea. Available from: https://dplga.gov.pg/wp-content/uploads/2018/03/Constitution.pdf (27 February 2019).

Government of Papua New Guinea, 2014. National Gazette No. G338: Declaration of the Establishment and Appointment of Conservation Area Management Committee Members for YUS Conservation Area, Its Rules, Fees and Penalties. Port Moresby.

Leverington, F., Peterson, A., Peterson, G., Jano, W., Sabi, J., Wheatley, A., 2017. Assessment of Management Effectiveness for Papua New Guinea's Protected Areas 2017. SPREP, Apia. Available from: http://www.pg.undp.org/content/papua_new_guinea/en/home/library/assessment-of-management-effectiveness-for-papua-new-guinea-s-pr.html (27 June 2019).

Pacific Island Legal Information Institute (PACLII), 1966. Papua New Guinea Consolidated Legislation: Fauna (Protection and Control) Act 1966. Available from: http://www.paclii.org/pg/legis/consol_act/faca1966290/> (27 June 2019).

Pacific Island Legal Information Institute (PACLII), 1978. Papua New Guinea Consolidated Legislation: Conservation Areas Act 1978. Available from: http://www.paclii.org/pg/legis/consol_act/caa1978203/ (27 June 2019).

Papua New Guinea Department of Conservation, 2014. Papua New Guinea's fifth national report to the Convention on Biological Diversity. Available from: www.cbd.int/doc/world/pg/pg-nr-05-en.pdf> (27 June 2019).

The World Bank, 2018. Rural Population (% of Total Population). Available from: https://data.worldbank.org/indicator/SP.RUR.TOTL.ZS?contextual=default&end=2017&locations=PG-XD&start=1960&view=chart (27 June 2019).

Tree Kangaroo Conservation Program, 2012. YUS Landscape Plan 2013–2015. Tree Kangaroo Conservation Program, Lae, Morobe.

Tree Kangaroo Conservation Program, 2017. Annual report. Woodland Park Zoo, Seattle, Washington.

UNEP-WCMC, 2020. Protected Area Profile for Papua New Guinea From the World Database of Protected Areas. April 2020. Available from: www.protectedplanet.net/country/PG (6 May 2020).

Wikramanayake, E., Dinerstein, E., Loucks, C., Olson, D., Morrison, J., Lamoreux, J., McKnight, M., Hedao, P., 2001. Terrestrial Ecoregions of the Indo-Pacific: A Conservation Assessment. Island Press, Washington, DC.

Woodland Park Zoo (WPZ), 2019. Conservation Documents: YUS-Community-Map. Available from: https://www.zoo.org/file/conservation-documents/YUS-Community-Map- - -webbw.pdf (27 June 2019).

Ziembicki, M., Porolak, G., 2016. *Dendrolagus matschiei*. The IUCN Red List of Threatened Species 2016: e.T6433A21956650. Available from: https://doi.org/10.2305/IUCN.UK.2016-2.RLTS.T6433A21956650.en (11 February 2019).

FURTHER READING

Dabek, L., Wells, Z., 2013. Establishing a conservation area in Papua New Guinea—lessons learned from the YUS conservation area. In: Beehler, B., Kirkman, A. (Eds.), Lessons Learned From the Field: Achieving Conservation Success in Papua New Guinea. Conservation International, Arlington, VA.

CHAPTER

14

Strengthening Community Conservation Commitment Through Sustainable Livelihoods

Trevor Holbrook

Tree Kangaroo Conservation Program, Woodland Park Zoo, Seattle, WA, United States

OUTLINE

INTRODUCTION

Throughout the 20th century the modern protected area movement spread across the globe, resulting in the establishment of tens of thousands of formal, government-owned and -controlled protected areas to ensure the long-term conservation of nature and the services it provides (IUCN, 2020a). In many cases, the creation of these protected areas was driven largely by top-down influence and based upon the interests and priorities of the government. Within this "classic" paradigm, protected areas generally served to "set aside" would-be productive resources in order to protect wildlife (i.e., game) and scenic beauty more so than functional ecosystems. In particular, the interest in protecting "wilderness" demanded the exclusion of human influence from such protected landscapes (Phillips, 2003). And while the immense conservation value of these protected areas—and their collective role in advancing the practice of conservation—is virtually incalculable, one must also recognize the absence of consideration or involvement of the local communities and indigenous peoples affected by their creation.

In recent decades, the scope of priorities and considerations for the establishment and management of protected areas has expanded substantially. In contrast with the more limited interests and objectives which drove the creation

Tree Kangaroos
https://doi.org/10.1016/B978-0-12-814675-0.00027-0

of protected areas throughout much of the 20th century, the paradigm has shifted to acknowledge and incorporate the perspectives, needs, rights, and knowledge of local and indigenous communities as well as the complex relationship between conservation and sustainable development. Whereas traditional protected areas had generally aimed to wholly guard against human activity and influence, this new paradigm has effectively expanded the definition as to what a Protected Area is, and what objectives they serve (Locke and Dearden, 2005). To support this expanded interpretation, the International Union for the Conservation of Nature (IUCN) established six categories within which all protected areas align (IUCN, 2020b):

IUCN Category Ia—Strict Nature Reserve: A strictly protected area set aside to protect biodiversity and also possibly geological/geomorphical features, where human visitation, use, and impacts are strictly controlled and limited to ensure protection of the conservation values.
IUCN Category Ib—Wilderness Area: A protected area which is usually a large unmodified or slightly modified area, retaining its natural character and influence without permanent or significant human habitation, which are protected and managed so as to preserve their natural condition.
IUCN Category II—National Park: A large natural or near natural area set aside to protect large-scale ecological processes, along with the complement of species and ecosystems characteristic of the area, which also provide a foundation for environmentally and culturally compatible, spiritual, scientific, educational, recreational, and visitor opportunities.
IUCN Category III—Natural Monument or Feature: A protected area set aside to protect a specific natural monument, which can be a landform, sea mount, submarine cavern, geological feature such as a cave or even a living feature such as an ancient grove. They are generally quite small protected areas and often have high visitor value.
IUCN Category IV—Habitat/Species Management Area: A protected area which aims to protect particular species or habitats with management that reflects this priority.
IUCN Category V—Protected Landscape/Seascape: A protected area where the interaction of people and nature over time has produced an area of distinct character with significant, ecological, biological, cultural, and scenic value: and where safeguarding the integrity of this interaction is vital to protecting and sustaining the area and its associated nature conservation and other values.
IUCN Category VI—Protected Area with Sustainable Use of Natural Resources: A protected area which conserves ecosystems and habitats together with associated cultural values and traditional natural resource management systems. They are generally large, with most of the area in a natural condition, where a proportion is under sustainable natural resource management and where low-level non-industrial use of natural resources compatible with nature conservation is seen as one of the main aims of the area.

While Categories I through IV largely encompass the common types of traditional protected areas which serve to "set aside" unmodified, uninhabited natural areas, Categories V and VI specifically recognize the integral role of people in relation to the value of the protected area. The recognition of protected areas adhering to IUCN's Category VI, in particular, has enabled a substantial increase in the establishment of protected areas encompassing large "living landscapes" in which both nature and human communities coexist. IUCN's Category VI allows for a much broader spectrum in terms of protected area governance and management, and calls for a much wider range of objectives and benefits relating to both conservation and sustainable development. Whereas traditional protected areas were meant to be free of human influence, this new paradigm acknowledges and embraces the view that people are an inextricable part of many landscapes and ecosystems—and therefore must participate in, and benefit from, the sustainable management of those areas.

By acknowledging and incorporating the roles and relationships of people within a protected landscape, and with the incorporation of sustainable development objectives into the scope of a protected area, its management plans must extend beyond protection to address both sustainable resource use and community well-being factors. In the case of protected areas under customary ownership or local control, in particular, management plans need to account for both the operating costs as well as the opportunity costs of foregoing other types of land use (FAO, 2014). While local and indigenous communities often assign substantial value to the inherent and intangible benefits of such protection, the long-term sustainability of a Category VI protected area requires consideration and action in response to the resource needs and livelihoods of the communities affected.

COMMUNITY-BASED CONSERVATION IN PAPUA NEW GUINEA

In Papua New Guinea, a participatory, community-based approach to the creation of protected areas is essential to ensuring the protection of the country's vast biodiversity and the health of its ecosystems. With over 95% of the land under customary ownership and control, the local communities collectively have tremendous influence over the management, use, and conservation of the country's land and resources. With a majority of the country's population living a rural subsistence lifestyle, there remains a strong, direct dependence on local natural resources and ecosystem services. Because local communities both control and depend heavily upon their natural resources, the long-term viability and success of Category VI protected areas in Papua New Guinea relies in large part on the communities' ongoing commitment, support, and involvement.

Papua New Guinea's YUS Conservation Area is comprised of over 400,000 acres (162,000 hectares) of customary land, owned and controlled by the clans living throughout the landscape encompassing the Yopno, Uruwa, and Som river watersheds. The YUS Conservation Area was established in 2009 to protect the endangered and endemic Matschie's tree kangaroo (*Dendrolagus matschiei*) and its cloud forest habitat, along with many other unique and threatened species of flora and fauna (TKCP, 2009). Situated on the north side of the Huon Peninsula and stretching from coastal reefs to 4000-meter peaks, the area is extremely rugged and steep. Because of its difficult terrain, YUS is one of the most remote and inaccessible areas in Papua New Guinea. With no road access, transportation and movement within the YUS landscape is limited to walking trails which connect the area's 50 villages. Six grass airstrips provide for once-weekly flights over the mountains to the Provincial capital of Lae, while boats provide transport between coastal villages and neighboring Madang via a six-hour journey. As is the case throughout much of Papua New Guinea, the economy of YUS is highly localized and subsistence-based. In large part, the communities' livelihoods are directly dependent on the area's natural resources through subsistence-level hunting and farming and through the use of local forest products for fuel and shelter.

Throughout the consultation process with landowners to secure commitment and land pledges for the creation of the YUS Conservation Area (Chapter 10), the benefits of protection were largely framed in economic terms to recognize the community's reliance on their landscape. The protection of forests and other habitats was presented as the creation of a "Wildlife Bank"—in which their investment would yield long-term interest via robust wildlife populations for sustainable hunting outside of the core protected area. Given the community's recognition and concern regarding the diminishing availability of vital hunted species in their forests, this perspective underscored one important element of the YUS Conservation Area's contribution to local livelihoods and well-being.

Given the remoteness of the YUS landscape and its limited economic interaction beyond its borders, the area offers few opportunities for its residents to earn cash income. While the production of food crops, livestock, and poultry is largely intended for subsistence use, most families throughout YUS rely primarily on coffee (in higher-elevation areas) and cocoa (in lower-elevation areas) as cash crops (TKCP, 2012). Having been introduced and heavily promoted throughout many rural areas of Papua New Guinea in the 1950s and 1960s, the two crops represent a sizeable share of the country's non-mineral exports and serve as substantial sources of income for as much as 50% of the population (Coffee Review Australia, 2009). As a result, these crops were planted among subsistence food gardens in many of the villages across YUS. However, as compared with other areas of Papua New Guinea with somewhat better road infrastructure, the high cost of transporting goods out of YUS presents additional challenges for the profitability of such products (Fig. 14.1).

FIGURE 14.1 Coffee grows among native shade trees and subsistence crops in YUS, Papua New Guinea. *Source: Tree Kangaroo Conservation Program.*

While Papua New Guinea and YUS benefited from high global coffee prices and improving infrastructure during the 1960s and 1970s, the following decades led to the industry's significant decline—including a particularly large drop in the sale and export of smallholder-produced coffee, due in part to high transport costs (Coffee Review Australia, 2009). During this time, many families in YUS shifted their focus away from coffee production as a cash crop. Coffee trees were often left unpruned in the gardens, and without active care, yields decreased over time. And although some of the product continued to be harvested and processed in smaller quantities, most families ceased to invest time and resources into coffee in favor of essential subsistence crops. So, as yields and incomes from coffee production dropped substantially, so too did efforts and expectations for its ability to provide an adequate livelihood for the people of YUS.

In preparation for the establishment of the YUS Conservation Area, the landowners and communities of YUS expressed the importance of strengthening livelihoods, including opportunities for income generation. Recognizing that the protection and sustainable management of the YUS landscape largely provides for local subsistence needs in terms of food security, the communities emphasized their need for cash income to support family health costs, children's education fees, and other household expenses. To address this, the Tree Kangaroo Conservation Program (TKCP) and YUS community leaders undertook a review of potential income-generating opportunities which would align with, leverage, and support the conservation objectives of the protected area—particularly the protection of the Matschie's tree kangaroo and its core cloud forest habitat. In itself, the creation of the YUS Conservation Area provided a number of opportunities for employment and income through management activities and research projects. Protected areas often also offer potential through the promotion of eco-tourism, which can support the development of a wide range of income-generating activities while underscoring the direct economic benefit of the protected wildlife and habitats. However, given the nascent tourism industry in Papua New Guinea as a whole, the lack of supporting infrastructure in Morobe Province, and the distribution of the YUS community throughout dozens of remote villages, the YUS Conservation Area could not be expected to generate sufficient visitor interest to support livelihoods throughout the landscape.

Importantly, TKCP and stakeholders in YUS paid particular attention to the existing resources and capabilities among the communities in YUS, seeking to build upon those opportunities rather than introducing new and untested ventures. In doing so, coffee emerged as a viable product among the higher-elevation inland villages due to the community's familiarity with its production and the large number of established trees in gardens throughout the landscape. Because coffee farming had faltered as an income-generating activity in YUS in the past, TKCP conducted a review to better understand the reasons for its collapse. While the high cost of transport is generally unavoidable, it is relatively predictable and can be factored into the cost of production. The market, however, is much less predictable and is driven by a variety of factors, both globally and within Papua New Guinea. The country's industrial infrastructure and market structures for coffee were also developed primarily for the production of commodity-grade coffee for export. Commodity-grade coffee is comprised of beans sourced from a large number of farms, and its value throughout the supply chain generally favors quantity above quality. As a result, the market for coffee parchment (the state of the coffee bean after wet-milling and drying, which is usually the point at which farmers in PNG sell their product) places little value on the unique origins, flavor profiles, or quality of the product. In fact, because coffee parchment is purchased based on weight, the market inadvertently incentivizes farmers to sell coffee that has not been fully dried and is therefore heavier. And, because coffee parchment buyers have come to expect this practice, it can be expected that the offered price also factors in the higher moisture content—creating a cycle which actually serves to discourage best practices for the production of high-quality coffee.

Despite these challenges, Papua New Guinea's climate, terrain, and volcanic soils are very well-suited for the production of high quality, specialty grade Arabica coffee. In contrast with commodity grade, the market for specialty coffee places significant value on the unique qualities and characteristics of the location in which the coffee was grown, the techniques applied throughout the fermentation and drying processes, as well as the social impacts and environmental sustainability of the coffee's production throughout its entire supply chain. Increasingly, the global specialty coffee market as well as Papua New Guinea's Coffee Industry Corporation have sought to develop and expand the production of specialty grade coffee in order to serve the rapidly-growing sector of the market. However, such a transition requires substantial changes to practices and standards throughout the coffee production process and supply chain as a whole. The specialty market's demand for higher quality and product consistency requires greater coordination and standardization among the participating farmers, and the market's need for differentiation and traceability requires a greater degree of management and record-keeping from the farm to the end-buyer (Fig. 14.2).

FIGURE 14.2 YUS Conservation Coffee farmer, Mr. Mangkera, oversees the coffee drying process in Dungket village, YUS, Papua New Guinea. *Source: Trevor Holbrook, Woodland Park Zoo.*

To further explore the potential of developing YUS coffee for the specialty market, TKCP sought the advice of experts through its partner, Seattle's Woodland Park Zoo. With Seattle's reputation as an epicenter for the world of coffee, TKCP and the YUS Conservation Area caught the interest of Caffe Vita—an independent coffee roasting company and a pioneer of the "Farm Direct" approach to sourcing specialty coffee. The "Farm Direct" or "Direct Trade" approach is built on the first-hand relationship between the coffee roaster (the end-buyer) and the coffee producer (the farmers), and suggests that the beans are not transferred through intermediaries or traded based on third-party certifications. Instead, the quality and the social-environmental impacts at origin are directly confirmed and assured by the roaster themselves. The relationship is intended to be mutually beneficial, enabling the roaster's direct influence and input into the coffee's production and ensuring that the producers receive a larger share of profits from the sale of their coffee. With the strong commitment of the YUS community in protecting the health and biodiversity of their landscape, and with the demonstrated potential for the production of unique, high quality coffee among the area's mountain villages, Caffe Vita agreed to provide support with the hopes of establishing a fruitful, long-term Farm Direct relationship with the coffee farming families of YUS. By emphasizing the importance of environmental sustainability and local conservation action to the premium market value of the coffee produced in YUS, this unique partnership clearly underscored the potential economic benefit of the YUS Conservation Area for participating communities throughout the area's 35 higher-elevation villages (Fig. 14.3).

Commencing in 2010, the project's early years focused heavily on quality improvement and control, with the close involvement and support of Caffe Vita's green coffee buyer Daniel Shewmaker as well as production specialist Karl

FIGURE 14.3 Caffe Vita's former Green Coffee Buyer Daniel Shewmaker inspects the quality of coffee during a technical training visit to YUS, Papua New Guinea. *Source: Trevor Holbrook, Woodland Park Zoo.*

Aglai and others from Papua New Guinea's Coffee Industry Corporation. Focusing initially on several villages within the Uruwa Zone of YUS, trainings focused on tree rejuvenation and husbandry practices, as well as quality management throughout each step of coffee harvesting, pulping, washing, fermenting, and drying. To improve consistency and production coordination, farmers were encouraged to process their coffee in groups as opposed to individually in order to ensure greater product uniformity and to facilitate collective quality management. Following the farmers' adoption and application of the quality improvement practices, Caffe Vita purchased its first lot of "YUS Conservation Coffee" beans in 2011. Aiming to provide farmers with a minimum of price parity with local market rates in addition to covering the high cost of transport, participating farmers received six kina per kilogram of parchment for coffee delivered to airstrips in YUS. The price effectively doubled farmers' profits as compared with prior years' earnings. Nearly 30 60-kilogram bags of the coffee arrived in Seattle, branded among Caffe Vita's signature line of Farm Direct Single Origin coffees and labelled "Papua New Guinea Yopno Uruwa Som." The branding prominently highlighted the coffee's unique origin, the powerful conservation impact of the YUS Conservation Area and the Tree Kangaroo Conservation Program, and the impact of the Farm Direct coffee trade on the livelihoods and well-being of the YUS communities. Several bags of the roasted coffee returned to YUS to celebrate the milestone achievement and to reiterate the direct connection between the community's conservation efforts and the high prices earned through the sale of YUS Conservation Coffee (Fig. 14.4).

The project has grown more than 20-fold over the subsequent 8 years, welcoming the participation of coffee farming families across virtually all of the inland villages among Yopno, Uruwa, and Som Zones. Average annual coffee production throughout YUS has increased to more than 35,000 kilograms, harvested from shade-grown coffee trees planted among subsistence food gardens within designated "livelihood zones." Caffe Vita's demand for its "PNG YUS" single origin coffee has increased as well, with annual shipments between 10,000 and 15,000 kilograms. With the support of TKCP partner Zoos Victoria in Melbourne, Australia, YUS Conservation Coffee farmers have established a second Direct Trade relationship with Melbourne's Jasper Coffee. Both Zoos Victoria and Jasper Coffee actively highlight and promote the unique story of the YUS Conservation Area, inviting socially- and environmentally-minded coffee consumers in Melbourne—an epicenter for the world of specialty coffee in its own right—to support

FIGURE 14.4 YUS Conservation Coffee is prominently branded and shipped internationally in burlap bags displaying the Tree Kangaroo Conservation Program and Caffe Vita logos. *Source: Tree Kangaroo Conservation Program.*

conservation efforts in Papua New Guinea through the purchase of Jasper's "YUS Kopi" single origin coffee (TKCP, 2018) (Fig. 14.5).

With the growth and development of the YUS farmers' conservation-based coffee enterprise, the Tree Kangaroo Conservation Program is shifting the focus of its livelihoods efforts toward sustainability and local leadership. To ensure the project's continued viability and success over the long-term, the farmers need to continue expanding market linkages and strengthening YUS Conservation Coffee's marketability among a broader range of specialty coffee buyers. Direct Trade partners Caffe Vita and Jasper Coffee will undoubtedly continue to be the primary buyers of YUS Conservation Coffee, providing a strong foundation both in terms of coffee sales as well as shared conservation values. However, as production capacity increasingly outpaces the needs of these two core buyers, YUS Conservation Coffee will need to appeal reliably and profitably to more buyers. In order to formulate a broader market strategy and to connect YUS Conservation Coffee with new specialty markets, TKCP partnered with Sucafina Specialty (formerly MTC International Coffee Group)—a global coffee processing and trading company with extensive market linkages and familiarity with Papua New Guinea's coffee production context. In addition to supporting the farmers' logistical solutions for the transport and delivery of YUS Conservation Coffee to market, Sucafina is well-equipped to provide continued technical guidance for the project regarding product development, branding, and marketing—with consideration given to industry demands and trends (Fig. 14.6).

Given the success and experience in developing long-lasting Direct Trade partnerships, TKCP and the YUS Conservation Coffee farmers have a strong interest in establishing additional relationships with coffee roasters who value the project's contributions to conservation and community well-being. However, recognizing that such relationships often take time and require sustained direct connection, as well as consideration for market diversification, the project will also explore other segments of the specialty coffee market. With guidance from Sucafina, TKCP and the farmers will revisit the potential value of certification schemes such as USDA Organic Certification, Fair Trade Certification, Rainforest Alliance Certification, and others. While such certifications may be of less interest to buyers engaged in Direct Trade relationships—due to the fact that those buyers have first-hand knowledge of the coffee's production processes and social-environmental impacts—they do have

FIGURE 14.5 YUS Conservation Coffee farmers examine their product, roasted and packaged for retail sale as "Kopi YUS" by Melbourne, Australia's Jasper Coffee. *Source: Trevor Holbrook, Woodland Park Zoo.*

FIGURE 14.6 MTC International Coffee Groups's Harrison Koch (right) and Caffe Vita's Mason Sager (left) evaluate samples of YUS Conservation Coffee at New Guinea Highlands Coffee Exports cupping lab in Goroka, Papua New Guinea. *Source: Trevor Holbrook, Woodland Park Zoo.*

the potential to provide YUS farmers with access to a significantly larger market segment. Many small- and mid-sized specialty coffee roasters are committed to sourcing responsibly-produced coffee, but lack the financial or technical capacity to establish and maintain fruitful direct relationships with producers. For these buyers, certification schemes provide assurance that the coffee has been produced and sold in accordance with certain standards. As TKCP and the YUS farmers move forward with guidance from Sucafina, the benefits and costs of various certification schemes will be considered along with other potential opportunities to increase market access (Fig. 14.7).

In order to establish and strengthen effective local leadership and management of the farmers' YUS Conservation Coffee enterprise, including its continued role in promoting local commitment to conservation, TKCP is supporting the farmers in forming the YUS Conservation Coffee and Cocoa Cooperative. With more than 700 coffee farming families now participating in the production of YUS Conservation Coffee throughout the landscape, the need for coordination, quality assurance, logistics, and business management surpasses TKCP's role and capacity, and must rely on functional local systems and structures. As participating farmers had already organized into production groups or "clusters" within their respective villages in order to improve quality management and localized production coordination, the Cooperative serves to coordinate among all of the clusters for the delivery and sale of its members' products. Led and managed by an elected Board of Directors, the Cooperative will assume increasing responsibility for the operational and financial management of the YUS Conservation Coffee initiative (Fig. 14.8).

FIGURE 14.7 The Tree Kangaroo Conservation Program's Danny Nane (2nd from right) discusses with YUS Conservation Coffee farmers about the presence of local wildlife in the coffee gardens to emphasize the value of conservation and sustainable resource management. *Source: Gemina Garland-Lewis.*

FIGURE 14.8 The Tree Kangaroo Conservation Program's Stephen Fononge (center) loads YUS Conservation Coffee onto a plane for transport out of the YUS landscape. *Source: Ryan Hawk, Woodland Park Zoo.*

In addition to TKCP's efforts to support the sustainability of the YUS Conservation Coffee initiative following its successful growth and expansion throughout the inland villages of YUS, the program is also seeking to replicate those successes among the coastal villages through their production of YUS Conservation Cocoa. Between 2011 and 2015, TKCP assisted coastal farmers to employ quality improvement techniques with technical guidance provided by Seattle-based chocolatier Theo Chocolate. As a result of those improvements, YUS Conservation Cocoa caught the interest of PNG-based Queen Emma Chocolates. In line with Queen Emma's selection of chocolate bars using single-origin cocoa sourced from communities throughout Papua New Guinea, the company expressed great interest in featuring TKCP and the YUS Conservation Area on its packaging. Queen Emma's "YUS Kakao" chocolate bars were distributed and sold across Papua New Guinea, providing YUS cocoa farmers with increased incomes and promoting the importance of protecting the country's unique wildlife. Unfortunately, the spread of the invasive cocoa pod borer throughout many of the farms in YUS during 2016 resulted in significant damage to both the harvest and the trees (TKCP, 2016). In response, TKCP has sought to assist the YUS cocoa farmers in recovery and rehabilitation of their livelihoods. Following a needs assessment and the co-development of a business plan together with the farmers and other stakeholders (led by conservation-based livelihood specialist Francis Hurahura), TKCP aims to facilitate collaboration among the District and Local Level Governments, the PNG Cocoa Board, the YUS Conservation Coffee and Cocoa Cooperative, and leading cocoa processing companies. Guided by the YUS Conservation Cocoa business plan, the farmers and other stakeholders will focus on re-developing cocoa production capacity designed to serve the specialty market. The membership criteria and expectations set by the YUS Conservation Coffee and Cocoa Cooperative will continue to reinforce the important role of conservation and sustainable resource management in relation to the branding and market value of its products (TKCP, 2018).

CONCLUSIONS

Ten years after its creation, the YUS Conservation Area is recognized by Papua New Guinea's Conservation and Environment Protection Authority as a national model for Category VI protected areas, due in part to its successes in engaging local community members in landscape management planning, decision-making, and conservation-based activities (Leverington et al., 2017). Among the YUS communities, TKCP's livelihoods initiatives are often credited with garnering local interest and support for conservation efforts. As a result, the Matschie's tree kangaroo serves as an icon representing the YUS communities' pride in protecting their landscape and heritage. Both community observations and TKCP's monitoring data suggest a stabilized and healthy tree kangaroo population in YUS, with increased sightings in core protected areas as well as near villages and gardens (TKCP, 2018). Into the future, TKCP aims to strengthen the resilience of the YUS communities through the diversification of local livelihoods, based upon local resources and expertise and rooted in the value of a healthy YUS landscape. As the number of Category V and VI protected areas around the globe continues to increase, and the interests and well-being of local communities within those areas are taken into careful consideration, the growing global market for responsibly-sourced products presents a significant opportunity for the development of such creative, scalable, market-driven initiatives.

REFERENCES

Coffee Review Australia, 2009. Papua New Guinea: Land of Opportunity. Available from: https://web.archive.org/web/20090226193312/http://coffeereviewaustralia.com/2009/02/22/papua-new-guinea-land-of-opportunity/ (28 July 2020).

Food and Agriculture Organization (FAO), 2014. Protected Areas, People, and Food Security. Available from: http://www.fao.org/3/a-i4198e.pdf (28 July 2020).

International Union for the Conservation of Nature (IUCN), 2020a. Protected Areas. Available from: https://www.iucn.org/theme/protected-areas/about (28 July 2020).

International Union for the Conservation of Nature (IUCN), 2020b. Protected Area Categories. Available from: https://www.iucn.org/theme/protected-areas/about/protected-area-categories (28 July 2020).

Leverington, F., Peterson, A., Peterson, G., 2017. Assessment of Management Effectiveness for Papua New Guinea's Protected Areas 2017. Available from: https://www.sprep.org/attachments/Publications/BEM/pngpa.pdf (28 July 2020).

Locke, H., Dearden, P., 2005. Rethinking protected area categories and the new paradigm. Environ. Conserv. 32 (1), 1–10. Available from: https://www.wild.org/wp-content/uploads/2010/01/Locke-Dearden-Rethink-PA-Categories-paper.pdf (28 July 2020).

Phillips, A., 2003. Turning ideas on their head: the new paradigm for protected areas. The George Wright Forum 20 (2), 8–32. Available from: https://notendur.hi.is/hjs11/Projects/Phillips%202003%20IUCN%20new%20paradigm.pdf (28 July 2020).

Tree Kangaroo Conservation Program (TKCP), 2009. 2009 Annual Report. TKCP, Lae, Papua New Guinea. Available from: https://www.zoo.org/document.doc?id=1723 (28 July 2020).

Tree Kangaroo Conservation Program (TKCP), 2012. YUS Landscape Plan 2013–2015. TKCP, Lae, Papua New Guinea. Available from: https://www.zoo.org/document.doc?id=904 (28 July 2020).

Tree Kangaroo Conservation Program (TKCP), 2016. 2016 Annual Report. TKCP, Lae, Papua New Guinea. Available from: https://www.zoo.org/file/visit-pdf-bin/TKCP-2016-Annual-Report-web.pdf (28 July 2020).

Tree Kangaroo Conservation Program (TKCP), 2018. 2018 Annual Report. TKCP, Lae, Papua New Guinea. Available from: https://www.zoo.org/document.doc?id=2537 (28 July 2020).

CHAPTER

15

Using a One Health Model: Healthy Village-Healthy Forest

Robert M. Liddell[a], Emily R. Transue[b], Joan Castro[c], Nancy Philips[d], W. Blair Brooks[d], Marti H. Liddell[e], Peter M. Rabinowitz[f], and Lisa Dabek[g]

[a]Center for Diagnostic Imaging Quality Institute, Seattle, WA, United States
[b]Washington State Health Care Authority, Olympia, WA, United States
[c]PATH Foundation Philippines, Inc., Makati, Philippines
[d]Dartmouth Hitchcock Medical Center, Lebanon, NH, United States
[e]The Polyclinic, Seattle, WA, United States
[f]Environmental and Occupational Health Sciences, University of Washington, Seattle, WA, United States
[g]Tree Kangaroo Conservation Program, Woodland Park Zoo, Seattle, WA, United States

OUTLINE

Tree Kangaroos
https://doi.org/10.1016/B978-0-12-814675-0.00023-3

INTRODUCTION

Papua New Guinea is facing a number of important health challenges, including the challenge of how developing human populations can coexist with sensitive natural ecosystems and unique wildlife species. The country faces multiple issues including climate change, the need to improve public health and medical services, agricultural development including expanded livestock farming, and how to preserve fragile rain forest and other natural habitat and populations of sensitive wildlife species. A national "One Health" model can be formed to balance these concerns about human, animal, and ecosystem health in a model of healthy coexistence.

The One Health concept considers the many interrelationships among the health of humans, animals, and the ecosystems they depend upon. The One Health approach has been endorsed by international health agencies including the World Health Organization (WHO, 2020), the World Organisation for Animal Health (OIE, 2020), the UN Food and Agriculture Organization (FAO, 2020) and the US Centers for Disease Control (CDC, 2020). On a national level, countries as diverse as Sweden and Kenya have developed national One Health strategies. While much of One Health activity is centered around infectious diseases that can pass between animals and humans (zoonotic diseases), the concept is broader than this and encompasses other aspects of health (Rabinowitz et al., 2018).

One Health provides a model for considering ways to maximize and monitor the health of human populations living in close proximity with wildlife and domestic animals in a sensitive ecosystem. It provides a platform for transdisciplinary cooperation and collaboration between government and professional organizations that may have separate missions involving one or more of the sectors of human, animal, or environmental health. For example, in Papua New Guinea, the Ministry of Agriculture typically deals with livestock health and disease issues, while the Ministry of Health manages human health affairs, and a number of non-governmental agencies (NGOs) are involved in conservation of natural resources. Under the umbrella of One Health, these agencies would work together to develop mutually beneficial strategies for moving forward to encourage healthy humans, animals, and ecosystems.

A key aspect of the One Health model is the integrated assessment of human, animal, and ecosystem health indicators in a particular region. This can involve obtaining and sharing data about the health of the human populations, the animal populations, and the ecosystem of a region, and through such data sharing developing a better understanding of shared environmental health risks across species. In addition to such integrated assessments, One Health stresses the importance of integrated interventions (using the same transdisciplinary approach) that consider and track the effect of any health interventions on humans, animals, and the environment (Rabinowitz, 2017).

This chapter describes a One Health approach to sustainable human, animal, and environment health within a remote region of Papua New Guinea. This local One Health model can be expanded to other remote regions of Papua New Guinea.

THE PEOPLE OF PAPUA NEW GUINEA

Eighty-five percent of Papua New Guineans live in rural villages, many of which are situated within steep mountainous and forested terrain. In these rural areas, there are few roads and little government-provided infrastructure such as electricity, water, or waste utilities. Many homes are constructed of thatched grass and most are without running water or toilets. Cooking is done over open wood fires within homes, and these indoor open fires also provide the only source of heat. Subsistence farming, hunting and foraging provide most of the food staples.

These conditions cause significant health implications for people living in remote areas of PNG. Lack of safe water leads to diarrheal diseases and other infections. The use of open wood fires in poorly ventilated homes results in smoke inhalation which causes chronic lung disease and increased susceptibility to lung cancers and respiratory infections, such as tuberculosis. Inadequate nutrition leads to generalized poor health and stunting of growth. Maternal and neonatal health are adversely affected by poor nutrition and lack of safe water. Tobacco use and betel nut chewing lead to poor dentition and oral cancers. Other human health issues include sexually transmitted infections, vaccine-preventable diseases, and domestic violence.

Subsistence living places significant pressure upon the forest, which is a source of food and wood. Cultivation and livestock grazing practices also contribute to degradation of the forest and pollution of water resources. Further, hunting of wild animals has led to extinction pressures upon endangered endemic species such as tree kangaroos (*Dendrolagus* spp.), echidnas (*Zaglossus* spp.) and cassowaries (*Casuarius* spp.).

The pressures of remote living in mountainous regions with limited resources increase the necessity of understanding the human impact upon the landscape and has led to taking a unified One Health approach of keeping the health of humans, animals, and their environment in balance.

TREE KANGAROO CONSERVATION PROGRAM AND HUMAN HEALTH

The Tree Kangaroo Conservation Program (TKCP) is a nongovernmental conservation organization based in the Yopno Uruwa Som (YUS) Conservation Area of Papua New Guinea's Huon Peninsula (Chapters 10, 13, and 16). The people of YUS have consistently engaged TKCP to help the communities of YUS improve their general health and improve access to health care. TKCP responded by committing to assess and improve healthcare in the area.

Initial assessment began in 2005 when TKCP brought a team of volunteer physicians and conservation scientists to eight mountain villages (Fig. 15.1). The team of physicians and scientists also met with government health authorities in the provincial capital of Lae and the national capital of Port Moresby (TKCP, 2005). In the villages, TKCP physicians spoke with local healthcare providers and villagers and held clinics in each village. In Lae and Port Moresby, the team met with government officials including political leaders and persons responsible for managing and distributing healthcare throughout the country.

The physicians and scientists found that infrastructure as well as supplies and medication were lacking. Refrigeration for vaccines and radiocommunication in emergencies were nonexistent due to lack of electricity. Vaccination patrols were significantly delinquent. Critical medications were also deficient. There was no capacity to test for infectious diseases which resulted in inconsistent and perhaps both under and over treatment. Dental and eye care were nonexistent.

Additional factors were identified that affected the health of the YUS population including:

- Indoor wood fires used for cooking and heating and the resultant smoke exposure contributed to acute respiratory infections and chronic lung disease (Fig. 15.2).
- There was no indigenous midwife tradition; in fact, women's blood is considered dangerous and women were often expected to deliver alone. Other women did not assist

FIGURE 15.1 Two physicians examine a young child during a health assessment survey in 2005. *Source: Lisa Dabek.*

FIGURE 15.2 Smoke rises through the thatched roof from an open wood fire within the hut. *Source: Robert Liddell.*

FIGURE 15.3 A pig eats food left behind in an area shared by humans and livestock, creating an opportunity for spread of infections between humans and animals. *Source: Robert Liddell.*

out of fear of contamination and becoming ill. Despite undergoing training to be Village Birth Attendants, the lay midwives were not generally utilized by villagers and most had stopped offering their services.
- There was a strong tradition of elected officials directing funding to their own area; a small, less populated area such as YUS saw little government resources for health care as a result.
- Livestock animals, including chickens and pigs, lived in yards shared by people, leading to potential spread of diseases between people and animals (Fig. 15.3).
- Sexual assaults and polygamy were tolerated and resulted in widespread risk of sexually transmitted diseases to include hepatitis B and cervical cancer caused by human papilloma virus.
- Domestic violence was common and there were no programs to address this and little recourse for the persons involved.

The 2005 assessment found that the fundamental resources needed for a healthy population were largely present, including adequate shelter, sanitation, and clean water, though agricultural runoff threatened safe drinking water. Subsistence farming provided basic nutrition supplemented by occasional chickens and pig livestock raised in the villages. However, there were multiple barriers to achieving optimal human health. In addition, local health care workers had only rudimentary training and financial support for health care from the national and provincial governments. Overhunting of wildlife within the forest was leading to local extinction pressures upon wildlife (TKCP, 2005). The information collected led to the realization that a unified approach was needed to address the inter-connected needs of the people living within their remote environment, giving rise to TKCP's Healthy Village, Healthy Forest Initiative.

ONE HEALTH MODEL: THE HEALTHY VILLAGE, HEALTHY FOREST INITIATIVE

The Healthy Village, Healthy Forest Initiative has the goal of connecting the communities of the YUS Conservation Area with improved access to government health resources, while developing local capacity to provide for basic health needs, and helping show the connections human, environmental, and animal health such as clean water, clean air, and healthy forests. This human connection to the environment has been critical to the overall success and mission of TKCP (Ancrenaz et al., 2007; TKCP, 2016).

Activities TKCP Facilitates Through the Healthy Village, Healthy Forest Initiative

- The Initiative facilitates connections between the YUS communities and provincial and national governmental agencies and NGO health resources to support quality health facilities, medication, and vaccine access, appropriately trained community health workers, and use of routine health patrols.
- The Initiative provides essential training to Community Health Workers and Village Birth Attendants (midwives) regarding the prevention and treatment of common illnesses, safe pregnancy and delivery for women and babies, healthy behaviors, and PNG's referral process for the treatment of serious medical conditions.
- Population Health and Environment (PHE) Educators trained by TKCP teach the connections between human health and environmental health to local communities (see section Population Health and Environment" below).
- TKCP and volunteer health professionals lead Health Workshops and Clinics to provide health care and health and environment education to aid residents of YUS in planning their own healthy communities and address issues such as water quality, family health, disease prevention, and nutrition (see sections Health Workshops," Health Education Program," and Health Care Clinics" below).
- Veterinarians and physicians cooperate in the assessment of potential zoonotic diseases of wild tree kangaroos and humans (see section Cooperative Assessment of Zoonotic Diseases in Wild Tree Kangaroos" below).
- TKCP connects with other NGOs such as Marie Stopes International to facilitate contraception and family planning resources for women.
- TKCP staff promote local and national publicity of the success of the YUS Conservation Area in order to increase support from local, national, and international entities and raise awareness of the needs of the YUS community.
- TKCP encourages full staffing of community health facilities by offering to sponsor training scholarships and continuing medical education for health workers who commit to staffing health facilities in their community.

Population Health and Environment (PHE)

TKCP strengthens its Healthy Village, Healthy Forest Initiative by training a network of Population-Health-Environment (PHE) Peer Educators to serve as a valuable source of information within the communities (Fig. 15.4). The volunteer Peer Educators of PHE draw attention to the inter-relationships among populations, health, and environmental dynamics, with the goal of enhancing the well-being of people living in critical biodiversity areas, while also

FIGURE 15.4 Small group session for a Population Health and Environment Education includes a physician, a conservation scientist, and YUS adults. *Source: Ryan Hawk.*

improving the health of the ecosystems upon which these people depend. The volunteers are also responsible for referring community members to the appropriate health facilities for further information and care in the absence of trained health workers.

The growing human population, high maternal death rate, low rates of immunization, low level of health awareness, high prevalence of infectious disease, and poor maternal and child health are the greatest health-related burden in the YUS area. A high percentage of non-functioning medical aid posts, lack of health providers, and the remoteness of YUS contribute to challenges in accessing health services and commodities. In addition, community discussions highlight the issue of food insecurity due to limited ability to hunt tree kangaroos and other animals within the Conservation Area. Women's involvement is lacking in conservation and livelihood activities. Issues impacting women include widespread high-risk pregnancies, in particular teenage pregnancy, large family size, and older mothers having babies. Community members have prioritized the need to address family planning and reproductive health concerns in YUS. The Healthy Village, Healthy Forest Initiative increases awareness among YUS families on family planning/reproductive health and its role in health, nutrition, and environmental conservation. The initiative seeks to improve access to and use of family planning methods and services alongside TKCP's work in conservation, education, livelihood and health.

While the PHE program has been implemented within YUS, the TKCP model and the lessons from implementation present an opportunity for replication in similar areas across PNG. The scale-up is necessary to achieve a meaningful impact that would benefit more people and the ecosystem. Continuous advocacy and promotion of the PHE integrated approach to donors and governments to drive resource allocation and influence policies will enable implementation of similar initiatives in other areas.

Promote Healthy Practices

The remoteness of YUS leads to practices that affect the health of the physical environment, which directly and closely impacts human and animal health and well-being. TKCP promotes healthy practices to maintain a healthy environment for both humans and animals.

An example is the YUS Conservation Area. This was created by the residents of YUS with the guidance of TKCP and the approval of the national legislature (Chapters 10 and 13). Hunting of all animals is prohibited within the defined boundaries of the Conservation Area. Over time, this practice has allowed previously over-hunted game animals such as tree kangaroos to increase in number. In addition to maintaining a strong population within the Conservation Area, this leads to a steady flow of animals venturing outside of the Conservation Area where they can be legally hunted by the residents of YUS, and used as a much needed dietary protein source as well as use of fur for clothing and in traditional cultural celebrations.

Another example is the use of cooking fires. The large majority of structures in the region are thatched-roof huts with a central open fire that is used for cooking as well as for warmth and as a center of social activity. Homes lack ventilation; smoke rises through the thatched roofs, and smoke residue helps to seal the roof against water. Indoor smoke also serves as protection against infestation by insects. While these benefits are significant, the near continuous use of low-efficiency open fires indoors without ventilation results in serious consequences for human health, including high rates of asthma, chronic cough, chronic emphysema, and increased susceptibility to tuberculosis. The fires also impact outdoor air quality for both humans and the animal populations in surrounding areas; and supplies of wood for burning are limited in some areas, so harvesting for fires impacts vegetation. TKCP teams help to educate the local population about the health and environmental impacts of fires and smoke and are investigating opportunities to connect to NGOs working on clean-burning, high-efficiency alternatives to open fires. It is understood that any solution would need to be culturally acceptable and take into account the secondary roles of wood fires on social structure, roof sealing, and insect control.

Other areas where healthy practices are addressed by the TKCP team include management of sanitation, human waste, garbage, clean water, and farming practices.

Health Workshops

Beginning in 2014, TKCP volunteer health teams have visited YUS every 2 years to conduct health education and clinical workshops. Each workshop consists of a combination of health education, provision of health care, and health needs assessments. Frequently, patients travel by foot for as long as several days to present for care, given the scarcity of trained providers and absence of physicians working in this region. International volunteers in the health program have included internal medicine physicians, emergency medicine physicians, family medicine physicians, musculoskeletal specialists, radiologists, and public health specialists from PNG, the United States, and the Philippines (Fig. 15.5). Planned expansion of the program includes the addition of obstetrical specialists, dental providers, nurses and nurse-practitioners, and physical therapists.

Health Education Program

The health education component of each workshop includes presentations on a variety of topics with significant impact on the health of the YUS population. Topics include respiratory health and disease, gastrointestinal issues, sexually transmitted infections (STIs),

FIGURE 15.5 Volunteer physicians, nurses and scientists teach basic human and environmental health concepts to community health workers and volunteer birth assistants. *Source: Robert Liddell.*

neurological issues, nutrition, dental health including impacts of betel nut chewing and tobacco use, smoking cessation, family planning and reproductive health including pregnancy, delivery, and contraception, and others. Education is focused on practical knowledge that can be applied with existing local resources and make a significant population impact. Health education includes sessions for the general village population, focusing on general health knowledge and disease prevention, as well as specialized sessions directed toward health workers, volunteer birth assistants, and volunteer peer health educators. A combination of teaching methods is used including interactive sessions and video presentations as well as didactic teaching.

The volunteer health care providers also teach elementary school children in villages where the Health Workshops are held. Discussions center on how individual health is a part of family and village health, which is directly related to viable forests, safe drinking water and plentiful wild animals for hunting (Fig. 15.6). Students are encouraged to teach others in their family about what they have learned.

Health Care Clinics

Health care clinics are managed in coordination with local health workers and other local volunteers, who help organize the clinic and serve as translators for the volunteer providers. To avoid starting medication courses which cannot be completed locally, the volunteers use

FIGURE 15.6 Elementary school students are taught basic human health concepts by volunteer physicians. *Source: Lisa Dabek.*

medications that are available in the local aid stations rather than bringing medications with them. However, diagnostic equipment including battery operated ultrasound and X-ray are brought to enable improved diagnosis during the clinics. For conditions not amenable to medication management with the aid station formulary, the focus is on diagnosis and referral, with facilitation of referral and transport to a population center when needed. Common diagnoses managed at the clinics include musculoskeletal disorders such as chronic and acute knee and back pain; respiratory conditions including asthma and emphysema related to indoor smoke exposure from cooking fires and tobacco smoking; infectious diseases including tuberculosis, with a variety of presentations including tuberculous lymphadenitis as well as pulmonary symptoms; abdominal and pelvic pain and sexually transmitted infections; menstrual irregularities and contraception management; dental issues including abscesses; and vision and hearing abnormalities including cataracts and ear wax impaction (Fig. 15.7).

Diagnoses made by ultrasound include cirrhosis likely related to chronic hepatitis B, hepatocellular carcinoma, and other cancers. Prenatal ultrasounds have been a prominent feature of every clinic, and often the patients volunteer to allow other villagers to watch the procedure; the opportunity to provide education about fetal development through real-time ultrasound as well as to give information to the patient about her pregnancy has been extremely well received (Fig. 15.8).

Assessment of General Health Needs

In addition to providing education and delivering care, the clinician volunteers use the clinics as well as their other interactions

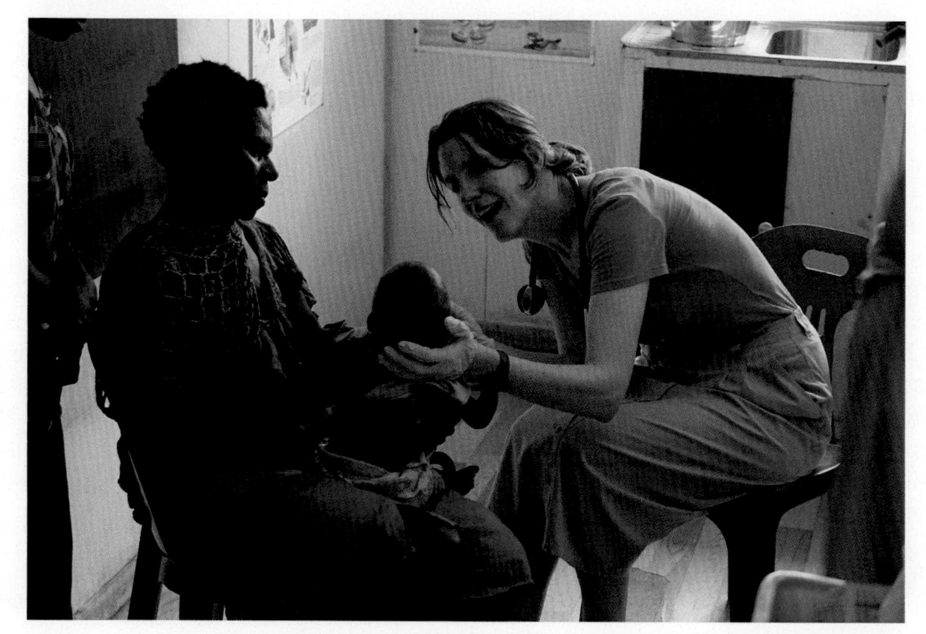

FIGURE 15.7 Volunteer physician assesses a small child in the health clinic within the YUS Conservation Area. *Source: Hartmut Frenzel.*

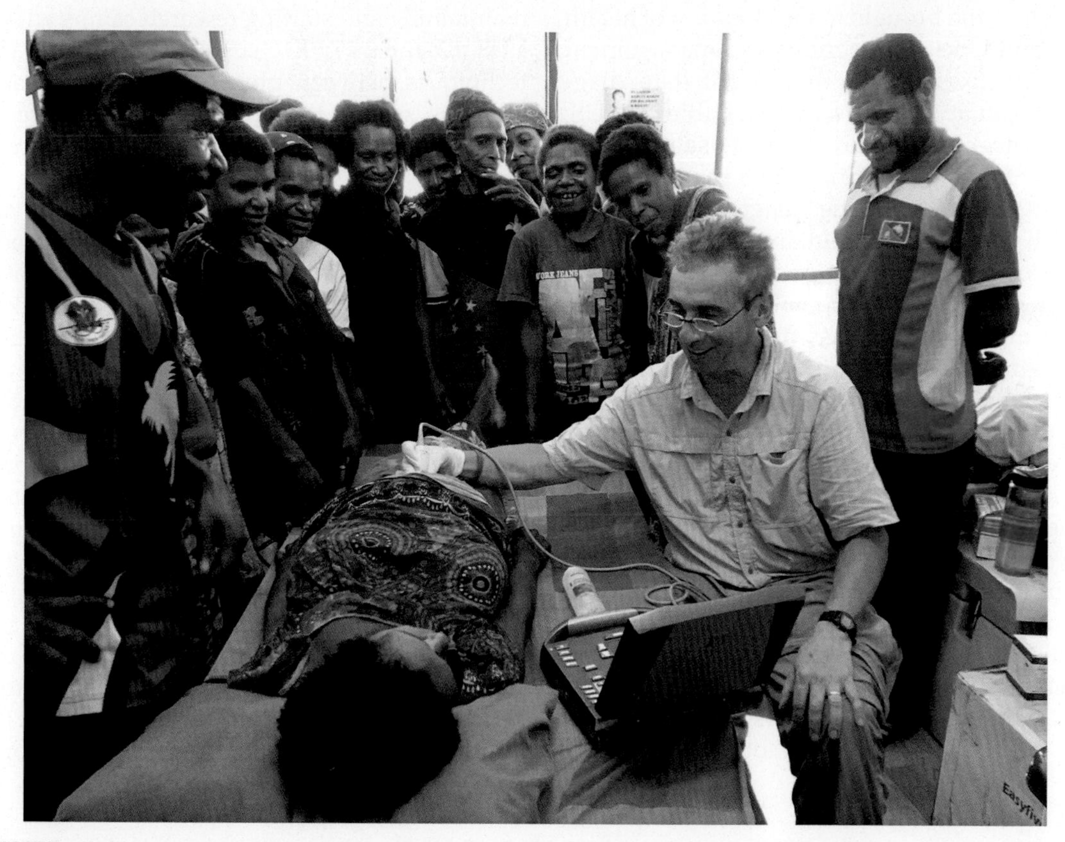

FIGURE 15.8 Volunteer physician uses ultrasound to teach obstetric anatomy to volunteer birth assistants. *Source: Lisa Dabek.*

with local villagers to assess the general health needs of the population (TKCP, 2016). Records are kept regarding presenting symptoms and assessments, to allow analysis of local health patterns. Observations are made regarding the nutritional status and diet patterns of the population, general health status and specific complaints, and opinions and experiences around health and wellness are gathered from villagers. The local experience of health care is also discussed, yielding information about staffing and effectiveness of aid stations, in addition to direct observation of maintenance and supplies (medications, vaccines, vaccine refrigerators with adequate solar batteries for consistent power). This enables general assessment of local health status and needs, which is then conveyed to regional and national health authorities to assist in directing resources and responding to gaps in care.

Particular areas of concern which have been identified and brought to the attention of health authorities include diagnosis and management of tuberculosis (TB). In the current state, diagnostic testing for TB is not available in YUS; prevalence is high; and the expense and difficulty of travel to an urban center prevents most patients from accessing diagnostic services. Within the region, certain areas have access to tuberculosis medication, and staff may initiate empiric treatment of tuberculosis based on clinical symptoms. However, the appropriateness of treatment is not confirmed and courses of treatments are often interrupted or incomplete, raising the risk of developing antibiotic resistance which can further threaten the health of the local population as well as spread TB via travel to cities and more distant populations. The need for local availability of appropriate diagnostic testing, and for the application of treatment and monitoring standards to patients in this region, has been discussed and emphasized with regional and national authorities. TKCP international volunteers are also exploring the possibility of other resources (nonprofits, global health programs) to improve TB care in the region.

Another major issue has been vaccine delivery in YUS. In theory, PNG has a national program ensuring availability and delivery of standard vaccines. In practice, there have been several challenges to maintaining equipment and consistently provide vaccines. TKCP health volunteers have worked with regional health authorities to advocate for resolving these issues. After the initial assessment in 2005, TKCP provided solar refrigerators in the main health centers in YUS as part of a collaborative effort with the province to support the immunization program.

Cooperative Assessment of Zoonotic Diseases in Wild Tree Kangaroos

TKCP supports research into the biology and health of Matschie's tree kangaroos, to include maintaining and staffing research camps within YUS (Chapter 25). Exchange of knowledge and techniques between physicians and veterinarians has served to further the knowledge of the biology and health of wild tree kangaroos assessed at TKCP research camps.

Matschie's tree kangaroos living in zoos are susceptible to a mycobacterium which infects lungs, other soft tissues, and bones (Travis et al., 2012). As humans living in Papua New Guinea have a significant incidence of infection by the tuberculosis mycobacterium (Aia et al., 2018), it has been hypothesized that a zoonotic relationship of mycobacterial infections could be shared between humans and tree kangaroos in YUS. Zoonotic mycobacterial infections have been demonstrated between several animal species and humans (Vogelnest, 2013).

Radiographic and ultrasound equipment brought to YUS to assess for TB in humans was transported to the tree kangaroo research camp where temporarily-captured wild tree kangaroos were radiographed to assess for pulmonary infections, soft tissue infections,

FIGURE 15.9 Veterinarian and physician team uses battery-powered digital radiography to assess tree kangaroo health in the Wasaunon TKCP research camp. *Source: Lisa Dabek.*

and bone infections (osteomyelitis) (Fig. 15.9). Of eight tree kangaroos assessed, none had any evidence of past or ongoing mycobacterial infections involving the lungs, soft tissues, or bones (Fig. 15.10). No evidence of infection was found in wild tree kangaroos, and therefore there is no proof from these assessments of a zoonotic relationship between humans and tree kangaroos for mycobacterial infections.

One of eight animals had ultrasound findings of small bowel endoparasitic worms (helmiths spp.), though the type of helminthic species is unknown at this time. Further, it is

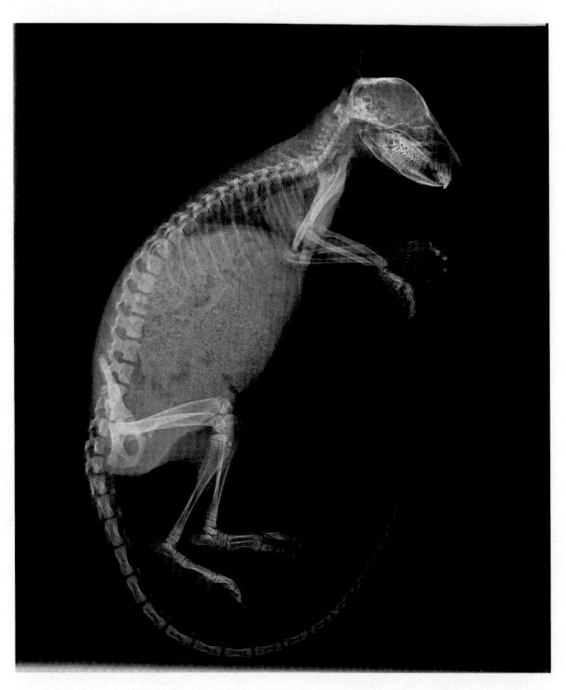

FIGURE 15.10 Lateral radiograph of a Matschie's tree kangaroo obtained at Wasaunon camp in YUS. The exam is normal and reveals no evidence of a pneumonia or osteomyelitis (bone infection). *Source: Robert Liddell.*

unknown at this time if the symbiotic relationship is a parasitic, commensal, or mutual relationship.

Other Programs Linking Human Health to Wildlife and Environmental Health

Health in Harmony, an Indonesian-based health program, directly connects the health of people to the health of animals and forests (Health in Harmony, 2020). Former illegal logging of the forest has been replaced with sustainable farming practices, encouraged by the availability of on-going health care provided by both Indonesian and international health care providers. The health care facility is a center-piece of the community, not only providing health services, but also serving as a place of education of conservation best practices. TKCP has learned from Health in Harmony's success by understanding how human health care is vitally important to the well-being for individuals, families and entire communities.

Blue Ventures is based in remote and underserved Madagascar fishing communities, where there is a severe lack of access to basic health care. The rapidly growing human population driven by coastal migration and unmet family planning needs contributes to food insecurity and worsening poverty. Long term family planning taught by peer educators helps couples to meet immediate needs of their families, allowing them to work and earn more while learning sustainable fishing practices and alternative livelihoods. Restriction of fishing in selected areas has led to improved fishing yields in adjacent waters which previously had produced very little harvest (Blue Ventures, 2020). This practice is similar to that enacted within the YUS Conservation Area where animals are not hunted, allowing a resurgence of game animals to become available for hunting once the animals migrate out of the Conservation Area. TKCP also uses the peer education model to help people understand the connections between the food sources and human health.

LOOKING AHEAD

The key path to success for One Health in the YUS Conservation Area and throughout Papua New Guinea is for those living in remote regions to work with local, provincial, and national governments to ensure appropriate supplies, infrastructure, and personnel to provide health care and to link human health to the health of animals and their shared environment.

TKCP plans to expand the Healthy Village, Healthy Forest Initiative by expanding collaborations with established providers of health care, to include medical and nursing schools. The need to understand nutritional needs will require collaboration with nutritionists as well as expanded expertise in agricultural practices. Durable mechanisms to store and deliver safe drinking water will decrease water-borne infectious diseases. Alternatives to open wood fires in closed spaces will be made available to prevent inhalational diseases.

It is hoped that the concepts of the Healthy Village, Healthy Forest Initiative will be implemented by people in other remote regions in Papua New Guinea. Successful implementation will require the support of local, provincial, and national government agencies. The national government will best serve the country by creating a unifying One Health agenda to optimize the health of its people and animals within the country's complex and beautiful landscape.

The people of YUS have learned that a One Health approach has improved their own health, forest health, safe water resources, and increased the abundance of tree kangaroos and other wild animals. Understanding and living within the complex interrelationships of humans, animals and the landscape has led to a sense of pride for the people of YUS, and brings deserved admiration from others in Papua New Guinea and the rest of the world.

ACKNOWLEDGMENTS

This chapter is dedicated to Dr. Blair Brooks, 1953–2019, one of the founders of the TKCP health program. His passion for serving the people of YUS inspired the involvement of many others, and his infectious joy, curiosity, and enthusiasm permeate the spirit of the health program.

REFERENCES

Aia, P., Wangchuk, L., Morishita, F., Kisomb, J., Yasi, R., Kal, M., Islam, T., 2018. Epidemiology of tuberculosis in Papua New Guinea: analysis of case notification and treatment-outcome data, 2008–2016. Western Pac. Surveill. Response J. 9 (2), 9–19. Available from: https://pubmed.ncbi.nlm.nih.gov/30057853/.

Ancrenaz, M., Dabek, L., O'Neill, S., 2007. Recognizing a role for local communities in biodiversity conservation. PLoS Biol. 5 (11), 2443–2448. Available from: https://www.ncbi.nlm.nih.gov/pmc/articles/PMC2039772/pdf/pbio.0050289.pdf.

Blue Ventures, 2020. Available from: www.blueventures.org (17 July 2020).

Centers for Disease Control and Prevention (CDC), 2020. One Health. Available from: https://www.cdc.gov/onehealth/index.html (17 July 2020).

Health in Harmony, 2020. Available from: https://healthinharmony.org/ (17 July 2020).

Rabinowitz, P.M., 2017. One health for Papua New Guinea. Unpublished manuscript. .

Rabinowitz, P.M., Pappaioanou, M., Bardosh, K.L., Conti, L., 2018. A planetary vision for one health. BMJ Global Health 3, e001137. Available from: https://www.ncbi.nlm.nih.gov/pmc/articles/PMC6169660/.

Travis, E.K., Watson, P., Dabek, L., 2012. Health Assessment of free-ranging and captive Matschie's tree kangaroos (*Dendrolagus matschiei*) in Papua New Guinea. J. Zoo Wildlife Med. 43, 1–9.

Tree Kangaroo Conservation Program (TKCP), 2005. 2005 Annual Field Report. Available from: https://www.zoo.org/document.doc?id=1719 (17 July 2020).

Tree Kangaroo Conservation Program (TKCP), 2016. Annual Report 2016. Available from: https://www.zoo.org/file/visit-pdf-bin/TKCP-2016-Annual-Report-web.pdf (17 July 2020).

United Nations Food and Agriculture Organization (FAO), 2020. AGA News – FAO as One Health: Moving Forward. Available from: http://www.fao.org/ag/againfo/home/en/news_archive/2011_FAO_as_One_Health.html (17 July 2020).

Vogelnest, L., 2013. Tuberculosis: an emerging zoonosis. NSW Public Health Bull.. 24 (1) Available from: https://www.publish.csiro.au/NB/pdf/NBv24n1#page=32 (17 July 2020).

World Health Organization (WHO), 2020. One Health. Available from: https://www.who.int/news-room/q-a-detail/one-health (17 July 2020).

World Organisation for Animal Heath (OIE), 2020. One Health. Available from: https://www.oie.int/en/for-the-media/onehealth/ (17 July 2020).

FURTHER READING

Lerner, H., Berg, C., 2017. A comparison of three holistic approaches to health: One Health, EcoHealth, and Planetary Health. Front. Vet. Sci. 4, 163. https://doi.org/10.3389/fvets.2017.00163 . Available from: https://www.ncbi.nlm.nih.gov/pmc/articles/PMC5649127/.

CHAPTER

16

Building Conservation Leadership in Papua New Guinea for Tree Kangaroo Conservation

Mikal Eversole Nolan[a], *Danny Samandingke*[a], *and Timmy Sowang*[b]

[a]Lae, Papua New Guinea

[b]Tree Kangaroo Conservation Program, Lae, Papua New Guinea

OUTLINE

INTRODUCTION

In Papua New Guinea (PNG), 97% of the land is owned and controlled by the customary landowners (AusAid, 2008). This land is locally governed by traditional rules for use, access, and the transfer of rights. As the Constitution provides for the protection of property rights "as a special right of citizens", customary land cannot be sold or transferred to external entities. As such, the success of conservation initiatives depends upon the continued commitment and stewardship by the local communities.

Rural men and women living near forests depend heavily on resources derived from these areas for subsistence and livelihood opportunities (Fig. 16.1). This reliance impacts the sustainability of fauna and flora, as evidenced by the primary threats to biodiversity being localized pressures such as over-hunting and habitat destruction through forest clearing (Papua New Guinea Department of Conservation, 2014). With over 85% of the PNG population living in rural areas (The World Bank, 2018), coupled with a growing population, the pressures on local natural resources will likely only

Tree Kangaroos
https://doi.org/10.1016/B978-0-12-814675-0.00030-0

FIGURE 16.1 Traditional village life for rural men and women living within the YUS Conservation Area Landscape.

intensify. In order to mitigate the negative impacts to biodiversity, developing strong local leadership and conservation advocates must be a priority.

Few conservation efforts in Papua New Guinea have had an enduring positive outcome. The history of failed conservation projects may be rooted in external interests dominating the planning and decision-making, negatively affecting the sense of ownership of the initiative (Beehler and Kirkman, 2013). Understanding the local context and influence of the formal and informal institutions within the project area increases the likelihood of successful conservation efforts. Inclusion of local institutions, from inception and throughout the project lifecycle, will likely have positive outcomes. The formal laws, rules, and regulations supporting conservation perform best when developed to reflect the local culture and informal rules of society. Importantly, the strategic engagement of community leaders provides key entry points for accessing and influencing the appropriate individuals and groups for driving local decision making that will best conform to the local institutions. As community members are engaged and empowered as decision makers, they will assume leadership roles that benefit both the community and the conservation program.

Building local leadership to guide and direct institutions has proven to be a critical tool for exerting influence for successful conservation programs. As with global initiatives, achieving long-term conservation outcomes in Papua New Guinea must be rooted in the exploration and understanding of the different roles, responsibilities, and challenges that men, women, and youth face within each community. External partners should take into consideration the various ways community members use their natural resources, societal roles and responsibilities, and the future impact they may have upon the landscape. The identification of values, by whom and for what purpose, represent prerequisite knowledge for appropriate planning and development of conservation programs. By understanding the local context, the project can help to alleviate threats, implement appropriate land-use planning, and effectively facilitate a sustainable conservation initiative.

The Tenkile Conservation Alliance (TCA) and Tree Kangaroo Conservation Program (TKCP) represent two organizations in Papua New Guinea that leverage enhanced biodiversity protection and long-term conservation outcomes through the attention afforded to threatened species identified on the IUCN Red List (IUCN, 2020), specifically tree kangaroos. Established as independent non-governmental organizations, the TCA and TKCP similarly incorporate and strive to develop local leadership capacities as part of their respective missions. Funded through contributions from a variety of partners, including core support from a collection of global zoos (Chapter 22), both organizations implement holistic programing to foster enhanced biodiversity protection as well as address the well-being of the communities who serve as environmental stewards. Having been individually founded in the late 1990s, TCA and TKCP have accumulated decades of experience working directly with rural communities (Chapter 13). The continued presence of these organizations is attributed to having sought first to understand and then to cultivate local institutions and people. Their experiences demonstrate that by developing local leadership through the strengthening of organizational capacity, providing educational and capacity building opportunities, addressing gender disparities, and enhancing local competencies to advocate for conservation outcomes, locally championed conservation efforts possess the ability to make an enduring impact for tree kangaroo conservation.

STRENGTHENING OF ORGANIZATIONAL CAPACITY

Successful community conservation programs require effective and transparent organizations which understand the institutional environment that shapes the community's behavior, perception, and choices that will ultimately sway performance and outcomes of the conservation goals. Recognizing the various kinds of formal and informal institutions, as well as the diverse leadership required, remains central to effective conservation initiatives. Local organizations necessitate home-grown leadership and empowered communities to formulate preferences and assume decision making roles. The acknowledgment of the indigenous cultural beliefs and norms, role of local interest groups and political entities, and existing internal conflicts (such as land rights, clan disputes, etc.) cannot be underestimated as these will influence the mechanisms and entry into organizational strengthening. The TKCP seeks to build a strong network of actors including a non-government organization (NGO), community-based organization (CBO), agricultural cooperative, and government partners. Each of these entities require investment to foster effective and transparent operations with strong local leadership and conservation advocates.

Horwich and Lyon (2007) have observed that central to successful community conservation projects, are the CBOs and institutions managed by empowered local people and communities. The TKCP, based in the Yopno-Uruwa-Som (YUS) area of Huon Peninsula in Morobe Province, encourages conservation leadership through various local outlets including the YUS Conservation Organization (YUS CO), a community-based organization whose representative body functions as environmental stewards and advocates for biodiversity within their communities. The YUS CO was established in 2009 to collectively organize landowners from various clans, wards, and regions across the

YUS Conservation Area Landscape, a 158,271 ha area that is co-managed as a protected area and an area for sustainable use (Chapter 10 and Chapter 13). The TKCP's support during development and establishment of the YUS CO has been influential in cementing their relationship for achieving common goals within the YUS Conservation Area Landscape.

The role of the TKCP is to facilitate community engagement, provide technical assistance and training, secure funding for implementation of the landscape management plan, and perform the functions of monitoring and evaluation. CBOs, such as the YUS CO, provide a platform for local people to self-organize and affect decisions, which stimulates overall effectiveness of the conservation program (Eklund and Cabeza, 2017). The YUS CO also fulfills an advisory function ensuring conservation and community projects are relevant and appropriate. As the two organizations each serve important roles in the management and governance of the protected area, strong and collaborative leadership is critical. Sustaining a culture of respect, transparency, and trust, between and within the two organizations, guides constructive co-management and implementation of the landscape management plan.

Contemporary conservation approaches acknowledge the influence of governance as it signifies who exercises power and decision-making authority (Borrini-Feyerabend and Hill, 2015). Societies throughout Papua New Guinea exhibit a preference toward consensus decision making. Strong conservation leaders thus serve as critical influencers within governance models. In order to exert the decision making role, the YUS CO holds one Annual General Meeting (AGM) where representatives gather to discuss management, program implementation, monitoring and challenges of the protected area (Fig. 16.2). Prior to the AGM, members hold local (ward-level) and regional (zone-level) meetings to collect input from various stakeholders who may not be able to travel the distances required to personally attend the AGM. This model promotes the inclusion of good governance principles such as participation, openness, and responsiveness. Under the scope of institutional strengthening, the advancement of these principles of good governance occupy a realm equivalent in value to that of the actual outcomes of the decision-making process. Once the YUS CO realizes the full capacity to exercise authority, the TKCP hopes their leadership will be best positioned to champion and influence consensus for the benefit of biodiversity and sustainable development.

Beyond NGOs and CBOs, local livelihood and socio-economic institutions can be utilized to maximize the value of natural resources and biodiversity to landowners (Chapter 14). Livelihood and socio-economic advancement can produce a negative impact on biodiversity if not addressed as an integrated component of the conservation project. Neglect of sustainable livelihood options perpetuates poverty and may led to increased pressure on local natural resources. Unless natural resources are managed by and for the benefit of local people, they risk being replaced by less sustainable alternatives. Engaging local people in effective home-grown institutions mitigates the threat of local people being marginalized from their local resources.

The TKCP model supports the removal of obstacles for the quality production of coffee and cocoa in livelihood zoned areas, market integration, and handover of management to local producer organizations who maintain control over their natural resources. The TKCP has assisted the local producer organizations to unite under The YUS Conservation Coffee and Cocoa Cooperative (Chapter 14). As a component of institutional capacity building, the cooperative leadership participated in a process to establish inclusive membership guides, appropriate structures, systems, and policies. Sound financial models assist leaders in making informed decisions and demonstrate transparency to stakeholders. A clear linkage between biodiversity and economic incentives encourages local leaders to assume responsibility for

FIGURE 16.2 Women perform a traditional welcome ceremony for the YUS Conservation Organization's Annual General Meeting (AGM).

on-the-ground monitoring and sustainable management of natural resources. Leadership from within the producer cooperative led a successful bid to utilize the project as an additional mechanism to enforce the YUS Conservation Area Bylaws. These nationally gazetted bylaws prohibit activities within the protected area such as the hunting of endangered species, the destruction of plants, and igniting fires amongst others (Chapter 13). Under their remit any person found in violation of a YUS Conservation Area Bylaw will be ineligible to sell their commodities under the conservation program until the associated penalty is satisfied. Across the YUS Conservation Area Landscape, leadership under the producer groups have facilitated greater support for conservation efforts, enforce protected area rules and penalties, and enhanced the productivity of already-cleared land, thereby mitigating the need to clear existing forests for agriculture which threatens the sustainability of the protected area.

The Papua New Guinea Conservation Areas Act (PACII, 1978) imparts oversight of protected area management to a Conservation Area Management Committee (CAMC), who is mandated to represent the interests of local landowners, as well as the local and provincial government authorities. In the case of the YUS Conservation Area Management Committee, the landowner representatives, as elected from the membership of the YUS CO, possess the strongest voice with one-third of the voting rights. Other members of the YUS CAMC include a representative from local, provincial, district, and national authorities, as well as TKCP. The devolution of authority to local landowners allowed the communities to establish the rules and penalties of the protected area which were endorsed by the CAMC and ultimately supported by PNG's Minister for the Environment. Under the Conservation Areas Act (PACII, 1978), the CAMC is empowered to undertake the following functions:

(a) to manage the conservation area; and
(b) to make recommendations to the Minister on the making of rules applicable within the conservation area; and
(c) to advise the Minister in respect of coordination of development within the conservation area; and
(d) to prepare a management plan for the conservation area outlining the manner in which land use will be managed and features of special significance conserved; and
(e) to direct the work of rangers.

The co-management of the YUS Conservation Area has proven beneficial in furtherance of wildlife and habitat protection through the various levels of authority, providing specific technical knowledge and as an integrated mechanism for monitoring. While the CAMC oversees the implementation of the management plan, the customary landowners continue to serve as the direct day-to-day managers of the protected area. The landowners believe that the powers imparted by the legal and formal institutions have elevated the recognition and support of their custodianship of the YUS Conservation Area.

The YUS Conservation Area Rangers blends the formal legal requirement for monitoring the protected area with a process of selection by the traditional leadership of the local community (Chapter 12). With Rangers recruited from clans who pledge land for indefinite protection, their leadership supports the development of strong environmental custodians to assume control over protection of biodiversity such as the tree kangaroo. While the activities of the Rangers are formally mandated to the CAMC, the TKCP undertakes the role to build their capacity to effectively implement their roles and responsibilities. The TKCP implements responsibility through on the job training for developing a supported and confident work force. Rangers must demonstrate the skills to understand specific scientific knowledge, analyze and synthesize conclusions from the information they collect. In addition to the scientific information, soft skills are equally important. Written and verbal communication, interpersonal skills, teamwork, listening, problem solving, time management, and ability to exert influence are examples of essential soft skills. On the job, Rangers use technology such as GPS (Global Positioning System) and SMART (Spatial Monitoring and Reporting Tool) enabled devices to monitor and collect information regarding tree kangaroos and other target species. An effective Ranger also requires the capacity to accurately share monitoring data and patrol information with community members and institutions to make informed decision for best conservation outcomes. The ability to earn and keep community trust, especially within the PNG context, serves as an essential skill when working with diverse stakeholders, cultures, and traditional clans.

EDUCATION AND YOUTH DEVELOPMENT

Papua New Guineans embrace a strong cultural history of environmental awareness, respect, and protection of their natural resources. Many rural people continue to value the traditions and customs passed down from generation to generation with the hope that the young people within their communities become empowered individuals who will lead through future conservation stewardship and sustainable community development. Yet, as young people leave their rural homes to pursue educational opportunities or careers in the larger urban areas, they risk losing this part of their culture. Experience has shown that when these young people return to their rural communities, there is a possibility that they bring back negative influences that undermine community values and change existing systems of local leadership (TKCP, 2016).

From an anthropological perspective, Slotta (2014) suggests the Yopno peoples' social interactions and standing within the community has been significantly influenced through the engagement with international partners and institutions. He describes how churches, schools, and non-governmental organizations have come to be seen as sources of foreign, "esoteric knowledge" that confers those with access enhanced status and power within the community. This has led to changes in the traditional power relations at the village level. Young people with access to sources of foreign knowledge, now assume greater authority over decision making within traditional institutions. As the rural communities become more globalized, it is important to acknowledge societal changes, seek to mitigate unintended negative consequences, and appropriately engage young people.

Importantly, the young people of today hold the key to tomorrow's biodiversity. Early engagement in conservation discussions set their perceptions and expectations. With 97% of land nationally deemed customary, clans maintain rights over their natural resources in perpetuity with inheritance predominantly through patrilineal descent (AusAid, 2008). Customary land cannot be "locked-up" by the government or other privatized means for conservation as with traditional "fortress" conservation models. In the YUS Conservation Area, land is voluntarily pledged and remains under the clan's ownership with community members participating in the governance and management of the protected area. Thus, for the long-term conservation of nature, the future landowners and custodians must continue to support the conservation efforts and continue to hold regard for the land pledges of the clan elders.

Recognizing the need to support community youth education to build future capacity, the TCA and TKCP have sought to incorporate a variety of approaches, both through formal educational institutions and informal capacity development programs. Their approaches seek to blend scientific knowledge with local relevance. This aligns with the global experience giving indigenous knowledge equal importance for environmental decision making. Müller et al. (2015) describe the relevance of conveying the science of phenomena such as climate change to local people whose traditional knowledge systems may not be equipped with management systems.

With a focus on the formal education system, the TCA has developed a supplemental teaching manual to provide local teachers with "sufficient background information to deliver conservation education within their classroom both now and in the future, with the aim of empowering the students to help 'make a difference' in preventing the decline of PNG's unique flora and fauna" (TCA, 2006). As science and environmental education is under-developed across much of Papua New Guinea, better support for these program areas may positively contribute to conservation efforts and ultimately result in the improved status of tree kangaroos.

From conversations with hundreds of children, youth, and community members, the TKCP has heard a clear interest in opportunities that would foster the development of skills which would enable them to become positive members and leaders in their own communities. Yet as education levels in the communities of the YUS Conservation Area are overall lower than the national and provincial averages with some exceptions, educational institutions underperform in conveying these skills. Few students who commence their formal education continue through to completion. On average less than 9% of youth who complete grade 6 continue on to complete grade 10, a contrast to national (25.2%) and Morobe Province (24.3%) retention rates (TKCP, 2012). The disparity may be at least partially attributed to a lack of a high school within the area thus compelling students who have adequate grades and financial backing to attend secondary school in the urban centers. A number of other factors may also include an inadequate supply of teachers for primary school instruction, insufficient grades, and the lack of funds for school fees. The low level of education overall, coupled with the critical under-representation of girls in both primary and secondary school for various social and cultural reasons, restricts the effectiveness of educational institutions.

With limited access to formal education, young people are hungry for knowledge and skills to position them for future leadership opportunities. The TKCP's Junior Ranger Program aims to develop a strategic program to support positive youth development and leadership through a community driven and facilitated model (Chapter 22). By providing networked education, stewardship, and leadership opportunities for youth from pre-school through young adulthood, the program hopes to directly link outcomes to support the core vision of "a sustainable, healthy and resilient Huon Peninsula landscape which supports the area's unique biodiversity, human communities, and culture" (TKCP, 2012). The Junior Ranger Program trains community teachers, local volunteers, and mentors to support the suite of Junior Ranger programs to ensure that these opportunities will be supported and sustained by each community, while TKCP staff provides training and guidance to ensure their ongoing success (Fig. 16.3). This model seeks to fill the gaps left by inadequate formal education.

Lessons are designed to encourage Junior Rangers to directly think about leadership within their communities, as well as the community links to biodiversity. Activities are varied from map reading and linking locally relevant species with their habitats, to exploring career opportunities and conducting interviews with local leaders about traditional knowledge. The program seeks to expose young people to research taking place within the landscape and the monitoring of biodiversity being done by local YUS Conservation Area Rangers. These concepts are linked back to identify local threats and recognizing their personal role, present and future, in threat reduction. The Junior Ranger program seeks to balance western science with traditional knowledge. The TKCP is optimistic that this approach will positively contribute to building future environmental stewardship and strong conservation leaders.

ADDRESSING GENDER INCLUSIVITY

Women and men have distinctive needs and priorities, experience different threats, and hold a diverse array of interests regarding their natural resources. Their experiences, knowledge, and capacity not only influence conservation outcomes, they shape adaptive management strategies and effectiveness. The 2017 assessment for protected area management in Papua New Guinea revealed an under representation of women and youth in management decision making. Despite attempts to engage them, generally women's attendance, participation, and input during workshops and meetings was minimal (Leverington et al., 2017). Developing women's leadership potential and ensuring appropriate opportunities to participate in decision making bodies encourages better representation of community views.

A gender needs assessment (TKCP, 2017), conducted by the TKCP, confirmed that across the YUS Conservation Area Landscape, as throughout much of mainland Papua New Guinea, land is held under patriarchal dominated customary tenure which affords women limited rights and control over land use decisions. In many of the enduring traditional societies across the country, men are granted a higher status than women within their communities. This resulting

FIGURE 16.3 Future conservation leaders.

inequality not only influences the roles within households and the community, it leaves women more vulnerable to adverse economic and environmental changes. This may be exacerbated when women's voices are neglected in land-use planning and management decision making, thus failing to recognize their demand and need for natural resources. A lack of representation and participation can lead to underreported stress on the local environment, as well as aggravated poverty and negative impacts on the wellbeing of women and youth. Including women's knowledge and priorities in planning and decision making enhances the likelihood that the priorities of the entire population are represented, in turn fostering more sustainable resource use and long-term conservation outcomes (Fig. 16.4).

Across the YUS Conservation Area landscape, women commonly experience greater obstacles than men in engaging in monitoring and research activities due to their traditional household and family roles and responsibilities, safety and security concerns, perceived lack of knowledge of more remote areas of the landscape, and potential conflict in relationships with their husbands (TKCP, 2017). These are in addition to the technical and education barriers that both men and women face. Women who hold an interest in monitoring and research activities can add unique value to the project. As women are often community knowledge keepers of indigenous medicinal plants and other non-timber forest products, neglecting to engage them in monitoring activities could result in valuable information and biodiversity being lost. Alleviating obstacles to women's participation will positively contribute to responsive and adaptive management of local biodiversity.

FIGURE 16.4 Women resource users must be engaged in environmental decision making. *Source: David Gillison.*

Even when women's knowledge and priorities are recorded, their inclusion in decision making is not guaranteed. Experience from within the global community suggests that the inclusion of women on decision making bodies diminishes the likelihood of neglecting their knowledge and priorities. Furthermore, while one or two women may have negligible impact on outcomes, the participation of three or more women on decision making bodies may substantially increase the probability of adequate representation of women's priorities factored into decisions (TKCP, 2017). Understanding local customs and institutions will assist in establishing equitable and appropriate participation.

The TCA has structured the program's stakeholder and advisory committees to achieve equal representation of men and women. The TCA Representative Committee is comprised of one male and one female from each of the 50 villages represented under the program's scope, in order to foster a bottom-up approach to conservation and development. Similarly equal, the TCA's Advisory Committee consists of 16 elected members, eight males and eight females, whose role is to bring community perspectives to the TCA Board of Directors supporting a representative governance model (TCA, 2016). These models represent an important step forward for the inclusion of women. Equal representation may afford women and men comparable opportunities for voicing priorities and concerns, thus supporting an inclusive approach to stakeholder engagement.

While such forms of representative governance models seek to establish equal membership from the community, participatory governance models may present greater opportunity for inclusion of marginalized women and youth. These models often involve whole communities thus fostering greater ownership of the program and outcomes. Program models should be adapted to ensure women have meaningful participation and that barriers are mitigated. Employment of culturally appropriate gender mainstreaming tools may assist in collecting women's voice and priorities. Naturally, leadership development and empowerment of women must transpire as a process, evolving over a period of time, as well as being sensitive to the local culture, customs, and institutions. Despite a lengthy and culturally specific process, the long-term benefits of providing space for women to have a voice and meaningful participation will certainly advance collective conservation goals.

ADVOCACY

Horwich et al. (2015) expresses that "when conservation practitioners act as catalysts rather than project owners, communities respond favorably and contagion and community activism can emerge." Leaders think, talk, and act in a way that pulls people together to exert influence over their situation. Developing local leadership strengthens the communities' ability to deepen strategic partnerships (NGO, CBO, socio-economic, government, etc.) and assume roles as advocates for improved services and enhanced protection of local natural resources.

Effective governance structures involve positive connections between decision makers at various institutional levels, from the local and regional to national policies and international conventions. The grassroots' voice cannot be under-valued as direct stewards of their environment. Under the TKCP model for the YUS Conservation Area, the YUS CO is strategically positioned as a forum for local and regional leaders to share lessons, discuss common concerns, and collectively seek solutions for challenges that occur beyond the individual reach. The power of YUS CO exists in collective action by and for local resource owners and users. The support of networks such as this reinforce a united community voice and call to action as direct custodians, decision makers, and beneficiaries of conservation programs.

As programs seek to empower and develop individual leadership, the group and community will also be continually strengthened. Face-to-face encounters set positive examples and propel others to action. While comparative estimates are challenging, it is arguable that the development of local leadership and advancement of strong advocates, may contribute to less need for external funding in the long term. As conservation in Papua New Guinea has demonstrated to be a costly endeavor due to poor infrastructure and challenging transportation, mitigating reliance on external partners while supporting local action may lower costs and encourage sustainable long-term programs. Collaborative networks and advocacy provide communities with the ability to relate and attract greater support and resources.

CONCLUSIONS

In Papua New Guinea, the longevity of tree kangaroo conservation depends upon the local stewards and landowners of conservation programs. Investing in local leadership development strengthens both the individual and the institution. Developing capacity and opening doors to greater involvement of men, women, and young people to engage in decision making fosters a holistic approach to natural resource management, representing a variety of needs and priorities. For decades, the TCA and TKCP have fostered locally championed conservation efforts with strong advocates for the protection of biodiversity. The long-term protection of tree kangaroos depends on continued partnerships and leadership by the indigenous communities entrusted as the local custodians.

REFERENCES

AusAid, 2008. Land Tenure Systems in the Pacific. Making Land Work. Reconciling Customary Land and Development in the Pacific, vol. 1. Australian Government, Canberra.

Beehler, B., Kirkman, A. (Eds.), 2013. Lessons Learned From the Field: Achieving Conservation Success in Papua New Guinea. Conservation International, Arlington, VA.

Borrini-Feyerabend, G., Hill, R., 2015. Governance for the conservation of nature. In: Worboys, G.L., Lockwood, M., Kothari, A., Feary, S., Pulsford, I. (Eds.), Protected Area Governance and Management. ANU Press, Canberra, pp. 169–206.

Eklund, J., Cabeza, M., 2017. Quality of governance and effectiveness of protected areas: crucial concepts for conservation planning. Ann. N. Y. Acad. Sci. Available from: https://nyaspubs.onlinelibrary.wiley.com/doi/pdf/10.1111/nyas.13284.

Horwich, R.H., Lyon, J., 2007. Community conservation: practitioners' answer to critics. Oryx 41 (3), 376–385. Available from: https://doi.org/10.1017/S0030605307001010.

Horwich, R.H., Shanee, S., Shanee, N., Bose, A., Fenn, M., Chakraborty, J., 2015. Creating modern community conservation organizations and institutions to effect successful forest conservation change. In: Zlatic, Miodrag (Ed.), Precious Forests—Precious Earth. IntechOpen Available from: https://www.intechopen.com/books/precious-forests-precious-earth/creating-modern-community-conservation-organizations-and-institutions-to-effect-successful-forest-cohttps://doi.org/10.5772/61133.

IUCN, 2020. The IUCN Red List of Threatened Species. Version 2020-1. Available from: https://www.iucnredlist.org (4 June 2020).

Leverington, F., Peterson, A., Peterson, G., Jano, W., Sabi, J., Wheatley, A., 2017. Assessment of Management Effectiveness for Papua New Guinea's Protected Areas 2017. SPREP, Apia.

Müller, E., Appleton, M.R., Ricci, G., Valverde, A., Reynolds, D., 2015. Capacity development. In: - Worboys, G.L., Lockwood, M., Kothari, A., Feary, S., Pulsford, I. (Eds.), Protected Area Governance and Management. ANU Press, Canberra, pp. 251–290.

Pacific Island Legal Information Institute (PACII), 1978. Papua New Guinea Consolidated Legislation: Conservation Areas Act 1978. Available from: http://www.paclii.org/pg/legis/consol_act/caa1978203/ (27 June 2019).

Papua New Guinea Department of Conservation, 2014. Papua New Guinea's fifth national report to the Convention on Biological Diversity. Available from: www.cbd.int/doc/world/pg/pg-nr-05-en.pdf (27 June 2019).

Slotta, J., 2014. Revelations of the world: transnationalism and the politics of perception in Papua New Guinea. Am. Anthropol. 116 (3), 626–642. Available from: https://doi.org/10.1111/aman.12114.

Tenkile Conservation Alliance (TCA), 2006. Teacher Training Manual: Science and Conservation. Available from: www.tenkile.com/documents/TCA-Teacher-Training-Manual.pdf.

Tenkile Conservation Alliance (TCA), 2016. Tenkile Conservation Alliance annual report. Available from:https://www.tenkile.com/documents/TCA%202016%20Annual%20Report.pdf.

The World Bank, 2018. Rural population (% of total population). Available from: https://data.worldbank.org/indicator/SP.RUR.TOTL.ZS?contextual=default&end=2017&locations=PG-XD&start=1960&view=chart (9 August 2018).

Tree Kangaroo Conservation Program (TKCP), 2012. YUS Landscape Plan 2013-2015. Tree Kangaroo Conservation Program, Lae, Morobe. Available from: https://www.zoo.org/document.doc?id=904.

Tree Kangaroo Conservation Program (TKCP), 2016. YUS Youth Junior Ranger report. .

Tree Kangaroo Conservation Program (TKCP), 2017. A Gender Strategy: Conservation and Sustainable Development in the YUS Conservation Area. .

CHAPTER

17

Status of Tree Kangaroo Science and Conservation in Indonesian New Guinea

Freddy Pattiselanno[a,b,d], *Johan F. Koibur*[a,b], *and Agustina Y.S. Arobaya*[b,c]

[a]Faculty of Animal Science, University of Papua, Manokwari, West Papua, Indonesia
[b]Biodiversity Research Centre, University of Papua, Manokwari, West Papua, Indonesia
[c]Faculty of Forestry, University of Papua, Manokwari, West Papua, Indonesia
[d]Australasian Marsupial and Monotreme Species Specialist Group, Brisbane, QLD, Australia

OUTLINE

INTRODUCTION

Indonesia is an archipelago in Southeast Asia made up of over 14,7000 islands. The islands can be grouped into the Greater Sunda Islands of Sumatra (Sumatera), Java (Jawa), the southern extent of Borneo (Kalimantan), and Celebes (Sulawesi); the Lesser Sunda Islands (Nusa Tenggara) of Bali and a chain of islands that runs eastward through Timor; the Moluccas (Maluku) between Celebes and the island of New Guinea; and the western extent of New Guinea (generally known as Papua).

Indonesia is second after Brazil in terms of ecosystem diversity among 12 "mega-diversity" countries. Its strategic location between the Indomalayan biodiversity of tropical Southeast Asia in the west and the Australasian species in the east together with its species-rich forest environment contributes to the high diversity of plants and animals present in the area (Pattiselanno and Arobaya, 2012). Indonesia is also a country of enormous cultural diversity. Among its more than 273.5 million inhabitants (World Odometers, 2020), there are more than 500 ethnic groups speaking over

Tree Kangaroos
https://doi.org/10.1016/B978-0-12-814675-0.00019-1

1000 languages and dialects (Pattiselanno and Arobaya, 2012).

After the Amazon, Indonesia contains the most densely forested areas on Earth. In spite of having the fourth largest human population in the world, Indonesia contains roughly 11% of the world's flowering plant species, 13% of its mammal species (including 46 primate species), 6% of its amphibians, 7% of its reptiles, 16% of its birds, and 14% of its fish (including freshwater and saltwater species) (Fauna and Flora International, 2019).

Indonesian New Guinea (both Papua and West Papua provinces) consists of mountainous areas, prominent lakes, and swamps and mangroves at the head of Bintuni Bay. This area is considered a complex piece of the planet formed by a convoluted tectonic history. Both provinces have a species-rich forest environment making them home to endemic Australopapuan flora and fauna including unique New Guinean species. Several studies indicate that Papua and West Papua provinces are home to 146 mammal species, 329 reptile and amphibian species, and also 650 bird species that represent 50% of Indonesia's terrestrial animals (Pattiselanno and Arobaya, 2013).

Marine areas of Papua are home to a high diversity of marine life. The Coral Triangle Marine Protected Area at the north western part of Papua is the habitat of the world's greatest diversity of coral-reef fish, containing more than 1650 species in eastern Indonesia alone. The coastline of Papua is magnificent, with a fringe of more than 1000 islands like Raja Ampat, Biak, Supiori, Yapen, and the satellite islands of Cenderawasih Bay. These astonishing features along the coast are currently considered icons of ecotourism of the Bird's Head Peninsula which makes up the northwest portion of the island of New Guinea (Arobaya and Pattiselanno, 2016).

Among the mammalian species richness in Indonesia, the tree kangaroo (*Dendrolagus*) is one of the marsupial genera found in Papua and protected under the Minister of Agriculture Decree 247/Kpts/Um/1979 dated April 5, 1979 (Yepasedanya, 2003). Tree kangaroos are mammal species classified in the family Macropodidae of the order Diprotodontia, an order of rare species of marsupials. There are six species of tree kangaroos that live in Indonesia, including the Ndomea tree kangaroo (*Dendrolagus dorianus*), Goodfellow's tree kangaroo (*Dendrolagus goodfellowi*), Wakera tree kangaroo (*Dendrolagus inustus*), Mbaiso tree kangaroo (*Dendrolagus mbaiso*), Seri's tree kangaroo (*Dendrolagus stellarum*), Vogelkop tree kangaroo (*Dendrolagus ursinus*), and Wondiwoi tree kangaroo (*Dendrolagus mayri*) (Table 17.1).

The six species of tree kangaroos endemic to Papua are found in the forests. Unlike other macropods such as kangaroos and wallabies in Australia who spend their lives on the ground, tree kangaroos spend most of their lives in the trees. Mbaiso, for example, have big feet like a cushion covered with rough skin and slightly curved legs, giving them a firm grip on the trunk and branches of the tree. Tree kangaroos in nature prefer young leaves/shoots, fruit and soft stems of *Ficus* sp., *Gnetum gnemon*, *Schuur mansia heningsii*, *Tetracera*, *Elatostema*, *Procris*, as well as weeds and some creeping plants (Flannery, 1995; Chapter 2). Lorentz National Park in Papua was recognized as the habitat of tree kangaroos from a long-term survey done in the era of Flannery and Petocz in the 1970s. Thus, a regular survey was conducted by the Worldwide Fund for Nature (WWF) starting in 2010 to monitor the presence of tree kangaroo and all fauna species around the park. In 2011, a similar survey was done to develop a conservation strategy for tree kangaroos. A report of [Worldwide Fund for Nature WWF—Indonesia Region Sahul Papua (2011)], describes the tree kangaroo species around the Lorentz National Park in Papua Province as follows:

The report states that tree kangaroos are listed as Vulnerable and Endangered (IUCN Redlist, 2019; Table 17.1). It also notes that in Indonesia, tree kangaroos (*Dendrolagus* sp.) in general and especially the grizzled tree kangaroo-Wakera

TABLE 17.1 Tree kangaroo species in Papua, local name, and conservation status.

Indonesian	Local term[a]	English	Scientific name	IUCN Red List conservation status 2019
Kanguru pohon doria	Ndomea (Moni, Papua); Naki (Amungme, Tembagapura); Weya (Dani, Kwiyawagi)	Doria's tree kangaroo	*Dendrolagus dorianus*	Vulnerable
Kanguru pohon hias		Goodfellow's tree kangaroo	*Dendrolagus goodfellowi*	Endangered
Kanguru pohon kelabu	Wakera (Lobo Bay)	Grizzled tree kangaroo	*Dendrolagus inustus*	Vulnerable
Kanguru pohon coklat	Dingiso (Moni); Mbaiso (Dani); Namenaki (Amungme)	Mbaiso tree kangaroo	*Dendrolagus mbaiso*	Endangered
		Seri's tree kangaroo	*Dendrolagus stellarum*	Vulnerable
Kanguru pohon Cenderawasih	Wangoirie (Triton Bay)	Vogelkop tree kangaroo	*Dendrolagus ursinus*	Vulnerable
Kanguru pohon Wondiwoi		Wondiwoi tree kangaroo	*Dendrolagus mayri*	Critically endangered

[a] *Flannery (1995) – name represents places.*

(*D. inustus*) have been protected by the Indonesian government based on the Minister of Agriculture Decree 247/Kpts/Um/1979 dated April 5, 1979 (Koibur, 2011) due to threats to survival from hunting, and forest conversion for other purposes including roads, logging, mining, and modern agriculture such as palm oil production.

Locally, in 2008, WWF in collaboration with local government, Wasur National Park Authority, and local communities in Merauke, conducted a campaign to protect tree kangaroos at the local community level (DetikNews, 2019). Although it was not known yet outside of local communities, each ethnic group in Papua have their own traditional wisdom related to the use of wild animals (Pattiselanno, 2008; Pattiselanno et al., 2016), traditions that are passed down from their ancestors. It is currently acknowledged that traditional ecological knowledge indirectly protects wildlife and is used for the conservation purposes of protected species. Local communities have their own way of ensuring the protection of wildlife, in this case the tree kangaroo.

DESCRIPTION OF INDONESIAN TREE KANGAROOS

Ndomea tree kangaroos (*Dendrolagus dorianus*) are known as the Tingkalong, Ndomea (Moni, Pogapa), Naki (Amugme-Tembagapura), Weya (Dani, Kwiyawagi) in the local dialect based on where they were found. This species is classified as Vulnerable (VU) on the IUCN

Red List (Leary et al., 2016a). The Ndomea tree kangaroo has a long, silvery tail less than 20% body length, with reddish limbs, and red or yellow at the base of the tail. The body length is 52–77 cm, tail length 40–66 cm (77–80% body length), foot length 9–12 cm, with an average body weight of 13.3 kg for males and 10.2 kg for females (Worldwide Fund for Nature WWF—Indonesia Region Sahul Papua, 2011; Flannery et al., 1996; Chapter 1).

Ndomea tree kangaroos live in trees that are generally found in alpine and tundra alpine pastures in the range of 600–3650 m above sea level (asl). Ndomea lives in small groups consisting of male and female adults with their offspring. Generally, they are active in the morning and evening, reducing activity at night and during heavy rain. They produce up to 6 (six) different sounds for different purposes of communication including contact for the mating season, danger warning (threats), and territory of the group (Worldwide Fund for Nature WWF—Indonesia Region Sahul Papua, 2011). Sexual maturity is reached around 21 months of age. They have long copulation bouts that last around 20 min. They prefer leaves such as *Asplenium* spp. and are sympatric to *D. ursinus* and *D. inustus*.

Goodfellow's **tree kangaroo** (*Dendrolagus goodfellowi*) or ornamental tree kangaroo, is classified as Endangered (EN) on the IUCN Red List (Leary et al., 2016d). This species has a pale abdomen color on the lower body with a darker color on its back. They have pink to brown faces. There is a trio of lines starting about half-way down the back reaching down to the tail, with two yellow lines on the outside and a brown to black line in the middle. The ears are not tufted, and the edge of the ear has a lighter color. The tail is yellow on the underside and has a unique number of brown rings on the dorsal side. Body length is 57–63 cm (male) and 56–64 cm (female), tail length is 56–64 cm (male) and 59–76 cm (female) or around 116–118% body length, foot length 11–12 cm, body weight 6.7–9.1 kg (male) and 7.7–8.6 kg (female). In Intan Jaya of Papua, Goodfellow's have previously been reported to live in pairs and being active in the morning or late at night at an altitude of 680–2865 m asl (Worldwide Fund for Nature WWF—Indonesia Region Sahul Papua, 2011; Chapter 1). Copulation is carried out on the ground. Females have a 54 day menstrual cycle. The gestation period is 44.5 days and the joeys remain in the pouch for 8–10 months. These animals are thought to occur in moss forests on the upper part of the lower montane forest and on alpine grassland vegetation but there are no recent sightings in Indonesia.

Wakera tree kangaroo (*Dendrolagus inustus*) or grizzled tree kangaroo, is listed as Vulnerable (VU) on the IUCN Red List (Leary et al., 2016e). Flannery (1995) described that on average, males (15.5 kg) are larger than females (11.4 kg), with small head and flat muscle (Ho, 2004). Body length of males is also longer (76 cm) compared with females (67.7 cm). Ho (2004) explains that the grizzled coloration of *D. inustus* makes this species differ from other tree kangaroos. They have slate gray to chocolate brown coats with medium length fur.

The food plant species recognized during Flannery's survey at the Torricelli Mountain in 1989–1992 were leaves and fruits of *Ficus* sp., leaves of *Gnetum gnemon*, and leaves of Araceae. Records of the reproduction in this species of tree kangaroo showed that young were born in March and June. From different data recorded during the survey, Flannery (1995) concluded that breeding of this species is non-seasonal, with the number of offspring ranging from 1 to 2 (Ho, 2004). Females attained sexual maturity at a weight range of 8.5–10.6 kg, and males about 12 kg (Flannery, 1995), at about 2 years (Ho, 2004).

Dingiso (*Dendrolagus mbaiso*) (Moni Tribe), also known as Bondegezon (Bibido), Mbaiso and Wanun (Dani), Itimili (Nduga), Mayamaya (Pagopa) and Nemenaki (Amungme), is a species of tree kangaroo endemic to Papua in Indonesia. This species was discovered by Flannery, Boeadi, and Szalay in 1994 through specimens obtained in the Tembagapura and Kwiyawagi regions (Flannery, 1995).

The **Dingiso** is one of the most threatened species in Indonesia and is classified as Endangered (EN) on the IUCN Red List (Leary et al., 2016c). The Dingiso tree kangaroo population is increasingly threatened as land clearing continues to shrink their forest habitat. The population is estimated to have decreased by more than 80% over the past 30 years. Besides loss of habitat due to land clearing, the Mbaiso tree kangaroo is threatened by hunting by indigenous communities in Papua for consumption and often used as a substitute for pork for their traditional ceremonies (WWF, 2019).

Dingiso has adapted to living in the trees usually above 2700 m asl. This species has a long tail (40–94 cm) that helps balance the body as it moves through the trees. The body length is about 52–81 cm and body weight is 6.5–14.5 kg (WWF, 2019). The color of the fur that stands out is black with a white line from the chin to the base of the tail and a white pattern on the muzzle and part of the forehead. The back color is generally blackish or black charcoal with variations in the ends of the hair on the cheeks and the body. The length of the back hair is around 60 mm. The ears, arms and legs, and the tail are black with straw yellow splashes. The yellow straw coloring is also found at the tip of the tail. The white color is found around the muzzle, forehead, chest, abdomen, and throat. Body length is 66–67 cm, tail length 42–52 cm (72% body length) for females and 46 cm for males with foot length 11 cm and body weights of 8.5–9 kg.

Seri's tree kangaroo (*Dendrolagus stellarum*) is listed as Vulnerable (VU) on the IUCN Red List (Leary et al., 2008). Information on this species was obtained from Tenkile Conservation Alliance (TCA, 2020). Seri's tree kangaroo was distributed between 2600 and 3200 m asl within the Central Cordillera in Indonesia and Papua New Guinea. Adults are brown with a dark tail and silver-tipped fur on their limbs while very young animals have a dark body color and a bright yellow tail. The average weight for males is 9.5 kg and for females, 8.6 kg. The population is suspected of a decline of 30% in the last three generations and threats continue including heavy hunting.

Vogelkop **tree kangaroo** (*Dendrolagus ursinus*) is listed as Vulnerable (VU) on the IUCN Red List (Leary et al., 2016f). Differing from other species, this tree kangaroo is black in color with pale or reddish cheeks and pale belly (Flannery, 1995). Despite occurring sympatric with *D. inustus* in the Arfak Mountains, *D. ursinus* is only found at higher elevations. The information about its body weight is limited or unknown. Flannery (1995) described the male (66 cm) as longer than the female (59 cm). There is very little information about reproduction in this species.

Wondiwoi tree kangaroo (*Dendrolagus mayri*), listed as Critically Endangered (CR) on the IUCN Red List (Leary et al., 2016b), is one of the most poorly known mammals in the world and has been rarely seen since the first sighting but not yet collected. Ernst Mayr was the first to see this species in the mountains of the Wondiwoi Peninsula in 1928. It is rare and elusive and disappeared for nearly a century; thus, it was assumed to be extinct. In 2017, Michael Smith, an amateur botanist from Farnham, England, with his team went on an expedition to explore the West Papuan Mountains for rare species including orchids, rhododendrons, tulips and the Wondiwoi. The species was seen at a height of 30 m above the ground and a few identification photos were taken to demonstrate the presence of the Wondiwoi tree kangaroo (Pickrell, 2018, Chapter 27). The Wondiwoi population is estimated to have less than 50 individuals and has rarely been seen (Leary et al., 2016b).

DISTRIBUTION

Tree kangaroos live in almost all parts of Papua (land and islands) and occupy 80% of Papua's land area (416,000 km^2). In some cases the island populations have been translocated (Martin, 2005). Tree kangaroos are arboreal macropods that are found from sea level to high mountain forests around 4000 m asl, with a temperature range of 10–27 °C and humidity 78.5–86.5% (Petocz, 1994; Yepasedanya, 2003). Especially in West Papua, tree kangaroos can be found in Manokwari, Bintuni, Sorong, South Sorong, Maybrat, Waigeo, Fak-fak, Kaimana, Wondama/Wasior and Raja Ampat (Petocz, 1994).

THREATS

Palm Oil Deforestation

Expansion of oil palm plantations is a priority of both the federal government and the government of Papua. The area of oil palm plantations has increased rapidly in Papua from 11,367 ha in 1991 to 50,000 ha in 2005 (GRM International, 2009). The Indonesian government is keen to develop oil palm plantations in Papua and is offering investors the opportunity to establish concessions of up to 200,000 ha. Over 50,000 ha of oil palm have already been planted in Papua and permits have been allocated to develop another 500,000 ha (GRM International, 2009). Abood et al. (2014) indicated that the total areas allocated for oil palm in Papua is approximately 500,000 ha or 3.3% out of the total forest areas allocated for industrial use. In just 25 years, 76 million acres of Indonesia's rainforests have disappeared - an area the size of Germany.

Embury-Dennis (2016) reported that the systematic bulldozing and burning of the endangered species' rainforest home to supply palm oil for use in a massive range of everyday products, is leading to a dramatic dwindling of tree kangaroo numbers. Environmental groups that work in Papua claim the palm oil company Korindo has been burning forests in Papua, knowing full well the practice is illegal. During the extreme dry season period in Indonesia, multiple wildfire outbreaks happen not only in palm plantations but also in natural forests. It is also assumed that locals who have gained easy access to the areas burned the forests for hunting purposes causing the decline of wild animal species in the remaining trees.

Wildlife Hunting

The Wakera tree kangaroo was among the native species considered a hunting target by local hunters along the coastal sites of the Bird's Head Peninsula of Papua (Pattiselanno and Lubis, 2014; Pattiselanno and Koibur, 2018). Koibur (2006) reported in Yapen Papua, that tree kangaroos were hunted by people using traditional weapons (machetes, spears, and arrows), modern weapons, and snares for consumption and trade, dead or alive. Selling prices ranged from 500,000 to 1,000,000 IDR (Indonesian Rupiah) per animal according to body weight. Those caught in good condition usually have been raised in captivity as pets (Fig. 17.1), and very often sold for cash.

As previously explained, WWF monitored the population and surveyed the conservation status of tree kangaroos in Intan Jaya, around the Lorentz National Park. In many parts of Intan Jaya in sites surveyed by WWF, Dingiso was often hunted using trained dogs. Local people acknowledged that Dingiso tend to be docile (tame) and thus easy to catch. Dingiso is hunted using both traditional and modern tools including bow and arrow, spear and air rifle. Dingiso, together with cuscus (*Spilocuscus* sp.) and long-beaked echidna (*Zaglossus bruijni*) are consumed as animal protein sources and their skins are used for decoration and warm clothes.

The probability of collecting Dingiso from their habitat is high because they live in family

FIGURE 17.1 A young Wakera tree kangaroo (*Dendrolagus inustus*) kept as a captive animal in the outskirts of Manokwari. *Source: O.R. Faidiban.*

groups, male, female and juvenile. Usually more than one adult female joins the group, so a group ranges from 5 to 6 animals. Thus, in one hunting session, hunters can collect up to 15 animals from the forest. When hunters have caught one animal, others will be easy to catch because they are hiding close to each other within the range (Worldwide Fund for Nature WWF—Indonesia Region Sahul Papua, 2011).

For special occasions including weddings or other traditional ceremonies such as stone burnt (bakar batu), a traditional ritual to express gratitude for an abundant blessing, this species is hunted extensively and served as part of traditional food. The tendency to collect this animal from its habitat also increases along with the demand from the market. Dingiso, together with long-beaked echidna (*Zaglossus bruijni*), bring the highest selling prices across the community. Both species are usually sold alive or dead with prices ranging between 1.5 million IDR and 2 million IDR.

Other threats to tree kangaroos include habitat loss from road development and forest conversion to other purposes including logging, mining and modern agriculture such as palm oil. The creation of new regencies and districts under the special autonomous law also threatens these animals due to forest conversion for new infrastructures.

STRATEGY FOR CONSERVATION

Although tree kangaroos are hunted, local people have their own traditional wisdom that strongly supports conservation purposes. For example, in Intan Jaya, if hunters catch an adult female with a juvenile, the juveniles are released into the forest after having tags on their tails or ears. If other hunters find the tagged animals during hunting, these animals are released again because of the tags. Tagged animals are not allowed to be hunted until they reach mature period and gain more weight to supply the need of meat. In other parts of Papua, people from other clans or ethnic groups who do not have land tenure rights are not allowed to hunt in the forest that they do not own.

All tree kangaroo species are protected by the Indonesia Government Regulation (PP RI No. 7/1999 and UU NO. 5/1990). Locally, tree kangaroo species that are caught alive are kept by hunters as pets (Fig. 17.2).

As there are no further studies to document the population status of tree kangaroos in Papua, information on these species is lacking, although they are considered a threatened species. These species were considered threatened due to habitat loss and illegal hunting that massively occurred across the land of Papua. Locally, the government of Papua and West Papua provinces also acknowledge the implementation of the regulations, but in reality, there is no clear action in protecting this species. Individually, scientists and students from the University of Papua conducted an ecology survey of tree kangaroos to obtain a better understanding of the future plans for the conservation of this species (Fig. 17.3).

FIGURE 17.2 Wakera tree kangaroos (*Dendrolagus inustus*) are raised as pets by local communities at Kasoneweja, a district of Mamberamo, Papua.

Fortunately, in collaboration with WWF Papua Program, the conservation program for tree kangaroos is conducted through intense population surveys around Lorentz National Park. A moral movement is now starting along with a declaration of West Papua Province as a conservation province. Provincial government is currently taking a lead to give more concern on the conservation of flora and fauna in Indonesian New Guinea. Plans to leave about 70% of the province areas for conservation purposes shows a good willingness to give more area for flora and fauna without any disturbance.

FIGURE 17.3 Wakera tree kangaroo (*Dendrolagus inustus*) placed in the Faculty of Animal Science of UNIPA facility for the purpose of study.

It is now important to follow up on specific conservation efforts for tree kangaroo species. Long-term work and commitment of WWF Papua Program in protecting this species should be taken into account for a future plan of tree kangaroo conservation action. In collaboration with students, WWF Papua is taking step by step action to study tree kangaroos within their habitats across the Papua and West Papua provinces. International collaboration needs to be established in order for future conservation efforts to be effective for these species.

ACKNOWLEDGMENTS

The authors would like to acknowledge WWF Papua Program for sharing the report of the Tree Kangaroo survey at The Lorentz National Park.

REFERENCES

Abood, S.A., Lee, J.S.H., Burivalova, Z., Garcia-Ulloa, J., Koh, L.P., 2014. Relative contributions of the logging, fiber, oil palm, and mining industries to forest loss in Indonesia. Conserv. Lett. 8 (1), 58–67.

Arobaya, A.Y.S., Pattiselanno, F., 2016. Follow up on West Papua as Conservation Province. Jakarta Post, 17 March 2016.

DetikNews, 2019. Di Pelukan Adat, Kanguru Papua Berlindung dari Kepunahan. Available from: https://news.detik.com/berita/d-3509923/di-pelukan-adat-kanguru-papua-berlindung-dari-kepunahan (28 March 2019).

Embury-Dennis, T., 2016. Tree Kangaroos 'on Brink of Extinction' due to Palm Oil Deforestation. Available from: https://www.independent.co.uk/news/world/asia/tree-kangaroos-extinct-palm-oil-deforestation-indonesia-asia-a7220731.html (28 Mar 2019).

Fauna and Flora International, 2019. Indonesia. Available from: https://www.fauna-flora.org/countries/indonesia (8 Nov 2019).

Flannery, T., 1995. Mammals of New Guinea. Cornell University Press, Cornell, New York.

Flannery, T.F., Martin, R., Szalay, A., 1996. Tree Kangaroos: A Curious Natural History. Red Books Australia, Port Melbourne, Victoria.

GRM International, 2009. Papua assessment USAID/Indonesia, final report. GRM International, Jakarta.

Ho, Y., 2004. Dendrolagus inustus. Animal Diversity Web. Available from: https://animaldiversity.org/accounts/Dendrolagus_inustus/ (11 April 2020).

IUCN Redlist, 2019. Dendrolagus. Available from https://www.iucnredlist.org/ (11 Nov 2019).

Koibur, J.F., 2006. Perburuan Kanguru Pohon Kelabu (*Dendrolagus inustus*) dan Pemanfaatannya Oleh Masyarakat Angkaisera Di Yapen Papua. Laporan Penelitian Dosen Muda Dirjen Dikti Jakarta.

Koibur, J.F., 2011. Karakteristik dan organ reproduksi betina Kanguru Pohon Kelabu *Dendrolagus inustus* di Papua. Bull. Anim. Sci. 35 (1), 17–23.

Leary, T., Seri, L., Flannery, T., Wright, D., Hamilton, S., Helgen, K., Singadan, R., Menzies, J., Allison, A., James, R., Aplin, K., Salas, L., Dickman, C., 2008. *Dendrolagus stellarum*. The IUCN Red List of Threatened Species 2008: e.T136812A4342630. Available from: https://doi.org/10.2305/IUCN.UK.2008.RLTS.T136812A4342630.en (12 April 2020).

Leary, T., Seri, L., Flannery, T., Wright, D., Hamilton, S., Helgen, K., Singadan, R., Menzies, J., Allison, A., James, R., 2016a. *Dendrolagus dorianus*. The IUCN Red List of Threatened Species, 2016: e.T6427A21957392. Available from: https://doi.org/10.2305/IUCN.UK.2016-2.RLTS.T6427A21957392.en (9 Apr 2020).

Leary, T., Seri, L., Flannery, T., Wright, D., Hamilton, S., Helgen, K., Singadan, R., Menzies, J., Allison, A., James, R., 2016b. *Dendrolagus mayri*. The IUCN Red List of Threatened Species: 2016 e.T136668A21956785. Available from: https://doi.org/10.2305/IUCN.UK.2016-2.RLTS.T136668A21956785.en (9 Apr 2020).

Leary, T., Seri, L., Wright, D., Hamilton, S., Helgen, K., Singadan, R., Menzies, J., Allison, A., James, R., Dickman, C., Aplin, K., Flannery, T., Martin, R., Salas, L., 2016c. *Dendrolagus mbaiso*. The IUCN Red List of Threatened Species 2016: e.T6437A21956108. Available from: https://doi.org/10.2305/IUCN.UK.2016-2.RLTS.T6437A21956108.en (11 November 2019).

Leary, T., Seri, L., Wright, D., Hamilton, S., Helgen, K., Singadan, R., Menzies, J., Allison, A., James, R., Dickman, C., Aplin, K., Flannery, T., Martin, R., Salas, L., 2016d. *Dendrolagus goodfellowi*. The IUCN Red List of Threatened Species 2016 e.T6429A21957524. Available from: https://doi.org/10.2305/IUCN.UK.2016-2.RLTS.T6429A21957524.en (9 Apr 2020).

Leary, T., Seri, L., Wright, D., Hamilton, S., Helgen, K., Singadan, R., Menzies, J., Allison, A., James, R., Dickman, C., Aplin, K., Flannery, T., Martin, R., Salas, L., 2016e. *Dendrolagus inustus*. The IUCN Red List of Threatened Species 2016: e.T6431A21957669. Available from: https://doi.org/10.2305/IUCN.UK.2016-2.RLTS.T6431A21957669.en (9 Apr 2020).

Leary, T., Seri, L., Wright, D., Hamilton, S., Helgen, K., Singadan, R., Menzies, J., Allison, A., James, R., Dickman, C., Aplin, K., Salas, L., Flannery, T., Bonaccorso, F., 2016f. *Dendrolagus ursinus*. The IUCN Red List of Threatened Species 2016: e.T6434A21956516. Available from: https://doi.org/10.2305/IUCN.UK.2016-2.RLTS.T6434A21956516.en (12 April 2020).

Martin, R., 2005. Tree-Kangaroos of Australia and New Guinea. CSIRO Publishing, Melbourne, Australia.

Pattiselanno, F., 2008. Man-wildlife interaction: understanding the concept of conservation ethics in Papua. Tigerpaper 35 (4), 10–12.

Pattiselanno, F., Arobaya, A.Y.S., 2012. Biodiversity, Local Livelihood and the Nature's Conservation. Jakarta Post 6 November 2012.

Pattiselanno, F., Arobaya, A.Y.S., 2013. Managing tropical forest for Indonesian Papuan's livelihood. In: Proceeding of Institute of Foresters of Australia National Conference, 7–11 April 2013. Canberra, Australia, pp. 207–215.

Pattiselanno, F., Koibur, J.F.K., 2018. Returns from indigenous hunting in the lowland coastal forest of West Papua, benefits threatened wildlife species. J. Manaj. Hutan Tropika 24 (1), 46–50.

Pattiselanno, F., Lubis, M.I., 2014. Hunting at the Abun Regional Marine Protected Areas: a link between wildmeat and food security. Hayati J. Biosci. 21 (4), 180–186.

Pattiselanno, F., Koibur, J.F., Yohanes, C.H., 2016. Traditional ecological knowledge (TEK) in hunting: from culture to nature. In: ICSBP Conference Proceedings. International Conference on Social Science and Biodiversity of Papua and Papua New Guinea (2015), 2016, pp. 66–70.

Petocz, R.G., 1994. Mamalia Darat Irian Jaya. Gramedia, Jakarta.

Pickrell, J., 2018. Rare Tree Kangaroo Reappears After Vanishing for 90 Years. National Geographic. September 2018. Available from: https://www.nationalgeographic.com/animals/2018/09/rare-wondiwoi-tree-kangaroo-discovered-mammals-animals (28 Mar 2019).

Tenkile Conservation Alliance (TCA), 2020. Seri's Tree Kangaroo. Available from: https://www.tenkile.com/seris-tree-kangaroo.html (12 Apr 2020).

World Odometers, 2020. Population of Indonesia. Available from: https://www.worldometers.info/world-population/indonesia-population/ (8 April 2020).

Worldwide Fund for Nature WWF, 2019. Kangguru Pohon. Available from: https://www.wwf.or.id/?69772 (1 Apr 2019).

Worldwide Fund for Nature WWF—Indonesia Region Sahul Papua, 2011. Laporan Survey Identifikasi Jenis dan Strategi Konservasi Kangguru Pohon di Kabupaten Intan Jaya. WWF Indonesia Program.

Yepasedanya, S., 2003. Tingkah Laku Harian Kanguru Pohon Kelabu (*Dendrolagus inustus*) Dalam Penangkaran di Kampung Famboaman Distrik Yapen Selatan Kabupaten Yapen waropen. Skripsi Sarjana Peternakan FPPK UNIPA, Manokwari.

SECTION V

CONSERVATION SOLUTIONS: ROLES OF ZOOS

CHAPTER

18

Tree Kangaroo Populations in Managed Facilities

Jacque Blessington[a], *Judie Steenberg*[b], *Karin R. Schwartz*[c], *Ulrich Schürer*[d,*], *Brett Smith*[e], *Megan Richardson*[f], *Razak Jaffar*[g], *and Claire Ford*[h]

[a]Association of Zoos and Aquariums Tree Kangaroo Species Survival Plan® Program, Kansas City, MO, United States

[b]Association of Zoos and Aquariums Tree Kangaroo Species Survival Plan® Program, Maplewood, MN, United States

[c]Association of Zoos and Aquariums Tree Kangaroo Species Survival Plan® Milwaukee, WI, United States

[d]Wuppertal Zoo, Wuppertal, Germany

[e]Port Moresby Nature Park, Port Moresby, Papua New Guinea

[f]Melbourne Zoo, Parkville, VIC, Australia

[g]Wildlife Reserves Singapore, Singapore

[h]Taronga Conservation Society Australia, Mosman, NSW, Australia

OUTLINE

* Retired.

Tree Kangaroos
https://doi.org/10.1016/B978-0-12-814675-0.00020-8

INTRODUCTION

With tree kangaroo populations declining in the wild, global managed facilities have taken on important roles in the conservation of several species. These roles vary in capacity supporting in situ (in the wild) and ex situ (in managed facilities) conservation. Raising awareness and education about these rare animals is essential to supporting the species. Research, funding and technical support for in situ conservation programs are critical to their success. Scientific and collaborative breeding management programs that work towards establishing ex situ populations that maintain genetic diversity and demographic stability are the cornerstone for achieving sustainable populations (IUCN/SSC, 2014; Chapter 22).

Currently, there are five species in managed facilities in four regions of the world: Oceania, North America, Europe, and Asia. These five species include populations of Goodfellow's (*Dendrolagus goodfellowi*) and Matschie's (*D. matschiei*), plus a few individuals of Lumholtz's (*D. lumholtzi*), Grizzled (*D. inustus*), and Doria's (*D. dorianus*) tree kangaroos. With limited captive populations, it is crucial for facilities and regional programs to work together with careful planning and monitoring of the breeding programs to minimize inbreeding and maximize genetic diversity for healthy, sustainable populations.

This chapter gives an overview of zoological associations involved in tree kangaroo population management with a description of the associated conservation management programs for threatened species, and the history of tree kangaroos in global managed facilities. Information on regional and global population management strategies involving the international collaboration of zoos follows.

ZOO ASSOCIATIONS AND CONSERVATION BREEDING MANAGEMENT PROGRAMS

Zoological facilities are organized into global and regional zoological associations that manage conservation programs for the threatened species in their collections. The World Association of Zoos and Aquariums (WAZA) is the umbrella organization that brings together 400 institutions and organizations worldwide including regional associations, national federations, and zoos and aquariums. Their mandate is to facilitate collaborative efforts in animal health and welfare, environmental education, and global conservation (WAZA, 2019a). In 2015, WAZA developed The World Zoo and Aquarium Conservation Strategy (Barongi et al., 2015) which promotes species conservation strategies that incorporate both in situ and ex situ efforts (Chapter 22). WAZA oversees Global Species Management Plans (GSMP) that incorporate collaboration between regional associations for species programs that benefit from global coordination to maintain population characteristics that support the conservation value of the species. The GSMP for Goodfellow's tree kangaroo was approved by WAZA in 2013 and by 2019 was one of nine programs (WAZA, 2019b).

Each population management program is designed to implement strategies to maintain healthy populations that are genetically diverse, demographically stable, and behaviorally competent. In addition, the programs include components that link the ex situ programs to conservation of tree kangaroos in the wild, such as research, information exchange, funding, and technical support for in situ programs (Chapter 22).

There are three regional zoo associations that maintain the largest populations of tree kangaroos under conservation breeding programs. Approximately 100 zoos and aquariums and related organizations within the Oceania region (Australia, New Zealand, Papua New Guinea [PNG], and New Caledonia) are overseen by the Zoo and Aquarium Association Australasia (ZAA) (ZAA, 2019). ZAA manages 115 threatened species under Australasian Species Management Plans (ASMP) of which the Goodfellow's and Lumholtz's are identified ASMP programs.

In North America, the Association of Zoos and Aquariums (AZA) is a nonprofit organization with a membership of more than 230 zoos and aquariums in the United States, Canada, Mexico, and a small number of international member zoos. AZA zoos focus their mission on conservation, education, science, and recreation (AZA, 2019a,b). AZA oversees conservation management programs called Species Survival Plans® (SSP). The Matschie's Tree Kangaroo SSP (TKSSP) is one of over 500 SSPs for threatened species (AZA, 2019b).

The European Association of Zoos and Aquaria (EAZA) is the largest zoological association in the world and consists of over 377 member institutions in 43 countries throughout Europe and the Middle East (EAZA, 2019). The Goodfellow's tree kangaroo is one of over 400 species managed through EAZA's European Endangered Species Programmes (EEP) (EAZA, 2018).

Additionally, there are small populations of tree kangaroos maintained in Asia under the auspices of the South East Asian Zoo Association (SEAZA) (Singapore, Malaysia, and Indonesia) and the Japanese Association of Zoos and Aquariums (JAZA). These associations have limited threatened species conservation management programs, but many participate in global management programs such as the Goodfellow's Tree Kangaroo GSMP.

RECORDS-KEEPING AND POPULATION MANAGEMENT ANALYSIS TOOLS

The population management programs incorporate standardized records, population management, and analysis tools to establish breeding goals and recommendations for all of the individuals of that species in the region. Accurate records on origin and parentage are crucial for the development of studbooks, which are used as the basis for analysis of the populations' genetic and demographic composition. Studbooks are generally developed for one species or subspecies and contain a record of the pedigree and history of each individual, ideally traced back to wild founders of the population. There are regional studbooks (containing data on individuals within one region) and international studbooks (containing data on individuals found in zoological facilities worldwide). The data in the studbook comes from the animal records for each individual maintained within records-keeping programs such as the Species360 Zoological Information Management System (ZIMS) or other database systems.

Species360 (previously the International Species Information System) is the governing authority over a centralized database system that compiles animal records from over 1100 global zoos, aquariums, research facilities, and other wildlife organizations (Species360, 2019). Species360 members utilize ZIMS (Species360 ZIMS, 2019), a global web-based application for maintaining animal records including collection inventories, life history, physiology, reproduction, behavior, and health. Records from across the world can be accessed in real time by all Species360 members, facilitating collaborative animal husbandry and breeding management processes. The ZIMS Studbook module allows studbooks to be compiled within the same software as institutional animal records. Studbook data can be directly exported from ZIMS or other studbook programs such as PopLink (Faust et al., 2018) into PMx (Ballou et al., 2011), a population management program that uses a mean kinship strategy to analyze the demographic and genetic composition of the whole population and facilitate breeding management (Ballou and Foose, 1996; Lacy et al., 2012).

Studbooks are the foundation of most ex situ breeding programs. Studbook analysis enables breeding recommendations that promote pairing of least related individuals to maximize retention of genetic diversity while minimizing inbreeding. ZIMS, as a global information system, facilitates collaboration and information exchange for zoological facilities that rely on accurate records not only for breeding programs but also for husbandry, health management, and collaborative research projects.

TREE KANGAROOS IN MANAGED FACILITIES

Five species of tree kangaroos (*Dendrolagus* spp.) were described between the 1820s and the late 1880s: Black (Vogelkop) (*D. ursinus*), Doria's, and Grizzled tree kangaroos in New Guinea; Lumholtz's and Bennett's (*D. bennettianus*) tree kangaroos in Australia (Martin, 2005; Chapter 1). The London Zoo was the first facility to exhibit a living tree kangaroo. A female Grizzled was acquired in 1848 and it died in 1852 (Flannery et al., 1996). In 1890 the Adelaide Zoo became the first Australian zoo to maintain tree kangaroos when a pair of Grizzled was imported from PNG (Crook and Skipper, 1983). In 1893, four Bennett's tree kangaroos, captured in the wild in Queensland, Australia, became the first tree kangaroos at Melbourne Zoo in the state of Victoria, Australia. More than sixteen additional Bennett's tree kangaroos were added to the collection at the Melbourne Zoo in the next few years. However, the population did not thrive, possibly due to climate differences in Melbourne compared to their natural habitat in tropical northern Queensland. The collection eventually died out although it is believed that the last survivors may have been released in a national park in an early conservation attempt. Between 1898 and 1936, additional species or subspecies were discovered in Australia and New Guinea, yet tree kangaroos were scarce in zoological facilities (Martin, 2005).

Jones (1963) reported Black tree kangaroos in eight locations in South Africa, North America, Japan, and Europe from 1911 to 1958. Doria's tree kangaroos were reported at the Bronx Zoo (New York) in 1930, the Taronga Zoo (Sydney) in 1935, and in Sir Edward Hallstrom's private collection in Mona Vale near Sydney, Australia in 1962 (Jones, 1963). Between 1967 and 1982, Baiyer River Sanctuary (BRS), Mount Hagen, PNG exported a total of 33 Matschie's tree kangaroos to other zoological facilities (Schürer, 2019).

Census of Tree Kangaroos in Ex Situ Populations (1959–1994)

An historical account of the global populations of tree kangaroos (*Dendrolagus* spp.) in zoological facilities over four decades is derived from the annual *Census of Rare Animals in Captivity* and *Mammals Bred in Captivity and Multiple Generation Births* by institution in International Zoo Yearbook (IZY), Volumes 1–27 (1960–1986), from tree kangaroo studbooks, reports from noted tree kangaroo specialists, and literature.

R. L. Smith (1988a) reported that substantial numbers of tree kangaroos were taken into captivity in the 1970s to 1980s. According to IZY (Vols. 1–26), the following numbers were reported for seven species of tree kangaroos, in various global zoological facilities:

105—Grizzled Grey (*D. inustus*)
87—Goodfellow's (*D. goodfellowi*)

77—Matschie's (*D. matschiei*)
38—Black (Vogelkop) (*D. ursinus*)
34—Doria's (*D. dorianus*)
21—Bennett's (*D. bennettianus*)
9—Lumholtz's (*D. lumholtzi*)

Census data from IZY (Vols. 1–27) reported only a few Lumholtz's and Bennett's in zoological facilities, globally. Less than a dozen Black tree kangaroos were reported until 1975 when 17 (6.10.1—representing 6 males. 10 females. 1 unknown sex) were held in seven facilities. Doria's tree kangaroos were held at the Melbourne Zoo from 1969 to 1975. In 1981, there were 40 Doria's at seven zoos, which included 12.10 at BRS (reported as Mount Hagen), PNG with 3.3 born there that year. In 1977, IZY (Vol. 19) listed 76 (30.45.1) Grizzled tree kangaroos at 20 zoological facilities world-wide. At that time, Oklahoma City Zoo had the largest population of Grizzled (2.12), of which 0.4 were captive born. By 1994, the Grizzled population was considered hopeless and could not be sustained without the additional recruitment of tree kangaroos from the wild. At that time there were only 1.6 in three North American facilities and the lone male was 11 years old.

Fourteen facilities reported a total of 65 (31.31.3) Goodfellow's tree kangaroos in 1974. Mount Hagen (BRS) had the largest population at 8.18 in 1975, with 4.10 captive born. Throughout the 1970s and early 1980s Mount Hagen (BRS) consistently had the highest rate of reproductive success with several *Dendrolagus* species. By 1985 only 39 (20.19) Goodfellow's were reported at 11 facilities. By 1994, the global population of 38 (21.13.4) Goodfellow's was at a critical point and it was imperative that there be a cooperative effort among the 11 facilities housing them if the population was going to continue to exist in captivity.

Matschie's tree kangaroos once represented the largest on-going number of tree kangaroos in global collections. The first entry in IZY's (Vol. 1) *Census of Tree Kangaroos in Captivity* was in 1959 with a birth of a joey of unknown sex at Melbourne. In 1961 five zoos reported a total of 13 Matschie's (5.6.2), and by 1969 the population had increased to 40 (11.27.2) with seven born in zoological facilities. The number peaked in 1977 with 96 (29.58.9) at 15 global facilities; 42 were reported as captive born that year. Throughout the 1970s, Mount Hagen (BRS) had the largest collection of Matschie's, with several joeys born every year. By 1985 there were still 91 (34.57) in 19 locations globally. The 1988 International Studbook (ISB) for Matschie's Tree Kangaroo (Collins, 1990–1993) reported 76 (39.37) animals living in 26 different collections. The 1991 historical listing from the ISB recorded a total of 277 (111.130.36) individuals in zoological collections from 1967 to 1991. The 1994 ISB for Matschie's, Goodfellow's, and Grizzled Tree Kangaroos (Collins, 1994–1996) reported that the Matschie's tree kangaroo was the only species of the three that continued to show a gradual increase in the global population size and had increased to 95 individuals (43.46.6) at 30 facilities.

In 1992 at the first Master Plan Session (Steenberg, 1993) of the AZA TKSSP, there was discussion about concentrating the Goodfellow's population in Europe, Australia, and Asia while focusing on Matschie's in North America. This agreement was finalized at the 2005 Tree Kangaroo Summit Meeting in the Atherton Tablelands, Australia.

For any managed species, it is imperative to maintain the highest level of genetic diversity possible for the health and sustainability of the populations. Both Goodfellow's and Matschie's tree kangaroo ex situ populations would benefit greatly by the influx of "new blood" which can only be achieved by importing animals from the range country of PNG. Ideally, new founders (bloodlines) will be achieved through managed breeding programs at zoological facilities participating in rescue and rehabilitation of wild caught animals.

History and Current Status by Region

Oceania: Papua New Guinea

From 1968 to 1985 (IZY Vols. 10–26) Doria's, Goodfellow's, Grizzled, and Matschie's tree kangaroos were documented at Mount Hagen, PNG which was originally the Hallstrom Park Bird of Paradise Sanctuary and was later renamed Baiyer River Sanctuary (BRS) (Schürer, 2018).

The history of tree kangaroos (*Dendrolagus* spp.) from the BRS and those kept on Sir Edward Hallstrom's private property at Nondugl is reported in the article "Tree Kangaroos at Baiyer River Sanctuary, Papua New Guinea – Personal Recollections from Ulrich Schürer" (Schürer, 2018). During the time period from 1973 to 1979 the BRS census (under Mount Hagen) reported an average of 20 Matschie's per year, of which 60–70% were females. In 1973 they reported second, or subsequent generation births. In the 1975–1980 annual reports of the BRS, Roy Mackay reports that their collection of tree kangaroos had continued to breed well, even to the third generation. They hoped to breed a fourth generation before long (Mackay, 1981).

From 1970 to 1981, Mount Hagen (BRS) was the primary source for three species of tree kangaroo exported from PNG: Doria's, Goodfellow's, and Matschie's. These species were exported to various zoological facilities in Australia, Europe, and North America with the majority of these exports being Matschie's. Prior to that time period, Matschie's tree kangaroos had been received from Nondugl, where they had been bred while others had been obtained from the Lae area on the Huon Peninsula (Schürer, 2018).

According to Schürer (2018), although the BRS export records for the years 1976 to 1980 were not available, there could have been many more exports. After 1981, there was a decline in the population of tree kangaroos at BRS; the number of animals was greatly reduced with surplus stock exported to various approved zoos. A census of animals in the BRS collection at 1 January 1981 lists the following (Schürer, 2018):

- Doria's Tree Kangaroo 10.8.1
- Matschie's Tree Kangaroo 6.9.8
- Goodfellow's Tree Kangaroo 8.3.0

The last Goodfellow's export from BRS was in 1990, when 2.2 individuals were sent to Melbourne Zoo. Brett Smith (personal communication, 30 April 2019) reported that the BRS went through a lot of turmoil in 1993 with local unrest between two warring tribes in the area. Unfortunately, the main building was burned down and the sanctuary was abandoned. The BRS grounds continued to stay open to the public in various forms, but all of the animals in the collection and its facilities had perished. In 2016 the local government allocated some money towards the redevelopment of the facilities in the hope that it would draw tourists to the area. To date, very little infrastructure has been developed. Shortly after the export of the Goodfellow's tree kangaroos and the closure of BRS in 1993, PNG heavily restricted its borders to any further export of its fauna. This was mainly due to the lack of capacity to closely monitor its borders and it was easier to close them completely. Additionally, remaining institutions in the country did not have the capacity or generations of offspring to fulfill the requirements of government export permits.

The export of PNG fauna is covered under the 1982 International Trade (Fauna and Flora) (Fauna) Regulation (ITFFR, 1982) which was generated under the 1979 International Trade (Fauna and Flora) Act and is closely adhered to. In many ways this has preserved the fauna of PNG in comparison to the fauna in West Papua (formerly Irian Jaya), Indonesia, which is the western half of the island of New Guinea. Many Papuan tree kangaroo species are under immense pressure from forest clearing, mining, and legal and illegal animal trade. The

Indonesian government sets quotas for the number of animals that can be caught and exported but little enforcement and false declarations of breeding has resulted in many animals being illegally traded to zoos and the animal pet trade in Asia, and for medicinal purposes in China.

In PNG there have been a number of facilities that housed tree kangaroos since BRS was abandoned. They include the Wau Ecology Institute, The National Museum, and The Rainforest Habitat in Lae (which housed 4.1 Matschie's and 0.2 Goodfellow's in January 2019). Unfortunately, Rainforest Habitat is now very run down. Various other private collections and hotels display tree kangaroos that have been purchased from locals selling them in markets. The National Capital Botanical Gardens (NCBG) housed a number of different species throughout the years (Matschie's, Doria's, Goodfellow's, and Lowland [*D. spadix*]). In 2000, in partnership with the National Museum, NCBG sent Goodfellow's tree kangaroos (1.1) to Zoorasia, a zoo in Japan, as part of government goodwill for opening up new airline routes to that country.

In 2011, the Port Moresby Nature Park (formerly the NCBG) was established and is a well-recognized ZAA accredited institution. According to Brett Smith, in January 2019, three species of tree kangaroos were on display: Doria's (4.3), Goodfellow's (2.1), and Matschie's (1.1). In 2016 the park hosted the International Tree Kangaroo Conference and 2nd GSMP Master Plan session and continues to set the standard on the welfare for managed animals in PNG. The Adventure Park, located just outside of Port Moresby, has four species of tree kangaroos on display: Matschie's (4.2), Goodfellow's (8.9), Doria's (6.7), and Grizzled (1.3). It is privately owned and in 2013 it exported (4.4) Doria's tree kangaroos to Xiangjang Safari Park, a private zoo in China.

The Conservation, Environment Protection Authority (CEPA) of PNG, the government agency responsible for permitting exports of fauna from PNG, has indicated that they are agreeable to once again export animals, provided the animals are captive bred and not taken from the wild.

Oceania: Australia

The following is an excerpt from the Australian journal: Thylacinus: *Tree Kangaroos in Australian Collections* (Crook and Skipper, 1983):

Specimens of tree-kangaroos have been exhibited in Australian collections for many years. Adelaide Zoo obtained a pair of *D. inustus* from New Guinea in 1890 and in the following ninety years a singleton *D. bennettianus* and colonies of *D. m. matschiei*, *D. m. shawmayeri* and *D. dorianus* have been exhibited. Most of the other members of the genus have been represented at some time in Australia.

The Adelaide Zoo and Perth Zoo held the largest collections of Matschie's at that time. In 2017, Matschie's tree kangaroos were phased out of Australian zoos when the last animal died at the Adelaide Zoo.

The Melbourne Zoo, formerly the Zoological and Acclimatization Society, is the oldest zoo in Australia and one of the first zoos to import tree kangaroos. The 31st Annual Report of the Society lists Bennett's tree kangaroo (Fig. 18.1) in its inventory in 1894 (Melbourne Zoo, 1895). The Lumholtz's tree kangaroo was the second species brought to the collection in 1905 followed by Matschie's in 1929. During the late 1950s to 1960s, Doria's and Grizzled tree kangaroos were displayed in their collection. From 1978 to 1979, Goodfellow's were the last species to be added to Melbourne's collection (Richardson, personal communication, 2019). In early 1990, Melbourne Zoo received two males and two females from BRS. A trio was immediately set up and the additional young male was transferred to Currumbin Sanctuary. These animals are founders of the current Goodfellow's population.

Sam Bennett (2013) presented a history of tree kangaroos at the Taronga Zoo at the 2013

FIGURE 18.1 Bennett's Tree Kangaroo (*D. bennettianus*) from the Melbourne Zoo, 31st Annual Report of the Zoological and Acclimatization Society, 1894.

International Tree-Kangaroo Workshop held in Melbourne. Lumholtz's tree kangaroo was the first species kept at Moore Park, the former site of the Taronga Zoo, in 1909. Over the course of nearly 100 years the zoo has managed five species: Lumholtz's, Bennett's, Doria's, Matschie's, and Goodfellow's.

In the 1994 Goodfellow's draft management plan of the ASMP (Kingston, 1994), it was reported that Goodfellow's tree kangaroo have been recorded in the Australian region from the early 1960s. They, like other species, were probably held many years prior to this but clear records were not established until the 1970s. All animals were imported into Australia from PNG either from the BRS or private dealers and collections were kept at a number of facilities.

ZAA is a founding partner and host region for the Goodfellow's GSMP. The Melbourne Zoo was the host site for the 1st GSMP Master Planning session in 2013 (Richardson, 2018d). According to the 2018 ISB (Richardson, 2018a) there were 12.11.0 (23) animals in nine facilities in Australia.

The Lumholtz's studbook was first compiled by Rosie Booth in 2005 at the David Fleay Wildlife Park (DFWP), West Burleigh (Booth, 2005). Studbook records date back to 1979 from wild caught animals transferred to the Pallarenda Research Facility, Townsville. Several of these animals were later transferred to the DFWP in the late 1990s and early 2000s, including the last living representative from Pallarenda who died at 18 years old. This animal holds the record for oldest Lumholtz's tree kangaroo in captivity.

While Lumholtz's tree kangaroos have been held in captive collections for decades, it is only since 2013 that they have become a recognized ASMP managed species (Fig. 18.2). The first animals under this program were transferred to the following facilities: Dreamworld, Wildlife Habitat, and DFWP. The primary goals of the program are to secure quality long term housing for nonreleasable rescued animals and create zoo-based ambassador programs to raise awareness and financial support for conservation initiatives.

The February 2019 Lumholtz's Studbook recorded 8.5 animals housed in six ZAA accredited facilities (Legge, 2019). It should be noted that in 2019, approval was given to move the first animals to a facility outside of Queensland under the program. Oakvale Wildlife Park in New South Wales was the receiving facility. The Lumholtz Lodge and the Tree Roo Rescue and Conservation Centre are rescue and rehabilitation facilities that reported 11.12 animals in 2019 (Chapter 6).

Europe

European zoos have been exhibiting tree kangaroos for over 16 decades. Dr. Ulrich Schürer, retired Director of the Wuppertal Zoo in Wuppertal, Germany has compiled a history of tree kangaroos in Europe that provides information on dates and locations of animal acquisitions. The following historical information is based on selections from Schürer's writings (Schürer, 2019).

Grizzled tree kangaroos were the first reported in Europe when a female arrived at the London Zoo in 1848, where she lived for four years. The second arrived in 1865 from Zoo Rotterdam (Schürer, 2019) and a third one in 1877 (Flower, 1929). The Artis Zoo in Amsterdam acquired a female tree kangaroo in 1856 that did not live long. The two historically important Dutch zoos of Amsterdam and Rotterdam received many specimens from the former colony "Dutch East Indies" (Indonesia). Zoo Rotterdam was a turntable for tree kangaroo imports and distribution in the late 19th century.

FIGURE 18.2 Lumholtz's tree kangaroo (Wildlife Habitat). *Source: Terry Carmichael.*

It kept Grizzled tree kangaroos as early as 1862 and Black tree kangaroos by 1864 (Jones, 1963; Schürer, 2019). Rotterdam was the first zoo in Europe which kept both of these species simultaneously.

Berlin Zoo also had Black tree kangaroos in 1910 (Schürer, 2019). A pair purchased by the London Zoo in 1909 had a joey which was discovered on 10 April 1912, but unfortunately it died nine weeks later. This was the first documented breeding record for any tree kangaroo in Europe (Flower, 1929). In the 1940s, breeding groups of this species existed in the Zoological Gardens of Berlin and Leipzig (Steinmetz, 1940; Schneider, 1954; Gewalt, 1965). Collins (1973) reported a longevity record of 14+ years for a Black tree kangaroo from Leipzig Zoo transferred to Frankfurt Zoo.

Two pairs of Bennett's tree kangaroos were received at the Zoological Society of London in 1894 from the Melbourne Zoo (Flower, 1929). In August 1895, Berlin Zoo purchased a pair according to the "Berliner Börsenzeitung". According to "Naturkundemuseum Berlin", between 1895 and 1910 Berlin Zoo had at least ten Bennett's (Schürer, 2019). However, neither European institution published a breeding record for this species.

The Matschie's tree kangaroo was discovered in the former German colony "Kaiser-Wilhelmsland" in former German New Guinea (encompassing the Huon Peninsula) and is described by Förster and Rothschild (1907) honoring the German zoologist Paul Matschie, whose name is well known to mammal taxonomists. London Zoo had Matschie's tree kangaroos and bred them in 1932 and 1934 (Zuckerman, 1953). Frankfurt Zoo had a female in 1936, the first on the European continent (Zukowsky, 1937). Berlin Zoo had Matschie's since 1959 (Gewalt, 1965) and later bred this species as did numerous other institutions in Europe. Probably the best breeding group in Europe was kept by Zurich Zoo (Schürer, 2019).

Doria's tree kangaroos were introduced into zoological gardens later than the other species. They were reported for Blijdorp Zoo (Rotterdam) in 1961 (Jones, 1963). The Wilhelma Zoo in Stuttgart received Doria's in 1969 and started breeding in 1973. Karlsruhe Zoo in Germany started with a group in 1976 and had several offspring in the late 1970s. The British zoos of Twycross and Blackpool also kept groups of Doria's in the 1970s.

A male and two female Goodfellow's arrived at the London Zoo in 1963 and another male arrived in 1969 (Richardson, 2015). In 1969 the Berlin Zoo had 1.2 Goodfellow's tree kangaroos in their collection (Klös, 1969).

The first and probably only Lumholtz's tree kangaroo in Europe was kept in the Wilhelma Zoo in Stuttgart in 1967. It was a young, not yet fully weaned male that did not survive very long (Schürer, 2019).

German zoos have by far been the largest holders of tree kangaroos in Europe during the 20th century (Dressen, 2018). Data from IZY Volumes 1–27 show that 27 institutions in eight countries in Europe housed tree kangaroos. Germany alone had 15 institutions. The Zurich (Switzerland) and Rotterdam (Netherlands) zoos had Matschie's in their collections for many years. Up to four *Dendrolagus* species were reported in collections of Zoo Berlin and at three zoos in the Netherlands. Three zoos in Great Britain and two Italian zoos also reported tree kangaroos in their collections during this time period. In all, seven species of tree kangaroo were reported in European collections.

Similar to zoos in North America, European breeding programs did not show much success until the 1970s and 1980s. European zoos have been participating in global breeding programs for many years and have been participating partners of the Goodfellow's GSMP since its inception in 2013 (Ford, 2013). As of February 2019, Europe exhibits Goodfellow's and a few Matschie's tree kangaroos (1.2 at two institutions) (Fig. 18.3). Per the 2005 Atherton

FIGURE 18.3 Goodfellow's tree kangaroo (Krefeld Zoo). *Source: Vera Gorissen.*

Tablelands Summit agreement, Matschie's are slated to be phased out of Europe. The population of Goodfellow's in Europe as of February 2019 was: 10.14.1 in ten institutions in four countries (Species360, 2019).

Continued education and international participation are critical for the success of tree kangaroos in Europe and for the European Endangered Species Programme. Zoo Krefeld has led the community by hosting two international meetings (2009 and 2018) and served as the location for the 3rd GSMP meeting and master planning session in 2018 (Richardson, 2018b).

North America

Tree kangaroos have been kept in North American (NA) zoological facilities since 1927. Records in Species360 ZIMS show that four wild born Lumholtz's tree kangaroos of unknown sex were transferred to the San Diego Zoo approximately 3 January 1927 (Legge, personal communication, 2019). The Philadelphia Zoo had a Black tree kangaroo from 1929 to 1939 and the National Zoological Park (NZP) in Washington, D.C. reported a longevity record of over 20 years for a Black tree kangaroo which they received in 1937 and died in 1957 (Collins, 1973). Records indicate there were other *Dendrolagus* species in NA zoos in the 1950s such as Bennett's, Matschie's, Doria's, and Grizzled tree kangaroos although most breeding groups were established in the 1970s and 1980s. (Collins, 1973; Crandall, 1964; Smith, 1988a, 1988b).

R.L. Smith reported that during the early 1970s there had been a substantial number of tree kangaroos taken into captivity. For at least three species (Goodfellow's, Grizzled, and Matschie's), there was a nucleus of potential founders which could have insured long term species survival in NA (Smith, 1988a). However, from 1960 to 1986 only six zoos reported Doria's in their collections and the last Doria's died in 1991 at the San Diego Zoo.

Smith (1988b) also reported that 52 Grizzled tree kangaroos were imported into NA facilities in the 1970s and 1980s. Due to the lack of a specific breeding program there were only 12 remaining in 1989 and by December 1999, there were only three female Grizzled left. The TKSSP (Steenberg, 1992) tried to identify one facility to house all three animals in order to better monitor and understand the species. Unfortunately, a single facility was not identified, and no behavioral or biological data were collected. The last Grizzled tree kangaroo died in 2006 at 17 years of age at the San Antonio Zoo (Rodden, 2006–2008).

Goodfellow's tree kangaroos were reported in five NA zoological facilities from the early 1970s to 1986 (IZY Vols. 12–27). In 1994, the

ISB reported 38 (21.13.4) Goodfellow's in 11 institutions (Collins, 1994-1996). By 1999, the Goodfellow's population numbered 18 and had declined to 11 animals by 2006. Transfers recommended at the 2005 Tree Kangaroo Summit in the Atherton Tablelands Australia, as well as age-related mortalities, contributed to this decline. The last Goodfellow's tree kangaroo died in 2017 at 17 years of age at the San Diego Zoo. While the region no longer holds Goodfellow's tree kangaroo, AZA is a founding partner and participant in the GSMP.

In 1961, three NA facilities reported Matschie's tree kangaroos in their collections and by 1985 the number had increased to ten with most facilities housing 1.1 animals (IZY Vols. 1–27). The 1988 ISB for Matschie's tree kangaroo reported 43 (22.21) Matschie's at ten NA facilities (Collins, 1990-1993). Records indicate that the NZP acquired captive born Matschie's tree kangaroos from BRS in 1970 and breeding behavior was observed in 1971 (Collins, 1973). In 1982, Woodland Park Zoo (WPZ) (Seattle) would receive the last import of Matschie's from BRS (Fig. 18.4).

The original success of the Matschie's tree kangaroo in NA facilities is directly related to the dedication of Larry Collins, formerly the Mammal Curator at the NZP's Conservation and Research Center (NZP-CRC). Collins started a breeding program with six (1.5) Matschie's tree kangaroos at the NZP–CRC in 1977. Three additional Matschie's were received from BRS in 1978, and several other individuals were acquired from a variety of sources. By 1990 the NZP-CRC population was 18 (7.9.2), with an additional 13 Matschie's out on breeding loan. Sixty births had been recorded but not all joeys survived (Heath et al., 1990). Research and careful records-keeping confirmed an average gestation period of 44.2 days for Matschie's and the need to separate male and female tree kangaroos, resulting in a significant increase of joey survival rate as reported by Benner and Collins (1988).

Due to the reproductive research conducted at NZP-CRC and subsequent behavioral research at WPZ, tree kangaroo husbandry improved. In 1997, the population of Matschie's tree kangaroos reached an all-time high of 90 (39.46.5) animals, at 27 institutions (Rodden, 1997). This was the result of cooperative husbandry management and reproductive efforts throughout NA zoos. Most importantly, there was a steady increase in the survival of tree kangaroo joeys.

This population peak occurred after a husbandry survey, conducted in 1989 (Mullett

FIGURE 18.4 Last Matschie's imported into North America. "Huen" was transferred from Wildlife Reserves Singapore to Woodland Park Zoo, Seattle WA in 2009. *Source: Ryan Hawk.*

et al., 1990a), helped identify housing and reproductive issues which were then addressed through the updated publication of the Tree Kangaroo Husbandry Manual (Mullett et al., 1990b), the establishment of the TKSSP (Steenberg, 1992) and the first TKSSP Master Plan (Steenberg, 1993). However, from 1998 to 2007 (Rodden, 1998–2005, 2006–2008) there was a steady decrease in the Matschie's population. The population had declined to 78 within three years and for the next six years it continued a downward trend. By December 2007 there were only 49 Matschie's tree kangaroos remaining in NA. Several factors contributed to the decline:

- The birth rate and survival of joeys continued to be lower than needed for sustainability.
- The population was aging, and mortalities were exceeding births on a regular basis
- Noncompliance of institutions to the TKSSP protocols and recommendations.

Over the next 10 years, numbers fluctuated but remained close to 50, which is significant in regard to how an SSP is classified. The AZA Animal Programs are divided into four categories. Table. 18.1 provides an overview of those categories. The TKSSP maintained Yellow SSP status for many years. Unfortunately, in 2017 with the population in decline, the program status changed to a Red SSP.

Throughout the past decade, the TKSSP has taken many strides to help address the decline in the population in the hopes of once again building numbers to a Yellow SSP:

- Regular Breeding and Transfer planning sessions with the Population Management Center (PMC)
- Publication of the updated 2007 Tree Kangaroo Husbandry Manual (Blessington and Steenberg, 2007)
- Publication of Medical and Necropsy reports 2007 (Blessington and Steenberg, 2007)
- Joey Developmental Milestones database created in 2008
- Institutional Diet Survey in 2015
- Development of a Body Condition Score Chart in 2017
- Nutrition analysis and diet reformulation in 2019
- Organization of and participation in 10 workshops, of which four were international
- Fecal hormone assay testing in females to assess female reproductive cycles

TABLE 18.1 Association of Zoos and Aquariums (AZA) Species Survival Plan® (SSP) Animal Program Designations. Taxon Advisory Groups (TAG) first determine if program is recommended for management; population size, number of AZA holding institutions and projected gene diversity determine program level.[a]

Criterion	Green SSP Program	Yellow SSP Program	Red SSP Program	Candidate Program
TAG recommended for cooperative management	Yes	Yes	Yes	Yes
Population Size (N)	50 and greater	50 and greater	49–20	19 and fewer
# AZA member institutions	3 and above	3 and above	3 and above	2 or fewer
Projected gene diversity (%GD) at 100 years or 10 generations	90.0% or above	Less than 90.0%	Less than 90.0%	NA
Institution participation	Mandatory	Voluntary	Voluntary	Voluntary

[a] *Components of table from: Association of Zoos and Aquariums (AZA), 2020. Species Survival Plan® (SSP) Program Handbook. AZA, Silver Spring, MD. Table 1. Applying Sustainability Criteria to Designate Animal Program Management Levels.*

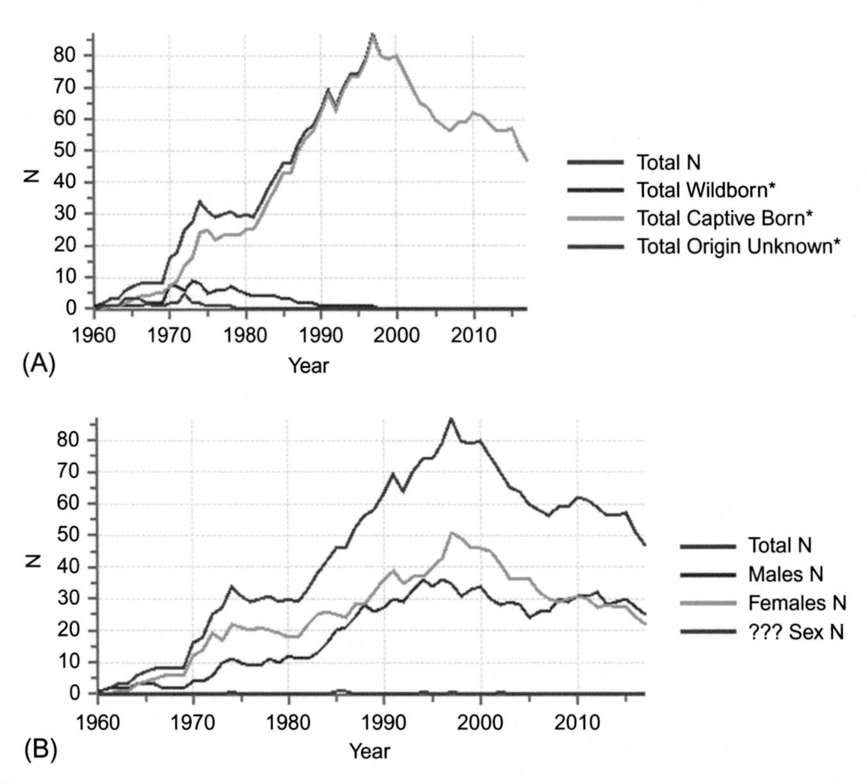

FIGURE 18.5 Census of Matschie's tree kangaroos in the North American population from 1960 to 2018, by origin (A) and sex (B). *Source: Association of Zoos and Aquariums (AZA), 2018. Population Analysis & Breeding and Transfer Plan, Matschie's Tree Kangaroo* (Dendrolagus matschiei)*. AZA Species Survival Plan (SSP) Red Program. AZA Population Management Center, Lincoln Park Zoo, Chicago, IL.*

Despite these efforts, the NA population continues to remain below 50 individuals (Fig. 18.5), primarily because the birth rate remains lower than needed for sustainability. In February 2019, the ISB recorded that the population was 43 (23.19.1) at 21 institutions with an additional 1.2 in two European institutions (Norsworthy, personal communication, 2019). Brett Smith reported an additional 5.3 in PNG (personal communication, 2019).

A Matschie's Tree Kangaroo Population Viability Analysis (PVA) was performed by the AZA Population Management Center in January 2019 (Che-Castaldo et al., 2019). Using the 13 December 2018 ISB data, the summary indicated that the population on average decreased by 2.5% annually and had 3.8 births per year in the past decade. The population currently retains 85.3% of founding gene diversity and has an average inbreeding coefficient (F) of 0.069. The 2018 Matschie's population represented by sex and age ratio in managed facilities is depicted in Fig. 18.6.

The PVA also projects population models which compare a baseline scenario (current management with no changes) to alternate scenarios that reflect potential changes in population management. For tree kangaroos, these alternate scenarios might include importation, varying breeding rates (probability of breeding),

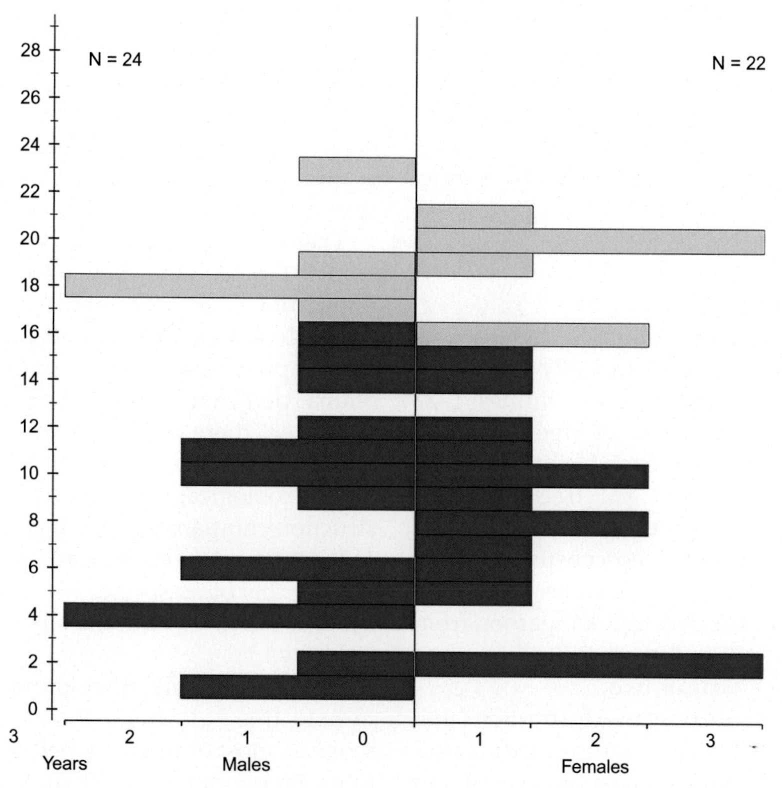

FIGURE 18.6 Age structure of the Matschie's tree kangaroo Species Survival Plan (SSP) population. N = number of individual animals. Dark blue indicates individuals of reproductive age. Light blue indicates individuals of postreproductive age (females) or nonreproductive status (males). *Source: Association of Zoos and Aquariums (AZA), 2018. Population Analysis & Breeding and Transfer Plan, Matschie's Tree Kangaroo (*Dendrolagus matschiei*). AZA Species Survival Plan (SSP) Red Program. AZA Population Management Center, Lincoln Park Zoo, Chicago, IL.*

and/or genetic management strategies. Under current breeding rates and without imports, the SSP population would decline and have an average extinction probability of 78% over the next 100 years. If possible, regular and sustained imports would improve the long-term viability of the TKSSP. Importing two individuals at least every four years could help the population avoid extinction and have an average size of ~48 individuals in 100 years by producing ~5 births/year. It will also help retain 90% genetic diversity (GD) and lower inbreeding (F = 0.05) (AZA, 2019b).

The TKSSP is working with the Port Moresby Nature Park (PMNP) in PNG on potential, future exports of captive bred animals from their facility to aide in managed population sustainability. Larry Collins and WPZ-CRC proved that with minimal new genes added to a managed population, it could become sustainable. The NZP-CRC breeding program for Matschie's started with 1.6 in 1977. Through careful management and working cooperatively with other zoological facilities over a 20-year span, the NA population steadily increased from 29 in 1977 and by 1997 peaked at 90 individuals

(Rodden, 1997). Reproductive and behavioral research contributed to this success.

Matschie's tree kangaroos are a long-lived species in managed programs. The 2018 Matschie's Tree Kangaroo ISB reported that the longest living male was 23 years four months and the longest living female was 27 years six months. (Norsworthy, 2018).

Asia

There are numerous countries and regions within Asia that have historically held tree kangaroos. Data from the IZY Volumes 1–27 (1960–86) gives a snapshot of tree kangaroo species that were reported from facilities in Asia. During this time period six tree kangaroo species were reported in four countries at 12 separate locations; there was no census report of Doria's being held in Asian facilities. While Singapore reported Grizzled tree kangaroos from 1977 to 1985, most of the other facilities had animals for only a year or two.

Currently, there are only two institutions that participate in global tree kangaroo breeding programs: Wildlife Reserves Singapore (WRS) and Zoorasia in Japan. Razak Jaffar reported the history of the last three decades of *Dendrolagus* captive management in WRS (Jaffar, personal communication, 30 January 2019). The first Goodfellow's tree kangaroo, a female, came into the collection from BRS in 1986. WRS has had the opportunity to work with three different species of tree kangaroos: Grizzled, Matschie's, and Goodfellow's. From 1986 to January 2019, a total of 18 tree kangaroos were at one-point part of the WRS collection.

In terms of reproduction, WRS has had successes in breeding both Matschie's and Goodfellow's tree kangaroos. WRS has also supported global breeding programs by providing exports to Hong Kong, Sydney, Japan and North America. In 2019, WRS sent a female Goodfellow's to France as recommended by the GSMP.

WRS exhibits Goodfellow's tree kangaroos and have a young pair with which they are trying to replicate breeding success of the past. Regionally, within SEAZA, WRS is the only holder of Goodfellow's tree kangaroos. WRS, as a founding partner of the GSMP, is focused on working with and contributing to the program. Within JAZA, Zoorasia currently has 1.1 animals that are a part of the GSMP.

There are a number of zoos in Indonesia that display Grizzled tree kangaroos, with the largest population held at Taman Safari Indonesia in Bogor. In March 2019, Taman Safari maintained 1.2.1 captive-born and 1.0 wild born Grizzled (Guha, personal communication, 17 Mar 2019). However, there are no specific details of the exact number currently being held in other Indonesian zoological facilities. As logging and construction companies move into the remote areas of West Papua (western half of New Guinea), it is likely that other species will start turning up in Indonesian zoos and the illegal animal trade industry.

China is the only other known region in Asia to hold tree kangaroos. Records are limited in scope as most of them are held in private collections. Brett Smith shared that Adventure Park in PNG exported eight Doria's tree kangaroos to a private zoo, Xiangjang Safari Park in 2013. In January 2019, the population in China was 3.2.1. There are talks that more species of tree kangaroos could be exported to China in the future, but nothing has come of it to date (Smith, personal communication, 2019).

GLOBAL SPECIES MANAGEMENT PLANS (GSMPs)

GSMPs are programs overseen by the Committee for Population Management (CPM) within WAZA (Ford, 2013). GSMPs were conceived in 2003 "...out of growing concerns about the long-term sustainability of wild animal populations in human care" (WAZA, 2019b).

Particular taxa have been identified where the strategic alliance between multiple regions can demonstrate greater sustainability of conservation results than a single regional approach. For many years regional zoo associations have been coordinating regional programs and even inter-regional programs. According to WAZA (2019b), "The current shift from a regional to a global population management framework is a change as significant as the move from institutional to regional species management in the 1980s and 1990s".

Global Management provides:

- An opportunity to link a number of isolated unsustainable populations improving demographic stability and managing inbreeding and gene diversity more effectively.
- The ability to strategically subdivide a global population by region and manage restricted migration between subpopulations to benefit genetic diversity.
- A mechanism for strategically distributing important founder lines in expanding populations primarily held in one region but sought after by others. This will avoid the scenario of over-represented lines exported, therefore reducing genetic potential and the conservation value of those populations.

The Goodfellow's tree kangaroo was endorsed by WAZA for management under a GSMP in 2013 and was one of nine species designated for these programs. The regional partners of the GSMP are: ZAA, AZA, EAZA, SEAZA, and JAZA. ZAA is the current host region of the Goodfellow's Tree Kangaroo GSMP program (Richardson, 2018d). The primary goal of the GSMP for Goodfellow's tree kangaroos is to enhance the sustainability of the zoo population globally. The population struggles with low numbers and gene diversity, which consequently leads to increased inbreeding coefficients (genetic similarity). Without a potential to acquire new founders, the genetic diversity will continue to decline. The population had a sex ratio highly skewed towards females, thus management recommended the establishment of trios (1.2) to maximize breeding opportunities. This recommendation has been successful, and the sex ratio is now leveling off (Richardson, personal communication, 2 June 2019).

The Goodfellow's tree kangaroo is one of two priority PNG tree kangaroo species for which there are international cooperative breeding programs. It is expected that the GSMP will expand to include Matschie's tree kangaroos, the other priority species, as the species programs share many common challenges and opportunities.

Bringing together the ex situ community, field conservationists, researchers and local people is integral to the GSMP to contribute to the One Plan Approach (WAZA, 2019b) for tree kangaroo conservation. Cooperative management at the global level is essential to enhance the sustainability of the species, genetically, demographically, and as well as to share husbandry expertise. It also provides a coordinated approach for supporting PNG tree kangaroos in the wild through prioritizing support for field conservation projects, namely the Tenkile Conservation Alliance and the Tree Kangaroo Conservation Program (Richardson, 2018d) (Chapters 13 and 22).

Goodfellow's have been documented in historical collections since the early 1960's. Fig. 18.7 displays a census of the population thru 2018. Data were obtained from the Goodfellow's Tree Kangaroo ISB (Richardson, 2018a).

In 2018, there were 55 individuals (26.28.1) across 21 institutions managed under the Goodfellow's GSMP in Europe, Australia, Japan, and Singapore. While individuals are known to be held in PNG at PMNP (2.1) and Adventure Park (8.9), they were not included in the genetic analysis of the managed population. The genetic summary is at 85.5% of founding gene diversity (GD) (Fig. 18.8) and has an average inbreeding coefficient (F) of 0.09 (Richardson, 2018c). Goodfellow's tree kangaroo, like the Matschie's, have potential for longevity in managed facilities. The

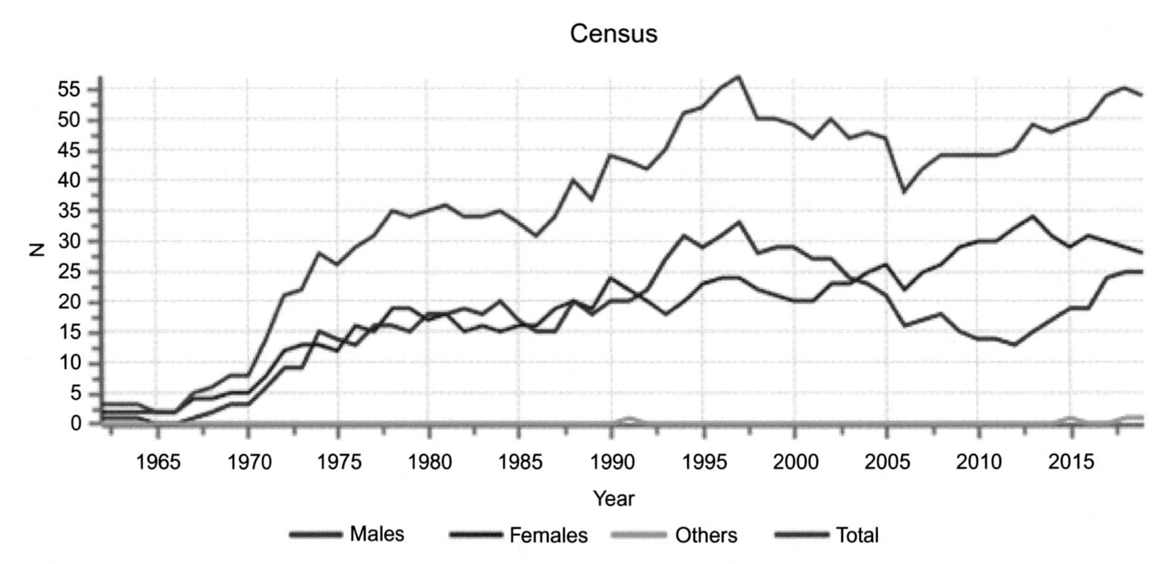

FIGURE 18.7 Goodfellow's tree kangaroo census in global zoos from the Goodfellow's Tree-kangaroo International Studbook (Richardson, 2018a) (updated January 23, 2019).

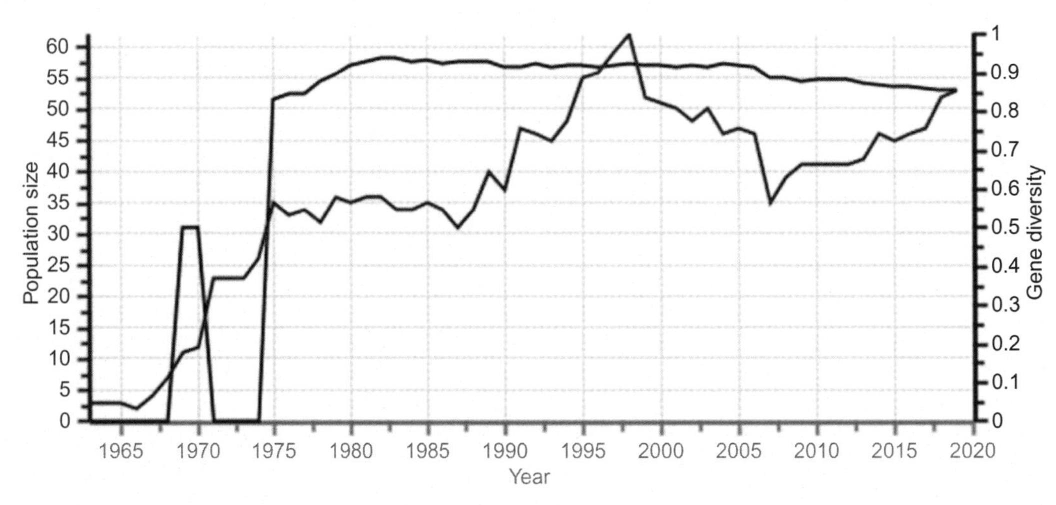

FIGURE 18.8 Population trends for Goodfellow's tree kangaroos in global zoos. *Data was obtained from the Goodfellow's Tree-kangaroo International Studbook (Richardson, 2018a) (updated January 23, 2019).*

2018 ISB records the oldest male living to ~19 years of age. The oldest female is recorded as living to ~28 years of age. However, because this female was wild born, the date of birth is estimated. She is also the oldest reported tree kangaroo of all species in managed populations.

The projections for this global population are similar to that of the Matschie's population as indicated in the PVA summary. Without the influx of new genes, the Goodfellow's population will see a similar decline in gene diversity and increase of inbreeding coefficients. As with the TKSSP, the GSMP and its regional associations would like to establish a working relationship with the PMNP in PNG on the potential, future export of captive bred animals from their facility to aide in the captive population sustainability.

CONCLUSIONS

Zoos, parks, and private collections have been exhibiting and holding tree kangaroos for well over 16 decades and have a rich history spanning the globe. Like many other species, tree kangaroo populations in the wild and in managed facilities have decreased over time. All facilities that choose to care for tree kangaroos must understand the importance of their role in supporting in situ and ex situ conservation. Dedicated efforts and contributions for raising awareness and education about these rare animals are key to public understanding and support. Participation in research and collaborative breeding programs are essential to the sustainability of all managed tree kangaroo species.

ACKNOWLEDGMENTS

The authors would like to acknowledge the contributions of TKSSP Coordinator Kathy Russell (Santa Fe College Teaching Zoo) for her review and recommendations for this chapter; Wolfgang Dressen (Krefeld Zoo) for current and historical information on Goodfellow's tree kangaroos in the European Association of Zoos and Aquaria collections; Josef Lindholm (Tulsa Zoo) for confirming historical International National Yearbook citation data and to Ken Kawata (formerly Staten Island Zoo) for comments and information regarding the reporting of historical zoo data.

REFERENCES

Association of Zoos and Aquariums (AZA), 2019a. Available from: https://www.aza.org/about-us. [9 March 2019].

Association of Zoos and Aquariums (AZA), 2019b. Available from: https://www.aza.org/ssp-population-sustainability [11 March 2019].

Ballou, J.D., Foose, T.J., 1996. Demographic and genetic management of captive populations. In: Kleiman, D.G., Allen, M., Thompson, K., Lumpkin, S., Harris, H. (Eds.), Wild Mammals in Captivity. University of Chicago Press, Chicago, IL, pp. 263–283.

Ballou, J.D., Lacy, R.C., Pollak, J.P., 2011. *PMx*: Software for Demographic and Genetic Analysis and Management of Pedigreed Populations (Version 1.0). Chicago Zoological Society, Brookfield, Illinois. Available from: http://www.vortex9.org/PMx.html.

Barongi, R., Fisken, F.A., Parker, M., Gusset, M. (Eds.), 2015. Committing to Conservation: The World Zoo and Aquarium Conservation Strategy. WAZA Executive Office, Gland, Switzerland.

Benner, K.S., Collins, L.R., 1988. Reproduction and Development in Matschie's Tree Kangaroos (*Dendrolagus matschiei*) in Captivity. National Zoological Park Conservation and Research Center, Front Royal, VA.

Bennett, S., 2013. The History of Tree-Kangaroos at Taronga Zoo and Current Management. In: Presentation at the International Tree-Kangaroo Workshop. Melbourne, Australia.

Blessington, J., Steenberg, J., 2007. Tree Kangaroo (*Dendrolagus* spp.) Husbandry Manual, Tree Kangaroo Species Survival Plan, third ed. Association of Zoos and Aquariums, Silver Spring, MD.

Booth, R., 2005. Lumholtz's Tree Kangaroo (*Dendrolagus lumholtzi*) Studbook. In: David Fleay Wildlife Park. Australasian Regional Association of Zoological Parks and Aquaria, West Burleigh, Australia.

Che-Castaldo, J., Russell, K., DeBo, D., Norsworthy, D.A., Blessington, J., 2019. Matschie's Tree Kangaroo (*Dendrolagus matschiei*) AZA Animal Program Population Viability Analysis Report. Lincoln Park Zoo, Chicago, IL.

Collins, L., 1973. Monotremes and Marsupials. Smithsonian Institution Press, Washington, DC, pp. 250–255.

Collins, L., 1990-1993. 1988–1992 International Studbook for Matschie's Tree Kangaroo, *Dendrolagus matschiei*. National Zoological Park Conservation and Research Center, Front Royal, VA.

Collins, L., 1994–1996. 1993–1995 International Studbook for Matschie's, *Dendrolagus matschiei*, Goodfellow's, *Dendrolagus goodfellowi*, and Grizzled Grey, *Dendrolagus inustus*, Tree Kangaroos. National Zoological Park Conservation and Research Center, Front Royal, VA.

Crandall, L.S., 1964. Family Macropodidae. In: Management of Wild Mammals in Captivity. The University of Chicago Press, Chicago, pp. 33–34.

Crook, G., Skipper, G., 1983. Tree kangaroos in Australian collections. Thylacinus 8 (7), 5–9.

Dressen, W., 2018. Historical perspectives of tree-kangaroos in captivity. In: Presentation at the International tree kangaroo meeting, Krefeld, Germany.

European Association of Zoos and Aquaria (EAZA), 2018. Overview of EAZA's Population Management Programs. Available from https://www.eaza.net/conservation/programmes.

European Association of Zoos and Aquaria (EAZA), 2019. About Us. Available from: https://www.eaza.net/about-us/. [10 Mar 2019].

Faust, L.J., Bergstrom, Y.M., Thompson, S.D., Bier, L., 2018. PopLink version 2.5. Lincoln Park Zoo, Chicago, IL.

Flannery, T.F., Martin, R., Szalay, A., 1996. Tree Kangaroos: A Curious Natural History. Reed Books, Port Melbourne, Victoria: Australia.

Flower, S.S., 1929. List of the Vertebrate Animals Exhibited in the Gardens of the Zoological Society of London 1828–1929. Mammals, vol. 1. Zoological Society of London.

Ford, C., 2013. Goodfellow's Tree-Kangaroo, *Dendrolagus goodfellowi* Global Species Management Plan, WAZA/ZAA. Taronga Zoo, Sydney, Australia.

Förster, F., Rothschild, W., 1907. Description of a new tree kangaroo. Novitates Zoologicae 14, 506.

Gewalt, W., 1965. Kleine Beobachtungen an selteneren Beuteltieren im Berliner Zoo II. Rotes Baumkänguruh *Dendrolagus matschiei* (Förster & Rothschild 1907). Der Zoologische Garten N. F. 31, 240–249.

Heath, A., Benner, S., Watson-Jones, J., 1990. Husbandry and Management of Matschie's Tree Kangaroo: A Case Study. In: AAZPA Regional Conference Proceedings. AAZPA, Wheeling, WV, pp. 518–527.

International Trade (Fauna and Flora) (Fauna) Regulation (ITFFR), 1982. Papua New Guinea Consolidated Legislation. Available from http://www.paclii.org/pg/legis/consol_act/itafr1982433/ [16 June 2019].

International Zoo Yearbook (IZY), Volumes 1–27, 1960–1986. Census of Rare Animals in Captivity, Mammals Bred in Captivity and Multiple Generation Births. The Zoological Society of London, London, UK.

IUCN/SSC, 2014. Guidelines on the Use of *Ex Situ* Management for Species Conservation. Version 2.0. IUCN Species Survival Commission, Gland, Switzerland pp. 1–15.

Jones, M., 1963. Mammals in captivity. In: Part 1 The Monotremata, The Marsupialia, 1–39. Private Publication.

Kingston, J., 1994. Draft Australasian Species Management Plan. Royal Melbourne Zoological Gardens, Melbourne, Australia, pp. 5–6.

Klös, H.G., 1969. Der Tierbestand des Zoologischen Gartens zu Berlin am Tage seines 125- jährigen Bestehens (1 August 1969). Berlin.

Lacy, R.C., Ballou, J.D., Pollak, J.P., 2012. PMx: software package for demographic and genetic analysis and management of pedigreed populations. Methods Ecol. Evol. 3 (2), 433–437. https://doi.org/10.1111/j.2041-210X.2011.00148.x.

Legge, A., 2019. Lumholtz's Tree Kangaroo (*Dendrolagus lumholtzi*) Studbook. Zoo and Aquarium Association (ZAA), Dreamworld. Gold Coast. Australia.

Mackay, R.D., 1981. Report of the Baiyer River Sanctuary, 1975–1980. Mount Hagen, Papua New Guinea.

Martin, R., 2005. Tree-Kangaroos of Australia and New Guinea. CSIRO Publishing, Australia.

Melbourne, 1895. 31st Annual Report of the Zoological and Acclimatisation Society of Victoria for 1894. Rae Bros Polygraphic Printers, Melbourne, Australia.

Mullett, T., Yoshimi, D., Steenberg, J., 1990a. Tree Kangaroo Husbandry Notebook 1989 Survey Results. Woodland Park Zoological Gardens, Seattle, WA.

Mullett, T., Yoshimi, D., Steenberg, J., 1990b. Tree Kangaroo Husbandry Notebook. Woodland Park Zoological Gardens, Seattle, WA.

Norsworthy, D.A., 2018. International Studbook Matschie's Tree Kangaroo (*Dendrolagus matschiei*). Lincoln Children's Zoo, Lincoln, NE.

Richardson, M., 2015. International Studbook Goodfellow's Tree-Kangaroo (*Dendrolagus goodfellowi*). Melbourne Zoological Gardens, Melbourne, Australia.

Richardson, M., 2018a. International Studbook Goodfellow's Tree-Kangaroo (*Dendrolagus goodfellowi*). Melbourne Zoological Gardens, Melbourne, Australia.

Richardson, M., 2018b. 3rd Goodfellow's Tree-Kangaroo (*Dendrolagus goodfellowi*) Global Species Management Plan (GSMP) Master Plan Session. Zoo Krefeld, Krefeld, Germany.

Richardson, M., 2018c. Australian species management program. In: Goodfellow's Tree Kangaroo (*Dendrolagus goodfellowi*) Annual Report and Recommendations, Melbourne Zoological Gardens. Zoo and Aquarium Association (ZAA), Melbourne, Australia.

Richardson, M., 2018d. Goodfellow's tree-kangaroo (*Dendrolagus goodfellowi*), Global Species Management Plan (GSMP). In: Melbourne Zoological Gardens. Zoo and Aquarium Association (ZAA), Melbourne, Australia.

Rodden, M., 1997. North American Regional Studbook for Matschie's Tree Kangaroo (*Dendrolagus matschiei*). National Zoological Park Conservation and Research Center, Front Royal, VA.

Rodden, M., 1998–2005. 1997–2004 North American Regional Studbook for Matschie's (*Dendrolagus matschiei*), Goodfellow's (*D. goodfellowi*), and Grizzled Grey (*D. inustus*), Tree Kangaroos. National Zoological Park Conservation and Research Center, Front Royal, VA.

Rodden, M., 2006–2008. 2005–2007 North American Regional Studbook for Matschie's (*Dendrolagus matschiei*), Goodfellow's (*D. goodfellowi*) and Grizzled Grey (*D. inustus*), Tree Kangaroos. National Zoological Park Conservation and Research Center, Front Royal, VA.

Schneider, K.M., 1954. Vom Baumkänguruh (*Dendrolagus leucogenys matschiei*). Der Zoologische Garten N. F. 21, 63–106.

Schürer, U., 2018. Tree Kangaroos at Baiyer River Sanctuary, Papua New Guinea. Zoo Grapevine & International Zoo News 48, 20–22 Independent Zoo Enhusiasts Society. Todmorden, Lancashire, UK.

Schürer, U., 2019. Zur Geschichte der Baumkänguru-Haltung in europäischen und einigen andern Zoos. Bulette (Berlin) 7, 7–45.

Smith, R.L., 1988a. New Studbook—Grizzled Tree Kangaroos. San Antonio Zoological Gardens, San Antonio, TX.

Smith, R.L., 1988b. Tree kangaroos: Vulnerable in the wild, threatened in captivity. In: Dresser, B.L., Reese, R.W., Maruska, E.J. (Eds.), Proceedings of the Fifth World Conference on Breeding Endangered Species in Captivity, Cincinnati, OH, 650.

Species360, 2019. About Us. Available from: https://www.species360.org/about-us/about-species360/ [12 March 2019].

Steenberg, J., 1992. Tree Kangaroo Species Survival Plan (TKSSP), American Association of Zoological Parks and Aquariums (AAZPA). Woodland Park Zoological Gardens, Seattle, WA.

Steenberg, J., 1993. Tree Kangaroo Species Survival Plan (TKSSP) Master Plan 1993–1994. American Association of Zoological Parks and Aquariums (AAZPA), Woodland Park Zoological Gardens, Seattle, WA.

Steinmetz, H., 1940. Nachrichten aus Zoologischen Gärten, Berlin. Der Zoologische Gärten N. F. 12, 335–343.

World Association of Zoos and Aquariums (WAZA), 2019a. Available from: https://www.waza.org/about-waza/ [9 March 2019].

World Association of Zoos and Aquariums (WAZA). 2019b. Available from: https://www.waza.org/piorities/conservation/conservation-breeding-programmes/global-species-management/plans/ [2 June 2019]

Zoo and Aquarium Association (ZAA), 2019. Who we are. Available from: https://www.zooaquarium.org.au/index.php/who-we-are/. [10 March 2019].

Zoological Information Management System (ZIMS), 2019. Available from: https://www (web archive link, 11 March 2019). zims.Species360.org. [11 March 2019].

Zuckerman, S., 1953. The breeding seasons of mammals in captivity. Proc. Zool. Soc. London 122, 827–950.

Zukowsky, L., 1937. Nachrichten aus Zoologischen Gärten, Frankfurt a. M. vom 1.I.1936 bis 15.IX.1936. Der Zoologische Gärten N. F. 9, 59–64.

CHAPTER

19

Genetics and General Husbandry of Tree Kangaroos in Zoos

Jacque Blessington[a], *Judie Steenberg*[b], *and Thomas J. McGreevy, Jr.*[c]

[a]Association of Zoos and Aquariums Tree Kangaroo Species Survival Plan® Program, Kansas City, MO, United States

[b]Association of Zoos and Aquariums Tree Kangaroo Species Survival Plan® Program, Maplewood, MN, United States

[c]Department of Natural Resources Science, University of Rhode Island, Kingston, RI, United States

OUTLINE

INTRODUCTION

In the past 40 years, the knowledge of tree kangaroo genetics and husbandry in managed facilities has improved immensely. While anecdotal reports still have a place in capturing information about species and individuals, research projects and data collection have become regular components of improving tree kangaroo husbandry around the world. Genetics research has been conducted on managed Matschie's (*Dendrolagus matschiei*) tree kangaroos in North American (NA) facilities to help determine founder population demographics. These data emphasized the importance of carefully planned pairings for reproductive success while maintaining genetic diversity.

Husbandry techniques are continually being evaluated and adjusted for optimal quality animal care. Through research and data

https://doi.org/10.1016/B978-0-12-814675-0.00003-8

collection, more questions will be answered which will continue to improve tree kangaroo husbandry in managed facilities. All facilities that have tree kangaroos in their collection are encouraged to acquire and make use of the Tree Kangaroo Species Survival Plan (TKSSP) 2007 Tree Kangaroo Husbandry Manual (TKHM) (Blessington and Steenberg, 2007). The TKHM contains detailed and useful information regarding the health and welfare management of tree kangaroos. The 2018 Population Analysis & Breeding and Transfer Plan (PABTP) from the Population Management Center (PMC) contains updated information from the TKHM (AZA, 2018). These materials can also be obtained from the TKSSP (AZA, 2019a). The Association of Zoos and Aquariums (AZA) oversees the development of Animal Care Manuals (ACM), which will replace husbandry manuals as a means for standardization and are considered works in progress (AZA, 2019b). The 2007 TKHM data and updates will be converted to the new ACM format and will become available thru the TKSSP. Through research and data collection, more questions will be answered which will continue to improve tree kangaroo husbandry.

GENETICS IN MANAGED FACILITIES

The use of genetic tools can inform a wide range of management decisions for captive animals, which include resolving taxonomic uncertainties, guiding captive breeding pairing and wild release decisions, monitoring genetic diversity and inbreeding levels, and measuring their adaptive potential (Frankham et al., 2010). An essential question when bringing an animal into captivity to start a managed population is: do animals from different geographic areas represent one species or subspecies? Confirming the animal's species identity is also critical, especially if a phenotypically similar species is sympatric. Genetics, along with other morphological and ecological analyses, can be used to answer these questions and prevent the inadvertent hybridization of two species. The subspecies versus species status of several tree kangaroo (*Dendrolagus spp.*) taxa is still in question and it would be critical to resolve this issue before starting a new captive breeding program for one of the unresolved taxa (e.g., Doria's tree kangaroo *D. dorianus*).

Below the species level, genetic tools can be used to determine if animals from different populations should be interbred. If samples can be collected from throughout a species range, genetic analyses can determine if the samples represent one population or numerous populations. If the samples are grouped into numerous populations, additional genetic analyses can determine if the differences among the groups are meaningful (i.e., adaptive) or mainly driven by random chance, such as genetic drift. The interbreeding of animals from isolated geographic areas can potentially degrade local adaptations that have developed over time and potentially cause an outbreeding depression, but on the positive side it can lead to increased genetic diversity or genetic rescue (Frankham, 2015). Interbreeding would be a factor to consider when importing animals from home-range countries.

Genetic tools can be used to determine how well a captive breeding strategy has preserved the captive population's genetic diversity and to avoid inbreeding. This information can be used to assess the genetic health of the captive population and help determine if the genetic diversity needs to be increased to avoid the negative impacts of inbreeding. Analyses of captive populations have shown that inbreeding can decrease juvenile survival. Inbreeding can also have effects on a number of other aspects which might include: elevated incidence of recessive genetic diseases; reduced fertility in sperm viability; and loss of immune system function (Ralls et al., 1988).

Unfortunately, captive populations are often founded with a limited number of animals that originated from only a portion of their full geographic range (Frankham et al., 2010). This was the case with the Matschie's tree kangaroo. The captive population was founded by the uneven genetic contribution of 19 animals, mainly from Baiyer River Sanctuary, with the vast majority of genetic contribution from only four founders (McGreevy et al., 2009). A mitochondrial deoxyribonucleic acid (mtDNA) analysis of samples from Matschie's tree kangaroos held at AZA facilities and wild Matschie's tree kangaroos from Papua New Guinea (PNG) revealed a paucity of mtDNA diversity in AZA Matschie's tree kangaroos. However, this result was not surprising because the AZA population only had four maternal lineages that contributed to the population (McGreevy et al., 2009). An additional nuclear DNA analysis by McGreevy et al. (2011) compared the amount of genetic diversity the AZA population has retained in comparison to animals from the wild and surprisingly they have retained similar levels of genetic diversity and do not show the genetic impact of a bottleneck. The genetic analysis of Matschie's tree kangaroo has been limited to the analysis of neutral markers, which are not influenced by natural selection or other factors.

Future studies could leverage the power of next-generation sequencing technologies to quantify their adaptive genetic diversity and determine if this also has been retained in captivity. Animals in captivity are negatively impacted by the selective forces of being in captivity, which can reduce their ability to be successfully reintroduced to the wild (Frankham, 2008).

One goal of a captive population is to preserve the highest level of genetic diversity possible to retain the option of releasing animals back into their natural habitat (Frankham et al., 2010). Genetic diversity is the raw material that natural selection acts upon and gives animals an increased chance of surviving a challenging environment (Markert et al., 2010). Current captive populations of Goodfellow's (*D. goodfellowi*) and Matschie's tree kangaroos are not considered genetically sustainable, which is crucial in captive management. PNG, the country of origin, has not requested any reintroduction programs and trying to recover wild populations by captive means has never been used by PNG. Traditionally, the PNG government, landowners and conservation groups such as Tree Kangaroo Conservation Program (TKCP) and Tenkile Conservation Alliance (TCA) have worked cooperatively to protect and leave these species alone to recover naturally (Smith, personal communication, 2 April 2019; Chapter 13). Lumholtz's (*D. lumholtzi*) tree kangaroos are currently part of a rescue/rehabilitation program in the Atherton Tablelands, Australia (Chapter 6). After appropriate health care and soft-release training for joeys, animals are released back into the wild. Most of the remaining Lumholtz's habitat is protected and current populations are relatively stable, negating a need for a captive breeding program for reintroductions (Valentine, personal communication, 27 March 2019).

All three tree kangaroo species held in captivity (Matschie's, Goodfellow's, and Lumholtz's) will encounter the same challenges due to their small population sizes. Even with the addition of founder animals, genetic diversity needs to be closely monitored to ensure healthy and sustainable captive populations. The future of these species in zoological facilities will require working together globally through breeding programs, education and awareness.

GENERAL HUSBANDRY

The welfare of animals in managed care has become an ever-increasing focus. In general, welfare is the condition of an animal at any given point of time and includes physical, behavioral, and psychological measures. The concept of welfare is often subjective and takes into account many variables. The AZA Animal Welfare Committee has described animal welfare using the following definition:

Animal welfare refers to an animal's collective physical, mental, and emotional states over a period of time, and is measured on a continuum from good to poor. Good welfare is typically experienced when an animal is physically and emotionally healthy, is experiencing physical comfort, and is well fed and safe from predation or other threats. Behaviorally, an animal experiencing good welfare should be able to perform behaviors that are highly motivated and typical for the species in the wild, and able to exhibit control over its environment. As physical, mental, and emotional states may be dependent on one another, and can also vary day to day, it is important to consider these states in combination with one another and over time when assessing an animal's overall welfare status (AZA, 2019c).

It is vital to the tree kangaroos' well-being that the natural conditions under which these animals survive in the wild are considered when they are housed in captivity. Arboreal habitat, social interactions in the wild, humidity, temperature, shade, and continuity of care are some of the important factors to consider when setting up an enclosure, whether for temporary or permanent housing. This will serve to provide a setting in which tree kangaroos can thrive. Their reproductive success is paramount in securing a future for tree kangaroo species in zoological facilities.

FIGURE 19.1 Matschie's tree kangaroo face and ear markings as individual identifiers (Miami Zoo). *Source: Yosanni Torres.*

Individual Identification

Tree kangaroos should be individually identified by physical characteristics as well as semi-permanent forms of identification, i.e. transponders. While most facilities house animals individually or in breeding/offspring pairs, they all require a form of identification for purposes of record keeping and transfer between facilities.

Many species can be identified from physical markings. Matschie's tree kangaroos are easily identified by facial and ear markings whereas tail markings are reliable identification cues for Goodfellow's tree kangaroos (Figs. 19.1 and 19.2). Other species such as Grizzled (*D. inustus*) or Doria's are not as easy to identify but could have individual characteristics that define them. Drawings and/or photographs of markings can be useful tools and a detailed description is essential to complete the specimen record.

Ear tagging, a common identification marker for terrestrial kangaroos and wallabies, is not recommended for tree kangaroos because tags often fall out or are torn out (Crook and Skipper, 1983). Tattooing is a reliable form of permanent identification but is only visible when the tree kangaroo is in hand. The TKSSP recommends the use of the transponders for

FIGURE 19.2 Goodfellow's tree kangaroo tail markings as individual identifiers (Healesville Sanctuary). *Source: Amie Hindson, Zoos Victoria.*

semi-permanent IDs. The capsule should be implanted between the scapulae, not behind the ear or any other location where grooming occurs.

Social Interactions

Tree kangaroos are mainly solitary animals in the wild, and therefore it is recommended that mature males be housed separately. The only exceptions are pre-reproductive males (Thompson, personal communication, n.d.). If necessary, these males can be housed together temporarily with very careful monitoring. Three young male Matschie's, all less than 24 months old, had to be separated when two were injured (Collins and Steenberg, personal observations, 1990). In general, so little is known about male social interactions in the wild that until researchers know more, it is best to be cautious and keep all male tree kangaroos in separate quarters.

Special consideration should also be given for potential male tree kangaroo aggression towards staff. Some individuals have been known to demonstrate aggression on a regular basis, others when the female is in estrus (Chapter 21). Exhibits and holding areas should allow for 'protected contact' of these individuals so staff do not have to enter their areas when they are present.

Another housing consideration for tree kangaroos is accommodation for pouch-gravid females. The safest way to avoid harassment from other animals is to separate conspecifics prior to parturition. A pouch-gravid female should remain in her original enclosure. The conspecifics should be moved to quality quarters since it will most likely be a long-term separation (Thompson, 2000). The separation of conspecifics is important because they have been known to pull joeys from the female's pouch (Collins, 1986). Males should be moved out of olfactory range, preferably into another area. Space must also be allocated for the permanent separation of joeys from females (Chapter 21).

Exhibits and Housing

Tree kangaroos live in rainforest habitats in Northeastern Australia and the island of New Guinea. It is vital to the tree kangaroos' well-being that the natural conditions under which these animals survive in the wild are considered when they are housed in captivity. Natural history is important to consider when setting up an enclosure.

The TKSSP has a review process for institutions that are housing tree kangaroos for the first time, designing new exhibits, renovating current facilities, or for temporary holding such as quarantine. The Facility Review Checklist and Exhibit Checklist are forms provided by the TKSSP as guidelines. These checklists can be found in the PABTP (AZA, 2018) or obtained by contacting the TKSSP through AZA (2019a). These checklists can be utilized by any facility planning to hold tree kangaroos.

Tree kangaroos need spacious enclosures with extra height for climbing because of their arboreal nature. The Australian species are reported to be capable of jumping huge distances: 9 m downward to another tree and 18 m to the ground. The New Guinea species are less agile, but their acrobatic abilities should not be taken lightly either (George, 1982). The TKCP has documented Matschie's tree kangaroos leaping down 20 m (Porolak et al., 2014).

The 1993 TK-SSP Master Plan (Steenberg, 1993) "Basic Care and Housing Standards" document suggests a 6 × 3 × 3 m minimum enclosure space for one animal. Enclosure size should also take into consideration for social housing needs (Figs. 19.3 and 19.4).

The TKSSP recommends tree kangaroo enclosure substrate be layered to a minimum of 20–30 cm (Heath et al., 1990). Layered substrates might include sand, hay/straw, crushed limestone, pea gravel, wood chips, leaf litter, peat, pine bark, white cedar mulch, natural soil and grass. This method serves to primarily cushion an animal if it falls, and can also add support for trees and poles, filter particulate matter,

FIGURE 19.3 Indoor tree kangaroo exhibit (Roger Williams Park Zoo). *Source: Christine MacDonald.*

FIGURE 19.4 Outdoor tree kangaroo exhibit (Melbourne Zoo). *Source: Megan Richardson, Zoos Victoria.*

and create a more natural looking exhibit for the public. A cushioned substrate is critical for young joeys that tend to be unsteady when first exiting the pouch. If a joey falls from a perch or platform, this extra padding increases their chance of survival.

Tree kangaroos will climb on cage mesh and screening material on enclosures should be of adequate size to avoid nail sheaths from getting stuck in the mesh and being pulled off. The mesh should be a sturdy fabric with openings approximately 2.5 × 2.5 cm. A mesh fencing material, 5 × 10 cm, has also been successfully used.

Inhabiting dense rainforests, tree kangaroos need shade and cannot tolerate prolonged direct sunlight or dry heat. Enclosures should be built to provide plenty of shade with some areas of either natural or artificial light. Natural light can be from a complete or partial outdoor exhibit, skylights, or other windows. Fluorescent or incandescent lights, or other sources of artificial light, can be used in indoor exhibits. Artificial light can also be used to increase the length of the photoperiod during short winter days.

Tree kangaroos are considered crepuscular (mainly active during the periods of dawn and dusk) and should not be managed as nocturnal animals (Collins, personal communication, n. d.). Tree kangaroos also display levels of cathemeral behavior (sporadic and random intervals of activity during the day or night in which food is acquired). Alana Legge (personal communication, 21 June 2019) shared that infrared cameras were utilized at Dreamworld, Australia on Lumholtz's tree kangaroos to monitor activity. On many occasions, different individuals browsed intermittently through the night and day. Further data collection on tree kangaroo activity will help to improve welfare in managed facilities.

The temperature of an indoor tree kangaroo enclosure should range between 16 and 27 °C. While they appear most comfortable at temperatures between 16 and 24 °C, temperature in the cloud forest of PNG can drop as low as 10 °C. In captive settings, it has been found that tree kangaroos can easily tolerate temperatures in the upper range of −1 to 4 °C for short periods with access to warm areas. An outdoor tree kangaroo enclosure may be as warm as 30 °C (though not recommended) provided that the humidity is not excessively high (>60%) at the same time (Collins, personal communication, n. d.). Tree kangaroos must be allowed access to shade and cooler areas when temperatures reach 30 °C. Heat stress seems to be common in tree kangaroos, regardless of species and can lead to death.

Tree kangaroos should be checked for signs of heat discomfort which include forearm licking for cooling effect (Fig. 19.5), decreased appetite, reduced activity, and increased water consumption. However, if it gets too hot, tree kangaroos can become noticeably uncomfortable, and the cooling effect of licking their forearms can be disrupted. Equipment such as fans, air conditioners and misters can aid in maintaining preferred temperatures. Direct, heavy misting on tree kangaroos can have an adverse effect if the fur coat becomes saturated; a water-logged coat can actually increase heat retention. Dehumidifiers can be beneficial in exhibits or holding areas where humidity becomes excessive.

FIGURE 19.5 Goodfellow's tree kangaroo with wet forearms (Melbourne Zoo). *Source: Megan Richardson, Zoos Victoria.*

To prevent scaly tail, dry lusterless coats, brittle hair and flaky skin, the relative humidity should be 50–55%. Features such as shallow pools (less than 10 cm deep), plants, moist substrate, misters or a combination of these can help maintain humidity in dry environments (Chapter 20).

Furnishings are important because they help replicate an animal's natural environment. In the wild, tree kangaroos utilize vines or small branches as they climb up into the canopy. Vertical and horizontal branches (10–15 cm diameter) for climbing and platforms for sleeping and feeding (at least 0.5 m wide by 0.9–1.8 m long) should be included in any holding or exhibit area (Blessington and Steenberg, 2007). Branches should be arranged to allow one animal to get out of the way of another, and to avoid dead-end situations. In addition, branches placed high over the heads of staff may provide the animal with a sense of security. The importance of platforms in a tree kangaroo enclosure is emphasized by the fact that they are the areas [where] the joeys are usually first seen out of the pouch (Steenberg, 1988). Rough-barked trees are preferred over smooth bark to ensure good footing. If bark becomes worn or smooth, branches can be scored or wrapped with natural rope. Platforms should also have a non-slip surface and spacing between boards should be minimized to prohibit joeys from falling through or getting stuck (Thompson, 1996).

Logs, rocks, nontoxic plants and trees, and other furnishings can also enrich a tree kangaroo enclosure. Tree kangaroos will browse on live trees and plants if accessible and palatable. If needed, protect parts of the trees with guards to effectively restrict the animals to certain areas of the trees or to prevent the animals from stripping bark and leaves.

Both wire mesh and dry moats have been used as successful perimeter barriers. Water moats are not recommended as tree kangaroos have drowned. Some facilities do have shallow pools and it is recommended that pools not exceed depths of 5–10 cm. Enclosing walls can also serve as successful barriers. In an open exhibit, it is important to remember that a high perch or structure should be a minimum of 2–3 m from the enclosing wall because of a tree kangaroo's ability to leap great distances. Walls should have an internal overhang and be high enough so animals cannot jump out. Surfaces of the wall should be smooth in texture and free from 'grab holds' to prevent tree kangaroos from climbing out (Blessington and Steenberg, 2007).

Electric fencing has been used to keep tree kangaroos out of planted areas or to separate animals. If an animal jumps onto the fence and is not touching the ground, there is no shock effect. Animals have been known to climb over without effect in this fashion (Woodland Park Zoo and Kansas City Zoo). However, it has also been noted that a male tree kangaroo standing on the ground grabbing the fence appeared to be 'frozen' as the fence was pulsing which resulted in burn wounds on the paws and the animal appearing dazed for several hours (Blessington, personal communication, 2001). Because of these situations, electric fence is not recommended in tree kangaroo enclosures.

Mixed-Species and Proximity Species Exhibits

In some cases, mixed-species exhibits can be used to better simulate the natural conditions under which an animal survives in the wild. A major concern when involving tree kangaroos in mixed species exhibits is the problem with Mycobacterium Avian Complex (MAC) and other associated *Mycobacterium* spp. infections (Chapter 20). The TKSSP recommends that birds not be housed with tree kangaroos to minimize the risk of infection. Also, tree kangaroos have been known to harass, stalk or even kill other animals in their exhibits (Drake, 1984; Mullett et al., 1988). It is important that they be watched closely for these behaviors when housed with other animals.

The success of mixed species exhibits depends on the amount of adequate cover for other species, the size of the exhibit and, of course, the individual animal personalities. There is no guarantee that any animal will be compatible with tree kangaroos. Another concern with mixed species exhibits is the potential interference from other animals during tree kangaroo copulation, which occurs primarily on the ground (Steenberg, personal communication, n. d.). A breeding pair of tree kangaroos should be carefully observed in a mixed species exhibit to see if there is any interference with their breeding activity.

The TKSSP has reported the following animals to be compatible based on surveys from AZA institutions: Rodrigues fruit bats (*Pteropus rodricensis*), small wallabies (*Macropus* spp., *Dorcopsis* spp.) or pademelons (*Thylogale* spp.) and echidnas (*Tachyglossus aculeatus*) (Mullett et al., 1990) (Fig. 19.6). A variety of tortoises have been successfully exhibited with tree kangaroos, but salmonella occurrence is often a concern with reptiles.

Consideration should be given when determining which other species to house or exhibit near tree kangaroos. Tree kangaroos do not have many natural predators and their adaptations for living near predatory species are diminished. Elevated stress levels over a short period can be resolved, but those that are long-term or continuous can have a detrimental effect on overall health (Chapter 20). Fecal assay testing for glucocorticoid levels was done on a female tree

FIGURE 19.6 Mixed species exhibit (Saint Louis Zoo): Matschie's tree kangaroo and short-beaked echidna. *Source: Saint Louis Zoo.*

kangaroo that was housed temporarily near a lion. The tree kangaroo showed a temporary rise in glucocorticoid production, with concentrations returning to baseline after approximately one week (Kozlowski, 2019).

Behavioral Husbandry

An animal's captive environment should be dictated by its native habitat and natural history (Forthman and Ogden, 1992; Forthman-Quick, 1984; Hediger, 1950, 1969). The concept and principles of enriching or enhancing an animal's environment are not new. In the 1920s, Robert Yerkes introduced the idea of enrichment when designing play items for the primates in his lab. Since then, others have recognized the need for perceptual input for captive animals (Shepherdson, 1992, 1999). Providing for enhanced mental well-being and improved welfare are now common principals for animals in captive settings. The AZA Behavior Scientific Advisory Group's (BAG), mission is to:

> promote positive welfare and science-based approaches to animal training, enrichment, and behavior-based husbandry programs. The BAG serves the AZA community of accredited institutions by supplying tools to support and enhance institutional programs, providing technical advice to meet the husbandry, social, and behavioral needs of animals, and promoting behavioral strategies which can be applied to the conservation of species (AZA, 2019d).

AZA also requires that accredited zoos have a staff member, committee, or volunteer appointed to oversee their enrichment programs.

Enrichment programs should have goals that solicit species-typical behaviors, promote positive social interactions, increase activity budgets, and decrease stereotypic behaviors. Enrichment programs should also include a framework that includes setting goals for the program, planning and implementing of enrichment, documenting what is accomplished, evaluating results, and making adjustments as needed.

FIGURE 19.7 Tree kangaroo painting as enrichment (Zoo Miami). Painting as enrichment fulfills multiple aspects such as sensory stimulation, item manipulation, and behavioral training. *Source: Ron Magill.*

Enrichment options should be goal oriented and include aspects related to feeding/foraging, sensory stimulation, social interaction, manipulable items, exhibit modification, and behavioral training (Fig. 19.7). These options should be approved by individual institutions prior to implementation. Safety is a primary concern when providing enrichment and items must be monitored on a regular basis. Caution should be taken to avoid causing undue stress on animals. It is also recommended that all food enrichment be included as part of the daily nutritional requirements.

The following items have been used by various institutions for tree kangaroo enrichment:

- Boxes – Non-wax coated, staples and tape removed, monitor for ingestion.
- Grain bags – Paper and burlap, monitor for ingestion.
- Cardboard tubes – Monitor for ingestion.
- Kong Toys™ – available at local pet stores.
- Browse – local, non-toxic, Veterinary Department approved species.
- Swimming pools with leaves, straw, water, wood chips. NOTE: Tree kangaroos can, and have, drowned in water moats. The depth of the water in the pool should not be more than a few inches and, is not recommended for joeys
- Crates – also useful to crate-train for shipping.
- Platforms – non-skid surfaces are important.
- Boomer Balls™ – assorted sizes.
- Scents – spices, extracts, perfumes – monitor and adjust for responses. NOTE: some scents can elicit an adverse reaction.
- Hanging canvas bags stuffed with straw.
- Cargo nets – monitor each individual for ease of movement.
- Rope ladders – same as for cargo nets.
- Plastic barrels – five-gallon pails with the wire handled removed, large 20-gal plastic drums with ends removed (pop syrup containers) and other non-toxic plastic containers
- Stumps – non-toxic species.
- Misters – can be used for cooling during high temperatures; however, do not directly spray or saturate animal as it can have adverse effects.
- Radio/tapes/CD's – for use of playing natural sounds and/or background noise for something different in the animal's environment stimulating a different sense. NOTE: avoid loud noise and observe to see if animal is agitated by sounds offered.
- Grapevine – can be hung and incorporated into other enrichment.
- Ferret logs – stuffed with computer paper or straw and treats; smear inside with peanut butter.
- Frozen treats – any type of approved food items frozen individually or in water.
- Sod and/or seed grown grass – must be chemical free; watch for netting in commercial sod. NOTE: Fescue grass is not recommended.
- Sprouts – check for freshness; they can mold quickly if they get too warm.

When enrichment is carefully planned to suit the needs of the species and individual animals, and is well documented to record its effectiveness, it can provide both activity for an animal and help reduce husbandry problems such as stereotypic behaviors.

Training programs should also have goal-based results. Programs might include training for husbandry or medical needs, solicitation of natural behaviors, behavioral enrichment, socialization, and animal presentations. Operant conditioning through positive reinforcement can be of great benefit when considering training techniques for tree kangaroos. When starting a training program several things must be taken into consideration including individual behavior. There can be great variances in personalities of individual animals in a collection. While one individual may be very laid back and tractable, others may be quite nervous and skittish. Some individuals, particularly males, can be aggressive and need to be handled with a different approach. Other considerations might include age of animal, health and capabilities of animal, enclosure/exhibit parameters (where will training occur), and activity patterns (when will training occur) to name a few.

Since tree kangaroos are animals that can become easily stressed, all external factors should be taken into consideration. Animals that are in quarantine or are being introduced to a new exhibit are exposed to a lot of stress. Training may need to be delayed during adjustment to new environments unless the animal is comfortable with the trainer(s). Behaviors solicited may also be limited to those that are reliable to ensure that the animal is set-up for success. The introduction of new cage mates can also cause additional stress. Training may be considered a potential tool for decreasing stress levels during introductions to new areas and cage mates. Offering 'rewards' or reinforcement for approximations to new areas and positive interactions with cage mates may decrease stressful situations.

Training is considered 'voluntary' participation by the animal. If an animal chooses not to participate in a training session, they should not be forced to perform a behavior. Rather, evaluate what might be going on with the animal or their environment and adjust accordingly.

FIGURE 19.8 Tree kangaroo target training for acceptance of a pouch check (Saint Louis Zoo). *Source: Saint Louis Zoo.*

Target training is the most commonly used training technique for tree kangaroos (Fig. 19.8). Favored food items should be identified and used for reinforcement. As with enrichment, it is recommended that all foods consumed as a part of a daily diet must be included in the overall nutrition requirements and caloric intake. The following is a list of behaviors that tree kangaroos might be trained for:

- Shifting across or through a space
- Standing on a scale
- Entering and remaining in a crate
- Standing for voluntary pouch checks
- Nail trims
- Eye, ear, mouth exams
- Body palpations
- Voluntary radiographs
- Voluntary ultrasounds
- Stationary urine collection
- Hand injections
- Application of medications to body parts
- Natural behaviors for presentations

There are many resources for training and enrichment available through various websites and organizations. For example, the American Association of Zoo Keepers, Inc. (AAZK) Enrichment Notebook, 4th Edition 2018 not only provides enrichment ideas with guidelines, but also provides a list of website links and locations for resources (AAZK, 2019). Two training resources commonly used in zoos and aquariums are:

Don't Shoot the Dog: The New Art of Teaching and Training (Pryor, 1999)
Animal Training: Successful Animal Management Through Positive Reinforcement (Ramirez, 1999).

Positive animal welfare is the goal for all animals in captive situations. Through research, observations and information sharing, facilities will continue to improve welfare. Husbandry techniques, enrichment, and training must be held to the highest standards for all animals in zoological facilities. Considering the natural conditions under which these animals survive in the wild are vital to tree kangaroos' well-being in captivity.

ACKNOWLEDGMENTS

The authors would like to acknowledge the contributions of TKSSP Coordinator Kathy Russell (Santa Fe College Teaching Zoo) for her review and recommendations for this chapter: Megan Richardson (Melbourne Zoo) for reviewing the chapter and providing Goodfellow's tree kangaroo data and photographs; Alana Legge (Dreamworld) for providing materials and information regarding the Lumholtz's tree kangaroo.

REFERENCES

American Association of Zoo Keepers (AAZK), 2019. Available from: https://www.aazk.org (01 March 2019).

Association of Zoos and Aquariums (AZA), 2018. Population Analysis & Breeding and Transfer Plan Matschie's Tree Kangaroo (*Dendrolagus matschiei*). AZA Species Survival Plan (SSP) Red Program. AZA Population Management Center. Lincoln Park Zoo, Chicago, IL.

Association of Zoos and Aquariums (AZA), 2019a. Tree Kangaroo Species Survival Plan (TKSSP). Available from: https://www.aza.org/contact-us (01 March 2019).

Association of Zoos and Aquariums (AZA), 2019b. Tree Kangaroo Animal Care Manual. Available from: https://www.aza.org/animal-care-manuals (01 March 2019).

Association of Zoos and Aquariums (AZA), 2019c. Animal Welfare Committee. Available from: https://www.aza.org/animal_welfare_committee (March 25, 2019).

Association of Zoos and Aquariums (AZA), 2019d. Behavior Scientific Advisory Group (BAG). Available from: https://www.aza.org/behavior-scientific-advisory-group (March 3, 2019).

Blessington, J., Steenberg, J., 2007. Tree Kangaroo (*Dendrolagus spp.*) Husbandry Manual. In: Tree Kangaroo Species Survival Plan. third ed. Association of Zoos and AquariumsSilver Spring, MD.

Collins, L., 1986. Big Foot of the Branches. Zoogoer, July-AugustNational Zoological Park, Washington, DC, pp. 35–40.

Crook, G., Skipper, G., 1983. Tree kangaroos in Australian collections. Thylacinus 8 (7), 5–9.

Drake, B., 1984. Response: (tree kangaroo predation). Thylacinus 9 (3), 15–16.

Forthman, D.L., Ogden, J.J., 1992. The role of applied behavior analysis in zoo management: today and tomorrow. J. Appl. Behav. Anal. 25, 647–652.

Forthman-Quick, D., 1984. An integrative approach to environmental engineering in zoos. Zoo Biol. 3, 65–77.

Frankham, R., 2008. Genetic adaptation to captivity in species conservation programs. Mol. Ecol. 17, 325–333.

Frankham, R., 2015. Genetic rescue of small inbred populations: meta-analysis reveals large and consistent benefits of gene flow. Mol. Ecol. 24, 2610–2618.

Frankham, R., Ballou, J.D., Briscoe, D.A., 2010. Introduction to conservation genetics. Yale J. Biol. Med. 83 (3), 166–167.

George, G.G., 1982. Tree kangaroos (*Dendrolagus spp.*) their management in captivity. In: Evans, D.D. (Ed.), The Management of Australian Marsupials in Captivity. Zoological Board of Victoria, Melbourne, pp. 102–107.

Heath, A., Benner, S., Watson-Jones, J., 1990. A case study of tree kangaroo husbandry at CRC. In: AAZPA Regional Conference Proceedings. American Association of Zoological Parks and Aquariums (AAZPA), Wheeling, WV, pp. 518–527.

Hediger, H., 1950. Wild Animals in Captivity. Butterworths Scientific Publications, London.

Hediger, H., 1969. Man and Animal in the Zoo. Delacorte Press, California.

Kozlowski, C.P., 2019. [tree kangaroo endocrinology]. Unpublished raw data. Saint Louis Zoo. St. Louis, MO.

Markert, J.A., Champlin, D.M., Gutjahr-Gobell, R., Grear, J.S., Kuhn, A., McGreevy, T.J., Roth, A., Bagley, M.J., Nacci, D.E., 2010. Population genetic diversity and fitness in multiple environments. BMC Evol. Biol. 10, 205.

McGreevy Jr., T.J., Dabek, L., Gomez-Chiarri, M., Husband, T.P., 2009. Genetic diversity in captive and wild Matschie's tree kangaroo (*Dendrolagus matschiei*) from Huon Peninsula, Papua New Guinea, based on mtDNA control region sequences. Zoo Biol. 28, 183–196.

McGreevy Jr., T.J., Dabek, L., Husband, T.P., 2011. Genetic evaluation of Association of Zoos and Aquariums Matschie's tree kangaroo (*Dendrolagus matschiei*) captive breeding program. Zoo Biol. 30, 636–646.

Mullett, T., Yoshimi, D., Steenberg, J., 1988. Tree Kangaroo Husbandry Notebook. Woodland Park Zoological Gardens, Seattle, WA.

Mullett, T., Yoshimi, D., Steenberg, J., 1990. Tree Kangaroo Husbandry Notebook 1989 Survey Results. Woodland Park Zoological Gardens, Seattle, WA.

Porolak, G., Dabek, L., Krockenberger, A., 2014. Spatial requirements of free-ranging Huon tree kangaroos, *Dendrolagus matschiei*, (Macropodidae) in upper montane forest. PLoS 9 (3), e91870. https://doi.org/10.1371/journal.pone.0091870.

Pryor, K., 1999. Don't Shoot the Dog: The New Art of Teaching and Training. Bantam Books, New York, NY (revised edition).

Ralls, K., Ballou, J.D., Templeton, A.R., 1988. Estimates of lethal equivalents and the cost of inbreeding in mammals. Conserv. Biol. 2, 185–193.

Ramirez, K., 1999. Animal Training: Successful Animal Management Through Positive Reinforcement. Shedd Aquarium, Chicago, IL.

Shepherdson, D., 1992. Environmental enrichment: an overview. In: American Association of Zoological Parks and Aquariums (AAZPA) Annual Conference Proceedings, pp. 100–103.

Shepherdson, D., 1999. Introduction: tracing the path of environmental enrichment in zoos. In: Shepherdson, D., Mellon, J.D., Hutchins, M. (Eds.), Second Nature Environmental Enrichment for Captive Animals. Smithsonian Institution Press, Washington, DC.

Steenberg, J., 1988. The management of Matschie's tree kangaroo (*Dendrolagus matschiei*) at Woodland Park Zoological Gardens. In: American Association of Zoological Parks and Aquariums (AAZPA) Western Regional Conference Proceedings. Wheeling, WV, pp. 94–111.

Steenberg, J., 1993. Tree Kangaroo Species Survival Plan (TKSSP) Master Plan 1993-1994. Woodland Park Zoological Gardens, Seattle, WA.

Thompson, V., 1996. Tree Kangaroo Species Survival (*Dendrolagus spp.*) Master Plan 1995-1996. San Diego Zoological Society, San Diego, CA.

Thompson, V., 2000. Tree Kangaroo Species Survival (*Dendrolagus spp.*) Master Plan 1999-2000. San Diego Zoological Society, San Diego, CA.

CHAPTER

20

Biology and Health of Tree Kangaroos in Zoos

Jacque Blessington[a], *Judie Steenberg*[b], *Terry M. Phillips*[c], *Margaret Highland*[d], *Corinne Kozlowski*[e], *and Ellen Dierenfeld*[f]

[a]Association of Zoos and Aquariums Tree Kangaroo Species Survival Plan® Program, Kansas City, MO, United States

[b]Association of Zoos and Aquariums Tree Kangaroo Species Survival Plan® Program, Maplewood, MN, United States

[c]Association of Zoos and Aquariums Tree Kangaroo Species Survival Plan® Program, Richmond, VA, United States

[d]Kansas State University, Manhattan, KS, United States

[e]Saint Louis Zoo, St. Louis, MO, United States

[f]Association of Zoos and Aquariums Tree Kangaroo Species Survival Plan® Program, St. Louis, MO, United States

OUTLINE

Tree Kangaroos
https://doi.org/10.1016/B978-0-12-814675-0.00011-7

INTRODUCTION

Animal welfare can be measured by many aspects including health and psychological well-being. Adequate nutrition, exercise, and other environmental conditions can have direct or indirect effects on health. The absence of disease or poor physical condition indicates an animal is generally in good health. When given choice and opportunity to respond/behave in a species appropriate manner, animals generally handle stress better which leads to overall improved welfare (Irwin et al., 2013).

Over 30 years of managing tree kangaroo health has been well documented in managed facilities for Matschie's (*Dendrolagus matschiei*), Goodfellow's (*D. goodfellowi*), and Lumholtz's (*D. lumholtzi*) tree kangaroos. Protocols for exams, diagnosis, treatment, and animal transfers have been written to facilitate good practices and procedures regarding tree kangaroo health and reproduction. This chapter will address aspects of tree kangaroo health and biology in managed facilities.

IMMUNOLOGY OF CAPTIVE TREE KANGAROOS (*DENDROLAGUS* SPP.)

Immunological research of tree kangaroos was conducted at the National Zoological Park-Conservation Research Center (NZP-CRC) from 1990 to 2000 by Dr. Terry Phillips, Chief of the Ultramicro Analytical Immunochemistry Resource, National Institutes of Health (NIH) (Phillips, 2000) (Montali et al., 1998). This was the first, and to date, only research conducted on the immunological status of Matschie's (*D. matschiei*) and Goodfellow's (*D. goodfellowi*) tree kangaroos in captivity; preserved blood samples were provided by North American (NA) zoological facilities This data was later compared to extracts from blood spots from wild Matschie's tree kangaroos that were provided by the Tree Kangaroo Conservation Program (TKCP) (Chapter 25).

Mycobacterium avium complex (MAC) bacteria cause opportunistic infections in humans and mammals, indicating functional problems with the host's immune system. MAC infections have been documented in NA zoo collections of tree kangaroos indicating a possible immune dysfunction in these animals (Buddle and Young, 2000; Griner, 1978; Montali et al., 1998; Travis et al., 2012). In light of this evidence, a series of studies was undertaken to determine the status of the tree kangaroo immune system and why MAC is a serious problem in these animals.

To help the reader understand these studies, a simplistic overview of the immune system is given in Delves et al. (2011). Mammals rely on two interlinked immune systems – the innate and the acquired. The innate system is the "front-line" defense, requiring no priming or prior exposure for recognition and protection against pathogen invasion. It is comprised of physical barriers (skin and epithelia lining the respiratory and digestive tract), associated chemical barriers (tears, sweat, saliva), multiple leukocytes in tissues and circulating in blood (eosinophils, basophils, neutrophils, natural killer cells, macrophages/monocytes/dendritic cells/Langerhans cells) that can phagocytose and/or release chemicals to kill pathogens. The innate system acts to control and guide the development of the acquired system, which is comprised of T and B lymphocytes that are primed to react rapidly and robustly to specific pathogens that the animal has been previously exposed to, either naturally or through vaccination. T cells can become cytotoxic T cells having T cell receptors that recognize specific proteins/antigens in acquired cell-mediated immunity, while B cells are primed to become plasma cells, the producers of antibodies that function in humoral acquired immunity. For intracellular pathogens such as mycobacteria, cell-mediated immunity is much more important than humoral immunity. Control of immune systems is maintained by a complicated network of molecules called cytokines. The two systems work together to protect the host from disease. Immune control of intracellular infections, like MAC, requires an effective innate immune response and an acquired cell-mediated T cell response (Th1 response), rather than a humoral or antibody (immunoglobulin) response (Th2 response), to protect the host from infection (Romagnani, 1992).

Studies were conducted to assess immunity in a large cohort of captive tree kangaroos at the NZP-CRC in Front Royal, VA, and other NA zoos. Assessment of innate immunity was performed by isolating total white blood cells from tree kangaroo blood (Boyum, 1984; Montali et al., 1998) and studying their reactivity to heat-inactivated mycobacterium organisms. To study tree kangaroo cell-mediated immunity, non-specific tests such as lymphocyte stimulation (LS) and mixed lymphocyte culture (MLC) (Montali et al., 1998) were utilized. These tests were chosen because they are used to assess immune function in human patients and animal models and are well suited for studying marsupials (Wilkinson et al., 1992). Additionally, immunoglobulins (Ig) were isolated and measured physio-chemical techniques as indicators of B cell function. Three Igs (IgM, IgG1, and IgG2) were isolated which compared to the findings reported by Bell (1977). Cytokines (interleukin [IL]-1,IL-2, IL-4, IL-6, IL-10 and tumor necrosis factor alpha) were isolated from serum by a modification of the techniques described for measuring cytokines in pygmy rabbits (Harrenstien et al., 2006). These cytokines were used as indicators of T and B cell activity as well as indicators of the presence of inflammation.

Tree kangaroos were found to have a vigorous innate system, where all the assays (reactive oxygen production, phagocytosis, and chemotaxis) showed a marked increase when compared to other mammals. This may indicate that tree kangaroos rely on innate immunity as their prime line of defense. This finding was reinforced when results from lymphocyte studies demonstrated that there was a lower lymphocyte activity when compared to other mammals (Montali et al., 1998). Both T and B cell proliferation were 3–6 times lower in healthy captive tree kangaroos when compared to other mammalian species. While this may indicate an inherently lower immune response, it may also be related to how tree kangaroos react to plant mitogens. When T lymphocytes were stimulated with mycobacterial antigens, MAC positive animals showed T cell reactivity against *M. avium* but no other mycobacterial species, thus indicating that some T cell recognition was present in clinically infected animals. However, this activity was quite low by mammalian standards and cell surface antigen recognition by T cells was depressed, indicating that captive tree kangaroos are unable to mount normal mammalian cell-mediated responses to MAC antigens. The results from these lymphocyte assays showed evidence of a fairly inactive adaptive cell-mediated immune response which greatly differed from the vigorous activity of the innate immune response (Table 20.1).

Studies on tree kangaroos Igs showed that they had lower concentrations of all three Igs when compared to those of the other mammalian species. It was not surprising that the IgM concentrations were low as this Ig is a "first responder" and only associated with early infections. Surprising was the finding that the IgG concentrations were also low and although it is unknown what roles the different IgG subclasses play in tree kangaroos immunity, IgG1 and IgG2 are the immunoglobulins in mammals associated with active B cell-mediated immunity. Antibody-mediated immunity plays no effective role in combating Mycobacterial infections in other mammals but the finding that all Igs were significantly reduced in captive tree kangaroos when compared to Igs from other mammalian species (Fig. 20.1) further suggests the presence of a weak or less functional B cell acquired immunity.

TABLE 20.1 Overview of tree kangaroo (TK) immune parameters (Montali et al., 1998; Phillips, 2000).

	Captive TK					
Immune system arm	**Healthy**	**MAC[a] infected**	**Wild TK**	**Human**	**Mouse**	**Dog**
Innate						
Oxygen production (% activity)	62–69%	22–33%	NT[b]	21–27%	31–35%	18–24%
Phagocytosis (% activity)	38–55%	5–18%	NT	26–35%	52–60%	23–27%
Chemotaxis (% activity)	53–66%	6–10%	NT	31–45%	28–36%	17–23%
Acquired						
T cell Stimulation (SI[c])	62–65	12–20	NT	180–212%	316–370	243–250
B cell Stimulation (SI)	39–46	15–25	NT	161–185	237–301	228–264
Mixed lymphocyte Stimulation (SI)	34–61	18–32	NT	176–193	275–387	180–250
M. avium T cell stimulation (SI)	5–11	15–20	NT	2–12	NT	NT
M. avium B cell stimulation (SI)	3–16	33–51	NT	0–9	NT	NT
Immunoglobulin G (mg/dL[d])	169–180	118–124	NT	900–1400	685–1100	580–225
Cytokine production						
Th1[e] (pg/mL[f])	11.5	6.3	201.2	NT	NT	NT
Th2[g] (pg/mL)	12.7	28.5	145.5	NT	NT	NT
Inflammatory (pg/mL)	2.1	350.9	66.3	NT	NT	NT

[a] *MAC:* Mycobacterium avium *complex.*
[b] *NT: Not tested.*
[c] *SI: Stimulation Index.*
[d] *mg/dL: milligrams/deciliter.*
[e] *Th1: T helper cell type 1.*
[f] *pg/mL: picogram/milliliter.*
[g] *Th2: T helper cell type 2.*

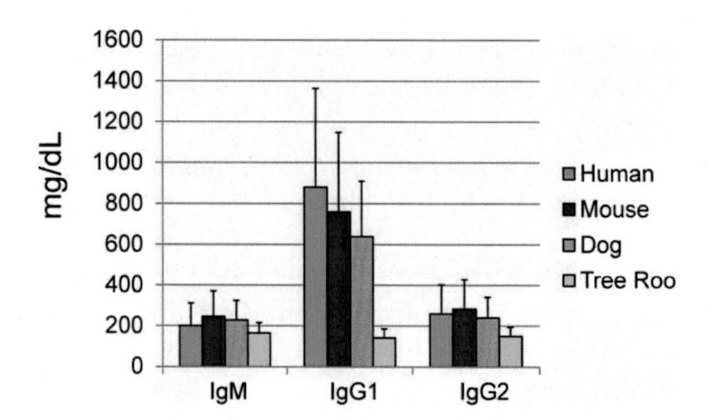

FIGURE 20.1 Immunoglobulin comparisons between tree kangaroos and other mammals (Phillips, 2000).

Both T (Th1) and B (Th2) cytokines were isolated and measured in captive tree kangaroos and in extracts from dried blood spots from wild tree kangaroos supplied by the TKCP team in Papua New Guinea (PNG). While the wild tree kangaroos had Th1 and Th2 cytokine concentrations similar to those found in other mammals, the captive tree kangaroos had lower concentrations. In MAC-infected captive animals there was a significant decrease when compared to both healthy captive and wild tree kangaroos in Th1 cytokines indicating that cytotoxic T cell functions were impaired or nonfunctional. Th2 cytokines from captive animals were also depressed in healthy animals although some increase was seen in MAC-infected animals. Clearly, these finding suggest that factors exist that alter Th1 and Th2 cytokine production in captive tree kangaroos. Phagocyte-associated cytokines from both non-infected captive and wild animals were found to be comparable to cytokine concentrations found in other mammals thus indicating that phagocytic functions were not impaired. But high concentrations of IL-1, TNF-alpha and IL-6 cytokines found in MAC-infected captive animals indicated that active inflammatory processes were present (Fig. 20.2 and Table 20.1).

Quantification of stress-associated hormones including cortisol, norepinephrine, and serotonin, revealed each to be higher overall in captive tree kangaroos as compared to their wild counterparts, with especially high concentrations found in captive animals infected with MAC. Considering the known immunosuppressive effects of these hormones on immune function in mammals, it is reasonable to consider that stress can play a part in determining the effectiveness of captive tree kangaroo immunity (Buddle et al., 1992; Griffin, 1989).

These studies indicate that tree kangaroos rely primarily on their innate immune system, as compared to the adaptive cell-mediated immune system which is required for controlling MAC infections. Additionally, both innate and adaptive immune function may be negatively influenced by stress. The effects of stress have been tied to depression of immunity in both humans and other mammals (Griffin, 1989; Kiecolt-Glaser et al., 2002). The immunosuppressive effects of stress on T cell immunity has been reported in brushtail possums (*Trichosurus vulpecula*) (Buddle et al., 1992) and other captive species (Hing et al., 2016). Although tree kangaroos may visually appear unstressed, the effects of subliminal stress will

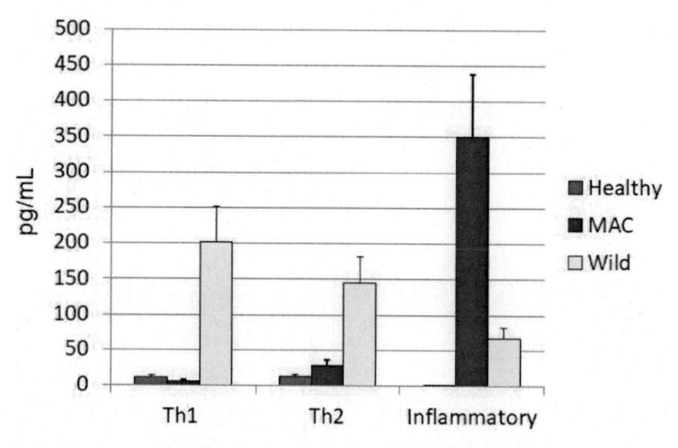

FIGURE 20.2 Inflammatory response to *Mycobacterium avium* complex (MAC) between healthy, infected, and wild tree kangaroos. Th1 response = innate immune response and an acquired cell-mediated T cell response. Th2 response = humoral or antibody (immunoglobulin) response.

have some effect on their immune function, but the extent of this effect is unknown at this time and warrants further investigation.

More recent studies have indicated that additional work needs to be done on the identification and characterization of the diversity of the major histocompatibility complex (MHC) and T-cell receptor (TCR) proteins, the proteins involved in specific antigen recognition in acquired cell-mediated immunity in tree kangaroos. A lack of appropriate MHC and TCR types may be responsible for susceptibility to specific intracellular pathogens such as those belonging to MAC. Skin testing in MAC-positive tree kangaroos produces ambiguous results, generally consistent with a low or inadequate cell-mediated immune response (Burns et al., 1994). Therefore, skin testing for MAC is not recommended in tree kangaroos.

Because marsupials are significantly divergent from eutherian mammals, their cytokine genes and resultant proteins have diverged. This provides major challenges with availability of tools for immune assays. For example, recombinant brushtail possum TNF-alpha has been shown to stimulate brushtail possum lymphocytes much more than human TNF-alpha; the converse was true for murine lymphocytes (Wedlock et al., 1999). Cross-reactivity of antibodies to proteins from different hosts decreases with genetic distance (Nollens et al., 2008); assays using anti-eutherian mammal cytokine antibodies to measure cytokines in marsupials would be expected to significantly underestimate their concentrations in marsupials.

MANAGEMENT OF MAC AND MYCOBACTERIOSIS IN MANAGED FACILITIES

Historically, the term avian tuberculosis has been used to define the disease that most affects tree kangaroos in captivity. The original description was a tuberculosis-like disease caused by a combination of mycobacterial species, but primarily arising from a bird isolated organism called *Mycobacteria avium*. Two species, *M. avium* and *M. intracellulare*, were initially cultured from infected tree kangaroos.

Avian tuberculosis is now known to be an inappropriate term and has been replaced with MAC. The term "tuberculosis" is a disease of humans or other animals that is caused by species of mycobacterium that are distinct from MAC bacteria and are part of the *M. tuberculosis* complex (Daley, 2017). MAC currently refers to any of 10 species of *Mycobacteria*, that have been identified to date, which includes *M. avium*.

Mycobacteriosis in captive tree kangaroo collections is typically caused by environmental exposure. MAC organisms are commonly isolated from aquatic biofilms, dust, and soil (Daley, 2017; Whiley et al., 2012) and can infect and cause disease in immunocompromised hosts. Hosts infected with MAC are not typically considered a source of infection to others. While exposure of tree kangaroos to MAC may be high, some animals may develop infections while others do not.

The effects of stress on the immune system may contribute to an increased susceptibility to mycobacteriosis and overall decrease in animal health. Signs may include, but are not limited to, coughing with increased mucus production, green/yellow/tan nasal discharge, lethargy, decreased appetite and weight loss. A change in mobility can be caused by MAC-related osteomyelitis. If any of these signs are present or suspect, additional testing for MAC infection should occur. It should be noted that tree kangaroos testing positive for MAC may show little to no signs of infection.

Tree kangaroos with MAC infections are not typically considered infectious to other tree kangaroos. However, coughing animals may shed MAC organisms into the environment through sputum, which may be a concentrated dose of mycobacterium, and are best managed away from direct contact with other tree kangaroos. Another concern would be contact from an animal with a draining wound. Moving the infected animal to the other end of the building is considered adequate. Staff working with these animals should also take precautions with proper personal protective equipment to prevent zoonotic disease risk, particularly staff that may be immunocompromised.

Full body radiographs are considered first step testing to identify if lesions are in the lungs or long-bones; these are common sites for infection. If lesions are seen radiographically in the lungs, a tracheal wash should be performed. This method is likely to pick up mycobacteria that are free-floating in the respiratory tract. Tracheal washes should be performed with caution because anesthesia is typically needed, and lung function may be compromised in tree kangaroos with mycobacteriosis. Repeat washes can also cause damage to the trachea. If the lesions are encapsulated in the lungs or are indicated in other sites, then a biopsy is recommended. Both tracheal washes and biopsies are considered invasive and should be performed under veterinary consultation.

Lymphocyte stimulation tests are the best prognostic assessment for an effective immune response and can help monitor an animal's reaction to treatment. Information on testing tree kangaroo immune profiles suspected of mycobacterial disease can be found in the 2007 Tree Kangaroo Husbandry Manual (TKHM) (Blessington and Steenberg, 2007). Lymphocyte stimulation tests are not currently being performed, but they will resume once a new testing lab is identified.

The Association of Zoos and Aquariums (AZA) Tree Kangaroo Species Survival Plan® (TKSSP) recommends combination drug therapy (amikacin, rifabutin, ethambutol and azithromycin) for initial treatment of MAC while antibiotic sensitivity testing is pending. Other species of *Mycobacterium* may be better treated with different drug combinations. Single drug therapy is likely to fail and result in resistant organisms. There has been an increase over time in drug resistant isolates (Wang et al., 2010). The best recommendation is for an initial course of 6–12 months of combination drug therapy, followed by reevaluation as to whether therapy needs to be continued; courses of 2–3 years or more may be indicated. Decisions on continuation of treatment are based upon diagnostic tests, including blood tests, radiographs, and examination of each individual animal. Signs and overall health may improve following removal of stressors in conjunction with treatment. Once therapy for MAC is started, institutions may need to continue some form of therapy for the rest of the animal's life. MAC can be tolerated reasonably well by infected animals, and those that are still eating will typically respond to treatments and can live for years. Early diagnosis is paramount.

Much of the aforementioned information has been referenced from the TKHM (Blessington and Steenberg, 2007) and the Population Analysis & Breeding Transfer Plan (PABTP) (AZA, 2018).

GENERAL HEALTH AND BEHAVIOR

In December 2003, a survey request for Medical History Reports (MHRs) and Necropsy Reports (NRs) was sent to NA institutions that had tree kangaroos (*Dendrolagus* spp.) in their collection from 1993 to 2003. The survey was later expanded to include all deaths (NRs) from 1993 to May 2006. The survey included three species: Matschie's, Goodfellow's, and Grizzled (*D. inustus*) tree kangaroos (Steenberg et al., 2007).

The survey results represent 436 documents:

1993–May 2006 NRs:
111 Matschie's
18 Goodfellow's
8 Grizzled

1993–2003 MHRs:
116 for 61 living Matschie's
20 for 16 living Goodfellow's
10 for 4 living Grizzled
153 for animals that died (all three species)

The purpose of the survey was to gather the following information:

1. What was the cause of death (mortality) for tree kangaroos in captivity in NA?
2. What were the major causes of disease (morbidity)?
3. Were there husbandry issues that needed to be improved?

One hundred percent of the available necropsy reports on the three species listed in their respective studbooks were received and approximately 95% of the available medical history reports.

The results of the survey indicate that the leading causes of mortality continued to be diseases of bacterial origin, primarily MAC, and the loss of pouch young. These results are similar to previous surveys which had documented the status of primarily the NA populations of Matschie's, Goodfellow's, and Grizzled tree kangaroos in 1989 (Mullett et al., 1992), and from 1989 to 1993 (Steenberg and Blessington, 2001). Mycobacterial disease persists as a major concern for this genus that shows a lowered cellular immune response (see "Immunology" section) as compared with eutherian (placental) mammals and other marsupials. No cases of *Mycobacterium* have been reported for Lumholtz's tree kangaroos (A. Legge, personal communication, 14 March 2019).

Reproductive and perinatal disease, which includes abandoned or traumatized pouch young, along with conspecific aggression among adults also continued to be a cause of morbidity and mortality. Other significant causes of disease were reflective of inadequate husbandry and also included gastrointestinal abnormalities, such as trichobezoars (a mass formed from the ingestion of hair), diarrhea, and degenerative musculoskeletal disease, such as osteoarthritis. Traumatic injuries to the tail, toenails and teeth were also documented.

Preventative Medicine

The following information on health and behavior is general in nature but has been developed primarily based on experiences with Matschie's tree kangaroos.

Medical Data Reference

Many zoological facilities that care for tree kangaroos maintain health records for their animals in Species360 Zoological Information Management System (ZIMS) Medical Module (Species360 ZIMS, 2019) range of functionalities includes:

- Anesthesia records
- Pathology
- Treatments and prescriptions
- Pharmacy inventory
- Samples
- Tests and results
- Physiological measurements
- Clinical notes
- Diagnoses and procedures
- Necropsy and Biopsy records
- Medical resources:
    - Anesthesia summaries
    - Drug usage extracts
    - Morbidity and mortality analysis
    - Hematology, chemistry/fluid analysis, endocrinology/reproduction reference values

- Medical term, test and treatment dictionaries
- Reports generated for complete medical histories, caseload management, daily medical summaries, aesthesia summaries, samples, tests and results, active prescriptions/treatments, pharmacy reports, and unresolved medical issues

Physical Exam and Diagnostic Procedures

Routine medical examinations are recommended for the early detection of disease and to help prevent mortality in tree kangaroos (Steenberg et al., 2007). These exams will not only aid in the health maintenance of the individual but also the collection. They will also provide a source of physiological and developmental data from which normal values can be developed.

Routine exams for pouch-gravid females should be postponed until the joey is permanently out of the pouch. Females can become stressed and expel the joey during induction, or under general anesthesia as the dam relaxes control of the pouch, a joey may fall out. A joey in the pouch can also be affected by the anesthesia if using an induction chamber and some anesthetic drugs may be passed in the milk. Joeys are generally not examined until they are permanently out of the pouch and it is time to separate them from their dams.

In situations that a pouch-gravid female must be examined, caution should be administered. Younger joeys may fall out and simply need to be placed back in the pouch. A joey-at-foot may be removed for exam and then returned to the dam once she has recovered. When reintroducing a joey to the dam they should be carefully monitored. Occasionally, a dam will initially seem to reject the reintroduced joey and although the rejection is usually temporary it can be confusing and distressing for the joey (Steenberg, 1987).

Routine examination includes a dental/oral exam, pouch check, fecal samples, rectal cultures, microchip installation, blood collection, weight documentation, and body condition score assessment. Nail length should be assessed and trimmed if needed. Measurements should be documented for all ages, including the length of feet, tail, and from crown to rump. It is important to specify age and sex of each animal at the time it is measured.

Full body radiographs, focusing on the thorax and long bones are recommended to look for pulmonary lesions, osteomyelitis which is associated in many cases with mycobacteriosis, and for documentation of gastrointestinal trichobezoars (Zdziarski and Bush, 1991). Intradermal TB testing has not been shown to be a reliable detection test for the exposure to *Mycobacterium* species in tree kangaroos. Immunological blood testing can be performed to determine infection (see "Immunology" section). Some mycobacteria are very difficult to culture, and it can be challenging to interpret what a positive culture result means for the animal. If the animal has a positive MAC culture, and an associated lesion, then treatment is indicated. MAC continues to be isolated from various systems and attempts to treat the infections have met with limited success (Burns et al., 1994; Bush and Montali, 1995; Montali et al., 1998). This subject has been covered in depth in the preceding section on Immunology.

Performing an electrocardiogram and echocardiogram are important for establishing baseline cardiac information. Tree kangaroos have been diagnosed with cardiomyopathy (Steenberg et al., 2007). Hypertrophic cardiomyopathy has been reported in a Matschie's tree kangaroo (Fredholm et al., 2015).

Fecal examinations for ova and parasites should be conducted at least twice a year. Treatment is usually contingent upon isolation and identification of a parasite and related symptoms. In general, gastrointestinal parasites are not a common problem for tree kangaroos.

Blood samples can be obtained from the lateral tail base, jugular, or cephalic veins for routine CBC, serum chemistry, vitamin E/

selenium levels, frozen serum banking, and for other disease survey protocols requesting samples, as approved by the TKSSP Veterinary Advisory Group.

Inoculations

According to the Lumholtz's Tree Kangaroo Husbandry Manual (Gow, 2006), several vaccines are administered to this species. In NA, inoculation of tree kangaroos is not routinely performed, however, killed rabies (Imrab), clostridial, and tetanus toxoid have been used. Rabies vaccine should only be used as required by regulating authorities since marsupials are fairly resistant to the disease. Members of the AZA TKSSP Veterinary Advisory Group have more information on immunological stimulants research (AZA, 2019).

Illness and Injury; Indications and Treatment

Stress plays an important role in the welfare of captive animals. Physiological and psychological stress can lead to increased production of the hormone cortisol by the adrenal glands. Elevated cortisol levels can lead to a number of physiological changes that are adaptive in the short-term. However, when cortisol is elevated for a prolonged period, these changes are associated with negative impacts on the neurological, cardiovascular, immune and reproductive systems (Nelson, 2011).

In general, long-term increases in cortisol production may be evidence of compromised welfare. However, adrenal responses also occur during beneficial behaviors that require physical activity, including mating behavior and copulation, and responses to environmental enrichment. This can make it challenging to differentiate between adaptive responses and those signaling genuine stress. In addition, individuals of the same species may also differ in their responses, as a result of differences in temperament and previous experiences. Therefore, measuring a range of behavioral, health, and physiological responses is recommended to ensure that endocrine data are interpreted correctly.

To date, very little research has been conducted to investigate the factors that may contribute to elevated cortisol production by tree kangaroos. However, findings from work on captive Matschie's tree kangaroos in NA zoos suggests that institutional transfers and moving to a new exhibit can lead to short-term increases in cortisol production (Kozlowski, 2019).

Behavioral Manifestations of Illness

Tree kangaroo behavior is often subtle and can be difficult to interpret. It is imperative that staff can identify what is 'normal' for each individual animal. Daily activity patterns and individual characteristics must be studied closely. The ability to recognize changes in normal behavior is essential for the early detection of illness, and continuity of care is critical to tree kangaroo welfare.

Stress can reveal itself in many different behaviors in tree kangaroos. They can show these behaviors in response to environmental changes including high temperatures, unfamiliar sights, smells, sounds, or loud noises. It is crucial when housing tree kangaroos in captivity to maintain consistency in their routine. The introduction of new animals, a break or change in normal routine, or unfamiliar personnel can cause stress.

Healthy and sick animals can exhibit the same behavior and the significance of that behavior must be interpreted in context when observed. Behaviors that are normal for one individual may be signs of stress or illness for another. For example, the typical tree kangaroo sleeping position is a sit-curl or tucked posture (Fig. 20.3); however, some individuals may sleep in the prone position (Fig. 20.4). If an individual, who normally sleeps in a certain position, is observed in a different sleeping position, it may be indicative of an incipient health problem. For another

FIGURE 20.3 Goodfellow's tree kangaroo in sit-curl or tucked posture (National Zoo & Aquarium Australia). *Source: Megan Richardson, Zoos Victoria.*

FIGURE 20.4 Matschie's tree kangaroo in prone posture (Cologne Zoo). *Source: Megan Richardson, Zoos Victoria.*

individual (even in the same collection), the prone position might be normal behavior. A curled sleeping position is also the same as a high-stress withdrawal posture. Tree kangaroos generally have a preferred location for resting. A change in resting location, for example from a high location to a lower location, may indicate a sick or weak animal that may not be able to climb.

The sucking or licking of forearms can also have two meanings. It might indicate that the tree kangaroo is stressed or that it is over-heated and attempting to cool itself. Both situations and conditions should be remedied to aid the animal. Hyperactivity, climbing to the highest perch, looking upward, and vocalizing are also indicators of stress.

Behavioral changes such as a normally sociable animal that becomes aggressive or fearful can indicate it is in pain or that it has an illness. Animals should be bright, alert, and responsive. One that appears sleepy or is unresponsive to external stimuli ("out of it") may be an indicator of poor health, or extreme stress. Animals that are consistent with training may stop performing a behavior (e.g. pouch checking) or training completely if there is an underlying health issue. Any behavioral change should alert staff to problems. The earlier a problem is recognized and addressed, the higher the chance of resolution.

Physical Symptoms of Illness

The first signs of illness may be seen as a change in appetite or diet preference. An animal that normally consumes an entire diet might start leaving portions or all of the diet. A change to preferring soft food may indicate oral or dental problems, which can be exacerbated by an all soft food diet. While browse tannins can stain or change tooth appearance, discoloration may also indicate tooth injury or root abscess.

Tree kangaroos tend to defecate in designated areas and normal feces are pelleted. A change in fecal output frequency, composition, and location are all important to monitor. While this change may indicate illness, it can also be an indicator that a female is ovulating. Males tend to defecate along a barrier where they can see or smell a female that is in estrus (DeBo, personal communication, 2018).

A change in urine output frequency, appearance, and location are also important indicators of overall health. Dark reddish urine can be caused by the presence of blood or the addition of a pigmented dietary item. For example, certain varieties of browse can cause urine to change color.

Generally, tree kangaroos have little nasal discharge. If discharge is present, it is important to note the color and consistency. A stressed or over-heated animal will often have clear, continuous, nasal discharge, which can be quite spontaneous. A sick animal can have colored (yellow/green/tan), thick, intermittent discharge.

Coughing should always be taken seriously and monitored carefully. An occasional cough can occur, but a repeated, persistent cough is a definite concern and the underlying cause needs to be determined. It could be a sign of pulmonary disease such as *Mycobacterium* infection.

When humidity is too low, coat can become dry and lusterless with brittle hair and flaky skin. Relative humidity levels should be maintained above 50%, if not, a condition called scaly tail may develop which affects the tip of the tail (Fig. 20.5). If untreated, scaly tail can become a

FIGURE 20.5 Goodfellow's tree kangaroo 'scaly tail' (Melbourne Zoo). *Source: Megan Richardson, Zoos Victoria.*

serious condition, requiring surgical partial tail amputation. Some tree kangaroos may develop a bare patch of skin at the base of the tail due to sitting on their tail in the same manner repeatedly, causing a "wear spot", not to be confused with scaly tail.

Decreased body condition and weight can be an important and early indicator of disease. Obesity not only affects overall health but can also contribute to lowered reproductive success (Norman and Clark, 1998). Captive tree kangaroo weights for Matschie's should range from 7 to 9 kg (±1 kg) with a body condition score of 4/5 based on a 7-point scale (TKSSP Body Condition Score (BCS) Chart, 2015). Weight over 10 kg is considered obese for Goodfellow's, Lumholtz's, and Matschie's tree kangaroos. Grizzled tree kangaroos are considerably larger with an average weight for females of 11.4 kg and males 15.5 kg (TCA, 2019). This corresponds with information received on captive populations (Smith, 28 March 2019; Gutha, 28 March 2019, personal communications).

More information on weights and diets are found in Nutrition and Feeding. BCS values should be recorded with each weight. Physical scoring (body palpations) is more accurate, but visual assessment scoring is also valuable in daily animal observations. Weight and BCS may easily be obtained with training and conditioning. Minimum, monthly recordings will track trends and help improve animal care and welfare.

Trauma

Abandoned or traumatized pouch young continue to be the most common problem seen in captive tree kangaroos. The negative impact of social stress in the loss of joeys has been well-documented at the NZP-CRC, the Woodland Park Zoo, and the Adelaide Zoo (Collins, 1986; Dabek and Hutchins, 1990; Goonan and Arlidge, 1992; Hutchins et al., 1991; Steenberg, 1988; Wemmer, 1985). This validates the importance of minimizing stress for pouch-gravid females. Refer to Chapter 21 on Reproduction for more recent information on this topic and the abandonment and loss of joeys before one year of age.

Antibiotics

Tree kangaroos are foregut-fermenters and some oral antibiotics can have an adverse effect on their digestive systems. When administering oral antibiotics, it is very important to monitor the feces closely. If diarrhea occurs, feces become black (tar-like) in appearance or blood/mucus is found in the feces, the antibiotic should be reviewed and changed. Oral medications are most successfully given to animals that are used to being hand fed a favorite food item.

Anesthesia

Both injectable anesthesia and inhalant anesthesia are routinely used in tree kangaroos. A combination of both is used for longer and/or more invasive procedures. Tree kangaroos should be monitored closely during and following anesthesia, and not allowed to climb until they resume normal functions. Tree kangaroos should be kept in a covered crate, in a quiet place, until they have recovered from anesthesia. An alternative method is to suspend them in a cotton/canvas/hessian bag (similar to other macropods). This has been the safest and most effective method for anesthesia recovery, transport and short term holding for emergency maintenance at Dreamworld Wildlife Habitat and David Fleay Wildlife Park, Australia (Legge, personal communication, 17 March 2019).

Zoonosis

Zoonosis is the transfer of disease from animals to humans and should be taken into consideration when caring for captive animals. However, there are few documented zoonotic diseases of concern relative to tree kangaroos.

While tree kangaroos are very susceptible to toxoplasmosis, they do not shed infectious oocysts in their feces. Therefore, this route of transmission is not usually a source of infection for people when dealing with tree kangaroos.

While MAC infections are serious concerns in tree kangaroos, this can also be of concern for zoonosis in human caretakers if the immune system is compromised or there is an underlying pulmonary disorder (Blessington and Steenberg, 2007).

Guidelines and Protocols

Cortisol production increases during stressful events, such as institutional transfer and quarantine (Dembiec et al., 2004; Ji et al., 2013; Snyder et al., 2012; Wells et al., 2004). Concern regarding the impact of these procedures on animal well-being has been raised, particularly for sensitive species like tree kangaroos that have heightened susceptibility to disease such as mycobacteriosis (see "Immunology" section). Recommendations and guidelines for Preshipment, Transport, and Quarantine are provided by the TKSSP for NA institutions. These documents can be found in the PABTP (AZA, 2018) or contacting the TKSSP (AZA, 2019). This information may also be relevant for and utilized by tree kangaroo regions globally.

These recommendations are intended to help ensure that tree kangaroos are in good health prior to shipment, which also helps reduce stress for the animal once at the receiving institution. Results from the pre-shipment exam are available through the Species360 ZIMS Medical Module. The American Association of Zoo Keeper's (AAZK) (2019) Animal Data Transfer Form (ADT) should also be completed. All information should be sent prior to shipment, or with the animal.

Tree kangaroos are no more difficult to ship than other mammals of the same size. The requirements of the 2019 International Animal Transport Association (IATA) Live Animals Regulations 45th Edition, Container Requirement 83 (IATA, 2019) should be used to evaluate the proper shipping container. Domestic or international transfer, and size of the tree kangaroo being shipped, will determine the dimensions and materials needed (Fig. 20.6).

It is imperative that all tree kangaroos be necropsied, including neonatal carcasses. The TKSSP utilizes a postmortem protocol. After necropsy, the TKSSP requests that the carcass or portions of the carcass be sent to the Burke Museum at the University of Washington, in Seattle. The Burke Museum Mammalogy Collection (2019) supports research, education, and outreach related to mammals. They maintain and continue to expand a large archival collection of mammal specimens and make these specimens available to local educators and the global research community. Burke Mammalogy is committed to documenting and archiving mammalian diversity and natural history, while providing perpetual access to the specimens and information they hold (Chapter 22).

NUTRITION AND FEEDING

Tree kangaroos represent a unique group of arboreal macropod marsupials, distributed through the tropical forests of northern Australia and the island of New Guinea. Studies of feeding ecology (Dabek and Betz, 1998; Heise-Pavlov et al., 2014, 2018; Martin, 1992; Procter-Gray, 1985) provide lists of food plants utilized in various locales, with some limited indications of food preferences. Despite the feeding diversity noted in free-ranging animals (>90 spp. consumed; Dabek and Betz, 1998), few analytical data on foods eaten by tree kangaroos are available from which to extrapolate nutrient concentrations for applied captive feeding programs. Nor have specific nutrient requirements been determined for the species. Zoo diets and feeding recommendations (Edwards and Ward, 2007) are based on a combination of practical

FIGURE 20.6 International shipping crate (Melbourne Zoo). *Source: Megan Richardson, Zoos Victoria.*

experience and information from species with similar digestive morphology and dietary habits.

Natural Feeding Habits

Considered browsing herbivores, tree kangaroos preferentially consume tree foliage (Drake, 1984); diet also includes a variety of ferns, fruits, bark, moss, flowers, and vines – eaten mostly above the ground (Archer, 1985; Dabek and Betz, 1998; Flannery, 1990; Troughton, 1965). Population densities appear to be linked with soil fertility and rainfall in locations studied, hence foliage quality, but detailed analyses of nutrient content are lacking. Tree kangaroos are found at multiple altitudes ranging from <700 to 1200 m (Lumholtz's) (Heise-Pavlov et al., 2018; Procter-Gray, 1985) and up to ~3000 m (Matschie's) above sea level (Flannery et al., 1996). Such variance in altitude and growing conditions may have implications regarding nutrient composition of forages consumed by tree kangaroos (Fimbel et al., 2001) and should be considered in future investigations of native food composition. Free-ranging tree kangaroos typically consume a moderately fibrous diet, anticipated to contain low levels of storage carbohydrates (simple sugars and starches), and possess physiologic and anatomic adaptations for processing this rather low-quality diet.

Macropod marsupials are considered foregut fermenters, with stomach modifications to support microbial populations capable of fiber fermentation; the tubiform or "midstomach" region of tree kangaroos comprises up to ~75% of stomach capacity (Hume, 1999).

Longitudinal muscles of the forestomach (for mixing gut contents as well as motility to the hindgut) differ anatomically among the *Dendrolagus* spp. studied. This may relate to differences in ecological feeding niches between species but has not been examined in detail. Teeth of tree kangaroos are also designed more for crushing, rather than shearing, with large surface area contact between upper and lower teeth. Tree kangaroos have larger, more cutting premolars compared with species that include grass in the diet, and their molars are flatter for precise occlusion, fine shearing and grinding capabilities which can reflect a heterogeneous diet. In addition to dentition, documented disparities in jaw musculature and even grasping ability may suggest subtle yet disparate feeding strategies even within the various tree kangaroo species (Iwaniuk et al., 1998).

Nutritional Status of Free-Ranging Tree Kangaroos

Circulating concentrations of fat-soluble vitamins, A and E as well as mineral values from a healthy population of *D. matschiei* in PNG were reported by Travis et al. (2012) as part of a larger health survey (also see Chapter 25). Although limited in number ($n = 4$ for free-ranging animals, $n = 5$ captive PNG animals), these data indeed provide useful parameters for aspects of diet assessment and may reflect normal values for this species.

Of note, vitamin E values were higher than might be anticipated for herbivores, more similar to values seen in omnivores (i.e. primates) (Dierenfeld and Traber, 1992). The significantly higher level in free-ranging compared to captive animals in PNG suggests that wild diets may provide a better source of this nutrient. As captive animals in this particular study weighed less and displayed lower body condition scores compared with free-range tree kangaroos, one might surmise that the captive diet (sweet potatoes and sweet potato greens) was potentially nutritionally suboptimal. Low levels of both selenium and copper in tree kangaroos from PNG and in human populations there (Travis et al., 2012) might suggest an overall deficiency of those nutrients; on the other hand, high molybdenum and heavy metals suggest that captive PNG diets may provide imbalanced mineral profiles. Except for phosphorus and potassium (which may have been compromised during handling), other values were within laboratory reference ranges. Three of four free-range females in this study had young, thus physiologic values reported can support reproduction in this species and provide interesting baselines for future comparisons with diet assessments of captive tree kangaroos in other regions.

Captive Feeding Management

Although specific nutrient requirements of tree kangaroos have not been determined, energetic studies with captive animals (*D. matschiei*, $n = 2$; McNab, 1988) determined that tree kangaroos possess a low metabolic rate (even contrasted with other marsupials), approximately 57% of similar sized eutherians – a value similar to that seen in other arboreal folivores (Hume, 1999). Daily basal metabolic requirements have been determined at 168 $kJ\ kg^{-0.75}\ day^{-1}$ for captive tree kangaroos, or approximately 900 kJ (equivalent to ~220 kcal/day for a 9.5 kg adult). With a low activity level (active <25% of day), maintenance requirements are thus calculated at ~245 to 350 kcal/day (or about 1000–1400 kJ per day). This information has direct implications for designing optimal diets for captive feeding programs. Captive tree kangaroos in NA are consistently reported as weighing more than free-living counterparts (9–10, even >12 kg versus 7–9 kg) (Blessington and Steenberg,

2007). Providing appropriate calorie content and nutrient densities is essential for maintaining appropriate weights, body condition, and ultimately, overall health and reproductive success.

Diets, suggested nutrient ranges, and feeding husbandry for managed populations of tree kangaroos in NA zoos have been previously recommended (Edwards and Ward, 2007). Adult animal diets offered in NA zoos have ranged between 300 and 600 g of fresh food daily (3–6% of body weight for an average 9.5 kg animal), supplying 2–3.5% of body weight in total dry matter (DM), or approximately 200–335 g DM. Calories provided by these diets, ranged from about 250 to >800 kcal/day per animal, which is excessive compared to caloric recommendations of 250–350 kcal/day. This allowed some animals the opportunity to self-select imbalanced diets, and/or over consume.

Global diet surveys conducted in 2015 and 2016 (Carlyle-Askew and DeBo, 2015; DeBo and Carlyle-Askew, 2018) reviewed ingredients offered but total quantities fed were not summarized. Differences were seen between diets offered to tree kangaroos in NA (n = 27 facilities) and Australian/European zoos (n = 14); both groups showed wide variety in the type of foods offered. Five of 14 zoos outside of NA fed commercially formulated dry feeds, 15–25% of diet weight, compared to 2–55% in US and Canadian zoos. Total weight of diets fed in NA zoos consisted mainly of vegetables (25–75%), various greens (10–65%), and fruits (0–35%). Other zoos in the survey offered vegetables (35–85%), fruit (1–60%), and less green plant material (5–45%). Zoos in PNG fed primarily flowers, leaves, and bark from native browse plants reported to be eaten by wild tree kangaroos.

To provide a balanced diet that supplies ~340 kcal, dietary amounts offered should be ~5% of body weight for a 9.5 kg animal or 475 g fresh food (~105 g DM) per day (Dierenfeld, personal observation, 2018). Percentages of recommended items fed include:

- browse and greens (45–50% or 215 g)
- vegetables (35% or 165 g) with no more than 10% (48 g) from root vegetables
- high fiber commercial primate or herbivore biscuits (15% or 70 g)
- minimal fruit (0–5%; up to 25 g)

The natural feeding habits of free-ranging tree kangaroos should be taken into consideration when providing diets, such as feeding on elevated platforms. Opportunities should be provided multiple times a day to simulate the activity of a browsing herbivore which in turn increases activity levels for individual animals. The type of greens and other produce within the above categories may be substituted daily to provide a varied diet. To minimize sudden alterations in gut microbiota, significant diet changes should be done on a gradual basis. Diets should be formulated by animal nutritionists and/or in consultation with veterinary and animal care staff to ensure nutritional balance as well as appropriate energy provision. Animal-based foods and/or cooking of vegetables are not recommended for tree kangaroos.

Daily provision of browse in tree kangaroo diets is highly recommended and has been linked to the prevention of loose stools. At least 45 species of locally available browses have been successfully fed to captive tree kangaroos in NA zoological facilities with no reported toxicity (Blessington and Steenberg, 2007). Browse materials should be presented in an upright fashion, such as secured to a branch or perch or placed in an upright canister (Fig. 20.7). The manner of presentation was shown to have a significant impact on consumption levels by captive animals (Mullett et al., 1988). When browse was placed on perching or branches, between 90% and 100% of the edible portion was consumed. When the same plant materials were "planted" in

FIGURE 20.7 Browse canister (Cologne Zoo). *Source: Megan Richardson, Zoos Victoria.*

sand at ground level, consumption decreased to 65–75%; only 50% of the browse was consumed when the portions were broadcast on the ground. Presentation should allow some individual selection of preferred species and/or plant parts by the animal.

Native browse species identified as tree kangaroo food plants (comprising trees, shrubs, vines, herbs, and flowers; n = 24 spp.) were collected in PNG and dried to determine water content (75.56 ± 10.14%). Dried samples were analyzed to determine proximate composition, fiber and mineral fractions. Summary data are reported in Table 20.2 and are compared with NA browses fed as well as nutrient recommendations for captive tree kangaroo diets (Dierenfeld, 2020; Dierenfeld et al., 2020; Edwards and Ward, 2007). Browses contained about twice the dietary fiber levels suggested by current TKSSP guidelines, as well as far less starch than would be provided by manufactured grain-based biscuits. Many of the minerals analyzed in native foods fell within recommended ranges; exceptions include Ca, K, Mn, P, Na, and Zn, with P and Na known to be limiting nutrients in many natural environments. Overall, locally available browses appear to provide suitable nutritional substitutes for native plants eaten by tree kangaroos, and should be considered essential dietary ingredients. Both fiber and starch can substantially impact gastrointestinal microbial

TABLE 20.2 Analysis of native tree kangaroo browse (mean ± SD) collected in Papua New Guinea compared with browses fed in North American zoos and recommended dietary nutrient concentrations (dry matter basis).

Nutrient	Units	Native browses (PNG; n=24 species)[a]	Locally available browses (NA zoos; n=18 species)[b]	TKSSP[c] nutrient recommendations[d]
Metabolizable energy (calculated)	kcal/g[e]	1.93 ± 0.25	1.99 ± 0.42	
Target calories provided by diet offered				250–350
ADF[f]	%	39.28 ± 9.97	33.45 ± 13.01	15.00
Lignin	%	15.13 ± 6.11	10.99 ± 5.46	10.00
NDF[g]	%	51.79 ± 12.87	48.47 ± 15.19	25.00
Crude fat	%	3.23 ± 1.88	3.09 ± 1.56	2.5–5.0
Crude protein	%	10.95 ± 4.80	11.89 ± 5.32	15.00
Starch	%	0.92 ± 0.92	1.20 ± 1.24	ND[h]
Vitamin A	IU A/g or RE/g	ND	ND	4.00
Vitamin D3	IU D3/g	ND	ND	0.80
Vitamin E	mg/kg[i]	ND	ND	100.00
Ash	%	7.57 ± 3.95	9.97 ± 6.48	10 (max)
Calcium	%	1.07 ± 0.99	2.24 ± 1.60	0.80
Copper	mg/kg	11.92 ± 12.79	10.74 ± 4.51	10.00
Iodine	mg/kg	ND	ND	0.30
Iron	mg/kg	47.34 ± 25.98	121.53 ± 111.09	60.00
Magnesium	%	0.32 ± 0.15	0.27 ± 0.18	0.20
Manganese	mg/kg	268.31 ± 225.23	51.26 ± 80.82	40.00
Phosphorus	%	0.19 ± 0.10	0.23 ± 0.09	0.40
Potassium	%	1.78 ± 0.90	1.53 ± 0.57	0.80
Selenium	mg/kg	ND	ND	0.20
Sodium	%	0.02 ± 0.02	0.03 ± 0.04	0.20
Zinc	mg/kg	33.85 ± 17.74	32.76 ± 17.52	80.00

[a] *Dierenfeld et al. (2020).*
[b] *Dierenfeld (2019).*
[c] *TKSSP = Tree Kangaroo Species Survival Plan.*
[d] *Adapted from Edwards and Ward (2007).*
[e] *kcal/g = kilocalories/gram.*
[f] *ADF = acid detergent fiber.*
[g] *NDF = neutral detergent fiber.*
[h] *ND = not determined.*
[i] *mg/kg = milligrams/kilogram.*

populations. However, tree kangaroos possess adaptations for processing and fermenting plant fiber fractions (Hume, 1999). These preliminary data can provide useful guidelines for improved captive diet development.

Future Nutrition Research

- Determine levels of dietary tannins, essential oils, and possible antioxidants for which free-ranging tree kangaroos may be adapted. Effects of sampled location (including altitude), growing conditions, and potential seasonal changes in nutrient levels should be considered.
- Investigation of microbial diversity, obtained from fecal, oral, and/or gastrointestinal samples may provide insight into digestive abilities of tree kangaroos, as well as possible alterations due to captive management and/or diet changes.
- Conduct digestibility studies with managed individuals of varying life stages to improve insight into optimal nutrition and further guide feeding management recommendations.
- Information from blood or tissue (urine, feces, hair, liver) samples will provide further guidelines for assessment of nutrient status related to animal health and welfare.
- Correlation studies among diet composition, body condition, reproductive outputs, and longevity will further inform improved captive management and conservation of tree kangaroo populations.

CONCLUSIONS

No one can dispute that both animals in managed facilities and those in the wild encounter stress and challenges, although the circumstances differ. These situations can contribute to health issues that can have direct impact on physical and psychological well-being which can in turn effect reproductive success. Great strides have been taken to study and improve the quality of care of tree kangaroos in managed facilities. These advances have led to the development of protocols and procedures for veterinary care and improvements in husbandry routines. MAC continues to be a disease of concern for captive tree kangaroos. Diagnosis and treatment of MAC should continue to be monitored and researched.

Nutrition also plays a direct role in the improved health and potential reproductive success of tree kangaroos. Obesity is a common concern with animals in captivity. Providing the appropriate nutrition for managed tree kangaroos is a complex process and is continually being evaluated. The analysis of plant samples of wild Matschie's tree kangaroo diets will be instrumental in the reformulation of captive diets.

ACKNOWLEDGMENTS

The authors would like to acknowledge the contributions of TKSSP Coordinator Kathy Russell (Santa Fe College Teaching Zoo) for her review and recommendations for this chapter: Megan Richardson (Melbourne Zoo) for reviewing the chapter and providing Goodfellow's tree kangaroo data and photographs; Alana Legge (Dreamworld) for providing materials and information regarding the Lumholtz's tree kangaroo; TKSSP Veterinary Advisors Drs Jim Wellehan (University of Florida) and Marisa Bezjian (Zoo Miami) for their review and contributions on the health of managed populations of tree kangaroos.

REFERENCES

American Association of Zoo Keepers (AAZK), 2019. Animal Data Transfer Form. Available from:https://www.aazk.org (1 March 2019).

Archer, M., 1985. The Kangaroo. Kevin Weldon Pty. Ltd. McMahon's Point, New South Wales.

Association of Zoos and Aquariums (AZA), 2018. Population analysis & breeding and transfer plan Matschie's tree kangaroo (*Dendrolagus matschiei*). In: AZA Species Survival Plan (SSP) Red Program. AZA Population Management Center. Lincoln Park Zoo, Chicago, IL.

Association of Zoos and Aquariums (AZA), 2019. Tree Kangaroo Species Survival Plan (TKSSP). Available from: https://www.aza.org/contact-us (01 March 2019).

Bell, R.G., 1977. Marsupial immunoglobulins: the distribution and evolution of macropod IgG2, IgG1, IgM and light chain antigenic markers within the sub-class Metatheria. Immunology 33, 917–924.

Blessington, J., Steenberg, J., 2007. Tree Kangaroo (*Dendrolagus* spp.) Husbandry Manual. In: Tree Kangaroo Species Survival Plan. third ed. Association of Zoos and Aquariums, Silver Spring, MD.

Boyum, A., 1984. Separation of lymphocytes, granulocytes and monocytes from human blood using iodinated density gradient media. Methods Enzymol. 108, 88–102.

Buddle, B.M., Young, L.J., 2000. Immunobiology of mycobacterial infections in marsupials. Dev. Comp. Immunol. 24 (5), 517–529.

Buddle, B.M., Aldwell, F.E., Jowett, G., Thomson, A., Jackson, R., Paterson, B.M., 1992. Influence of stress of capture on haematological values and cellular immune responses in the Australian brushtail possum (*Trichosurus vulpecula*). N. Z. Vet. J. 40, 155–159.

Burke Museum Mammalogy Collection, 2019. University of Washington, Seattle, WA. Available from: https://www.burkemuseum.org/research-and-collections/mammalogy/people-and-contac. Accessed 1 March 2019.

Burns, D.L., Wallace, R.S., Teare, J.A., 1994. Successful treatment of mycobacterial osteomyelitis in a Matschie's tree kangaroo *(Dendrolagus matschiei)*. J. Zoo Wildl. Med. 25 (2), 274–280.

Bush, M., Montali, R.J., 1995. Preliminary research proposal-diagnostic and therapeutic approach to tuberculosis in tree kangaroos. In: Steenberg, J. (Ed.), Tree Kangaroo *(Dendrolagus spp.)* Master Plan 1994-95. AZA, Silver Spring, MD.

Carlyle-Askew, B., DeBo, D., 2015. Tree Kangaroo (*Dendrolagus spp*) Captive Diet Survey and Reformulation. Poster presentation. AAZK Conference St. Louis, MO.

Collins, L., 1986. Big foot of the branches. In: Zoogoer. National Zoological Park, Washington, DC July-August, 35-40.

Dabek, L., Betz, W., 1998. 1998 field report Tree Kangaroo Conservation Program (TKCP). Association of Zoos and Aquariums. Tree Kangaroo Species Survival Plan.

Dabek, L., Hutchins, M., 1990. The social biology of tree kangaroos: implications for captive management. In: AAZPA Regional Conference Proceedings, pp. 528–535 Wheeling, WV.

Daley, C.L., 2017. *Mycobacterium avium* complex disease. Microbiol. Spectrum. 5 (2). https://doi.org/10.1128/microbiolspec.TNMI7-0045-2017 TNMI7-0045-2017.

DeBo, D., Carlyle-Askew, B., 2018. Nutritional Survey of Diets Fed to Matschie's Tree Kangaroos (*Dendrolagus matschiei*) in Human Care. Poster Presentation Association of Zoos and Aquariums Conference, Seattle, WA.

Delves, P.J., Martin, S.J., Burton, D.R., Roitt, I.M., 2011. Roitt's Essential Immunology, 12th ed. Wiley Blackwell Oxford, UK.

Dembiec, D.P., Snider, R.J., Zanella, A.J., 2004. The effects of transport stress on tiger physiology and behavior. Zoo Biol. 23, 335–346.

Dierenfeld, E.S., 2020. Dierenfeld, E.S. (Ed.), Tree Kangaroo Nutrition. Unpublished Raw Data. LLC, St. Louis, MO.

Dierenfeld, E.S., Okena, D., Oliver, P., Dabek, L., 2020. Composition of browses consumed by Matschie's tree kangaroo (*Dendrolagus matschiei*) sampled from home ranges in Papua New Guinea. Zoo Biol. 39 (4), 271–275.

Dierenfeld, E.S., Traber, M.G., 1992. Vitamin E status of exotic animals compared with livestock and domestics. In: Packer, L., Fuchs, J. (Eds.), Vitamin E in Health and Disease. Marcel Dekker, Inc., New York, pp. 345–360

Drake, B., 1984. Response: (tree kangaroo predation). Thylacinus 9 (3), 15–16 Australia.

Edwards, M.S., Ward, A., 2007. Nutrition, food preparation and feeding. In: Blessington, J., Steenberg, J. (Eds.), Tree Kangaroo (*Dendrolagus* spp.) Husbandry Manual. third ed. Association of Zoos and Aquariums, Silver Spring, MD.

Fimbel, C., Vedder, A., Dierenfeld, E., Mulindahabi, F., 2001. An ecological basis for large group size in *Colobus angolensis* in the Nyungwe Forest, Rwanda. Afr. J. Ecol. 39, 83–92.

Flannery, T.F., 1990. Mammals of New Guinea. Robert Brown and Associates, Carina, Queensland, pp. 89–109.

Flannery, T.F., Martin, R., Szalay, A., 1996. Tree Kangaroos: A Curious Natural History. Reed Books, Port Melbourne, Victoria, p. 417.

Fredholm, D.V., Jones, A.E., Hall, N.H., Russell, K., Heard, D.J., 2015. Successful management of hypertrophic cardiomyopathy in a Matschie's tree kangaroo (*Dendrolagus matschiei*). J. Zoo Wildl. Med. 46 (1), 95–99.

Goonan, P., Arlidge, J., 1992. Observations on the birth of a Matschie's tree-kangaroo *(Dendrolagus matschiei)* in captivity. Aust. Mammal. 15, 113–114.

Gow, K., 2006. Lumholtz's tree kangaroo (*Dendrolagus lumholtzi*) husbandry manual. (unpublished manuscript). David Fleay Wildlife Park, Queensland, Australia.

Griffin, J.F.T., 1989. Stress and immunity: a unifying concept. Vet. Immunol. Immunopathol. 20, 263–312.

Griner, L.A., 1978. Occurrence of tuberculosis in the zoo collection of the Zoological Society of San Diego, 1964–1975. In: Montali, R.J. (Ed.), Mycobacterial Infections of Zoo Animals. Smithsonian Institution, Washington, DC, pp. 45–49.

Harrenstien, L.A., Finnegan, M.V., Woodford, N.L., Mansfield, K.G., Waters, W.R., Bannantine, J.P., Paustian, M.L., Garner, M.M., Bakke, A.C., Peloquin, C.A., Phillips, T.M., 2006. *Mycobacterium avium* in pygmy rabbits (*Brachylagus idahoensis*): 28 cases. J. Zoo Wildl. Med. 37, 498–512.

Heise-Pavlov, S., Anderson, C., Moshier, A., 2014. Studying food preferences in captive cryptic folivores can assist in conservation planning: the case of the Lumholtz's tree-kangaroo (*Dendrolagus lumholtzi*). Aust. Mammal. 36 (2), 200–211.

Heise-Pavlov, S., Rhinier, J., Burchill, S., 2018. The use of a replanted riparian habitat by the Lumholtz's tree-kangaroo (*Dendrolagus lumholtzi*). Ecol. Manag. Restor. 19 (1), 76–80.

Hing, S., Currie, A., Broomfield, S., Keatley, S., Jones, K., Thompson, R.C.A., Narayanc, E., Godfreya, S.S., 2016. Host stress physiology and *Trypanosoma haemoparasite* infection influence innate immunity in the woylie (*Bettongia penicillata*). Comp. Immunol. Microbiol. Infect. Dis. 46, 32–39.

Hume, I.D., 1999. Marsupial Nutrition. Cambridge University Press, p. 434.

Hutchins, M., Smith, G.M., Mead, D.C., Elbin, S., Steenberg, J., 1991. Social behavior of Matschie's tree kangaroos *(Dendrolagus matschiei)* and its implications for captive management. Zoo Biol. 10 (2), 147–164.

International Animal Transport Association (IATA) Live Animals Regulations, 2019. Container Requirement 83, 45th ed. pp. 401–402.

Irwin, M.D., Stoner, J.B., Cobaugh, A.M. (Eds.), 2013. Zookeeping: An Introduction to the Science and Technology. University of Chicago Press, Chicago, IL, p. 55.

Iwaniuk, A.N., Nelson, J.E., Ivanco, T.L., Pellis, S.M., Whishaw, I.Q., 1998. Reaching, grasping and manipulation of food objects by two tree kangaroo species, *Dendrolagus lumholtzi* and *Dendrolagus matschiei*. Aust. J. Zool. 46 (3), 235–248.

Ji, S.N., Yang, L.L., Ge, X.F., Wang, B.J., Cao, J., Hu, D.F., 2013. Behavioural and physiological stress responses to transportation in a group of Przewalski's horses (*Equus ferus przewalskii*). J. Anim. Plant Sci. 23, 1077–1084.

Kiecolt-Glaser, J.K., McGuire, L., Robles, T., Glaser, R., 2002. Psychoneuroimmunology: psychological influences on immune function and health. J. Consult. Clin. Psychol. 70, 537–547.

Kozlowski, C.P., 2019. [tree kangaroo endocrinology]. Unpublished raw data. Saint Louis Zoo. St. Louis, MO.

Martin, R.W., 1992. An Ecological Study of Bennett's Tree-Kangaroo (*Dendrolagus bennettianus*). Project 116World Wide Fund for Nature, p. 67.

McNab, B.K., 1988. Energy conservation in a tree-kangaroo (*Dendrolagus matschiei*) and the red panda (*Ailurus fulgens*). Physiol. Zool. 61 (3), 280–292.

Montali, R.J., Bush, M., Cromie, R., Holland, S.M., Maslow, J.N., Worley, M., Witebsky, F.G., Phillips, T.M., 1998. Primary Mycobacterium avium complex infections correlate with lowered cellular immune reactivity in Matschie's tree kangaroos (*Dendrolagus matschiei*). J. Infect. Dis. 178, 1719–1725.

Mullett, T., Yoshimi, D., Steenberg, J., 1988. Tree Kangaroo Husbandry Notebook. Woodland Park Zoological Gardens, Seattle, WA (updated 1990).

Mullett, T., Yoshimi, D., Steenberg, J., 1992. Tree Kangaroo Survey. Woodland Park Zoological Gardens, Seattle, WA.

Nelson, R.J., 2011. An Introduction to Behavioral Endocrinology, fourth ed. Sinauer Associates, Inc., Sunderland, MA

Nollens, H.H., Ruiz, C., Walsh, M.T., Gulland, F.M., Bossart, G., Jensen, E.D., McBain, J.F., Wellehan, J.F., 2008. Cross-reactivity between immunoglobulin G antibodies of whales and dolphins correlates with evolutionary distance. Clin. Vaccine Immunol. 15 (10), 1547–1554.

Norman, R.J., Clark, A.M., 1998. Obesity and reproductive disorders: a review. Reprod. Fertil. Dev. 10, 55–63.

Phillips, T.M., 2000. Tree Kangaroo Immunology. Unpublished Raw Data. National Institutes of Health, Washington, DC.

Procter-Gray, E., 1985. The Behavior and Ecology of Lumholtz's Tree-Kangaroo (*Dendrolagus lumholtzi*) (Marsupialia: Macropodidae). (Ph.D. diss.)Harvard University Abstract in Dissertation Abstracts International 46(4B):1087.

Romagnani, S., 1992. Induction of TH1 and TH2 responses: a key role for the "natural" immune response? Immunol. Today 13, 379–381.

Snyder, R.J., Perdue, B.M., Powell, D.M., Forthman, D.L., Bloomsmith, M.A., Maple, T.L., 2012. Behavioral and hormonal consequences of transporting giant pandas from China to the United States. J. Appl. Anim. Welf. Sci. 15, 1–20.

Steenberg, J., 1987. Australasia Unit Records. Woodland Park Zoological Gardens, Seattle, WA.

Steenberg, J., 1988. The management of Matschie's tree kangaroo (*Dendrolagus matschiei*) at Woodland Park Zoological Gardens. In: American Association of Zoological Parks and Aquariums Western Regional Conference Proceedings. Wheeling, WV, pp. 94–111.

Steenberg, J., Blessington, J., 2001. Tree Kangaroo Husbandry Manual. Tree Kangaroo. Species Survival Plan. Association of Zoos and Aquariums.

Steenberg, J., Venold, F., McKnight, C., Collins, D., Blessington, J., 2007. Tree kangaroo species survival plan survey results: 1993-2003 medical history reports (living/dead) and 1993—May 2006 necropsy reports. 2007 Tree Kangaroo Husbandry Manual 10pp. 315–353.

Tenkile Conservation Alliance (TCA), 2019. Available from: https://wwwtenkile.com/grizzled-tree-kangaroo.html (18April 2019).

Travis, E.K., Watson, P., Dabek, L., 2012. Health assessment of free-ranging and captive Matschie's tree kangaroos (*Dendrolagus matschiei*) in Papua New Guinea. J. Zoo Wildl. Med. 43, 1–9.

Tree Kangaroo Species Survival Plan (TKSSP) Body Condition Score (BCS) Chart, 2015. Milwaukee County Zoo. Zoological Society of Milwaukee, WI.

Troughton, E., 1965. Furred Animals of Australia, eighth ed. Angus and Robertson, Sydney.

Wang, H.X., Yue, J., Han, M., Yang, J.H., Gao, R.L., Jing, L.J., Yang, S.S., Zhao, Y.L., 2010. Nontuberculous mycobacteria: susceptibility pattern and prevalence rate in Shanghai from 2005 to 2008. Chin. Med. J. 123 (2), 184–187.

Wedlock, D.N., Goh, L.P., McCarthy, A.R., Midwinter, R.G., Parlane, N.A., Buddle, B.M., 1999. Physiological effects and adjuvanticity of recombinant brushtail possum TNF-alpha. Immunol. Cell Biol. 77 (1), 28–33.

Wells, A., Terio, K.A., Ziccardi, M.H., Munson, L., 2004. The stress response to environmental change in captive cheetahs (*Acinonyx jubatus*). J. Zoo Wildl. Med. 35, 8–14.

Wemmer, C., 1985. A decade of research. In: Zoogoer. National Zoo, Washington, DC March/April 10-15.

Whiley, H., Keegan, A., Giglio, S., Bentham, R., 2012. *Mycobacterium avium* complex—the role of potable water in disease transmission. J. Appl. Microbiol. 113 (2), 223–232.

Wilkinson, R., Kotlarski, I., Barton, M., Phillips, P., 1992. Isolation of koala lymphoid cells and their in vitro responses to mitogens. Vet. Immunol. Immunopathol. 31, 21–33.

Zdziarski, J., Bush, M., 1991. Clinical challenges 3 (gastrointestinal trichobezoar in Matschie's tree kangaroos). J. Zoo Wildl. Med. 22 (4), 507–509.

Zoological Information Management System (ZIMS) Medical Module (previously ISIS—International Species Information System), 2019. Global Headquarters, Minneapolis, MN. Available from: https://zims.species360.org (01 March 2019).

CHAPTER

21

Reproductive Biology and Behavior of Tree Kangaroos in Zoos

Jacque Blessington[a], *Judie Steenberg*[b], *Megan Richardson*[c], *Davi Ann Norsworthy*[d], *Alana Legge*[e], *Deanna Sharpe*[f], *Margaret Highland*[g], *Corinne Kozlowski*[h], and *Gayl Males*[i]

[a]Association of Zoos and Aquariums Tree Kangaroo Species Survival Plan® Program, Kansas City, MO, United States

[b]Association of Zoos and Aquariums Tree Kangaroo Species Survival Plan® Program, Maplewood, MN, United States

[c]Melbourne Zoo, Parkville, VIC, Australia

[d]Lincoln Children's Zoo, Lincoln, NE, United States

[e]Dreamworld, Gold Coast, QLD, Australia

[f]Association of Zoos and Aquariums Tree Kangaroo Species Survival Plan® Program, Anchorage, AK, United States

[g]Kansas State University, Manhattan, KS, United States

[h]Saint Louis Zoo, St. Louis, MO, United States

[i]South Coast Environment Centre, Goolwa Beach, SA, Australia

OUTLINE

Tree Kangaroos
https://doi.org/10.1016/B978-0-12-814675-0.00017-8

INTRODUCTION

While good health is the cornerstone to maintaining individual animals, a successful breeding program is the cornerstone to population management. Assessment of reproductive health is essential and can determine the difference between a positive population trend and a negative one. Historically, the majority of reproductive biology and behavioral research has been conducted on Matschie's tree kangaroos (*Dendrolagus matschiei*) in captivity in North America (NA). Much of the following information has been extracted thru this research and data collection. Biology and behavior for Goodfellow's (*D. goodfellowi*) and Lumholtz's (*D. lumholtzi*) tree kangaroos are comparable to Matschie's and will be assumed as similar unless otherwise noted. It should also be acknowledged that much of the behavioral information on tree kangaroos has been acquired through experience and personal observation across global institutions.

Captive breeding populations face a variety of challenges. These challenges may include skewed male to female ratios, aging populations, and fertility complications. In addition, there are issues associated with effects of small population sizes on breeding success such as genetic drift and inbreeding. These can reduce genetic diversity and cause a reduction in reproductive fitness (Frankham et al., 2005). There are steps that can be taken to improve reproductive success. These include understanding reproductive physiology, tests to determine reproductive cycles in females and sperm viability in males, and knowledge of reproductive behavior. The 2007 Tree Kangaroo Husbandry Manual (TKHM) (Blessington and Steenberg, 2007) gives a wealth of information and examples of the reproductive process in *Dendrolagus* species. Documentation of reproductive behavior, fecal steroid assays, semen plugs exams, urine testing, electroejaculation, and physical exams have all been utilized to establish reproductive parameters.

Before starting a breeding program, it is paramount to research the species' needs. A Birth Management Plan (BMP) is a communication tool that can be utilized between the animal care and veterinary departments. Information in the plan should include both the species reproductive characteristics and those of the individual animal. Assessing housing and environmental parameters can help staff prepare for the physical needs of dams and joeys. While the birthing process might only take minutes, it is the hardest journey and most critical period for a joey. The BMP can be found in the 2018 Population Analysis & Breeding and Transfer Plan (PABTP) (AZA, 2018) or by contacting the Association of Zoos and Aquariums (AZA) Tree Kangaroo Species Survival Plan (TKSSP) (AZA, 2019).

Being prepared for a birth also means having the proper equipment and resources on hand in the event that a joey needs to be hand-raised. Because of the lack of physical development of macropods in early stages, hand-raising can be difficult when a joey is less than three months old and its best chance for survival is to be placed back in the pouch to be reared by the dam. As joeys develop, the outcome for successful hand-raising increases (see hand-rearing of rescued wild Lumholtz's tree kangaroos, Chapter 6).

PAIR MANAGEMENT

In the wild, Matschie's tree kangaroos are considered to be solitary. Research has shown that some female ranges overlap with each other (could be related females) and a male's range overlaps many females (Porolak et al., 2014). It has also been shown that young can stay with their mother for up to 24 months before dispersing (Dabek, personal communication, 2019). Due to their solitary nature; the TKSSP recommends that for Matschie's tree kangaroos, the male and female in a breeding pair should be housed separately except during the estrous period when the female is most receptive. Based on behavior, personality, and age, there may be exceptions to housing the male and female together. The decision to alter pair housing is determined by a discussion between the TKSSP and holding institution. Generally, Goodfellow's and Lumholtz's tree kangaroos are also managed as individuals. However, some facilities have successfully co-housed animals, but this should be approached cautiously and closely monitored. The Goodfellow's Global Species Management Plan (GSMP) (Richardson, 2018a) will help make determinations on pair management for Goodfellow's tree kangaroo (Chapter 18).

Considering their solitary and territorial nature, and potential heightened aggression, it is not recommended to house mature males together. While multiple females have been housed together in large exhibits, this practice is also not recommended as stress levels and pouch interference can affect reproduction. It was once thought (incorrectly) that Matschie's tree kangaroos in the wild lived in social groups of a male with several females. Animals that were managed in zoos using this social grouping did not reproduce well. Although breeding resulted in successful births, the dams would often desert their newborn joeys (Wemmer, 1985). There is ample evidence regarding the loss of joeys with group housing. The negative impact of social stress has been well-documented at the National Zoological Park-Conservation and Research Center (NZP-CRC), the Woodland Park Zoo (WPZ), and the Adelaide Zoo (Collins, 1986; Dabek and Hutchins, 1990; Goonan and Arlidge, 1992; Hutchins et al., 1991; Steenberg, 1988). These results validate the importance of giving pouch-gravid females their own space. At WPZ, 100% of the joeys were lost ($n = 10$) when two pouch-gravid Matschie's females were in various social situations. In contrast, 100% of the joeys survived to maturity ($n = 7$) when conspecifics were removed from the pouch-gravid female's area.

ONSET OF FERTILITY, ESTROUS CYCLES, AND INTRODUCTIONS

It is recommended to wait until both the male and female have reached sexual maturity before putting them into breeding situations. Historical data from Collins (1993) and Smith (1988a) reported Matschie's and Grizzled (*D. inustus*) females giving birth at 17 months of age, however, later reports stated 24 months for Grizzled (Smith, 1988b) and 32 months for Matschie's (Heath et al., 1990) for the female's first birth. Dabek (1994) and studbook data indicate that the average age for first reproduction in Goodfellow's, Lumholtz's, and Matschie's male and female tree kangaroos is 2–4 years, based on the animal's actual date of birth (Legge, 2019; Norsworthy, 2019; Richardson, 2018b).

First reproduction may occur at earlier ages, but factors such as husbandry management decisions may influence reproductive output. Simply because an animal can breed at a younger age does not mean it should be encouraged. While they may be hormonally capable, they may not be physically or even behaviorally mature (Cameron et al., 2000).

Tree kangaroos generally produce one offspring. However, there have been two recorded cases in NA of a Goodfellow's and a Matschie's tree kangaroo giving birth to twins (Norsworthy, 2019, Richardson, 2018b) (Fig. 21.1).

Use of fecal steroid hormone profiles is a precise and reliable method for tracking estrous cycles in Matschie's (Dabek, 1994). Fecal steroid analysis in female Matschie's tree kangaroo established a mean estrous cycle length of 55 days based on the intervals between estrogen and progestin peaks. Dabek's Dabek's (1994) findings indicated the mean length of an estrous cycle based on estrogen levels to be 56.83 ± 3.12 days ($n = 6$), and on progestin levels to be 54.17 ± 5.74 days ($n = 6$). Franke (1995) also reported that one female Matschie's tree kangaroo estrous cycle was very regular at every 56–57 days. Matschie's tree kangaroos, estrous cycles were reported to range between 48 and 66 days (North and Harder, 2008) (Fig. 21.2).

According to Schreiner et al. (2015), Goodfellow's average estrous cycle is 54.3 ± 1.6 days ($n = 7$). Richardson (2018b) reported the estrous cycle for Goodfellow's as ~55 days. Johnson and Delean (2003) stated that the estrous cycles for Lumholtz's tree kangaroos ranged from 47 to 64 days. The four species of tree kangaroos: Doria's, Goodfellow's, Grizzled, and Matschie's in Franke-Gunther's (2001) study "…revealed a mean estrous cycle length of approximately 53 days (n=9)." Franke-Gunther's data, along with behavioral data, suggested a similar estrous cycle for all four *Dendrolagus* species in the study. The studies on these four species confirm tree kangaroos have a longer estrous cycle than all other known macropods (Dabek, 1994; Blessington and Steenberg, 2007).

Reproductive behavior and estrous cues can be quite subtle and hard to distinguish. Knowing what is normal and abnormal for a particular animal is instrumental. Housing the male and female alongside each other in adjoining enclosures provides visual and olfactory access, allowing observation of female receptivity indicators (Dabek, 1994). Conducting fecal steroid assays on females will also help determine the estrous cycle stage and predict appropriate introduction timing.

Female estrous behavioral indicators may include increased activity, descending to the ground more frequently, approaching the area where the male is located, pouch cleaning, shivering or head shaking, scent marking, and vocalizing. Female behavior can be subtle, and often, male behavior is a better indicator that a female is coming into estrus. Institutions that were not able to house a male and female in adjoining areas, but in proximity of each other, recognized that a female was in estrus simply by the behavior change of the male.

Male behavior may include descending to the ground, approaching the area where the female is located, pacing along fence lines, vocalizing, sniffing or licking anogenital regions or other body parts, scent rubbing, and rubbing browse on genitals. If the male has access to the female or her area, he may smell urine or feces from the

FIGURE 21.1 Matschie's tree kangaroo with twin joeys (Lincoln Children's Zoo). *Source: Davi Ann Norsworthy.*

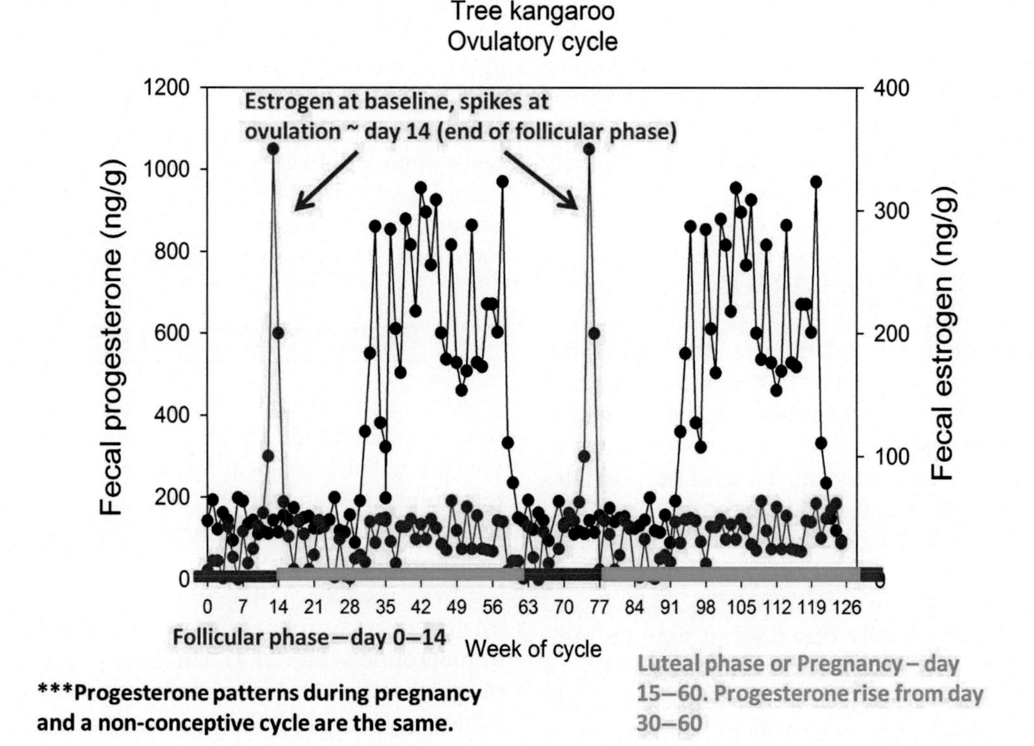

FIGURE 21.2 Expected patterns of estrogen and progesterone production during the reproductive cycle of Matschie's tree kangaroo. Ng/g = nanogram/gram. *Source: Kozlowski, C.P., 2019. [tree kangaroo endocrinology]. Unpublished data. Saint Louis Zoo, St. Louis, MO.*

female, groom the female's fur from behind, try to put his head in female's pouch, attempt to fully mount the female and copulate. The male may also be less focused on food or staff presence and appear dazed at times. It should be noted that some Goodfellow's, Lumholtz's, and Matschie's males may become aggressive toward staff when female tree kangaroos are in estrus.

It is best to introduce the male into the female's environment and preferably with adjoining enclosures where separation is possible. During an introduction, the male and female may vocalize with hissing and clicking sounds. Chasing often occurs including grabbing and cuffing of each other. Periods of rest will occur, as well as territorial marking of enclosure such as rubbing neck on furnishings. Flehmen response may be noted if the male tastes the female's urine or smells her cloaca. If the female is in estrus, and receptive, she will allow the male to approach and attempt copulation. If the female is not receptive, there is risk that injury can occur by either the male or female. Injuries can result from biting and/or grabbing/clawing with front paws. Once a pair has been introduced a couple of times throughout their breeding history, subsequent introductions may become less volatile, but this depends on the animals.

Estrus lasts only about 24–48 hours and the period of peak receptivity is quite short (Dabek, 1994; Blessington and Steenberg, 2007). If the initial introductions are positive, it is recommended to keep the pair together 24/7 until it has been determined behaviorally that the female is no longer cycling. If the female does not appear receptive and aggression has been noted, then the animals should be separated. Once estrus ends, the males usually resume their calm demeanor.

COPULATION

Copulation for Matschie's tree kangaroo occurs when the male "...repeatedly mounts the female with intermittent pelvic thrusting and intromission that occurs for up to 60 minutes, generally on the ground. Females may have thick mucoidal semen trailing from the cloaca following copulation, but this copulatory plug is often lost within 24 hours after copulation" (Heath et al., 1990) (Fig. 21.3). If copulation isn't visually observed, it may be indicated by a soiled cloaca on the female or finding a semen plug. Another indicator could include long drag or scratch marks evident in the substrate.

The fur on the female's back may have a disheveled appearance or be wet. She may also have marks or potential injury to the pouch from repeated grasping by the male's claws. Males may have wet forearms and chest. It has also been observed in Goodfellow's and Lumholtz's that the male may show indications of a sore penis by licking it and, in some cases, it may not retract immediately due to swelling.

Johnson and Delean (2003) have described Lumholtz's copulation as:

> ...male mounting from behind, clasping the thorax of the female with the forearms, while the female has the hindquarters raised and the forepaws on the ground. The investigation and the actual mating were accompanied by the female making a soft trumpeting sound and a shivering action of the head and neck. Mating occurred for 10–35 min with some pairs mating up to three times per day. This cycle lasted for 1–3 days with a semen plug usually present.

Males are generally removed after a confirmed mating or the estrous cycle has ended, and the female is no longer receptive. Occasionally the female will remain tolerant of the male's presence and they can remain together, but it is recommended to separate them prior to estimated parturition (birth). As previously mentioned, there are some instances where pairs have been deemed as suitable to leave together.

REPRODUCTIVE TISSUE

After copulation and ejaculation, the semen from most macropod males produces a copulatory plug. The biological function of the copulatory plug is believed to enhance success of fertilization and prevent mating by rival males, thus enhancing reproductive fitness. If protruding from the cloaca, the copulatory plug can be mistaken for a prolapse, a condition that is rare in macropods (Fig. 21.4). Unlike a prolapse, the plug usually disappears after a day or so (Veterian Key, 2019). Disappearance of the protruding plug occurs when the plug either falls out or is pulled out by the female and typically left behind in the enclosure. If intact when found, the copulatory plug may have an irregular "Y" shaped appearance, as it fills the two lateral vagina and urogenital sinuses; more often only fragments of the plug will be found. Copulatory plugs are typically firm and friable and vary from white to mottled tan and brown to dark red/maroon.

Very little histological data have been compiled on semen plugs or other reproductive tissue from tree kangaroos. In 2007, Dawn Fleuchaus, Milwaukee County Zoo, and Margaret Highland, Pathobiologist at Kansas State University, began studying tissue samples from Matschie's tree kangaroos in NA institutions. The samples were obtained from tree kangaroo enclosures. After gross examination, representative pieces of the tissue were selected and processed by standard methods into paraffin.

FIGURE 21.3 Matschie's tree kangaroo copulation (Zoo Miami). *Source: Yosanni Torres.*

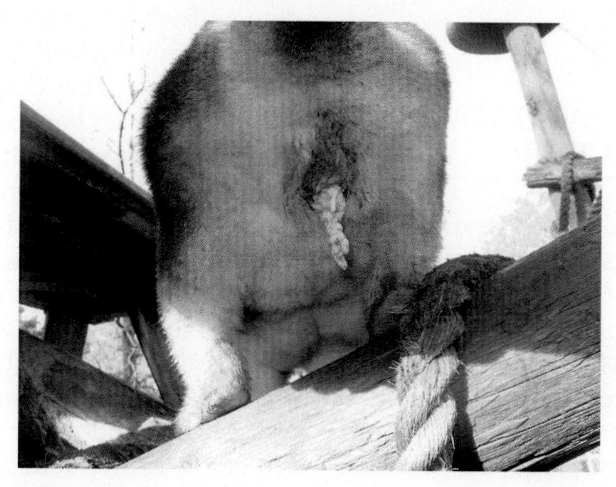

FIGURE 21.4 Goodfellow's tree kangaroo copulatory or semen plug (Melbourne Zoo). *Source: Megan Richardson, Zoos Victoria.*

Thin sections (3–6 μm) were placed on glass slides, H&E stained and examined by light microscopy to determine the type of tissue. To date four types of tissue samples have been identified through histology: semen plug, sloughed uterine lining, placenta, and intestinal/colonic mucous.

Histologic evaluation of copulatory plugs includes assessing the relative number of sperm (semiquantitative assessment) present as well as the amount of bacteria present internally (as opposed to bacteria that will grow along the superficial surfaces from environment contamination). Epithelial cells and hemorrhage are often observed at the surface regions of the plug, presumably due to the female manually removing the plug from the reproductive tract.

This long-term study will provide the TKSSP information on copulation among breeding pairs, and whether there are associations between sperm numbers and/or the overall composition of the plug and successful parturition. Examining all tissue found within enclosures may also aid in evaluating reproductive health (uterine lining, placenta) and potentially digestive health (intestinal/colonic mucosa/mucous). This information will assist with evaluating reproductive success of breeding pairs (Highland, 2019).

GESTATION PERIOD CALCULATION

Gestation should be calculated from the day breeding/copulation is observed or a semen plug is found. A study at the NZP-CRC by Benner and Collins (1988) concluded that Matschie's have a mean gestation period of 44 days. In 17 instances, the gestation period ranged from 43 to 45 days with an average of 44 days (Collins, 1986). In four examples of finding semen plugs, the gestation was estimated to be 39–45 days (Heath et al., 1990). A gestation period of 38 days (Lincoln Children's' Zoo) and 47 days (WPZ) have been reported. The TKSSP recognizes a gestational range of 38–47 days. Because of the variability in length, the TKSSP recommends that males must be separated from females by day 35 post copulation unless the pair has been approved to remain together.

Through the practice of pouch checking at the San Diego Zoo, Goodfellow's tree kangaroo were determined to have a mean gestation period of 44 days (Thompson, 2000). Richardson (2018b) reported a Goodfellow's average gestation period of 45 days, also based on pouch checking. According to Johnson and Delean (2003), Lumholtz's gestation is 42–48 days. Martin (1992) reported that tree kangaroo gestation is "...the longest recorded for a macropod, exceeding by 6 days the longest gestations of *Potorous tridactylus* and *Macropus giganteus*".

PARTURITION AND JOEY CONFIRMATION

Parturition (the action of giving birth) is the day the joey is born and climbs into the pouch. It is important to keep note and mark on a calendar the expected parturition date. All staff working with the female should be made aware of the date and what is expected of them during this time. During the birth process, a female will go into a birth position (sitting on the base of their tail, with the tail between their legs) (Fig. 21.5). They clean or groom their cloacal opening and might rock back and forth or pivot around. Legs can be stretched out and quivering. Females may strain or shake while having contractions. After parturition, the female may pouch clean. The cloaca may appear soiled after parturition and can be a good indicator that the event occurred. It is vital not to disturb the female during the birthing process. Disruptions or environmental stressors can result in the dam abandoning the birthing position and the joey not making it into the pouch. If the joey is observed on the ground, staff can intervene by placing it into the pouch (AZA, 2018; Sealy, 1999).

Historically, restrained pouch checks (where three people physically catch, manually restrain the female, and look in the pouch) were done to confirm the presence of a joey in the pouch. This was done on or around day 50, after seeing breeding/copulation behavior and/or finding a semen plug. Day 50 is the median between average gestation (44–45 days) and estrus (~55 days). If a joey was found during a restrained pouch check, the birth date was assigned as the mean gestational day of 44. Once a joey was confirmed, no further restrained checks were performed.

As husbandry and training have improved, more institutions are now able to perform standing pouch checks. This entails the female voluntarily allowing staff to open her pouch and look in or insert one finger to feel for pouch young. Depending on the female, this may be accomplished by one or more staff. A light source such as a borescope, pen light or phone light is needed to illuminate the pouch. Not only are standing pouch checks less stressful for the females, but the birth date can be more accurately determined.

If a joey is not confirmed by a restrained pouch check at day 50, staff should be prepared for the female to cycle again in 5–7 days (Dabek, 1994). Staff performing voluntary pouch checks should track the cycle days to determine when the next estrus will occur based on the absence of a joey. If a female gives birth but then loses the joey, she can return to estrus within a week and then be mated again, prior to the next cycle which will occur around day 55. In eight instances, female Matschie's returned to estrus an average of 7.3 days after losing a joey (Benner and Collins, 1988; Collins, 1986). George (1982) reported that a pouch would generally be occupied again within two months after either the removal or the weaning of a joey.

FIGURE 21.5 Goodfellow's tree kangaroo giving birth (Melbourne Zoo). *Source: Megan Richardson, Zoos Victoria.*

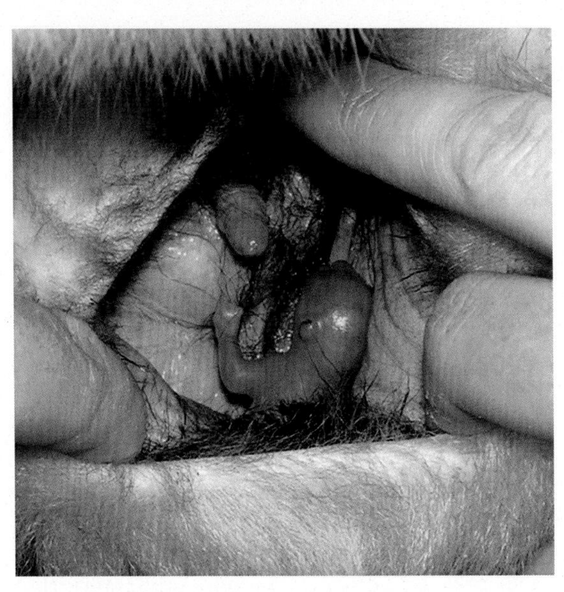

FIGURE 21.6 Determining developmental milestones through voluntary pouch checking. Goodfellow's joey at day 26 (Melbourne Zoo). *Source: Chandi de Alwis, Zoos Victoria.*

DEVELOPMENTAL MILESTONES

Developmental milestones (DM) are behaviors or physical characteristics observed as joeys grow and develop and begins at parturition. If the exact date of birth is not known, it should be estimated as close to the date as possible. The TKSSP has modified a DM chart based on Dabek (1991) that should be completed with every joey. This data is then entered into a database that is shared among institutions and can be used when studying tree kangaroos in the wild to help determine age estimates of joeys in the pouch. Another good resource for Goodfellow's DM is the Birth Date Determination in Australasian Marsupials (Richardson, 2012).

Data on the milestones can be collected in a couple of ways. The easiest method is to collect data points as they are observed. While this may require the least time involved, it may also not be as comprehensive because it only collects data points that can be seen outside of the pouch. Through training and standing pouch checks, developmental milestone data can be collected noninvasively. Milestones can be visually determined via standing pouch checks where the female allows staff to open her pouch and look in at the joey (Fig. 21.6). The use of a borescope, camera or smart phone can be used to video and record joey activity in the pouch during a standing pouch check. The video can then be viewed at a later date to record developmental and behavioral milestones.

Digital calipers were first used to record developmental data such as head, pes (foot) and tail measurements for captive Lumholtz's pouch young by Johnson and Delean (2003). Richardson (2012) was the first to obtain and report these measurements through voluntary pouch training of a Goodfellow's tree kangaroo (Fig. 21.7).

As with all training, it is essential that strong bonds are formed between staff and dam. Consistency in trainers is also beneficial for accurately recording measurements. Daily training and length of sessions should be determined by the acceptance of the dam. While daily pouch checks will gather the most accurate information

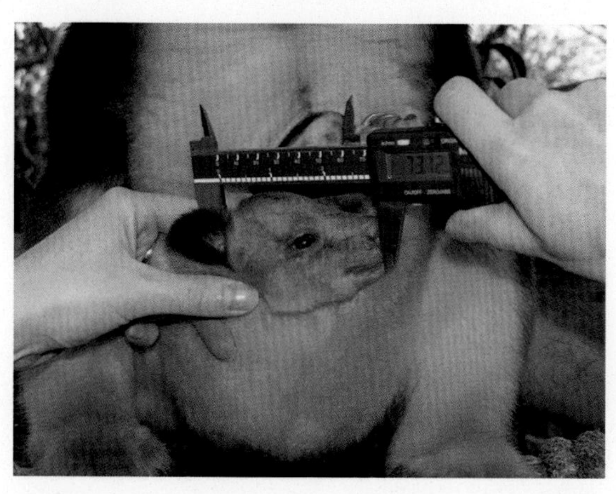

FIGURE 21.7 Using digital calipers to measure head growth in Goodfellow's tree kangaroo joey (Melbourne Zoo). *Source: Chandi de Alwis, Zoos Victoria.*

on development, intermittent checks will also have great value. Table 21.1 shows developmental milestone data for Matschie's, Goodfellow's, and Lumholtz's tree kangaroos.

Data points have been combined from various data sets. For consistency all data was converted to weeks in range. Data was collected in different methods which may contribute to a range in outcomes. Three methods were utilized to collect data: observations, restrained pouch checks, or trained/standing pouch checks. These were performed either daily, scheduled, or opportunistically.

Interpretation of data collected also varied when describing features such as signs of fur and fur coloration. Other milestones, such as weaned, varied in context. Weaned is typically described as the point in which young no longer nurse. In some cases, this was referred to as when the joey was separated from its dam. Data points for independence from the dam were also collected but are not included in this table. Facilities typically control when a joey is independent based on when it is physically separated from its dam which is not a true indicator when comparing to independence of animals in the wild. It would be important for future data collection to include an ethogram or description of data that is being documented.

Other notes to include in data collection are:

- Which of the four teats are they attached to: upper or lower, left or right?
- Note joey orientation when suckling
- Is recorder/observer/trainer doing standing pouch checks or is it observation only?
- Are pouch checks/observations scheduled or opportunistic?
- If scheduled, how often?
- What age of the joey did pouch checks/observations start?
- What age of the joey did pouch checks/observations stop?

JOEY SEPARATION

Dabek's field research indicates that joeys stay with their dam until 18–24 months of age Sometimes a young tree kangaroo is in proximity to its dam while a new joey is in the pouch (Dabek, personal communication, 2019). To follow this natural separation behavior, the TKSSP recommends leaving joeys with their dam until they are about 18 months of age. Some dams

TABLE 21.1 Developmental Milestones (DM) for Matschie's, Goodfellow's and Lumholtz's tree kangaroos. Samples were taken from a variety of sources and range was converted to weeks for consistency.

Developmental Milestone	Matschie's[a,b]		Goodfellow's[c,d]		Lumholtz's[e,f]	
	Range in weeks	Sample size	Range in weeks	Sample size	Range in weeks	Sample size
First Pouch movement	4–12	6				
Detachment from teat	10–16	6			11–15	Unknown
Eyes open	18–25	16	19–21	2	18–20	Unknown
Signs of fine fur (unspecified)	13–25	6	20–24	3	20–23	2+
Signs of fine fur—dorsum of forelimbs	3–4	12	25	1		
Signs of fine fur—face	11–22	8	21–22	2		
Signs of fine fur—rhinarium	15–16	7	21	1		
Signs of fine fur—cloacal vibrissae	16–21	9				
Signs of darker fur on arms and legs	24–31	8	24	1		
Signs of dark pigmentation on tail	23–29	8	20–23	2		
Fully furred	22–29	12	25–26	3	23–28	2+
Claws pigmented	4–8	12	6	1		
Whiskers Papillae apparent	8–13	3			11–14	Unknown
Teeth: Lower incisors	19–31	13	23			
Teeth: Upper incisors	19–31	6	27			
Ears: Free from head	11–23	12				
Ears: Becoming upright	16–24	5				
Ears: Fully upright	20–32	9			23–24	Unknown
Testes/penis visible	4–9	2				
Pouch visible	5–6	2				
Tail out of pouch	19–27	15			23–25	2
Limb out of pouch (may be one or both forelimbs/hindlimbs)	12–26	36	21–27	2	23–24	2
Head out of pouch	20–27	51	20–27	7	22–23	2
First pouch emergence	19–33	63	25–33	9	23–28 28–30	2+
At foot/head in pouch	29–34	6				
At foot/head in pouch/head bobbing (suckling)	30–32	4			38–39	2
Play behavior—Joey with dam						

TABLE 21.1 Developmental Milestones (DM) for Matschie's, Goodfellow's and Lumholtz's tree kangaroos Samples were taken from a variety of sources and range was converted to weeks for consistency—cont'd

	Matschie's[a,b]		Goodfellow's[c,d]		Lumholtz's[e,f]	
Developmental Milestone	**Range in weeks**	**Sample size**	**Range in weeks**	**Sample size**	**Range in weeks**	**Sample size**
Joey in pouch	26–36	9				
Joey out of pouch	30–32	6				
Permanent pouch exit	39–43	19	29–45	9	35–41	10
Sniffing foods	25–39	13				
Eating solid foods	26–41	23	27–36	3	28–36	2
Weaned (no longer nurses)	28–53	4			52–73	1+

[a] *Blessington and Steenberg, 2007, Tables 5.1 and 7.6.*
[b] *DeBo and Carlyle-Askew, 2019. [Tree Kangaroo SSP DM Database]. Unpublished raw data. Seattle, WA.*
[c] *Richardson, 2012.*
[d] *Blessington and Steenberg, 2007, Table 7.5.*
[e] *Johnson and Delean, 2003, Fig. 3, Table 2.*
[f] *Legge, personal communication, 15 May 2019.*

may encourage joeys to separate sooner while others will tolerate them longer.

Captive studies suggest that the dam begins to cycle again when her offspring is about 13 months old. Steenberg, (2000), in a two-year study, found that both the estrogen and progesterone levels were seen to rise slightly when the joey was about 11.5 months old. A more prominent rise was seen at around 13 months, 10 days with even higher levels around 15 months. These data show a pattern of a lactating female resuming full estrous cycle at approximately 15 months, though there can be variability between females and species.

The TKSSP recognizes that waiting to separate joeys at approximately 18 months of age will cause a missed opportunity for breeding. However, this correlates with observations in the wild that they would not breed with a joey present of less than 18 months of age (Dabek, personal communication, 2019). These data support that behavioral adjustment of the young tree kangaroo is more important than early separation for breeding. While the primary advantage of keeping pairs together is to allow breeding at the first return to estrus to maximize reproduction opportunities, this should only be done under consultation with the TKSSP or GSMP. Joeys may interfere or possibly become injured if present during the breeding process.

In preparation for permanent separation, the joey should be eating well on its own and from its own feeding station (Fig. 21.8). Ideally, institutions will have a separate holding area for the joey that is out of visual access to the dam and preferably in another building. A joey should be separated for a minimum of 30 days if shipping to another institution. It should be eating well and displaying normal behavior. This process can take up to six months before the joey is ready for a move. It is not advisable to return joeys to their dams after a long-term separation as it may result in serious injury or even death.

PLATEAU AND SENESCENCE

One reproductive situation that has not thoroughly been studied is the apparent tendency of some female Matschie's tree kangaroos to

FIGURE 21.8 Matschie's tree kangaroo joey eating solids (Woodland Park Zoo). *Source: Amanda Dukart.*

plateau during reproductively viable years. Many of the females that display this tendency produce joeys during younger ages, stop cycling for a few years and then resume cycling to produce one or two more joeys before reaching reproductive senescence. This phenomenon has been difficult to study because multiple factors may be contributing such as weight and infertility. Plateauing may also be confounded by the lack of a male being present or stress in their housing situation.

Reproductive senescence has not been thoroughly studied through fecal steroid assays but is primarily assumed thru studbook data and lack of joey production and rearing. These data may be incomplete, or inaccurate if the male or female was not provided a breeding opportunity.

A female Matschie's, captive born in 1974, was cycling in July 1991 at age seventeen. This observation was based on behavioral changes, copulation, and fecal steroid assays (Steenberg, 1991). Later it was found through fecal steroid assays that this same female was not cycling in March–July 1992 (Dabek, 1994); she had ceased to cycle between 17 and 18 years of age. A number of Matschie's and Goodfellow's females have been recorded at giving birth at 15–16 years of age, though some of those joeys failed to attach to the teat.

The oldest Matschie's recorded to give birth and successfully rear a joey was 16.5 years old at the Lincoln Children's Zoo (Nebraska) (Fig. 21.9). The oldest Goodfellow's female to give birth and successfully rear a joey was ~21 years of age at the Rotterdam Zoo (Richardson, 2018b). In Schreiner et al.'s study (2015), a 19-year-old female Goodfellow tree kangaroo was in an active reproductive cycle.

The International Tree Kangaroo Studbook (Collins, 1993) reported a male Matschie's tree kangaroo sired joeys until 15 years 9 months of age while another sired his last joey at 16 years and 4 months of age. A male Matschie's was electro-ejaculated at 17+ years of age and had good quality sperm. A male Goodfellow's sired a joey at 19 years of age (Richardson, 2018b). It is assumed that male tree kangaroos do not reach reproductive senescence (Kozlowski, personal communication, 2019).

EMBRYONIC DIAPAUSE

In Benner and Collins (1988), there were no indications of postpartum estrus in 11 years of breeding the Matschie's tree kangaroo. Martin (1992) reported that there was no evidence of embryonic diapause (delayed implantation) in a captive population of Lumholtz's.

FIGURE 21.9 'Milla', oldest reproducing Matschie's female and her joey 'Bek' at 15 months of age (Lincoln Children's Zoo). *Source: Davi Ann Norsworthy.*

Unlike other species of macropodids, postpartum estrus and embryonic diapause has not been observed in tree kangaroos (Heath et al., 1990; Johnson and Delean, 2003; North and Harder, 2008; Steenberg, 2000).

FUTURE REPRODUCTIVE BIOLOGICAL RESEARCH

Population sustainability for animals in managed collections depends primarily on the ability of holding institutions to get specific individuals to successfully breed. Reproductive sustainability within the AZA population is particularly important for Matschie's as there are currently no individuals available for importation from facilities in range countries (Chapter 18). The TKSSP has to manage the population for maximum breeding opportunities. Currently, the managed population is not sustainable as many females are nonreproductive due to age, health, or other unexplained factors. The TKSSP is working with colleagues in Papua New Guinea (PNG) on the potential for captive born animals to be exported from PNG to accredited institutions.

Contributing to a reduced reproductive output for the population is that important information on the general physiology, behavior, and reproductive biology of this species is lacking and needs further study. To address this shortage of information and to facilitate management decisions aimed at encouraging breeding and maximizing welfare, Koester (2019) is conducting a long-term fecal assay project titled "Examining potential factors affecting the reproductive endocrinology and breeding success of Matschie's tree kangaroos (*Dendrolagus matschiei*)".

This study will aim to:

(1) Measure ovarian and adrenal hormone concentrations in feces of female Matschie's tree kangaroos at AZA facilities through breeding attempts, confirmed pregnancy and lactation, and resumption of estrus.

(2) Compare ovarian and adrenal hormone concentrations between successful and unsuccessful breeding attempts.
(3) Compare female body weight and condition across successful and unsuccessful breeding attempts.
(4) Survey and compare the gut microbiome of tree kangaroos of different body conditions and on different diets.
(5) Analyze available, banked serum samples to establish captive reference ranges of the thyroid hormone thyroxine as a measure of metabolism.

Results of this study will provide, for the first time, hormonal and physiological information on female Matschie's tree kangaroos during breeding, pregnancy, and lactation, and shed light on potential factors impeding reproductive success for specific individuals.

SURROGATE PROJECT

The ability to cross-foster macropod pouch young from closely related species has a long history at several institutions within Australia. This capacity has allowed for an increase in the reproductive output of some endangered wallabies in conservation breeding programs (Taggart et al., 2010). As part of one of these programs, a group of Yellow-footed Rock Wallabies (YFRW) (*Petrogale xanthopus*) in Australia's Adelaide Zoo was maintained to act as surrogates for the critically endangered Southern Brush-tailed Rock-wallaby (BTRW) (*P. penicillata*) (Schultz et al., 2006; Taggart et al., 2005).

Due to trauma following an overnight accident, an adult female Goodfellow's tree kangaroo was found dead in her enclosure. Her 47-day old male unfurred pouch young was still alive in the pouch. Although detached from the teat, the joey, weighing 29 g, was still warm, bright, and active and the decision was made to attempt to cross foster using one of the YFRW surrogates. Although never previously attempted with an unrelated species with such different life strategies, the attempt was worth trying given the joey would not have survived at such a young undeveloped age without intervention.

Having identified a suitable surrogate YFRW carrying a pouch young of similar age and size, the routine procedure for cross-fostering was carried out (McLelland et al., 2015). The procedure is detailed fully in this paper. The joey began actively sucking on the new teat on the second attempt to introduce it into its mouth. As per normal practice, the surrogate was kept under light anesthesia for 30 min and the joey was checked every ten minutes to ensure it remained attached to the teat, before returning the dam to the enclosure.

After the initial anesthesia of the surrogate, monitoring was done by observation only, a minimum of twice a day. The dam was observed for pouch growth and movement. At 15 weeks, a limb was first seen protruding from the pouch. The first time the head popped out of the pouch was at 20 weeks, when fine fur was also seen on his limbs.

The surrogate started showing signs of discomfort when the tree kangaroo joey was five months old and had been in the pouch 3.5 months. While tree kangaroos and rock wallabies are similar in size and weight, individuals may vary which may have contributed to the surrogate's discomfort. After the joey was seen out of the pouch for the first time, there was potential risk of him being rejected due to the dam's discomfort. He was also at risk for injury if exiting the pouch while the surrogate was high on the rocky sections of the exhibit. A decision was then made to remove the joey for hand-raising; having the joey old enough to hand-raise was the ultimate goal. He weighed 865 g at the time and was hand-raised without any problem (Fig. 21.10).

The success of this procedure shows that even vastly different species can benefit from the

FIGURE 21.10 Goodfellow's tree kangaroo joey with Brush-tailed rock wallaby surrogate dam (Adelaide Zoo). *Source: Kate Fielder.*

techniques of cross-fostering which could potentially be utilized for genetically valuable individuals. This specialized procedure would require an institution to have established breeding programs to cross-foster macropods.

CONCLUSIONS

Over the last forty years, captive management of Matschie's, Goodfellow's, and Lumholtz's tree kangaroos has improved through behavioral observations and physiological testing of biological systems. Reproductive behavior in combination with well-established biological parameters for estrous cycles, gestation, parturition and rearing of young have been well studied and documented. However, questions still remain for each species and continued research is needed.

Careful management of current global populations is needed to maintain tree kangaroo species in captivity. The study of reproductive biology and behavior through observations, on-going data collection, and fecal steroid assays, continue to be major contributions to the success of populations. With cooperation among institutions, each of the managed tree kangaroo species and their respective programs has the potential to reach genetically sustainable populations in managed facilities in years to come.

ACKNOWLEDGMENTS

The authors of this chapter wish to acknowledge the contributions of Dr. Margaret Highland (Kansas State University) and Dawn Fleuchaus (Milwaukee County Zoo) for leading the reproductive tissue study; Dr. Donald Knowles (USDA, Pullman WA) for providing financial support of the tree kangaroo reproductive tissue analysis; Dr. Diana Koester (Cleveland Metroparks Zoo) for her work in reproductive endocrinology; Kate Fielder (formerly Adelaide Zoo) for providing review and edits to Surrogate Project section; TKSSP Coordinator Kathy Russell (Santa Fe College Teaching Zoo) for her review and recommendations to this chapter and to the TKSSP for the protocols and continued efforts to improve tree kangaroo reproductive success.

REFERENCES

Association of Zoos and Aquariums (AZA), 2018. Population Analysis & Breeding and transfer plan Matschie's tree kangaroo (*Dendrolagus matschiei*). In: AZA Species Survival Plan (SSP) Red Program. AZA Population Management Center, Lincoln Park Zoo, Chicago, IL.

Association of Zoos and Aquariums (AZA), 2019. Tree Kangaroo Species Survival Plan (TKSSP). [< https://www.aza.org/contact-us> 01 March 2019].

Benner, K.S., Collins, L.R., 1988. Reproduction and Development in Matschie's Tree Kangaroos (*Dendrolagus matschiei*) in Captivity. National Zoological Park Conservation and Research Center, Front Royal, VA.

Blessington, J., Steenberg, J., 2007. FormatTree Kangaroo (*Dendrolagus* spp.) Husbandry Manual. In: Tree Kangaroo Species Survival Plan. third ed. Association of Zoos and Aquariums, Silver Spring, MD.

Cameron, E.Z., Linklater, W.L., Stafford, K.J., Minot, E.O., 2000. Aging and improving reproductive success in horses: declining residual reproductive value or just older and wiser? Behav. Ecol. Sociobiol. 47, 243–249.

Collins, L., 1986. Big foot of the branches. In: Zoogoer, July–August. Washington, DC, National Zoo, pp. 35–40.

Collins, L., 1993. 1992 International Studbook for Matschie's Tree Kangaroo, (*Dendrolagus matschiei*). National Zoological Park Conservation and Research Center, Front Royal, VA.

Dabek, L., 1991. Mother-young relations and development of the young in captive Matschie's tree kangaroos (*Dendrolagus matschiei*). Master's thesis. University of Washington, Seattle, WA.

Dabek, L., 1994. The reproductive biology and behavior of captive female Matschie's tree kangaroos (*Dendrolagus matschiei*). Ph.D. diss., University of Washington, Seattle, WA. Abstract in *Dissertation Abstracts International* 55(1 IB): 4748.

Dabek, L., Hutchins, M., 1990. The social biology of tree kangaroos: implications for captive management. In: AAZPA Regional Conference Proceedings, Wheeling, WV, pp. 528–535.

Franke, M.C., 1995. Early developmental chronology of a Matschie's tree kangaroo (*Dendrolagus matschiei*) through daily standing pouch checks. Animal Keepers' Forum 22 (6), 226–233.

Franke-Gunther, M., 2001. Reproductive biology and behaviour in tree kangaroos (*Dendrolagus matschiei, D. goodfellowi, D. dorianus,* and *D. Inustus*) with applications for captive management and conservation. MS thesis, University of Kent at Canterbury (may 2001).

Frankham, R., Ballou, J.D., Briscoe, D.A., 2005. Introduction to Conservation Genetics. Cambridge University Press, Cambridge, UK, pp. 175–196.

George, G.G., 1982. Tree kangaroos (*Dendrolagus spp.*) their management in captivity. In: Evans, D.D. (Ed.), The Management of Australian Marsupials in Captivity. Melbourne, Zoological Board of Victoria, pp. 102–107.

Goonan, P., Arlidge, J., 1992. Observations on the birth of a Matschie's tree-kangaroo (*Dendrolagus matschiei*) in captivity. Aust. Mammal. 15, 113–114.

Heath, A., Benner, S., Watson-Jones, J., 1990. Husbandry and management of Matschie's tree kangaroo: A case study of tree kangaroo husbandry at CRC. In: American Association Zoological Parks Aquariums Regional Conference Proceedings, Wheeling, WV, pp. 518–527.

Highland, M., 2019. Tree kangaroo reproductive tissue. Unpublished raw dataWashington State University, Pullman, WA.

Hutchins, M., Smith, G.M., Mead, D.C., Elbin, S., Steenberg, J., 1991. Social behavior of Matschie's tree kangaroos (*Dendrolagus matschiei*) and its implications for captive management. Zoo Biol. 10 (2), 147–164.

Johnson, P.M., Delean, S., 2003. Reproduction of Lumholtz's tree-kangaroo (*Dendrolagus lumholtzi*) (Marsupialia: Macropodidae) in captivity, with age estimation and development of the pouch young. Wildl. Res. 30 (5), 505–512.

Koester, D., 2019. Examining Potential Factors Affecting the Reproductive Endocrinology and Breeding Success of Matschie's Tree Kangaroos (*Dendrolagus matschiei*). Cleveland Metroparks Zoo, Cleveland, OH Unpublished raw data.

Legge, A., 2019. Lumholtz's Tree Kangaroo (*Dendrolagus lumholtzi*) Studbook. Zoo Association of Australia, Dreamworld Wildlife Habitat, Goldcoast, Australia.

Martin, R.W., 1992. An ecological study of Bennett's tree-kangaroo (*Dendrolagus bennettianus*). In: Project 116, World Wide Fund for Nature 67 pp.

McLelland, D.J., Fielder, K., Males, G., 2015. Successful transfer of a Goodfellow's tree kangaroo (*Dendrolagus goodfellowi*) pouch young to a yellow-footed rock wallaby (*Petrogale xanthopus*) surrogate. Zoo Biol. 34, 460–462.

Norsworthy, D., 2019. International Matschie's (*Dendrolagus matschiei)* Tree Kangaroo Studbook. Lincoln Children's Zoo, Lincoln, NE.

North, L.A., Harder, J.D., 2008. Characterization of the estrous cycle and assessment of reproductive status in Matschie's tree kangaroo (*Dendrolagus matschiei*) with fecal progestin profiles. Gen. Comp. Endocrinol. 156 (1), 173–180.

Porolak, G., Dabek, L., Krockenberger, A., 2014. Spatial requirements of free-ranging Huon tree kangaroos, *Dendrolagus matschiei,* (Macropodidae) in upper montane Forest. PLoS. 9(3), e91870https://doi.org/10.1371/journal.pone.0091870.

Richardson, M., 2012. Herrmann, K. (Ed.), Birth date determination in Australasian marsupials. second ed. Zoo and Aquarium Association Australasia, Mosman NSW 2088 Australia.

Richardson, M., 2018a. Goodfellow's tree-kangaroo (*Dendrolagus goodfellowi*) global species management plan. Melbourne Zoological Gardens, Melbourne, Australia Zoo and Aquarium Association (ZAA).

Richardson, M., 2018b. International Tree Kangaroo Studbook for Goodfellow's (*Dendrolagus goodfellowi*). Melbourne Zoological Gardens, Melbourne Australia.

Schreiner, C., Schwarzenberger, F., Kirchner, W.H., Dressen, W., 2015. Behavioural and hormonal investigations in Goodfellow's tree kangaroos (*Dendrolagus goodfellowi*) (Thomas, 1908). Zoological Garden. 84 (1–2) January.

Schultz, D.J., Whitehead, P.J., Taggart, D.A., 2006. Review of surrogacy program for endangered brush-tailed rock wallaby (*Petrogale pencillata)* with special reference to animal husbandry and veterinary considerations. J. Zoo Wildl. Med. 37, 33–39.

Sealy, R., 1999. An amazing rescue of a Matschie's tree kangaroo joey. Animal Keepers' Forum 26 (2), 74–75.

Smith, R.L., 1988a. New Studbook—Grizzled (*Dendrolagus inustus*) Tree Kangaroos. San Antonio Zoological Gardens, San Antonio, TX.

Smith, R.L., 1988b. Tree kangaroos: Vulnerable in the wild, threatened in captivity. In: Dresser, B.L., Reese, R.W., Maruska, E.J. (Eds.), Proceedings of the fifth world conference on breeding endangered species in captivity, Cincinnati, OH. p. 650.

Steenberg, J., 1988. The management of Matschie's tree kangaroo (*Dendrolagus matschiei*) at Woodland Park zoological gardens. In: AAZPA Western Regional Conference Proceedings, Wheeling, WV, pp. 94–111.

Steenberg, J., 1991. Pilot project using fecal steroid assays with Matschie's tree kangaroos at National Zoo's Conservation and Research Center, Front Royal, Virginia. Unpublished raw data.

Steenberg, J., 2000. Matschie's tree kangaroo embryonic diapause: pilot research project. Tree kangaroo (*Dendrolagus* spp). Master plan 1999-2000. Thompson, V. (Ed.). San Diego, CA.

Taggart, D.A., Schultz, D., White, C., 2005. Cross-fostering, growth and reproductive studies in the brush-tailed rock wallaby, *Petrogale pencillata* (Marsupialia: Macropodidae): efforts to accelerate breeding in a threatened marsupial species. Aust. J. Zool. 53, 313–323.

Taggart, D.A., Schultz, D.J., Fletcher, T.P., 2010. Cross-fostering and short-term pouch young isolation in macropodoid marsupials: implications for conservation and species management. In: Coulson, G., Eldridge, M. (Eds.), Macropods: The Biology of Kangaroos, Wallabies and Rat Kangaroos. pp. 263–278.

Thompson, V., 2000. Tree Kangaroo (*Dendrolagus* spp.) Master Plan 1999-2000. San Diego Zoological Society, San Diego, CA.

Veterian Key, 2019. Macropods. Available from: <www.veteriankey.com/macropods> [Feb. 23, 2019].

Wemmer, C., 1985. A decade of research. In: Zoogoer. Washington, DC, National Zoo March/April 10–15.

CHAPTER

22

The Role of Zoos in Tree Kangaroo Conservation: Connecting Ex Situ and In Situ Conservation Action

Karin R. Schwartz[a], *Onnie Byers*[b], *Philip Miller*[b], *Jacque Blessington*[c], *and Brett Smith*[d]

[a]Association of Zoos and Aquariums Tree Kangaroo Species Survival Plan®, Milwaukee, WI, United States [b]IUCN SSC Conservation Planning Specialist Group, Apple Valley, MN, United States [c]Association of Zoos and Aquariums Tree Kangaroo Species Survival Plan® Program, Kansas City, MO, United States [d]Port Moresby Nature Park, Port Moresby, Papua New Guinea

OUTLINE

Tree Kangaroos
https://doi.org/10.1016/B978-0-12-814675-0.00026-9

INTRODUCTION

In 1872, William Hann, an Australian pioneer pastoralist, led an expedition for a mining survey in the Cape York Peninsula, a remote peninsula located in Far North Queensland of Australia. He encountered a strange, new animal that he described as a kind of kangaroo that climbs trees (Flannery et al., 1996). To Hann, the idea of a tree-climbing kangaroo was ridiculous. Yet he brought back the news and naturalists of that time were skeptical (Obituaries Australia, 2019). Fast forward to today, and even after 14 species of tree kangaroos have been described, many people within their native Australia, Papua, Indonesia, and Papua New Guinea (PNG), as well as other parts of the world, have never heard of a tree kangaroo much less seen one. Moreover, 12 of the 14 species are listed in the threatened categories on the International Union for the Conservation of Nature (IUCN) Red List (IUCN, 2019).

Tree kangaroo populations have declined due to overhunting and habitat loss (conversion to agricultural lands, devastation from mining practices, logging, and prevalence of forest wildfires during prolonged droughts due to climate change) (Chapter 3 – Australia; Chapter 4 – New Guinea). In order to reverse the trend of declining populations, immediate conservation measures are needed to sustain viable populations and prevent extinction. The global zoo community has been instrumental in partnering with governments, conservation organizations,and local communities in many areas within the natural ranges of tree kangaroos to work toward conservation of these threatened species.

The registered global tree kangaroo population (all *Dendrolagus* species) maintained in zoological facilities as of December 2019 totaled 108 individuals (52.52.4 [males.females.unknown sex]) in 49 zoos around the world (ZIMS, 2019). Five of the 14 known species of tree kangaroos are represented: Matschie's (*D. matschiei*; IUCN Endangered; Fig. 22.1), Goodfellow's (*D. goodfellowi* spp.; IUCN Endangered; Fig. 22.2), and Doria's (*D. dorianus*; IUCN Vulnerable; Fig. 22.3), and smaller populations of Lumholtz's (*D. lumholtzi*; IUCN Near Threatened: Fig. 22.4) and Grizzled (*D. inustus*; IUCN Vulnerable; Fig. 22.5) tree kangaroos (Chapter 18). Although the populations are very small, the participating zoos have the unique opportunity to introduce their visitors to this rare, little-known animal, learn about its biology and behavior, and advocate for conservation of these species in the wild.

Zoos have evolved from menageries that initially provided entertainment for private societies and later for the public, to organizations with an increased focus on conservation (Rabb, 1994; Minteer et al., 2018; Henson, 2018). The World Association of Zoos and Aquariums (WAZA) Conservation Strategy promotes animal welfare and conservation as the primary purposes for zoological institutions (Barongi et al., 2015). With a myriad of species

FIGURE. 22.1 Matschie's tree kangaroos at Sedgwick County Zoo, Wichita, Kansas, USA. *Source: Jan Nelson.*

FIGURE. 22.2 Goodfellow's tree kangaroo, Melbourne Zoo, Melbourne, Victoria, Australia. *Source: Zoos Victoria.*

FIGURE. 22.4 Lumholtz's tree kangaroo at Currumbin Wildlife Sanctuary, Currumbin, Queensland, Australia. *Source: Mel Spittall.*

FIGURE. 22.3 Doria's tree kangaroo at Port Moresby Nature Park, Port Moresby, Papua New Guinea. *Source: Port Moresby Nature Park.*

representing a large range of taxa (both native and foreign) in their collections, zoos globally are well positioned to work together for sustainable ex situ (under managed care in zoological facilities) populations, provide conservation education programs to promote awareness for visitors, and contribute to in situ (in the wild) conservation through research, funding support, capacity building, and involvement in conservation action planning and implementation.

Zoological facilities are organized into international and regional associations that maintain specialized animal programs for population management and collaborative conservation

FIGURE. 22.5 Grizzled tree kangaroo mother and joey at Taman Safari, Bogor, Indonesia. *Source: Taman Safari Indonesia.*

programs for threatened species (Chapter 18). Zoo staff with expertise in the husbandry and care of particular species contribute their knowledge not only to the care and welfare of their charges, but to conservation efforts of these species in the wild. It is clear that to mitigate the extinction risk of wild tree kangaroo populations, it is critical to utilize all available tools and resources and combine ex situ and in situ conservation efforts for a One Plan Approach for integrated tree kangaroo conservation. The One Plan Approach – a term originated by the IUCN Species Survival Commission (SSC) Conservation Breeding Specialist Group (CBSG) (in 2017, changed to Conservation Planning Specialist Group – CPSG). The One Plan Approach refers to integrated species conservation planning that considers all populations of the species (inside and outside the natural range), under all conditions of management, and engages all responsible parties and resources from the start of the conservation planning initiative (Byers et al., 2013).

This chapter discusses early and current involvement of zoos in ex situ and in situ tree kangaroo research and conservation action planning, the conservation role of zoological associations through population management programs and research, and the participation by ex situ facilities in conservation efforts in range countries. There is a discussion on additional conservation roles through education, awareness-raising, and the initiation, creation, and ongoing support and management of field conservation programs. This chapter emphasizes the importance of the collaborative One Plan Approach and working with government and local communities within the natural tree kangaroo ranges for effective, integrated conservation action.

INITIAL ZOO INVOLVEMENT IN TREE KANGAROO CONSERVATION

Zoo and Field Research

Tony Olds, Assistant Head Keeper and Larry Collins, Small Mammal Curator at the National Zoological Park's Conservation and Research Center (NZP-CRC) worked on initial research on the reproduction of Matschie's tree kangaroos in the 1970s and published this early work in the 1973 International Zoo Yearbook (Olds and Collins, 1973). This was the beginning of decades of research on tree kangaroos under managed care at the NZP-CRC. In the 1980s, in addition to the work of Larry Collins (1986), Graeme Crook and Gert Skipper, keepers at Adelaide Zoo in South Australia, described the care and behavior of the Matschie's tree kangaroos maintained at their zoo (Crook and Skipper, 1987). They reported on husbandry topics such as exhibit requirements, nutrition and food provision, social structure, and handling as well as reproduction and health care.

Benner and Collins (1988) furthered the work on identifying reproductive parameters for the Matschie's tree kangaroo at the NZP-CRC (Chapter 21). Steenberg (1988) identified the importance of social isolation of pouch-gravid females, noting that at Woodland Park Zoo (WPZ) in Seattle, Washington, all 10 joeys produced from two dams were lost when the females were housed with conspecifics. These studies were critical for understanding reproduction in the Matschie's tree kangaroo and contributed to improved husbandry practices for this species that continue today for ex situ breeding management. These seminal studies led to further research on biology and physiology of tree kangaroos, extending to work on tree kangaroos in the wild.

For Lisa Dabek, the mysterious Matschie's tree kangaroo of PNG brought great intrigue because, by the late 1980s, initial work had been done on reproduction but very little was known about social behavior between mother and young, and development of the joeys. Dabek, as part of a graduate program in Animal Behavior at the University of Washington, studied developmental milestones for this marsupial species (Dabek, 1991). The WPZ maintained Matschie's tree kangaroos and provided her with the opportunity to study the behavioral development of mother-young pairs. Dabek continued on with a PhD and devoted her research to reproductive biology and behavior of the Matschie's tree kangaroo at WPZ and at the NZP-CRC (now Smithsonian Conservation Biology Institute – SCBI). At the NZP-CRC, she worked with Larry Collins who had done leading research on gestation and reproduction on the NZP-CRC's colony of 20 Matschie's tree kangaroos (Olds and Collins, 1973). Her work in the reproductive endocrinology laboratory of Dr. Samuel Wasser established reproductive cycles through non-invasive hormone collection from dung (Dabek, 1994; Chapter 21). As part of her research, Dabek learned that the Matschie's tree kangaroo was considered endangered, and few studies had been done in the wild.

Dabek went to PNG in 1994 for the first time to determine a location to study Matschie's tree kangaroos in the wild. Will Betz joined Dabek in PNG as a Research Assistant in 1996 when the Tree Kangaroo Conservation Program (TKCP) was founded in Teptep village in the Yopno, Uruwa, and Som (YUS) area, Morobe Province, Huon Peninsula, PNG (Dabek and Betz, 1998). He and Dabek spent several years establishing relationships with the YUS villagers along with conducting research (Chapters 2 and 10). This work was important in leading to the formation of community conservation for tree kangaroos in PNG.

From 1998 until 2005, Dabek was supported by Roger Williams Park Zoo in Providence, Rhode Island in her role as Director of Research and Conservation. Dabek continued her tree kangaroo research and conservation work in PNG and brought the TKCP to Seattle, Washington in 2005 when she was appointed Field Conservation Director at the WPZ. The TKCP became the signature field conservation program for WPZ and the designated field program of the AZA Tree Kangaroo Species Survival Plan (TKSSP). This conservation program is supported by many zoos and aquariums in North America as well as in Australia, Asia, and Europe.

CONSERVATION PLANNING AND ACTION FOR DECLINING TREE KANGAROOS POPULATIONS IN PAPUA NEW GUINEA

New Guinea, a remote island north of mainland Australia, is one of the most biologically diverse islands in the world (Chapter 4). Due to the diverse and rugged mountainous topography, there is limited infrastructure and access to remote areas. Thus, little was known of the island's flora and fauna in the 1990s, their abundance and distribution, or the factors that threatened their existence. Papua New Guinea comprises the island's eastern half of the island

and is a country with low human population densities spread out in remote areas. The indigenous people own over 95% of the land and have a land tenure system for use of the natural resources (TKCP, 2014). They have a subsistence lifestyle, dependent on hunting (many species including tree kangaroos) and use of the land for agriculture. Village leaders started noticing that it was becoming harder to have successful hunts as they had to travel further to find tree kangaroos (Chapter 4). Dabek, working with Will Betz, discovered that there were increasing threats to tree kangaroos due to deforestation for agricultural use and over-hunting (Dabek and Betz, 1998).

Planning to Act – Transdisciplinary Collaboration for Tree Kangaroo Conservation

The status of wild tree kangaroos in PNG was not well understood in the 1990s, yet there was concern that the six known endemic species at that time (including three subspecies each for Goodfellow's and Doria's tree kangaroos) only existed as small populations. Three species occurred in very small geographic ranges totally within PNG (Matschie's, Lowlands [*D. spadix*], and Scott's or Tenkile [*D. scottae*]) while three other species (Doria's, Grizzled, and Goodfellow's) had somewhat larger distributions that extended into Papua Province and West Papua, Indonesia. Population declines and habitat loss had accelerated in the last three decades. It was clear that if PNG's tree kangaroos were to survive in viable populations beyond the beginning of the 21st century, a conservation plan had to be created and implemented.

Dr. Tim Flannery, a world expert on the fauna of New Guinea and principal research scientist at the Australian Museum in Sydney, explored New Guinea through 15 expeditions that led him to describe 20 new species (including two tree kangaroo species – Scott's and Golden-mantled [*D. pulcherrimus*]) and learn about the life and culture of indigenous peoples (Flannery, 1998). Flannery brought the threatened status of tree kangaroos in PNG to the attention of the conservation community.

In April 1992, there was an IUCN CBSG Conservation Assessment Management Plan (CAMP) Meeting on threatened taxa of Australasian marsupials and monotremes, held at the Australasian Regional Association of Zoological Parks and Aquaria (ARAZPA) Conference in Currumbin, Queensland (QLD), Australia. The purpose of the CAMP was to review the in situ and ex situ status of the threatened taxa and provide strategic guidance for management toward conservation goals. A post-ARAZPA Tree Kangaroo Symposium took the CAMP results into consideration and called for increased international cooperation for managing tree kangaroos in zoos as well as for conservation of these species in the wild. Six years later in 1998, CBSG facilitated a groundbreaking event held in Lae, PNG, supported by the global zoo community, that illustrated the long-term involvement and commitment of zoos in in situ and ex situ tree kangaroo collaborations for research, conservation action planning and implementation.

CBSG's reputation for balancing endangered species survival with the needs of local communities led the agencies responsible for tree kangaroo conservation to ask them to conduct a Population and Habitat Viability Assessment (PHVA) workshop and identify priority tree kangaroo species for conservation.

CPSG (previously CBSG) is a Disciplinary Specialist Group within the IUCN SSC whose mission is to "save threatened species by increasing the effectiveness of conservation efforts worldwide" (CPSG, 2019). CPSG follows three major tenets for all their conservation action planning workshops (CBSG, 2017):

- Science – effective planning depends on access to and analysis of the best available information in order to develop a shared

understanding of the species, habitat, and the impact of human communities' use of natural resources that threaten the species.
- Facilitation – social science skills are incorporated to ensure that all stakeholders from diverse groups are heard in defining and prioritizing problems, identifying goals, and coming to a consensus on strategies to meet those goals.
- Collaboration – this core philosophy of CPSG processes is the over-reaching principle that promotes global networking for conservation action. CPSG's One Plan Approach promotes integration of in situ and ex situ strategies, connecting zoos and aquariums to conservation of species in the wild.

The Conservation Planning Process for the Tree Kangaroos of Papua New Guinea

A PHVA workshop was held at the University of Technology, Lae, PNG from 31 August to 4 September 1998 (Bonaccorso et al., 1999). Forty-seven people attended the workshop which was hosted by the PNG National Museum, Rainforest Habitat, and the PNG Department of Environment and Conservation (DEC). The workshop was endorsed by the Marsupial and Monotreme Taxon Advisory Group (M and M TAG) of the Australasian Regional Association of Zoological Parks and Aquariums (ARAZPA; now Zoo and Aquarium Association – ZAA, of which PNG institutions are members), generously funded by zoos from Australia and the United States (Table 22.1), and facilitated by CBSG. Participants included representatives from PNG government (Department of Forestry and DEC), PNG museums and local zoo (Rainforest Habitat), local landowners, TKCP, and zoo specialists from ARAZPA M and M TAG and American Association of Zoological Parks and Aquariums (now Association of Zoos and Aquariums: AZA) TKSSP. Communication between the local landowners, PNG officials, and the non-PNG delegates was greatly enhanced by the participation of YUS landowner Mambawe Manono and Will Betz, representing TKCP and the TKSSP, who had spent several years in PNG working with and learning from the local villagers.

> "Much of the land in Papua New Guinea is privately owned, and wildlife is considered the property of landowners," said Peter Clark, Director of Life Sciences for Zoos South Australia. "Any planning or decisions made regarding the future conservation of tree kangaroos needed the landowners' input and agreement to have any meaning."

The primary aim of the workshop was to develop an action plan for the long-term conservation of genetically viable populations of tree kangaroos in PNG. Participants from PNG, Australia, and the United States specializing in population biology, captive management, reproduction, veterinary medicine, and human demographics, compiled and analyzed both published and unpublished information on all six tree kangaroo species of PNG. Most importantly, 13 local landowners, representing several regions of PNG where tree kangaroos are found, actively participated in the workshop, which had ongoing translation into pidgin (a common language in PNG).

For indigenous people living in the rugged terrain of the mountains and forests, tree kangaroos are an integral part of daily life, serving as a source of dietary protein and featuring prominently in local legends and customs (Chapter 10). Local people have intimate knowledge of tree kangaroos: where they move among the treetops, what they eat, and how they behave. Grassroots participation and approval by a spectrum of landowners is essential for the success of any conservation plan in PNG where 95% of all land is in private ownership. It is important to note that wildlife in PNG is the property of the landowner, not the government. In addition, social scientists need to be

TABLE 22.1 Hosts and sponsors of the 1998 Conservation Assessment and Management Plan for the Tree Kangaroos of Papua New Guinea and Population and Habitat Viability Assessment for Matschie's Tree Kangaroo (Bonaccorso et al., 1999).

In collaboration with IUCN/SSC Conservation Breeding Specialist Group		
Co-Hosted by	**Sponsored by**	
	Zoological Facility	**Country**
PNG Department of Environment and Conservation	Adelaide Zoological Gardens	Australia
	Royal Melbourne Zoological Gardens	Australia
PNG – National Museum & Art Gallery	Taronga Zoo	Australia
	Perth Zoological Gardens	Australia
Rainforest Habitat-University of Technology – PNG	Currumbin Sanctuary	Australia
	San Antonio Zoological Gardens and Aquarium	USA
	Roger Williams Park Zoo	USA
	Mill Mountain Zoological Park	USA
	Columbus Zoo	USA
	Smithsonian Conservation Biology Institute	USA

IUCN/SSC – International Union for Conservation of Nature/Species Survival Commission; PNG – Papua New Guinea; USA – United States of America.

involved to help translate human demographic and resource use trends into impacts on tree kangaroos and their habitat.

The Workshop Process

Effective conservation action is best built upon critical examination and use of available biological information, but also very much depends upon the actions of humans living within the range of the threatened species. At the beginning of each PHVA workshop, there is agreement among the participants that the general desired outcome is to prevent the extinction of the species and to maintain a viable population(s). The workshop process takes an in-depth look at the species' life history, population history, status, and dynamics, and assesses the threats putting the species at risk. The background information can be from many sources; the contributions of *all* people with a stake in the future of the species are considered. Information contributed by landowners, hunters, scientists, field biologists, and zoo managers all carry equal importance.

To obtain the entire picture concerning a species, all information gathered is discussed by the workshop participants with the aim of first reaching agreement on the state and status of current information. These data are then incorporated into a computer simulation model to determine: (1) risk of extinction under current conditions; (2) those factors that make the species vulnerable to extinction; and (3) which factors, if changed or manipulated, may have the greatest effect on preventing extinction. These computer-modeling activities provide a neutral way to examine the current situation and what needs to change to prevent extinction.

A successful PHVA workshop depends on determining an outcome where all participants coming to the workshop with different interests and needs, "win" in developing a management strategy for the species in question. Local solutions take priority. Workshop report recommendations are developed by, and are the property of, the participants.

Participants worked in stakeholder groups to identify the major issues and needs, related to tree kangaroos and their habitat, that they wanted the meeting to address. The landowner stakeholder group, working in pidgin came up with a list of eight issues and needs for reintroduction, public education, socio-economic concerns, law enforcement, and research. These, and the issues raised by the other stakeholder groups (biological/social scientists and captive managers/educators), then became the focus of four working groups: Life History and Modeling, Socio-economic Issues, Status and Distribution, and Government and Legislation.

The Life History and Modeling working group conducted a simplified Population Viability Analysis (PVA) to guide the development of targeted management recommendations designed to minimize the risk of extinction of the most threatened tree kangaroo species. PVA (Beissinger and McCullough, 1998; Reed et al., 2002) is a process that typically uses computer simulation modeling techniques to assess the relative magnitude of threats to wildlife populations, to evaluate the risk of decline and extinction of wildlife populations in the face of those threats, and to compare the relative predicted responses of threatened populations to proposed management alternatives. These models incorporate much of the fundamental knowledge of the dynamics of small population growth: that as population size decreases, genetic and demographic pressures cause a self-reinforcing spiral of decreased population stability, resulting in reduced abundance that ultimately puts a population at risk of imminent extinction from significant events like extreme weather events, disease outbreaks, etc. (Gilpin and Soulé, 1986). This concept can be readily applied to many tree kangaroo species – hence the urgency to hold the conservation planning workshop and to apply rigorous analytical methods to guide conservation action.

Experts in tree kangaroo biology were consulted before and during the workshop to derive a set of demographic and ecological parameters to use as base input values for models using Version 8.03 of the software package *Vortex* (Table 22.2) (Lacy, 2000). An important characteristic of this data assembly process was the responsible use of knowledge from the ex situ management community to fill in key gaps in the understanding of tree kangaroo demographics, breeding biology, etc. Another valuable highlight of this process was the use of rapid participatory appraisal tools before the workshop to conduct interviews in the field with local villagers for the purpose of gaining information on their interactions with nearby tree kangaroo populations – specifically, the rate of extraction of these animals through hunting (Nyhus et al., 2003). With this information, the PVA practitioner was able to estimate the amount of additional annual mortality to include in models designed to assess the impact of this practice. Moreover, the appraisal revealed a strong female bias among those tree kangaroos removed through hunting. Villagers identified females as both (1) easier to catch, owing to the frequent presence of a young joey in tow as they moved through the trees, and (2) a more highly sought-after food item because of higher fat content resulting from their frequent status as new lactating mothers. This knowledge was key in developing more realistic models of potential threats to local tree kangaroo populations – a clear objective of any PVA.

Models were constructed to better understand the biological drivers of tree kangaroo population growth, and the anticipated impacts of human-mediated threats to long-term species persistence. For example, sensitivity analysis revealed that both breeding success and survival of adult females were critical factors influencing population growth. In addition, the analyses also dramatically portrayed the inherent risks

TABLE 22.2 Input parameters used for *Vortex* models of tree kangaroo population dynamics[a]. Specific input parameter values used across range of models making up the population viability analysis can be found in Bonaccorso et al. (1999).

Model Category	Model Parameter
General population	Number of populations
	Current population-specific abundance, age structure
	Presence and extent of dispersal between populations
General habitat	Geographic range
	Current habitat carrying capacity across populations
Population reproduction	Breeding system
	Reproductive lifespan (first, last age of breeding)
	Mean annual breeding rate (% females reproducing) (SD[b])
	Offspring production (max, mean, sex ratio)
	Density dependence in reproduction, if documented
Population survival	Age- and sex-specific mean annual survival rate (SD)
	Additional impact of human hunting on mean age-specific survival
Population genetics	Inbreeding impacts on reproduction and/or survival (LE[c])
Catastrophes	Frequency (annual probability) of event(s)
	Impact of event(s) on reproduction and/or survival

[a] *Conservation Breeding Specialist Group (CBSG). 2010. Population and Habitat Viability Assessment (PHVA) Workshop Process Reference Packet. http://www.cbsg.org/sites/cbsg.org/files/PHVA_Reference_Packet_2010.pdf.*
[b] *SD, standard deviation in mean rate.*
[c] *LE, number of lethal equivalents quantifying severity of inbreeding depression.*

faced by small populations – risks that arise precisely because of the destabilizing stochastic demographic and genetic forces. Specifically, the models indicated that a small population of just 50 tree kangaroos, free from hunting pressure, could grow in abundance at an average annual rate of about 2.5% – yet that population had a 6.2% risk of declining to extinction through demographic instability, inbreeding, and other sources of variability (Fig. 22.6). A larger population, using the exact same input values, would show no such risk. Models like these demonstrate that small wildlife populations can become extinct simply because of bad luck.

More importantly, the systematic analyses conducted in the PHVA workshop demonstrated the considerable toll that the identified hunting strategy used by local villages – targeting females for a larger haul of tastier food with comparatively less effort – can exact on affected tree kangaroo populations (Fig. 22.6). A small population of just 50 tree kangaroos that is subject to additional removal of individuals through hunting (at the rates estimated at the workshop) faces a risk of extinction that is nearly seven times that of a population of the same size that is free from the same magnitude of hunting pressure. The consequences of this hunting threat are magnified when imposed on a smaller

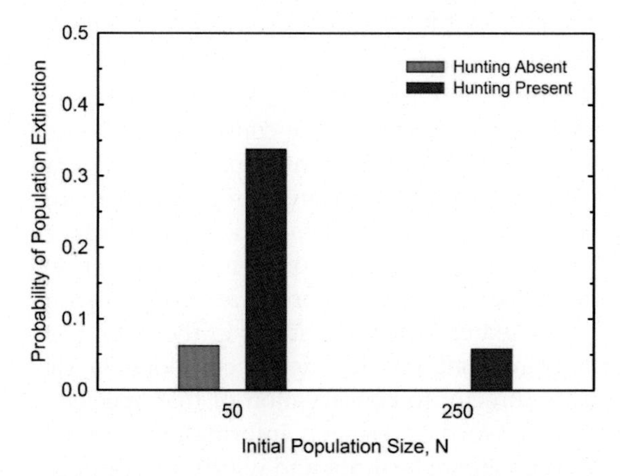

FIGURE. 22.6 Results of selected population viability models from the tree kangaroo Population and Habitat Viability Assessment (PHVA) workshop. Plot shows probability of extinction over 100 years for simulated tree kangaroo populations of different initial abundance in the absence (green bars) or presence (red bars) of female-biased hunting by local human villagers. For more information on these and other models used in the workshop, see accompanying text and Bonaccorso et al. (1999).

population that is already at risk from inherent destabilizing forces.

The PVA results were the basis for important discussions that took place during and after the workshop. Local villagers were not aware of the population biology principles that underpin the academic community's focus on threatened biodiversity conservation planning; seeing the models and their implications led them to see the current situation of their local wildlife in a dramatically different light. In order to facilitate this communication of knowledge to such a diverse assemblage of stakeholders, it was imperative that the workshop facilitators and technical practitioners could present and discuss often difficult, technical information in a way that was both accessible and non-threatening to those without the same professional background and experience.

"I thought this tree kangaroo was just another animal to be hunted," said Kausa Ilao from the Wau region. "But when I learned more about tree kangaroos from this workshop, I got excited. I will return to my village a happy man because in my wildlife area there are a lot of tree kangaroos. I am going to look after them."

Several recommendations were implemented during the workshop itself, including the development and translation into pidgin, of educational materials specifically requested by the landowners (Bonaccorso et al., 1999). TKCP was already established so landowners and team members were able to share what had been learned on the Huon Peninsula working to conserve the endangered Matschie's tree kangaroo (Nyhus et al., 2003). In addition, a rapid response team was formed at the request of landowners, to respond to the urgent need for conservation action directed toward the critically endangered Tenkile.

Action: The Rapid Response Team – Saving Tenkile Tree Kangaroos

The landowners' input revealed that one type of tree kangaroo, the Tenkile, was critically endangered: they believed that there were fewer than 100 left. Using this information, CBSG's risk assessment models confirmed that continued hunting of female Tenkile would edge the species closer to extinction – possibly within just a few years. The group agreed to send a small contingent of people from the workshop, including a local community member and a few zoo-based species experts, to travel to several villages in the Torricelli Mountains, learn from the communities there about the Tenkile's situation, and invite them to participate in the conservation effort for tree kangaroos. They subsequently visited several villages to gather more information on Tenkile and other tree kangaroos.

Leaders from some of the villages helped organize a regional meeting to discuss the priority needs of the villages and introduce the concept of a hunting moratorium on female Tenkile tree kangaroos in the area. At the meeting, the village leaders offered their opinions on the need for alternate protein sources and discussed their support for a proposed conservation field station in the area. Representatives of all 13 villages in attendance signed a two-year hunting moratorium and enthusiastically joined the conservation initiative for Tenkile in the region. A year later, Peter Clark (at the time Manager of Rainforest Habitat in PNG and later Director of Life Sciences for Royal Zoological Society of South Australia), and Gary Slater of Melbourne Zoo (Zoos Victoria) helped draft a community-based conservation plan out of which the nonprofit Tenkile Conservation Alliance (TCA) was formed. Thanks to the tireless work of its current directors, Jim and Jean Thomas, and the entire TCA staff, the TCA has become a successful conservation operation whose efforts have helped increase the wild population of Tenkile (Thomas, personal communication, 2019) (see section on the TCA under Role of Zoos in Creation of In Situ Conservation Programs).

TREE KANGAROO MANAGEMENT AND CONSERVATION THROUGH ZOO ASSOCIATIONS

Many zoos worldwide are involved in the One Plan Approach to tree kangaroo conservation through collaborative breeding management programs, sharing of information, conservation advocacy programs, capacity building with conservationists within tree kangaroo range countries, and provision of funding support for in situ programs. Ex situ populations of tree kangaroos are managed through regional zoological associations that have specialized animal management programs or globally through a collaborative population management program that oversees an international breeding program (Chapter 18). The focus of these programs is to maintain healthy populations that are genetically diverse, demographically stable, and behaviorally competent. Maintaining healthy, sustainable populations constitutes a conservation benefit by limiting the requirement to interject new genes from tree kangaroos taken from the wild to augment genetic diversity. In addition, these individuals act as ambassadors for their species, serving to educate visitors and promote awareness of their threatened conservation status in the wild. Each program contains additional components that relate directly to conservation of that species in the wild, such as research, information exchange between global facilities and wildlife researchers, contributing expertise in range country conservation programs, and funding support for conservation of the species in the wild.

Sharing of information facilitates the development of effective husbandry procedures for good health, successful reproduction, and welfare. The AZA TKSSP Tree Kangaroo Husbandry Manual (TKHM) (Blessington and Steenberg, 2007) is available to any facility that houses tree kangaroos, including those in other zoological regions (zoos in Oceania, Europe, and Asia). The manual, proposed to be updated to the new AZA Animal Care Manual format in 2021, provides best practices for tree kangaroo care and welfare (Chapter 19).

Global collaboration among in situ tree kangaroo programs in the different world regions for research in biology, reproduction, health, nutrition, comparison of genetic diversity between in situ and ex situ populations, and behavior (Chapters 19–21) not only benefit the management of tree kangaroos maintained in zoos, but elucidate life history parameters that are shared with those working on conservation of tree kangaroos in the wild. Through the use of standardized records-keeping processes such as the Species360 Zoological Information Management System (ZIMS) (ZIMS, 2019; Chapter 18), data from zoos maintaining tree kangaroos throughout the world can be compiled to facilitate breeding management processes, determine developmental milestones (Chapter 21; Table 21.1), share information on biology and behavior (Chapters 20 and 21), and link with conservation program practitioners in the wild. This next section describes the zoo associations that are involved and gives examples of individual institutions that carry on this important work.

Zoo and Aquarium Association (ZAA)

Species management programs – ASMP and GSMP

The Australasian Species Management Programs (ASMP) of the ZAA coordinate and facilitate management for priority species in the Oceania region (Australia, New Zealand, and Papua New Guinea). Each ASMP is assigned a primary species role such as Conservation Advocacy, Conservation Recovery, or Conservation Science and Research (although the management plan may include all three components). Each is categorized as a Conservation Program that is formally linked to a recovery effort, a Population Management Program (population sustainability of regionally relevant species to aid in conservation advocacy messages), or a Monitored Program (regionally significant species requiring general oversight rather than high-intensity management) (ZAA, 2019).

Although ZAA facilities had a history of managing five species of tree kangaroos (Lumholtz's, Bennett's, Doria's, Matschie's, and Goodfellow's) (Chapter 18), it was not until 1994 that a formal ASMP was developed for Goodfellow's tree kangaroo. In 2013, ZAA joined with the AZA, the European Association of Zoos and Aquaria (EAZA), the Japanese Association of Zoos and Aquariums (JAZA), and the South East Asia Zoo Association (SEAZA) to form the WAZA Goodfellow's Tree Kangaroo Global Species Management Plan (GSMP) (one of the first nine GSMPs formed by WAZA). This collaboration included 10 zoos in Australasia, 10 zoos in Europe and 2 zoos in Asia that maintained Goodfellow's tree kangaroos (Richardson, 2018).

The main objective of the GSMP, coordinated by staff at Melbourne Zoo in Victoria, Australia, is to effectively manage the global population of Goodfellow's tree kangaroos to maintain genetic diversity and demographic stability and to serve in the role of conservation advocacy to widely deliver key conservation messages concerning the threatened status in the wild (WAZA, 2019). The GSMP assists in the integration of conservation efforts between in situ conservation organizations and ex situ facilities which includes information exchange and education, working with the local communities living within the natural ranges, and support for priority conservation in situ programs such as the TCA and the TKCP in PNG.

Port Moresby Nature Park in Papua New Guinea

Port Moresby Nature Park (POMNP) was formed in 2012 following National Capital District Council's decision to cease operations of the National Capital Botanical Gardens and in its place, form a new charitable trust. It is an award-winning combined botanical, zoological, and cultural park located on 30 acres of tropical habitat in PNG. The Park features over 350 native animals, including Doria's, Goodfellow's, and Matschie's tree kangaroos. POMNP is the only zoo in PNG that has achieved membership in ZAA after passing assessments for international standards in animal care and welfare.

POMNP's vision is connecting people and nature through education, recreation, and conservation. POMNP has been successful and has won a number of awards in innovation and education programs. The park has established a sister-zoo partnership with Zoos Victoria (ZV) (see section below on Zoos Victoria) where both parties share knowledge and resources to train staff at both sites. At the time of writing, breeding tree kangaroos was not a priority as space for maintaining individuals was limited. Many wild animals that are injured or abandoned are offered to the Park, but most are not accepted due to lack of space and resources. Wholesale acceptance of such animals has the potential to contribute to illegal collection from the wild and the trade in wildlife. Stage two of the Park's Masterplan will see an extension of the POMNP through the establishment of a rescue/research/rehabilitation center to cater to the many species including tree kangaroos brought in as surrenders, rescues, and confiscations. Where possible, they will be rehabilitated and released but most of the tree kangaroos will require care for the remainder of their lives as the laws prevent animals being returned to the wild unless their precise origin is known.

The POMNP is currently negotiating with the government for the potential of breeding rescued tree kangaroos and provide animals for tree kangaroo populations in breeding programs under managed care globally. In February 2020, the Park opened stage two of a new tree kangaroo exhibit to enable the expansion of their tree kangaroo habitats (Fig. 22.7). Graphics share the conservation messages about the protected conservation areas in PNG and the steps the locals can take to establish these protected conservation areas on their own customary land. The TCA in the Torricelli Mountains and the TKCP in the Yopno, Uruwa and Som (YUS) Conservation Area in Morobe Province are highlighted at the exhibit to show the benefits of establishing these areas not only for the protection of fauna and flora species, but also the benefits they bring to communities within these areas.

One of the many roles of the POMNP is education of students and engaging the public to bring the message of wildlife and environmental protection in PNG. This is done through signage, public education days such as world wildlife day and world environment day, teaching of over 24,000 students annually that visit for formal education programs, hosting local television segments on PNG wildlife, and presenting weekly full page newspaper columns and posts on social media. The park is the most visited attraction in the country with over 140,000 people visiting each year.

In 2016, the POMNP hosted an international tree kangaroo summit where participants from overseas institutions maintaining tree kangaroos, PNG government representatives, and non-government organizations (NGOs) working on tree kangaroo conservation in PNG and Australia came together with local landowners to discuss the current and future global conservation of tree kangaroos.

Australian Zoos

Zoos Victoria

Zoos Victoria (ZV), the governing body for Healesville Sanctuary, Melbourne Zoo, and Werribee Open Range Zoo in Australia, has a long-term commitment to fight wildlife extinction. The Zoos Victoria Wildlife Conservation Master Plan 2019–2024 (Zoos Victoria, 2019) incorporates strategic conservation plans that integrate approaches that range across the biological and social sciences. This plan prioritizes 27 species native to Australia that are at risk of extinction and articulates two conservation partnerships for the Oceanian region. The latter includes the TKCP, with conservation outcomes for Matschie's tree kangaroo and other wildlife. As part of a new initiative included in the Plan, three of the international partnerships are linked with a respective sister zoo to "act as platforms for telling our conservation partners' stories in-country, enhance engagement with government, enable ZV to take our Fighting Extinction approach to a broader global audience, and significantly increase international participation of ZV staff" (C. Banks, personal communication, 2019).

ZV has a highly successful relationship with the POMNP as its Oceanian sister zoo. Since the partnership was launched in 2013, more than 30 staff have visited each other's zoos to share expertise in animal husbandry, records-keeping, veterinary care, environmental sustainability, master planning, and conservation education.

FIGURE. 22.7 New tree kangaroo habitat at Port Moresby Nature Park in PNG (February 2020). *Source: Port Moresby Nature Park.*

The partnership has facilitated the development of projects to engage the Port Moresby community in awareness of the conservation status and threat of species endemic to PNG (Banks, 2019). ZV has also been instrumental in the support of the TKCP's YUS Conservation Coffee project, an alternative livelihood program that benefits the local PNG village communities and wildlife (Banks, 2018; Chapter 14). This long term partnership has benefitted both TKCP and ZV. The tree kangaroo exhibit at ZV now features information about conservation coffee in YUS. ZV has also fostered a 10-year partnership with the TCA benefitting the Tenkile and Golden-mantled tree kangaroo (see section on the TCA below).

Currumbin Wildlife Sanctuary

Currumbin Wildlife Sanctuary (CWS), located on the Gold Coast of Australia, has housed tree kangaroo species for over 25 years with a focus on the Goodfellow's tree kangaroo endemic to PNG. In the past, keepers have worked in PNG with this species in situ and ex situ and gained valuable experience which has helped CWS to become one of the most successful breeding programs under managed care for Goodfellow's tree kangaroos (A. Molyneux, personal communication, 2020). CWS has seen 10 joeys born as part of the captive breeding program in its 25 years working with the species. In December 2017, CWS acquired their first male Lumholtz's tree kangaroo from Dr. Karen Coombes at the Tree Roo Rescue and Conservation Centre (TRRACC) in Malanda (Fig. 22.4). This male had been rescued from the wild because of vision issues. CWS continues to collaborate with the TRRACC in the care of rescued Lumholtz's tree kangaroos that are unable to be returned to the wild with the potential for beginning a managed breeding program to establish an ex situ population in collaboration with five other Australian zoological facilities (see section on TRRACC below).

Association of Zoos and Aquariums (AZA)

Matschie's tree kangaroos maintained in North American (NA) facilities are managed in 20 AZA institutions and one non-AZA institution through the AZA TKSSP (Che-Castaldo et al., 2019). The focus of AZA SSPs is to cooperatively manage the population of selected threatened species within AZA-accredited zoos and aquariums and to contribute to conservation of each species in the wild (AZA, 2018a). To be a part of the TKSSP, each facility is required to meet high standards for animal husbandry and health care. This includes following guidelines for proper habitats, appropriate social structure, breeding management, adequate veterinary coverage, attention to psychological well-being through behavioral enrichment, and nutrition. In addition, institutions that are considering adding tree kangaroos to their managed collections are encouraged to allocate funds to support field conservation programs for tree kangaroos in the wild.

The TKSSP, first approved by AZA in December 1991, is a collaborative program that enables participating institutions to responsibly manage their tree kangaroos through analysis of founder representation for each individual and through a mean kinship strategy, make pairing recommendations for breeding that would result in the sustainability of the overall population (Lacy et al., 1995; AZA, 2018b; Chapter 18). In addition, the TKSSP has a focus to support conservation of wild tree kangaroos. The TKSSP maintains the International Matschie's Tree Kangaroo Studbook to provide the pedigree data used to develop the breeding recommendations. The TKSSP originally covered Matschie's, Goodfellow's and Grizzled tree kangaroos, but a regional focus was adopted at the 2005 International Tree Kangaroo Summit on the Atherton Tablelands, Australia, with AZA concentrating on Matschie's tree kangaroos and Europe, Oceania, and Asia focusing on Goodfellow's tree kangaroos.

The TKSSP oversees collaborative research on reproduction, health, physiology, behavior, and husbandry processes to increase understanding of the biology of the species and to improve animal care (Chapters 19–21). There are established necropsy protocols that include body measurements, establishment of cause of death, and guidelines for carcass disposition to the Burke Museum of Natural History and Culture in Seattle, Washington. The Burke Museum, on the Seattle campus of the University of Washington, offers archival collections of specimens that are available to educators and global researchers. Jeff Bradley, Mammalogy Collection Manager at the Burke Museum, at the direction of the TKSSP, curates the collection and processing of tree kangaroo carcasses (skin, scull, body skeleton, and tissues) from TKSSP institutions and makes them available for research and education programs. A Mammalogy Database is accessible online for anyone interested in viewing the collection specimens of interest (Burke Museum, 2019).

The Burke Museum moved to a new building in October 2019, and a full body mount of a Matschie's tree kangaroo (that lived a long life at the WPZ) is featured in a prominent place in the Burke Museum's permanent exhibit, representing the Marsupial Mammal branch of the Tree of Life (Fig. 22.8). The exhibit is in the format of a cladogram – a diagram that groups species in "clades", showing organisms that have shared characteristics and can be traced back to a common ancestor. Although the museum is not a zoo, specimens in the collection are from ex situ institutions and these exhibits serve to educate people about tree kangaroos and links to conservation.

Due to the TKSSP's focus on maintaining a sustainable population of Matschie's tree kangaroos and linking to conservation, many member institutions support the TKCP for conservation of Matschie's tree kangaroos in PNG. Since the start of the TKCP, the TKSSP sponsored the field

FIGURE. 22.8 Matschie's Tree Kangaroo "Huen", representing the Marsupial lineage, looks over the rest of the Mammal branch of the Tree of Life, an exhibit at the Burke Museum, Seattle, Washington, USA. *Source: Jeff Bradley.*

conservation efforts and made TKCP their designated field program. TKSSP institutions have supported TKCP in many ways including funding, educational materials and resources, staff expertise, and collaborations on ex situ and in situ research. This support includes information exchange as well as directly contributing expertise in biological knowledge, veterinary medicine, field research (monitoring), animal husbandry, and records-keeping for conservation research in the YUS Conservation Area in PNG. For example, developmental milestones that were determined through documentation of tree kangaroo development in zoos assist to identify approximate age for those animals observed in the wild. Management and planning specialists from WPZ contribute to the administration of the TKCP and development of community conservation programs.

The TKSSP is looking forward to expanding and exploring new partnerships to facilitate tree kangaroo conservation in PNG. AZA launched SAFE: Saving Animals From Extinction, a collaborative initiative between ex situ and in situ partners to work together on field conservation efforts (AZA, 2020). Species-specific experts joining together in each SAFE program include AZA member zoos, scientists, non-government and government staff, and organizations working on threatened species conservation within the species' natural range. Currently (as of 2020), there are 28 endangered species covered by a SAFE program where partners work to identify the threats, develop action plans, obtain funding, and engage the local communities within natural ranges. The TKSSP in conjunction with WPZ SAFE program leaders Lisa Dabek and Beth Carlyle-Askew has been approved to conduct a Tree Kangaroo SAFE Program which will focus on Matschie's and Goodfellow's tree kangaroos. The Tree Kangaroo SAFE Program will partner with the TKSSP, the TKCP, local PNG community leaders, and organizations concerned with tree kangaroo conservation. This collaboration between ex situ and in situ conservation partners epitomizes the One Plan Approach with benefits to both sides as they share the common goal to prevent extinction of the species in their natural ranges.

European Association of Zoos and Aquaria (EAZA)

There were 10 EAZA zoos in four countries contributing to a European Endangered Species (EEP) program for Goodfellow's tree kangaroos, led by species program leaders at the Krefeld Zoo in Krefeld, Germany. The Goodfellow's Tree Kangaroo EEP had close ties to the TCA and supported them yearly beginning in 2007. In 2013, the EEP was integrated into the

Goodfellow's Tree Kangaroo GSMP along with zoos maintaining Goodfellow's tree kangaroos in Australia, Southeast Asia, and Japan. Of the 10 EAZA zoos participating in the GSMP, four are in Germany. Krefeld Zoo continues to be a leader in breeding Goodfellow's tree kangaroos. Research conducted on reproductive cycles by the four German zoos contributed to the primary goal of the GSMP of enhancing the sustainability of the GSMP population (Schreiner et al., 2015; Chapter 21). Krefeld Zoo and Rostock Zoo (Germany) continue to support the TCA as one of the designated recipients of in situ conservation support from the Goodfellow's Tree Kangaroo GSMP. Krefeld Zoo has hosted two International Tree Kangaroo Symposiums (2009 and 2018). The symposium in 2018 brought together global zoo professionals from the TKSSP and Goodfellow's Tree Kangaroo GSMP, wildlife researchers, veterinarians, and officials from the TKCP and TCA in PNG. The Symposium included a husbandry training workshop and exchange of information on health, population management, and research on tree kangaroos in the wild. Other European zoos have supported TKCP including the long term support from Zoo Parc de Beauval in Saint-Aignan, Loir-et-Cher, France.

Southeast Asia Zoo Association (SEAZA); Japanese Association of Zoos and Aquariums (JAZA)

Singapore Zoo (one of the facilities in the Wild Reserves Singapore Group), an active member of SEAZA, and Zoorasia (Yokohama Zoological Gardens) in Japan, an active member of JAZA, both maintain Goodfellow tree kangaroos as participants in the Goodfellow's Tree Kangaroo GSMP and support tree kangaroo field conservation efforts. Breeding recommendations and transfers are planned between countries to maintain genetic diversity for the global population. In 2014, a 47 day old joey, born at Adelaide Zoo in Australia, was orphaned when his mother was killed by a falling tree branch. He was too young to be hand-reared and in a world-first procedure, was cross-fostered (transferred to a lactating mother of a different species) by a yellow-footed rock wallaby (*Petrogale xanthopus*). At four months old, the joey was too big for the wallaby's pouch and hand-rearing was continued by his caretakers. At 2.5 years old, the youngster was sent to Singapore Zoo as part of the GSMP where he was paired with an unrelated female for future breeding (The Straits Times Singapore, 2020; Chapter 21). He recently successfully bred with a female (born at Taronga Zoo in Sydney, Australia and sent to Singapore) and a joey was born in February, 2020. Singapore Zoo is anticipated in the near future, to host an international tree kangaroo meeting, which will include a Goodfellow's GSMP Planning Session and a general tree kangaroo workshop. Participants will include field conservationists, researchers, and in situ program professionals as well as the global zoo conservation community that includes keepers to directors, veterinarians, regional officials, and wildlife caretakers.

Taman Safari Indonesia in Bogor, a member of SEAZA, is the only recorded ex situ facility that maintains Grizzled tree kangaroos (Fig. 22.5), a rare species endemic to the island of New Guinea. Taman Safari has the opportunity to bring awareness of this rare tree kangaroo species to its visitors through naturalistic habitats and graphics that show the fauna of Papua on the island of New Guinea (Fig. 22.9).

Role of Zoos in Creation and Support of In Situ Conservation Programs

Armed with expertise gained through long-term development of progressive husbandry, health, and welfare processes for tree kangaroos in zoological facilities, zoo professionals have initiated and contributed to conservation programs for tree kangaroos in the wild. Individual zoos as well as the conservation management

FIGURE. 22.9 Graphics outside the grizzled tree kangaroo habitat at Taman Safari Indonesia, on the fauna of Indonesian Papua on the island of New Guinea. *Source: Taman Safari Indonesia.*

programs of the global and regional zoo associations have stepped up to fully support development and sustainment of tree kangaroo conservation programs in the natural ranges of these species. This includes programs in PNG for Matschie's tree kangaroos on the Huon Peninsula, and for Tenkile in the Torricelli Mountains, and in Australia, two tree kangaroo rescue and rehabilitation programs in the Atherton Tablelands, Queensland, for Lumholtz's tree kangaroo.

Tree Kangaroo Conservation Program (TKCP) – The First Zoo-Based Tree Kangaroo Conservation Program

Inherent in any conservation effort is the necessity of local engagement and collaboration to create effective conservation action that is beneficial rather than detrimental to the most vulnerable people within the region. The most successful conservation action requires local participation with community-based strategies based on indigenous rights and cultural practices and investment by local government and international NGOs pointed toward local requirements for improving health, education, and livelihoods (Ellis, 2019). Dr. Lisa Dabek incorporated this conservation philosophy as she began field work on Matschie's tree kangaroo on the Huon Peninsula in PNG, and subsequently when she founded the TKCP in 1996.

The TKCP's mission is to "foster wildlife and habitat conservation and support local community livelihoods in Papua New Guinea through global partnerships, land protection, and scientific research" (Chapters 10, 13 and 16; TKCP, 2019). The TKCP began as a research program in collaboration with the local PNG community to identify the status of the Matschie's tree kangaroo in the wild. Working with local landowners, the project progressed into a holistic program to protect the endangered Matschie's tree kangaroo and other endemic species and their habitats while addressing local needs for livelihoods, health, and education (Chapter 13). The program increased its reach from the mountainous habitat of the tree kangaroos to encompass marine and coastal reef ecosystems, rainforests, alpine grasslands, agricultural areas, and villages for more than 50 settlements within the YUS watershed areas on the Huon Peninsula (TKCP, 2018b; Chapter 10 and Chapter 12).

In 2014, the TKCP became the umbrella name for two important partners: WPZ's TKCP and Tree Kangaroo Conservation Program – Papua New Guinea (TKCP-PNG), an independent PNG non-governmental organization. They work closely with the YUS Conservation Organization (YUS CO), a community-based organization representing the interest of local landowners and their communities (TKCP, 2019).

As the representative field conservation program for the TKSSP, the TKCP is supported by AZA member zoos and other international zoos as well as additional conservation partners and donors. These include the Global Environment Facility (GEF) facilitated by the United Nations Development Programme (UNDP), United States Agency for International Development (USAID) Biodiversity Project, Conservation International, and German Government BMU Lifeweb Initiative through KfW (a German state-owned bank). Such assistance enabled the TKCP to work with PNG landowners in 2009 to develop the first Conservation Area in PNG, offering protection for Matschie's tree kangaroos and other endemic species (Chapter 10; Fig. 22.10).

Zoo staff, with specialties in veterinary medicine, tree kangaroo husbandry and care, conservation research, marketing and education, and business management, have been instrumental in contributing their expertise to the TKCP processes in PNG. Dr. Erika Crook, a veterinarian at Utah's Hogle Zoo and a veterinarian advisor to the TKSSP, has assisted TKCP scientists with health assessments of wild Matschie's tree kangaroos in research to determine distribution, home range size, habitat use, and feeding ecology (Chapter 25). She and other zoo veterinarians including Dr.s Janet Martin, Louis Padilla, and Holly Reed, as well as field veterinarian Carol Esson and veterinarian technologist Trish Watson all have contributed to the success and animal well-being for the field research. Zookeepers from ZAA and AZA zoos, notably Beth Carlyle-Askew, have been assisting TKCP and the tree kangaroo research team since 1998. A collaboration among WPZ's Dr. Dabek, former TKCP Research and Conservation Manager Daniel Solomon Okena, and Microsoft engineer Doug Bonham resulted in new GPS tracking technology that automatically identifies a collared animal's location via satellite as well as contributing altitudinal and motion data (Fig. 22.11; Chapter 23). The GPS collars were first tested on Elanna, a WPZ Matschie's tree kangaroo before being used to track animals in the wild.

In 2010, National Geographic Society in collaboration with the TKSSP adapted their CritterCam device, a video camera that attaches to an animal, to fit around the neck of a tree kangaroo to document its daily life through video. Two of these CritterCams, were used by TKCP researchers to capture daily feeding behavior and activity of Matschie's tree kangaroos in the wild (Chapter 23).

PNG Forest Research Institute and PNG National Herbarium were co-collaborators with TKCP and TKSSP on a project to study nutrition in the tree kangaroos' diet. Complimented by local knowledge from hunters, TKCP staff and a PNG botanist collected samples of these plant species and sent them to the United States for a nutritional study headed by TKSSP Nutrition Advisor Dr. Ellen Dierenfeld, based in St. Louis, MO (Chapter 20; Dierenfeld et al., 2020; TKCP, 2016). The results of this nutritional study will inform researchers in PNG as well as assist in the improvement of diets for the animals under managed care.

The WPZ has supported full time staff dedicated to the development and management of TKCP programs to benefit the local PNG communities and tree kangaroo conservation. By

FIGURE. 22.10 YUS Community members in traditional dress during YUS Landscape Plan to celebrate the YUS Conservation Area in Isan Village. *Source: TKCP.*

2014, the TKCP had made many strides toward biodiversity conservation in PNG while at the same time improving the lives of people in the local communities through the formation of the YUS Conservation Area (Chapter 10), promoting and building capacity for alternative livelihoods (Chapter 14), and using a One Health approach to integrate health services and education into the conservation programs (Chapters 15 and 16, respectively). As the Program grew, there was a need to improve integration of the administrative teams in Lae and YUS, PNG and the team based at WPZ in Seattle, WA. Trevor Holbrook, with extensive experience in business management and organization development, is employed by WPZ as TKCP Program Manager to focus on livelihood program development, PNG staff professional development, grant management for the YUS Conservation Endowment Fund, and U.S. outreach projects (TKCP, 2015). A major responsibility is working with the TKCP's YUS Conservation Coffee Initiative, an alternative livelihood project where farmers in the YUS region grow and harvest coffee beans for export to coffee companies in Seattle, and Melbourne (Chapter 14).

In 2015, Eli Weiss, then WPZ Community Engagement Supervisor, tapped into his expertise in youth engagement and citizen science programs to assist in the development of the Junior Rangers program for the YUS Conservation Area in PNG (Chapter 16; Woodland Park Zoo, 2019) (Fig. 22.12). The program targets pre-kindergarten children to older youth and young adults and seeks to instill an appreciation for the environment, thus building

FIGURE. 22.11 A wild Matschie's tree kangaroo is fitted with a radio collar as the tree kangaroo capture and tracking team observe. Left to right, bottom row: TKCP Research and Monitoring Coordinator Daniel Okena, Field Veterinarian Carol Esson, and TKCP Director Lisa Dabek. *Source: TKCP.*

conservation stewards who would be future leaders for local conservation efforts. For a year prior to going to PNG, Eli mentored Danny Samandingke, former TKCP Leadership Training and Outreach Senior Coordinator, in planning and developing the program. Eli, supported by a WPZ ZooBright Fellowship, then traveled to YUS to work with Danny to facilitate a series of program design workshops, meeting with educators, parents, and youth to learn about their needs and challenges as well as their objectives for outcomes of the program. Through synthesis of the workshop results, the YUS Junior Ranger Program Model was developed with a Vision that "all youth will become empowered community members and lead through future conservation stewardship and sustainable community development across the YUS landscape" (E. Weiss, personal communication, 2019).

The Junior Ranger Program (JRP) has been a huge success. By 2019, there were over 500 Junior Ranger youth in the YUS area, participating in environmental education projects as diverse as forest stewardship, fishing, coffee farming, biological research on local fauna, and management of land to maintain local livelihoods while supporting biodiversity conservation (Woodland Park Zoo, 2019). The TKCP Education and Leadership Coordinator facilitates the Junior Ranger Program and TKCP supports JRP by providing training and materials for volunteer teachers.

Tenkile Conservation Alliance (TCA)

Dr. Tim Flannery's discovery of a new species of tree kangaroo – Scott's tree kangaroo (also known as Tenkile) in the early 1990s led to an investigation of the conservation status of the species and the early realization that the species was

FIGURE. 22.12 TKCP's Gibson Gala (right) leads YUS Junior Rangers in a song about health and their environment. *Source: Emily Transue.*

headed toward extinction (Flannery et al., 1996). The species was first assessed in 1994 and found to have an extremely restricted distribution with about 100 individuals occurring in only two locations – the Torricelli Mountain Range and Mount Menawa in the Bewani Mountain Range in Papua New Guinea (Leary et al., 2019). The seeds for the development of the TCA were sown after the CBSG PHVA on the tree kangaroos of PNG in 1998 (see the section on Action: The Rapid Response Team – Saving Tenkile Tree Kangaroos). In 2001, the TCA was formally incorporated and in 2003, Jim and Jean Thomas, zookeepers at Zoos Victoria, went to PNG to manage the TCA, based in Lumi, Sandaun Province. The Thomases cited their background at a zoological facility, experience with Australian endangered species, and a desire to do more for conservation in their decision to move to PNG. The TCA's Vision is as follows: The people of Papua New Guinea value and protect their natural resources, community, and culture in the context of advancing the overall wellbeing of their communities and their places (TCA, 2020a).

Conservation work was determined to have less to do with the animals and more to do with working with the local communities. Primary objectives were to facilitate processes to enable the rainforest communities to manage and protect their natural resources while maintaining their cultural heritage and traditions. This involved goals to improve health, provide education, and relieve poverty while protecting biodiversity. Inherent in the process was assistance in developing alternative protein sources to hunting to minimize the pressures on threatened species. This included programs for farming rabbits, chickens, and fish (the programs had varying success).

Initially working with 14 villages, the reach of the TCA has increased to 50 villages and over 12,000 people, with a goal to preserve over 185,000 ha of tropical rainforest (TCA, 2020a). The hunting moratorium covers not only Tenkile but also another critically endangered species, the Weimang or Golden-mantled tree kangaroo. The 20 villages in Tenkile territory are located in the western half of the Torricelli

Mountains while the 30 villages in Weimang territory are located in the eastern half of the Mountains. One of the objectives in the TCA conservation strategy is to establish the Torricelli Mountain Range as a Conservation Area. This protected area would have legal protection from commercial development (logging and mining) and safeguard species from local threats such as over-hunting and harvesting of natural resources (TCA, 2020b).

Tree Kangaroo Rehabilitation Centers in Australia

Lumholtz Lodge

In the early 1980s, Margit Cianelli tapped into her earlier experience with tree kangaroos in managed care at the Stuttgart Zoological Garden in Germany to work with the rescue and rehabilitation of wild Lumholtz's tree kangaroos on the southern Atherton Tablelands in the highland region of Queensland, Australia. Cianelli took in injured and orphaned joeys with the intent on returning them to the wild, if possible (Chapter 6). She developed successful hand-rearing techniques and identified development milestones as well as made the important discovery that special bacteria had to be introduced when the joeys were ready to eat solid food. This would enhance gut flora facilitating digestion of endemic plants in their natural diet that contained toxins. Cianelli has been quite successful in rehabilitating these joeys for a return to the wild at an appropriate stage in their development. She has made a huge contribution to the advancement in health and welfare of hand-raised Lumholtz's joeys by sharing the hand-rearing information with zoos that maintain this species and contributing to a husbandry manual for Lumholtz's tree kangaroos compiled by ZAA zoos.

Cianelli opened up her home as the Lumholtz Lodge, a bed and breakfast for visitors who are offered a unique experience in this lodge located in the middle of a tropical rainforest (Chapter 8). Nature lovers are treated to sightings of wild tree kangaroos and other forest denizens such as possums, wallabies, and a wide array of bird life. Visitors are also able to observe the care of the resident joeys and learn about this rescue and rehabilitation program, engaging them to care about wildlife conservation.

Tree Roo Rescue and Conservation Centre (TRRACC)

The Tree Roo Rescue and Conservation Centre (TRRACC), located in Malanda, Atherton Tablelands in Far North Queensland, Australia, is a non-profit organization dedicated to the prevention of extinction of Australian tree kangaroos through rescue and rehabilitation of orphaned, injured, displaced, or blind Lumholtz's tree kangaroos. Ultimately, the goal is to return rehabilitated tree kangaroos to the wild, if possible (TRRACC, 2020a). Some animals that are unable to return to the wild are placed in zoos accredited by ZAA as part of breeding management programs and as ambassadors in conservation education programs. Dr. Karen Coombes, Director and Chair, co-founded the TRRACC with her husband Neil McLaughlan in 2012. Coombes brought 20 years of experience specializing in the rescue, rehabilitation, and research of Lumholtz's tree kangaroos to head up the TRRACC (Coombes, 2005).

The TRRACC has a number of ex situ conservation partners that support its work to preserve the Lumholtz's tree kangaroo and educate the public on tree kangaroo status in the wild (TRRACC, 2020b). Dreamworld (a theme park and zoo on the Gold Coast) and the Dreamworld Wildlife Foundation are the main conservation partners. Dreamworld maintains Lumholtz's tree kangaroos that were rescued by the TRRACC but due to neurological blindness, could not be released. The blindness is neither hereditary nor contagious and TRRACC has been researching the cause. These tree kangaroos have become part of a successful breeding program at Dreamworld. TRRACC has been working closely with ZAA, the Lumholtz's Species Coordinator at Dreamworld and the Queensland government to place blind

Lumholtz's tree kangaroos into appropriate zoos including David Fleays Nature Park in Burleigh Heads, South East Queensland, Currumbin Zoo on the Gold Coast, Wildlife Habitat Port Douglas in Port Douglas, Snakes Down Under Wildlife Park at Childers, Wildlife HQ (formerly Queensland Zoo), Queensland, and now Oakvale Wildlife Park in New South Wales.

Originally, TRRACC, as a non-profit organization, received donations from the general public. Australian zoos soon contributed as they became aware of the conservation work of TRRACC (see the section on Tree Kangaroo Awareness Day below) and also took on the care of Lumholtz's tree kangaroos that could not be released back to the wild. Dreamworld Wildlife Foundation donates each year to TRRACC as part of their commitment to the species and their partnership with TRRACC.

ROLE OF ZOOS IN CONSERVATION EDUCATION, AWARENESS, AND FUNDING SUPPORT

Zoos and aquariums worldwide attract over 700 million visitors every year (Gusset and Dick, 2011), offering the opportunity to promote conservation messages about the animals in their care. WAZA has included in their conservation strategy, the tenet to engage visitors in a conservation ethic with a Vision that "Zoos and aquariums are trusted voices for conservation, and are able to engage and empower visitors, communities, and staff measurably to save wildlife" (Barongi et al., 2015). Gusset et al. (2014) conducted a large scale study of zoo and aquarium visitors to evaluate the understanding of biodiversity conservation with results showing that visitors increased their understanding of biodiversity issues and knowledge about actions that can be taken to protect biodiversity. Zoos and their staff play a vital role in incorporating aspects of the One Plan Approach by directly connecting the zoo visitor to tree kangaroos, raising awareness about their status in the wild, and exposing them to in situ community and culture. Zoos around the world, as conservation organizations, spend over 350 million US dollars on wildlife conservation every year (Barongi et al., 2015). Awareness campaigns go hand in hand to educate and also generate funding support for conservation of tree kangaroos in the wild.

In 2012, the TKSSP adopted the 4th Saturday of May as Tree Kangaroo Awareness Day and it has been adopted as a global event. While many facilities participate with activities on that day, others choose to raise awareness on different days throughout the year. Events typically incorporate graphics/posters that include information about tree kangaroo biology and behavior, natural habitat, and local culture. Many zoos promote awareness of the local community coffee growers program in PNG by sampling and selling coffee. TKCP facilitates the marketing of shade-grown YUS Conservation Coffee through the MTC International Coffee Group to sell the coffee to Seattle's Caffe Vita Roasting Company and Jasper Coffee in Melbourne (Chapter 14; TKCP, 2018a,b). The YUS farmers have conserved over 180,000 acres of forest for the protection of tree kangaroos and other species. YUS Conservation Coffee provides a vital source of income for the local communities serving as stewards of the rainforest in PNG (TKCP, 2019).

Awareness day activities are varied in nature, but all have an educational link back to tree kangaroos. The following is a list of activities with descriptions from some of the TKSSP facilities in the United States:

Milwaukee County Zoo, Milwaukee, Wisconsin, USA:

- Tree kangaroo face masks: visitors are educated on the fact that each individual Matschie's tree kangaroo has a unique face pattern; they are then able to create their own tree kangaroo face mask.
- Walk like a tree kangaroo: educate how tree kangaroos move through the forest canopy; visitors test their balance by walking on a log.

San Antonio Zoo, San Antonio, Texas, USA:

- Chutes and Ladders game with a tree kangaroo theme – a modification of the popular children's board game (Fig. 22.13).

Santa Fe Teaching Zoo, Gainesville, Florida, USA:

- Browse pick-up game: The object of the game is for participants to act like zookeepers, identify the browse items that tree kangaroos eat, and give them to the tree kangaroo.
- Tree kangaroo biodegradable pot craft: Participants can make a tree kangaroo biodegradable pot and plant their favorite plants with the pot right into their garden at home.
- Tree kangaroo Olympics: Participants can test their tree kangaroo skills by navigating through this simple obstacle course while carrying a tree kangaroo "joey" (plush toy) in a pouch (fanny pack).
- Bean bag board: large tree kangaroo with pouch cut-out allows participants to 'toss' bean bags into the pouch (Fig. 22.14).
- Photo opportunity with tree kangaroo photo board.

At TRRACC, support for Australian tree kangaroo conservation has expanded to include publicity around "Tree Roo Awareness Week", held from the 3rd weekend of May until the 4th weekend to coincide with the Tree Roo Awareness Day that has been held in zoos in the past. TRRACC holds a huge raffle using social media to involve the local community and to educate the public around the world about Australian tree kangaroo species.

Many zoos in Australia now run fundraising events for the TRRACC to help them continue their valuable work with this species and the

FIGURE. 22.13 Tree Kangaroo Chutes and Ladders game for Tree Kangaroo Awareness Day at San Antonio Zoo. *Source: San Antonio Zoo.*

FIGURE. 22.14 Celebrating Tree Kangaroo Awareness Day at Santa Fe Teaching Zoo in Gainesville, Florida, USA. *Source: Santa Fe Teaching College.*

number of zoos involved in "Tree Roo Awareness week" as a public awareness event is increasing each year. For example, the National Zoo and Aquarium in Canberra held a raffle and quiz night in 2019 and raised nearly $10,000 AUD. Some zoos in the USA have also held events and donated raised funds to the TRRACC.

The Port Moresby Nature Park in PNG holds an annual public event on World Wildlife Day, designated as the 3rd of March by the United Nations General Assembly. This event allows NGOs, government organizations, and communities working in wildlife conservation to showcase to the PMNOP visitors what is being done in the field. The event is structured to:

1. Draw attention to the threats to PNG wildlife
2. Provide practical actions to save PNG wildlife
3. Raise awareness of conservation groups in PNG
4. Generate as much media attention as possible

This event is followed by an education program that continues this theme over one month and reaches more than 2500 students (Fig. 22.15).

Visitor education and awareness can also be achieved from graphics at tree kangaroo displays/exhibits. Graphics might include information about tree kangaroo natural history and habitat, local community culture, and data about individual animals housed at the facility. Public relations, marketing, and animal staff often work together to enhance social media coverage. For the 2020 Tree Kangaroo Awareness Day in May, Zoorasia (Yokohama Zoological Gardens) in Yokohama, Japan, spotlighted their Goodfellow's tree kangaroos for three consecutive days on their Facebook page.

The TKSSP's Facebook page is focused on everything related to tree kangaroos including

FIGURE. 22.15 An animal staff member at Port Moresby Nature Park in PNG delivers a talk with a Doria's tree kangaroo during World Wildlife Month. *Source: Port Moresby Nature Park.*

photos from zoos, events going on in PNG, new research, and conservation connections. Promotions and posts might highlight births, developmental milestones, birthdays, holidays, and more. These charismatic animals not only bring awareness to the species and their conservation status but also have the potential to attract more visitors to the zoo which can in turn generate revenue that can be utilized to support in situ programs such as TKCP and TCA.

Several zoos maintain educational carts year round staffed by docents for direct interaction with visitors about tree kangaroo biology, ecology, and conservation. At the WPZ, docents meet annually with Dr. Dabek and Trevor Holbrook, TKCP Manager, for updates on TKCP achievements and activities. They staff the Tree Kangaroo Conservation Cart used on grounds and share these conservation messages with visitors. The cart includes photos, samples of the YUS Coffee bags, a tree kangaroo game, and a used radio collar (Fig. 22.16).

Zookeeper associations in North America and in Australia have events dedicated to contributing to conservation efforts worldwide. The American Association of Zoo Keepers (AAZK) chapters and zoo staff raise funds for tree kangaroo conservation in a number of ways. Products such as plush tree kangaroo toys have been sold and the YUS conservation coffee can be purchased not only at awareness day events but also throughout the year. A new event was debuted in May 2019 by the students at the Santa Fe Teaching College in Gainesville, FL. Students formed a partnership with a local brewery to host Brews for Tree 'Roos (Fig. 22.17). The local Swamp Head Brewery hosted the event and brewed special beers for the cause, one of which was canned with its own unique tree kangaroo label. The event was not only geared at generating funds, but to also encourage community support and involvement in conservation. A silent auction was held with items donated from local businesses.

FIGURE. 22.16 Tree Kangaroo Conservation cart staffed by docents at Woodland Park Zoo in Seattle, Washington. The cart contains information and photos about the Tree Kangaroo Conservation Program in Papua New Guinea, natural home of the Matschie's tree kangaroo. *Source: Woodland Park Zoo.*

FIGURE. 22.17 For the Brews for Tree 'Roos fundraiser for tree kangaroo conservation, zoo staff at the Santa Fe Teaching College in Gainesville, Florida teamed up with a local brewery. *Source: Santa Fe Teaching College.*

Informational posters and pamphlets provided information about tree kangaroos and in situ conservation. Educational presentations about the TKSSP and tree kangaroos were provided by staff. To round out the event, merchandise such as YUS coffee, t-shirts and buttons were sold. This event is intended to become an annual fundraiser with the possibility of being adopted by other zoos and AAZK chapters.

The Australasian Society of Zoo Keeping (ASZK) has hosted a number of yearly events to benefit TRRACC's rescue of Lumholtz's tree kangaroos. Tuesday Trivia for Tree Roos, hosted by the ASZK and Zoos Victoria, in April 2019 provided a fun night of a Trivia contest, auction, and raffles in Ascot Vale, Victoria. Also in April 2019, the ASZK hosted Trivia for Tree Kangaroos in Glebe, New South Wales to benefit the TRRACC. A Bowling for Tree Roo Rescue event was held by the ASZK and Taronga Zoo in Sydney, raising $22,000 AUD in funds for rescue, hand-rearing, and releasing Lumholtz's tree kangaroos as well as a project to control wild dogs in tree kangaroo areas.

CONCLUSIONS – TREE KANGAROO CONSERVATION SUCCESS DEPENDS ON EX SITU AND IN SITU COLLABORATION

Tree kangaroo populations face numerous threats in the wild due to anthropogenic factors including loss of habitat and hunting pressures, as well as natural catastrophes brought on by the effects of climate change (e.g., drought, wildfires). The global conservation community has been working together to ensure a future for these threatened species in a One Plan Approach calling for collaboration between ex situ and in situ forces. Conservation strategies with the goal of maintaining long-term viable tree kangaroo populations in healthy ecosystems incorporate a transdisciplinary process that involves local communities in range countries working with an international set of wildlife researchers and caretakers supported by zoos across the world.

Zoological institutions contribute to conservation of threatened tree kangaroo species and ecosystems on many levels. Global and regional zoo associations scientifically manage their tree kangaroo populations with the responsibility to maintain genetic diversity, demographic stability, and behavioral integrity. Zoo-based research has brought advancement in husbandry and breeding management, animal welfare and health care, and an understanding of biology and physiology that has been shared to increase the knowledge of the natural life of tree kangaroos in the wild. Alternatively, results of research on tree kangaroos in the wild assists in the understanding of the requirements for maintaining healthy ex situ populations. Tree kangaroos in global accredited zoos have a conservation role as ambassadors to raise awareness and engage visitors to care about their conservation status in the wild. Finally, zoos generate considerable funding to support overall One Plan Approach conservation strategies. Only through the global cooperation of ex situ and in situ conservation communities in close collaboration with local peoples within the natural ranges, will the future be ensured for tree kangaroo species in their natural habitats.

ACKNOWLEDGMENTS

The authors acknowledge Chris Banks for contributing information about Zoos Victoria's Conservation Strategy and for his careful review of the chapter; Peter Clark and Gert Skipper (Zoos South Australia) for historical information on initial conservation planning action; Megan Richardson for information on the GSMP and ASZK; and the following for their contributions of information and photographs: Anthony Molyneux (CWS); Dr. Karen Coombes (TRRACC); Jim Thomas (TCA), Biswajit Guha (Taman Safari Indonesia); Jeff Bradley (Burke Museum); Beth Carlyle-Askew (WPZ); Jan Nelson (Sedgwick County Zoo), Trevor Holbrook (WPZ-TKCP); and Eli Weiss (formerly WPZ). Special Acknowledgment goes to Judie Steenberg, TKSSP Historical Advisor, for careful reviews and edits, contributing information about the TKSSP and ex situ conservation programs. Dr. Lisa Dabek is at the forefront of the interconnection between ex situ and in situ conservation for tree kangaroos and provided valuable insight into the evolution of this chapter.

REFERENCES

Association of Zoos and Aquariums (AZA), 2018a. Population Analysis & Breeding and Transfer Plan Matschie's Tree Kangaroo (*Dendrolagus matschiei*). AZA Species Survival Plan (SSP) Red Program. AZA Population Management Center. Lincoln Park Zoo, Chicago, IL.

Association of Zoos and Aquariums (AZA), 2018b. Species Survival Plan® (SSP) Program Handbook. Association of Zoos and Aquariums, Silver Spring, MD.

Association of Zoos and Aquariums (AZA), 2020. AZA SAFE: Saving Animals From Extinction. Available from: https://www.aza.org/aza-safe (26 June 2020).

Banks, C., 2018. Saving species, one brew at a time. Wildlife Australia 55 (3), 24–25.

Banks, C., 2019. Partners in conservation. Wildlife Australia 56 (1), 22–23.

Barongi, R., Fisken, F.A., Parker, M., Gusset, M. (Eds.), 2015. Committing to Conservation: The World Zoo and Aquarium Conservation Strategy. WAZA Executive Office, Gland.

Beissinger, S.R., McCullough, D.R. (Eds.), 1998. Population Viability Analysis. University of Chicago Press, Chicago, IL.

Benner, K.S., Collins, L.R., 1988. Reproduction and Development in Matschie's Tree Kangaroos (*Dendrolagus matschiei*) in Captivity. National Zoological Park Conservation and Research Center, Front Royal, VA.

Blessington, J., Steenberg, J., 2007. Tree Kangaroo (*Dendrolagus* spp.) Husbandry Manual, Tree Kangaroo Species Survival Plan®, third ed. Association of Zoos and Aquariums, Silver Spring, MD.

Bonaccorso, F., Clark, P., Miller, P., Byers, O., 1999. Conservation assessment and management plan for the tree kangaroos of Papua New Guinea and population and habitat viability assessment for Matschie's tree kangaroo: final report. Conservation Breeding Specialist Group (SSC/IUCN), Apple Valley, MN. Available from: http://www.cpsg.org/content/tree-kangaroo-phva-and-camp-1998 (10 June 2019).

Burke Museum, 2019. The Burke Mammalogy Collection. Available from: https://www.burkemuseum.org/research-and-collections/mammalogy/people-and-contact (25 July 2019).

Byers, O., Lees, C., Wilcken, J., Schwitzer, C., 2013. The One Plan Approach: the philosophy and implementation of CBSG's approach to integrated species conservation planning. WAZA Mag. 14, 2–5.

Che-Castaldo, J., Russell, K., DeBo, D., Norsworthy, D.A., Blessington, J., 2019. Matschie's Tree Kangaroo (*Dendrolagus matschiei*) AZA Animal Program Population Viability Analysis Report. Lincoln Park Zoo, Chicago, IL.

Collins, L., 1986. Big foot of the branches. Zoogoer. July-AugustNational Zoological Park, Washington, DC, pp. 35–40.

Conservation Breeding Specialist Group (CBSG), 2017. Second Nature: Changing the Future for Endangered Species. Trio Bookworks, St. Paul, MN.

Conservation Planning Specialist Group (CPSG), 2019. Our Mission. Available from: http://www.cpsg.org/our-mission (7 June 2019).

Coombes, K.E., 2005. The Ecology and Utilisation of Lumholtz's Tree Kangaroos *Dendrolagus lumholtzi* (Marsupialia: Macropodidae), on the Atherton Tablelands, far North Queensland. (PhD thesis) James Cook University.

Crook, G.A., Skipper, G., 1987. Husbandry and breeding of Matschie's tree kangaroo (*Dendrolagus m. matschiei*) at Adelaide Zoological Gardens. Int. Zoo Yearb. 26, 212–216.

Dabek, L., 1991. Mother-Young Relations and Development of the Young in Captive Matschie's Tree Kangaroos *(Dendrolagus matschiei).* (Master's thesis)University of Washington, Seattle, WA.

Dabek, L., 1994. The Reproductive Biology and Behavior of Captive Female Matschie's Tree Kangaroos (*Dendrolagus matschiei*). (Ph.D. diss.)University of Washington, Seattle, WA Abstract in *Dissertation Abstracts International* 55(1 IB): 4748.

Dabek, L., Betz, W., 1998. Tree kangaroo conservation in Papua New Guinea. Endanger. Species Update 15 (6), 114–116.

Dierenfeld, E., Okena, D.S., Paul, O., Dabek, L., 2020. Composition of browses consumed by Matschie's tree kangaroo (*Dendrolagus matschiei*) sampled from home ranges in Papua New Guinea. Zoo Biol.. April 2020 online. Available from: https://doi.org/10.1002/zoo.21543 (1 June 2020).

Ellis, E.C., 2019. Sharing the land between nature and people. Science 364 (6447), 1226–1228.

Flannery, T.F., 1998. Throwim Way Leg: Tree-Kangaroos, Possums, and Penis Gourds. Atlantic Monthly Press, New York.

Flannery, T.F., Martin, R., Szalay, A., 1996. Tree Kangaroos: A Curious Natural History. Reed Books Australia, Port Melbourne, Victoria.

Gilpin, M.E., Soulé, M.E., 1986. Minimum viable populations: processes of extinction. In: Soulé, M.E. (Ed.), Conservation Biology: The Science of Scarcity and Diversity. Sinauer Associates, Sunderland, MA, pp. 19–24.

Gusset, M., Dick, G., 2011. The global reach of zoos and aquariums in visitor numbers and conservation expenditures. Zoo Biol. 30, 566–569.

Gusset, M., Moss, A., Jensen, E., 2014. Biodiversity understanding and knowledge of actions to help protect biodiversity in zoo and aquarium visitors. WAZA Mag. 15, 14–17.

Henson, P.M., 2018. American zoos: a shifting balance between recreation and conservation. In: Minteer, B.A., Maienschein, J., Collins, J.P. (Eds.), The Ark and Beyond: The Evolution of Zoo and Aquarium Conservation. University of Chicago Press, Chicago, IL, pp. 65–76.

International Union for Conservation of Nature (IUCN), 2019. Red List—Dendrolagus. Available from: https://www.iucnredlist.org/search?query=Dendrolagus&searchType=species (23 May 2019).

Lacy, R.C., 2000. Structure of the *Vortex* simulation model for population viability analysis. Ecol. Bull. 48, 191–203.

Lacy, R.C., Ballou, J.D., Princée, F., Starfield, A., Thompson, E.A., 1995. Pedigree analysis for population management. In: Ballou, J.D., Gilpin, M., Foose, T.J. (Eds.), Population Management for Survival and Recovery. Columbia University Press, New York, pp. 57–75.

Leary, T., Wright, D., Hamilton, S., Helgen, K., Singadan, R., Aplin, K., Dickman, C., Salas, L., Flannery, T., Martin, R., Seri, L., 2019. *Dendrolagus scottae*. The IUCN Red List of Threatened Species 2019. e.T6435A21956375. Available from: https://doi.org/10.2305/IUCN.UK.2019-1.RLTS.T6435A21956375.en (17 January 2020).

Minteer, B.A., Maienschein, J., Collins, J.P., 2018. Zoos and aquarium conservation: past, present, future. In: - Minteer, B.A., Maienschein, J., Collins, J.P. (Eds.), The Ark and Beyond: The Evolution of Zoo and Aquarium Conservation. University of Chicago Press, Chicago, IL, pp. 1–12.

Nyhus, P.J., Williams, J.S., Borovansky, J.S., Byers, O., Miller, P.S., 2003. Incorporating local knowledge: landowners and tree kangaroos in Papua New Guinea. In: Westley, F.R., Miller, P.S. (Eds.), Experiments in Consilience: Integrating Social and Scientific Responses to Save Endangered Species. Island Press, Washington, DC, pp. 161–184.

Obituaries Australia, 2019. Hann, William (1837-1889). Available from: http://oa.anu.edu.au/obituary/hann-william-3708 (23 May 2019).

Olds, T.J., Collins, L.R., 1973. Breeding Matschie's tree kangaroo *Dendrolagus matschiei* in captivity. Int. Zoo Yearb. 13 (1), 123–125.

Rabb, G.B., 1994. The changing roles of zoological parks in conserving biological diversity. Am. Zool. 34 (1), 159–164.

Reed, J.M., Mills, L.S., Dunning Jr., J.B., Menges, E.S., McKelvey, K.S., Frye, R., Beissinger, S.R., Anstett, M.-C., Miller, P.S., 2002. Emerging issues in population viability analysis. Conserv. Biol. 16, 7–19.

Richardson, M., 2018. International Studbook for the Goodfellow's Tree-Kangaroo (*Dendrolagus goodfellowi*). Melbourne Zoological Gardens, Melbourne.

Schreiner, C., Schwarzenberger, F., Kirchner, W.H., Dressen, W., 2015. Behavioural and hormonal investigations in Goodfellow's tree kangaroos (*Dendrolagus goodfellowi* Thomas, 1908). Der Zool. Garten 84 (1–2), 45–60.

Species360 Zoological Information Management System (ZIMS), 2019. Available from: https://zims.Species360.org (30 December 2019).

Steenberg, J., 1988. The management of Matschie's tree kangaroo (*Dendrolagus matschiei*) at Woodland Park Zoological Gardens. In: AAZPA Western Regional Conference Proceedings. Wheeling, WV, pp. 94–111.

Tenkile Conservation Alliance (TCA), 2020a. About TCA: Community. Available from: https://www.tenkile.com/about (1 June 2020).

Tenkile Conservation Alliance (TCA), 2020b. TCA Conservation Strategy, Component 4. Establishment of the Protected Area. Available from: https://www.tenkile.com/protect-biodiversity-culture (12 May 2020).

The Straits Times Singapore, 2020. Miracle Tree Kangaroo Moves to Singapore Zoo. Available from: https://www.straitstimes.com/singapore/miracle-tree-kangaroo-moves-to-singapore-zoo (2 June 2020).

Tree Kangaroo Conservation Program (TKCP), 2014. Tree Kangaroo Conservation Program annual report 2013. Woodland Park Zoo, Seattle, WA. Available from: https://www.zoo.org/document.doc?id=1727 (7 June 2019).

Tree Kangaroo Conservation Program (TKCP), 2015. Tree Kangaroo Conservation Program annual report 2014. Woodland Park Zoo, Seattle, WA. Available from: https://www.zoo.org/document.doc?id=1728 (16 August 2019).

Tree Kangaroo Conservation Program (TKCP), 2016. Tree Kangaroo Conservation Program annual report 2015. Woodland Park Zoo, Seattle, WA. Available from: https://www.zoo.org/file/visit-pdf-bin/TKCP-2015-Annual-Report.pdf (4 August 2019).

Tree Kangaroo Conservation Program (TKCP), 2018a. Tree Kangaroo Conservation Program annual report 2014. Woodland Park Zoo, Seattle, WA. Available from: https://www.zoo.org/document.doc?id=2537 (12 May 2020).

Tree Kangaroo Conservation Program (TKCP), 2018b. Tree Kangaroo Conservation Program annual report 2017. Woodland Park Zoo, Seattle, WA. Available from: https://www.zoo.org/document.doc?id=2407 (2 June 2020).

Tree Kangaroo Conservation Program (TKCP), 2019. Tree Kangaroo Conservation Program annual report 2018. Woodland Park Zoo, Seattle, WA. Available from: https://www.zoo.org/document.doc?id=2537 (2 June 2020).

Tree Roo Rescue and Conservation Centre Ltd. (TRRACC), 2020a. About Us. Available from: https://www.treerooresecue.org.au/about_us [6 February 2020].

Tree Roo Rescue and Conservation Centre Ltd. (TRRACC), 2020b. Other Organisations and Their Links. Available from: https://www.treerooresecue.org.au/faq-cik1 (6 February 2020).

Woodland Park Zoo, 2019. Junior Rangers of YUS: Conservation Heroes. Available from:https://blog.zoo.org/2019/06/junior-rangers-of-yus-conservation.html (11 August 2019).

World Association of Zoos and Aquariums (WAZA), 2019. Goodfellow's Tree Kangaroo GSMP. Available from: https://www.waza.org/priorities/conservation/conservation-breeding-programmes/global-species-management-plans/goodfellows-tree-kangaroo/ (13 December 2019).

Zoo and Aquarium Association (ZAA), 2019. 3.0 Policy—Australasian Species Management Program (ASMP). Available from: https://www.zooaquarium.org.au/wp-content/uploads/2011/10/3_0_P_ASMP_Policy.pdf (2 September 2019).

Zoos Victoria, 2019. Zoos Victoria Wildlife Conservation Master Plan 2019-2024. Available from: https://www.zoo.org.au/media/2183/48636_zoos-vic-wcs-master-plan-128pp_-final.pdf (2 September 2019).

SECTION VI

TECHNIQUES AND TECHNOLOGY FOR THE STUDY OF AN ELUSIVE MACROPOD

CHAPTER

23

Using Telemetry and Technology to Study the Ecology of Tree Kangaroos

Jonathan Byers[a], *Lisa Dabek*[b], *Gabriel Porolak*[c], *and Jared Stabach*[d]

[a]Perspective Solutions LLC, Missoula, MT, United States
[b]Tree Kangaroo Conservation Program, Woodland Park Zoo, Seattle, WA, United States
[c]University of Papua New Guinea, Port Moresby, Papua New Guinea
[d]Smithsonian National Zoo and Conservation Biology Institute, Front Royal, VA, United States

OUTLINE

INTRODUCTION

Understanding the habitat an animal uses is essential in designing effective conservation measures. Advances in remote sensing techniques, including animal telemetry and bio-logging, have facilitated previously inaccessible understanding of behavior and habitat for many difficult to study wildlife species, while simultaneously creating new challenges

https://doi.org/10.1016/B978-0-12-814675-0.00007-5

(Cagnacci et al., 2010; Ropert-Coudert and Wilson, 2005). Due to their low population numbers and density, cryptic behavior, and remote and rugged habitat, the Matschie's tree kangaroo (*Dendrolagus matschiei*) provides an ideal subject to apply remote sensing techniques to overcome the tremendous challenges of observational behavior research (Stabach, 2005). The Wasaunon Field Research Area (hereafter referred to as Wasaunon) has been the site of numerous studies including vegetation transects (Jensen and Fazang, 2005), very high-frequency (VHF) and global positioning system (GPS) collar home range and habitat studies (Porolak, 2008; Porolak et al., 2014; Stabach, 2005), and satellite land-cover classifications (Pugh, 2003; Stabach, 2005; Stabach et al., 2009). These studies have provided groundbreaking knowledge about the behavior and ecology of the Matschie's tree kangaroo in the wild and provide an ideal foundation for the refinement and testing of new research methods.

Some significant conservation and technological changes have occurred since some of these studies were conducted including: the establishment of a conservation area in 2009 which removed hunting pressure in the Wasaunon area (Wells et al., 2013); improvements in the accuracy, longevity, and reliability of GPS antennas and the addition of bio-logging sensors for altitude and motion on collars (Ropert-Coudert and Wilson, 2005); the development of more robust spatio-temporal movement and behavior investigation tools (Noonan et al., 2019; Lyons et al., 2013); and the advances in ultra-high resolution imaging using portable unmanned aircraft systems (UAS) and digital photogrammetry tools (Anderson and Gaston, 2013). This chapter summarizes previous research using telemetry and remote sensing to understand the habitat and behavior of the Matschie's tree kangaroos and discuss current research efforts and future directions of tree kangaroo research investigation.

BACKGROUND

Investigating Spatial Behavior and Home Ranges

Understanding tree kangaroo population densities, spatial requirements, and abundance has been a primary focus of the work of the Tree Kangaroo Conservation Program (TKCP) and is essential in planning effective conservation and land management strategies (Ancrenaz et al., 2007). The concept of home range is generally defined as "the area traversed by the individual in its normal activities of food gathering, mating, and caring for young" (Burt, 1943), and is commonly used to quantify the spatial behavior of individual animals. However, a multitude of home range estimation techniques exist and can lead to very different conclusions (Fleming et al., 2015; Kie et al., 2010; Powell and Mitchell, 2012). While the concept of determining an animal's home range is notoriously fuzzy, it does provide a starting point to conceptualize space use and compare with other animals. Research on tree kangaroos in Northern Australia reveal small home ranges from 0.96 ha (Newell, 1999) to 2.1 ha (Coombes, 2005) for Lumholtz's (*D. lumholtzi*) and 3.7–6.4 ha (Procter-Grey, 1985) for Bennett's (*D. bennettianus*). For the Matschie's tree kangaroo Flannery (1995) reported on a brief VHF collar study in 1991 by Ian Stirling that "suggested a home-range of 0.25 square km (25 ha)". Betz (2001) conducted extensive landowner interviews to identify habitat and food plant preferences of the Matschie's tree kangaroo and used distance sampling of scat along transects to estimate a density animal per hectare. The significance of these results, however, was reportedly limited by the difficulties of positive scat sample identification (McGreevy et al., 2009; Stabach, 2005).

To overcome the challenges of indirect or observational data collection, a series of fieldwork was conducted between 2004 and 2007 at

Wasaunon to deploy VHF and GPS telemetry collars on Matschie's tree kangaroos and improve vegetation and food plant monitoring. Porolak (2008) deployed 15 VHF collars resulting in the first understanding of home ranges and revealing interesting patterns of animal distribution. Porolak et al. (2014) analysis of VHF collar location data revealed large overall home ranges of 81.3 ± 16.9 ha that overlapped >50% with their neighbors at 90% Harmonic Mean (HM). Kernel home range analysis show a mean size of 98.7 ± 14.2 ha at 90% Utilization Distribution (UD) and 13.8 ± 2.9 ha at 50% UD. Their smaller core areas (50% UD) showed lower (5–10%) overlap between individuals and males overlapping female home ranges more than females overlapped the home ranges of other females.

Stabach (2005) presented a preliminary assessment of home range sizes from the 3 GPS collars deployed and found a mean fixed kernel home range of 29.9 ± 7.5 ha at 90% UD, and 5.8 ± 2.0 ha at 50% UD. These collars revealed very low GPS fix success rates of ~20% resulting in far less than one successful location per day, and less than half the amount of location data as the manual VHF tracking of the GPS collared animals produced. Stabach writes that their independence analysis reveals the points are "clustered and independent, indicating that animals do not use their habitat uniformly", however the cause of these different use patterns was undetermined.

Vegetation identification at VHF locations revealed tree kangaroos to be generalist herbivores and the animals were often found in *Dacrydium nidulum* trees, large canopy emergent species, (50% of observations) between 23 and 28 m above the forest floor (Porolak et al., 2014). A dominant forest type classification from Landsat imagery showed that *D. matschiei* are found in *Dacrydium nidulum* dominant forests in 61.7% of successful GPS fix locations (Stabach et al., 2009).

Because VHF animal location tracking requires manually locating animals, it inherently introduces human disturbance and habitat bias. GPS collars provide an advantage as they do not require manually locating the animal, and locations can be collected during any weather or time of day if the challenges of GPS signal attenuation in dense forest canopies can be overcome. The far higher frequency of GPS location collection without manual collection of habitat variables at those locations requires very high resolution habitat data from remote sensing to understand forest composition and structure which introduces its own challenges and limitations (Stabach, 2005).

Investigating Forest Structure and Composition

One of the biggest limitations in correlating high resolution location and motion data with habitat use is the lack of similarly high resolution habitat data (Hebblewhite and Haydon, 2010). In an effort to directly observe *D. matschiei* behaviors, a Crittercam™ video camera collar was used in partnership with National Geographic to record otherwise unobservable feeding, movement, and grooming behaviors high in the forest canopy (National Geographic, 2009). These video clips provided invaluable insight into feeding patterns, forage species, movement, and grooming behaviors, and have informed ongoing investigations into diet and food plant species. However, to be able to quantify the distribution of food plants and preferred habitat structure across the landscape, higher resolution forest composition and structure data were necessary.

Satellite imagery is the most widely used tool for understanding vegetation and habitat on a large scale but has always been limited by coarser spatial resolution. The 14.25 m Landsat 7 ETM+ data used by Pugh (2003) and Stabach et al. (2009) can be used to distinguish general forest classes, however small home range sizes, presence-only location data from collars, and complex forest structure, has limited the ability to make specific correlations between location and habitat from remote sensing. While the increasing availability of very high-resolution satellite data provides novel ways to investigate habitat use on larger scales, the near continuous presence of clouds in tropical regions, particularly at higher elevations like those found on the Huon Peninsula, hinder the regular acquisition of satellite imagery (Chambers et al., 2007). Despite these challenges, the increasing availability of free and very high resolution satellite data provides novel ways to investigate habitat use on larger scales in the future.

Light Detection and Ranging (LiDAR) and aerial imagery has been effectively used to understand habitat characteristics associated with difficult to study species (McLean et al., 2016) however applications of these techniques are limited in Papua New Guinea by the inaccessibility and cost of LiDAR and paucity of aerial imagery.

To overcome the limitations of satellite imagery and challenges of traditional aerial surveying, Unmanned Aircraft Systems (UAS) can be extremely useful tools to collect extremely high resolution imagery on-demand and at substantially lower costs than traditional aerial imagery (Anderson and Gaston, 2013). Advances in aerial photogrammetry from UAS and Differential GPS surveying allow the creation of canopy height models even in closed canopy forests (Isenberg, 2017; Mohan et al., 2017). The photogrammetric reconstruction of 3-D surfaces from UAS imagery gives the possibility to develop correlations between the forest structure and composition data and an animal's habitat use and behavior in novel ways (Davies et al., 2017; Goetz et al., 2010). Despite the incredible promise of these tools, implementation of these techniques in remote areas and rugged terrain remains challenging (Koh and Wich, 2012).

Current research continues to advance new techniques to increase the spatial and temporal resolution of our understanding of Matschie's tree kangaroo behavior and habitat by developing GPS collars with motion and barometric pressure sensors, using Unmanned Aircraft Systems (UAS) to develop very high resolution tree species classifications, and expanding to regional analyses with high resolution satellite imagery. These investigations are currently ongoing but show significant promise in revealing insights into previously unobservable behavior and habitat use of wild Matschie's tree kangaroos.

DATA AND METHODS

Animal Collar Telemetry

VHF and GPS Collar Animal Telemetry Research (2004–2007)

From March 2004 to November 2007, Matschie's tree kangaroos were captured at the Wasaunon Field Research Site and fitted with VHF collars (MOD-205 VHF Transmitter; Telonics Inc., USA). They were manually tracked for six months and locations were recorded with GPS units from which home ranges were calculated using Harmonic Mean (HM), Kernel (KM), and Minimum Convex Polygon (MCP) techniques (Porolak, 2008). Also in 2004, three adult female Matschie's tree kangaroos were captured and fitted with Televilt Posrec™ GPS collars (model C200). These collars collected two locations per day (6:00 am and 6:00 pm local time) for a five month study period. These collars had an additional VHF transmitter which was used to locate the animals daily. Data on the slope, aspect, temperature, tree species, canopy closure, tree height, and other habitat characteristics were collected at each location (Stabach, 2005). These studies investigated the effectiveness of GPS collars, provided a reference for VHF collar data, and laid the foundations for being able to understand habitat preferences using remote sensing (Stabach et al., 2009).

Results from these previous studies have served to provide information and guide the development of the current generation of GPS collars, specifically customized for Matschie's tree kangaroo research. Improvements in the accuracy and reliability of GPS antennas and the addition of motion and barometric altitude sensors can allow the study of feeding, movement, and energy budgets in 3-D while removing any observational disturbance (Brown et al., 2013).

GPS and Biologging Collars

In October 2017, a team including researchers, a field veterinarian, and local research assistants and trackers conducted a two-week field trip to Wasaunon Field Camp. Five Matschie's tree kangaroos were located and captured with two female and one male collared with custom GPS collars (Field Data Technologies, Essex, MT) (Fig. 23.1) and one male and one female collared with VHF radio telemetry collars (Telonics, Inc., Mesa, AZ). In October 2018 six more GPS collars were deployed (Field Data Technologies, Essex, MT). The GPS collars record locations with 4 h intervals and have barometric altitude and binary motion sensors summed hourly. Over the next six months, the GPS collar data were remotely downloaded regularly by local field researchers and assistants using 915 mHz telemetry radio to limit animal disturbance. Data were processed after they were sent from the field and required removing unused data, formatting, and preparation for input into R with a script in Python. Tracking Analysis functions in ArcMap 10.6 (ESRI Inc., Redlands, CA) were used to visualize movement patterns but data processing was primarily done in R.

FIGURE. 23.1 Matschie's tree kangaroo high in the forest canopy with the battery compartment of the GPS collar visible. *Source: Jonathan Byers.*

Because the objective of this study was to understand small scale habitat boundaries and use/nonuse patterns that can be derived from high resolution habitat data, a literature review identified Time Local Convex Hull (T-LoCoH), a nonparametric kernel home range technique in R (Lyons et al., 2013). The advantage of this method is that it is more sensitive to edge effects and boundaries than other models because it minimizes the utilization distribution (UD) volume that occurred outside the true home range boundary (Lichti and Swihart, 2011). Utilization hulls from T-LoCoH were exported as shapefiles to be compared to a Minimum Convex Polygon area in ArcMap 10.6 (ESRI Inc., Redlands, CA), and were visually assessed for representativeness of landscape structures by comparing with orthoimages and DSM from aerial imagery.

REMOTE SENSING OF HABITAT

Ground and Satellite Vegetation Classification

Satellite imagery is the most widely used tool for understanding vegetation and habitat on a large scale, and Landsat 7 ETM+ imagery (pansharpened to 14.25 m/pixel) was used by Pugh (2003) and Stabach et al. (2009) to distinguish general forest classes at Wasaunon. Because finding Landsat images that are relatively cloud free is extremely difficult for the montane areas of Papua New Guinea, both of these studies used the same image acquired on September 8, 2000. Stabach et al. (2009) found that the classification accuracy of dominant forest types using multispectral imagery from Landsat-7 Enhanced Thematic Mapper Plus (ETM+) imagery was 69.0% accurate and using Satellite Pour l'Observation de la Terre (SPOT)-4 imagery was 72.4% accurate. They suggest that classification of forest composition from satellite imagery does show some utility, "pixel-based classifications alone may be unreliable because of the complex structure of the vegetative communities" (Stabach et al., 2009).

Unmanned Aircraft Systems (UAS) and Photogrammetry

In 2017, Aerial mapping flights were conducted during a site visit in October 2017 using a DJI Mavic Pro (DJI Inc., Shenzhen, China) controlled by Pix4D Capture running on an iOS device with flight height varying between 80 and 120 m AGL and overlap between 50% and 80%. Pix4D Capture does not offer terrain following, so the Ground Sample Distance (GSD) and overlap varied depending on topography and flight altitude but averages ~8.5 cm/pixel. Strong daily cloud cover and dense forest canopy severely limited aircraft flight operations range and 6 flights over three non-consecutive days resulted in 864 photographs covering ~340 ha. The camera was a 1/2.3″ CMOS sensor with a rolling shutter that captures images of 4000 × 3000 pixels with a 78.8 degree Field of View. Shutter speed and focus were controlled automatically by Pix4D Capture. Images were processed using Pix4D Mapper (Pix4D SA) and Agisoft Photoscan (Agisoft LLC) to develop 3-D forest canopy surface models and orthophotos (Fig. 23.2).

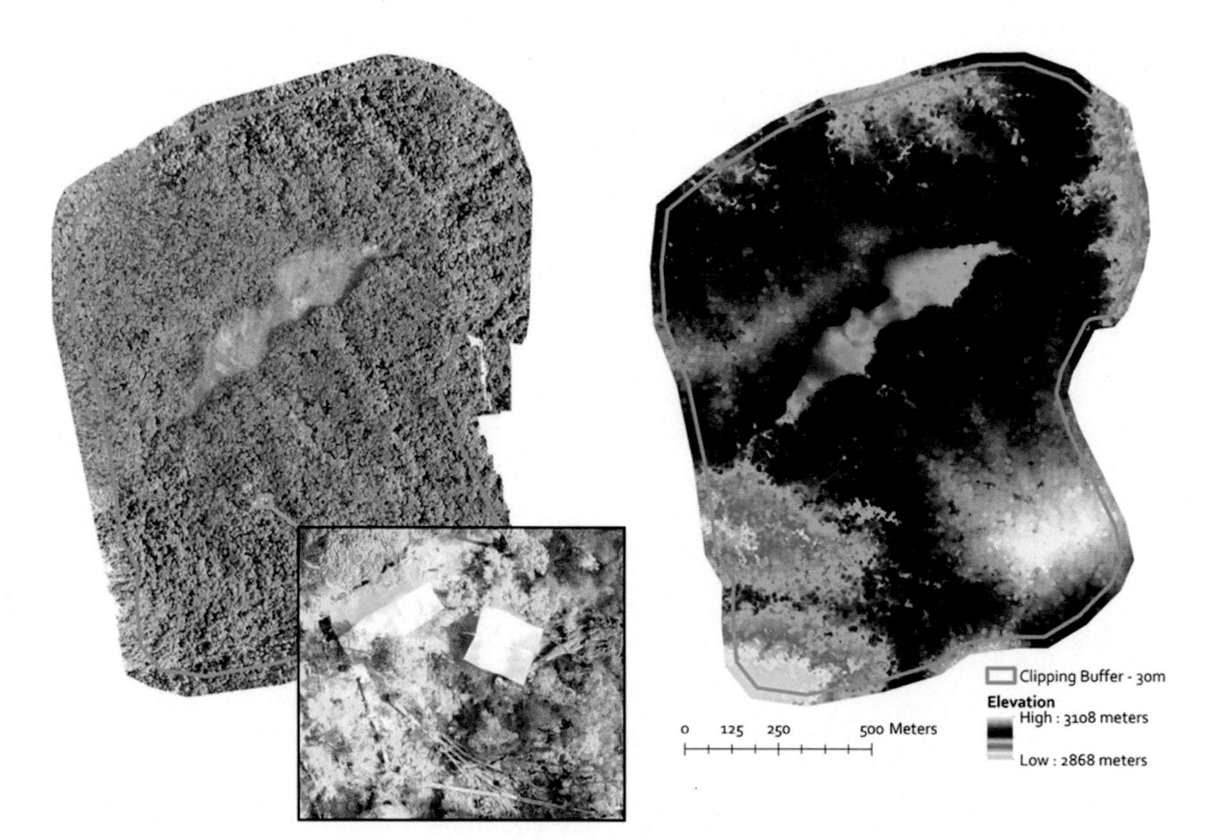

FIGURE. 23.2 Orthophoto of the aerial survey extent and inset of the research camp to illustrate resolution (left), and digital surface model (right) after processing. Insufficient overlap along the edges of the survey area reduced the accuracy of orthoimage merging and DSM and were trimmed (blue line). *Source: Byers, 2020.*

Canopy Structure Data Creation

Manual identification of canopy gaps was done using the photogrammetric DSM and orthoimage in ArcMap 10.6 (ESRI Inc.). This was a highly subjective process that relied substantially on local experience with forest canopy structure and terrain. A 13 m height difference between the surrounding canopy surface and the lowest point of the gap was used as a threshold for identifying gaps that reached fully through the canopy to the forest floor. This threshold was selected because the average canopy height is ~20 m, and the undergrowth on the forest floor often range from 2 to 5 m. There were several areas where depressions in the DEM aligned with gaps in the orthoimage but the difference to the surrounding canopy was <13 m indicating that the photogrammetry reconstruction had not successfully reconstructed down to the ground surface (Fig. 23.3). These points were not selected as canopy gaps.

The automated gap-detection process from the DSM used a process modified from Betts et al. (2005). First a fill (Hydrology toolbox) was applied so that larger canopy gaps did not lower the average height of their surroundings and create artifacts when identifying canopy emergent trees. A 25 × 25 m mean moving window average was applied to smooth the canopy surface. The smoothed surface was then subtracted from the DSM to yield a raster that showed areas lower and higher than the mean.

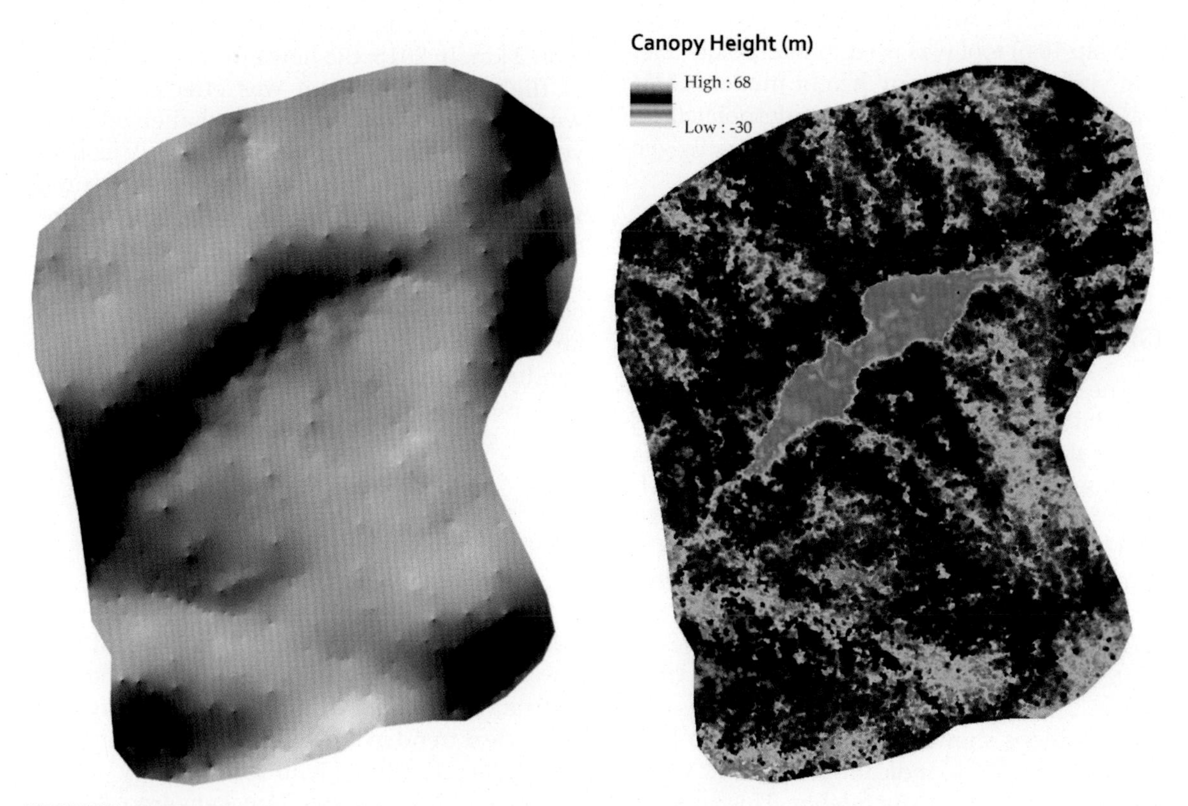

FIGURE. 23.3 Hillshade of the Digital Terrain Model from the ANUDEM terrain interpolation (left) and resulting canopy height model (CHM; right). *Source: Byers, 2020.*

The raster calculator tool was used with a threshold of ≤13 m below the average being identified as a gap, and ≥7 m above the average being identified as a canopy emergent tree. These raster classes were converted to polygons and the zonal statistics tool was used to place points on the lowest elevation areas from the DSM inside each gap polygon. A 10 × 10 m standard deviation of roughness modified from the Riley et al. (1999) Terrain Ruggedness Index is used as a metric to examine forest structural heterogeneity. A higher roughness value represents a larger difference between the highest and the lowest points in the moving window (e.g., canopy gaps, clearing edges, more vertical heterogeneity), a lower value represents less difference (e.g., consistent closed canopy forest, clearings). Once the ground surface points were established, the ANUDEM (Topo to Raster in ArcMap 10.6) tool was used to interpolate a terrain surface. The canopy height model (CHM) was created by subtracting the interpolated terrain surface from the photogrammetrically generated digital surface model (Fig. 23.3).

RESULTS

GPS Collar Performance and Home Range

The collars deployed in October 2017 and collected in April 2018 resulted in 76.8% fix success rate (Table 23.1). An MCP analysis of GPS collar locations shows an average 95% area of 21.72 ha ($N = 3$), with a smaller home range for females of 14.37 ha ($n = 2$) and 36.41 ha for the male. This same data processed using T-LoCoH a-method shows a 95% iso hull area average of 15.36 ha ($N = 3$), with a home range for females of 12.02 ha ($n = 2$) and 22.04 ha for the male (Table 23.2; Fig. 23.4). A temporal analysis of GPS locations shows an average Time to Independence of 72.6 h, indicating very slow movement and significant site fidelity. This supports the idea that these animals only use small parts, even only individual trees, in their home range at any given time. This suggests that even when the Minimum Convex Polygon home ranges overlap, there would be little spatial overlap.

AERIAL PHOTOGRAMMETRY PERFORMANCE

In 2017, six flights over three non-consecutive days resulted in 864 photographs covering ~340 ha with an average GSD of ~8.5 cm. In 2018, three flights collected 546 images covering 194 ha with a GSD of 4.9 cm. The aerial imagery collected in 2017 proved to have systematic problems (including bad focus and low overlap) which prevented successful surface reconstruction at a resolution allowing the identification of small canopy gaps. Additionally, signal interference limited flight range directionally to less than 2 km. In 2018, the new aircraft and change of flight control system was effective. While weather conditions were a challenge both years, the ability to opportunistically collect aerial imagery when it was clear in the morning was critical. Photogrammetric surface reconstruction using Pix4D proved to require a tremendous time investment as the available computer systems were not optimized for such large projects, and only imagery from flights conducted in 2018 was used for the remainder of these analyses.

CANOPY STRUCTURE PREFERENCES

The general findings of this comparison support the hypothesis that some variables of canopy structure are important for the habitat utilization of *D. matschiei*. There appeared to be no clear trend in their distance from manually or automatically identified canopy gaps although for Matschie's tree kangaroo (MTK) 3, the distance was lower using the automatically identified gap (Table 23.3). The significance

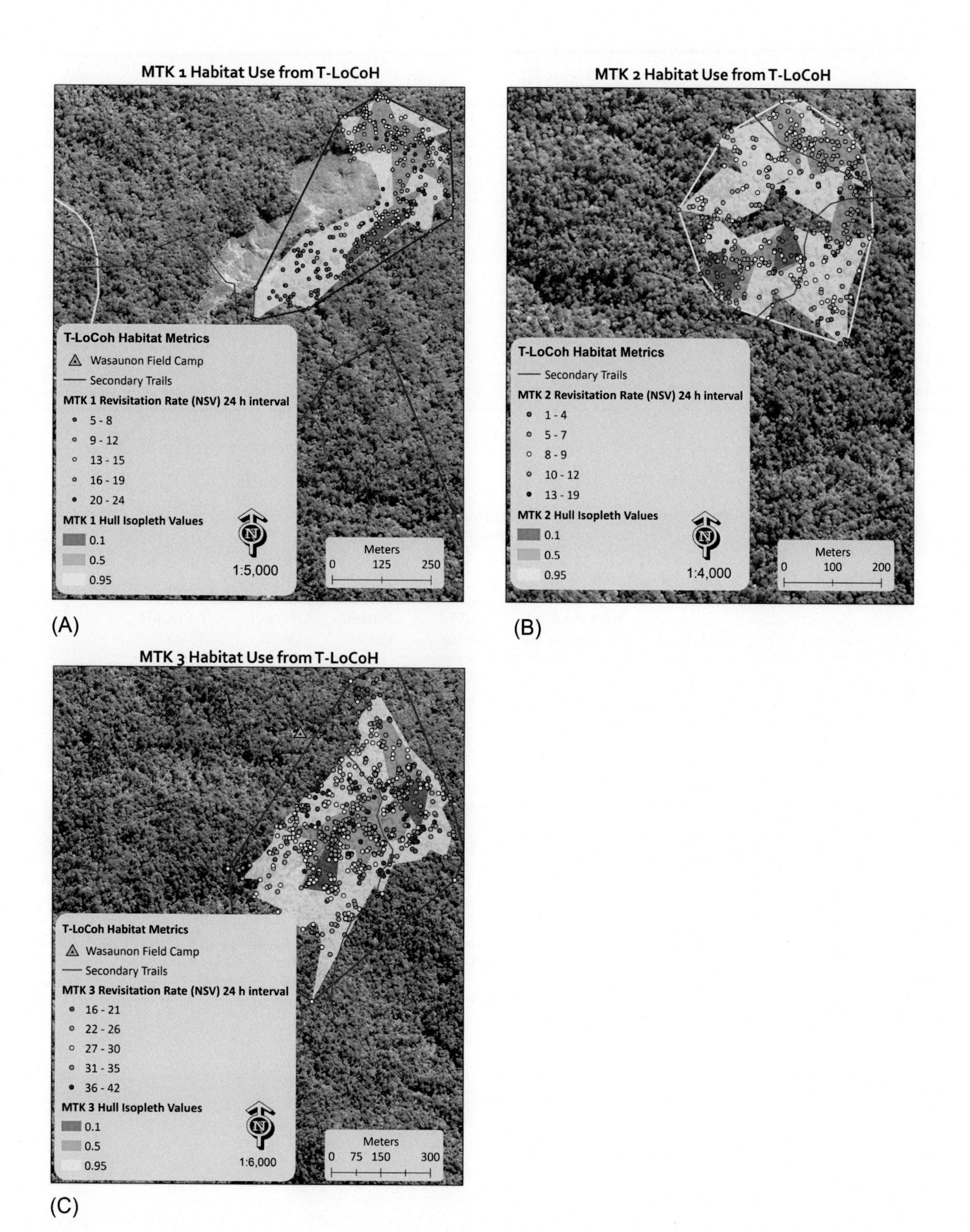

FIGURE. 23.4 Maps of location Revisitation rates from visits separated by 24 h windows using T-LoCoH for (A) MTK1, (B) MTK 2, and (C) MTK 3. Blue points represent areas visited infrequently and red the most frequent. Revisitation rates are dependent upon number of locations recoded by each collar and are not standardized. Isopleth values represent 10%, 50%, and 95% of points calculated using T-LoCoH. Colored polygon shows 100% MCP for comparison. MTK—Matschie's tree kangaroo; MCP—Minimum Convex Polygon; T-LoCoH—Time-Local Convex Hull. *Source: Byers, 2020.*

TABLE 23.1 Summary of tree kangaroo VHF and GPS collar deployment and performance in 2017–2018. MTK—Matschie's tree kangaroo.

ID	Collar Type	Start Date	# Days	# Locations	# of GPS Fixes Attempted	% Fix Success	Average Sampling Interval
MTK 1	GPS	9/27/2017	72.3	329	433	76.0%	4.9 h
MTK 2	GPS	9/30/2017	82.7	376	502	74.9%	4.8 h
MTK 3	GPS	10/2/2017	139.5	655	836	78.4%	4.2 h
MTK 4	VHF	12/9/2017	124	79	–	–	1.57 days
MTK 5	VHF	12/9/2017	124	80	–	–	1.55 days
Total	**GPS only**			1360	1771	76.8%	

Source: Byers, 2020.

TABLE 23.2 Mean home range area estimates from previous studies and this research. Location collection method and sample size are shown along with home range size in hectares from 100% Minimum Convex Polygon (MCP), and Harmonic Mean means calculated using 50% and 90% Kernels, and 95% Time Local Convex Hull (T-LoCoH) techniques.

Author	Method	Sample Size (# of animals)	MCP (100%)	Kernel (50%)	Kernel (90%)	T-LoCoH (95%)
Porolak (2008)	VHF	15	139.6 ± 26.5	13.8 ± 2.9	68.7 ± 14.2	–
Stabach et al. (2012)	GPS	3	–	7.3 ± 1.9	28.3 ± 2.3	–
Byers (2020)	GPS	3	21.7 ± 18.2	NA	NA	15.4 ± 6.7
	VHF	2	76.6 ± 19.3	–	–	–

TABLE 23.3 Mean values of canopy structure for each tree kangaroo compared to average values for the entire study area. *t*-test critical value of 1.96 for n-1 sample size. Bold values indicate significance lower than 0.01 *P*-value threshold. MTK—Matschie's tree kangaroo.

Mean Values	MTK 1	MTK 2	MTK 3	All Locations	Ortho-image	One tailed *t*-test	*P*-value
Distance from Manual Canopy Gaps (m)	23.9	19.2	27.9	26.0	25.6	0.99	0.32
Distance from Automatic Canopy Gaps (m)	22.9	19.3	18.4	20.0	22.7	**−6.39**	2.47^{-10}
Distance from Canopy Emergent Trees (m)	18.8	21.5	9.8	13.7	21.5	**−19.96**	2.56^{-74}
Canopy Height (m)	22.1	22.9	23.6	23.1	21.2	**6.80**	1.82^{-11}
Roughness	2.6	2.6	3.5	3.1	2.6	**8.08**	1.95^{-15}
Location Count	329	58	561	948			

Source: Byers, 2020.

of correlation between tree kangaroo locations and automatically generated canopy gaps is questionable since there is no significant correlation with manually identified gaps and the automatic method identified far higher numbers of gaps in some areas of the study site. The GPS-collared tree kangaroos were located significantly closer to canopy emergent trees than the mean within the study area, especially MTK 1 and MTK 3 (Table 23.3). Tree kangaroos were found at locations with taller canopy heights, and MTK 3 was also found in areas of more than mean canopy roughness.

Further investigation of these structural patterns with each animal's home range and using different utilization metrics would be interesting, however that will be better accomplished with additional collared animals and complete coverage of forest structure data. The fact that the CHM did not cover the entire areas traversed by the collared animals, particularly MTK 2, was disappointing but provides opportunities for synthesis with future high resolution height models or future aerial mapping efforts.

DISCUSSION

The overall home range areas assessed by Porolak et al. (2014) from VHF collar locations are nearly an order of magnitude larger than the utilization distributions of other tree kangaroo species (Coombes, 2005) and more than 2 × larger than those of the GPS collared animals from the same study (Stabach, 2005). The MCP overall home ranges identified from data from the 2004–2007 VHF collars are almost 5 × larger than those from the GPS collars deployed in 2017. One potential explanation for this variance is that up until 2009, tree kangaroos in this study area faced hunting pressure from local villagers. This would have lowered the population density and potentially increased home range area. Since the formation of the YUS Conservation Area and inclusion of the study site in the no-take zone, increased animal density and territorial behavior might be restricting home range sizes. Additionally, Porolak et al. (2014) used VHF collars which required locating the animals daily which could have caused a behavior shift that led to increased dispersion due to their sensitivity to disturbance. A larger sample size, comparing the data between the GPS and VHF collars, and studying home range size at different proximities to villages, as well as inside and outside the no-take zone would be required to understand this.

The variations in area between the Minimum Convex Polygon used by previous studies and T-LoCoH hulls illustrate why determining the entire area of landscape which an animal uses is very dependent on methodology. A visual assessment of habitat from UAS orthoimagery shows that T-LoCoH is better at excluding areas that are inside the kernel home range, but the animal does not use, such as the grassland (Fig. 23.4). T-LoCoh shows us that tree kangaroo habitat use is very sensitive to gaps in forest—whether grasslands or fallen down old trees. This is valuable insight in developing a species distribution model because previously it was unknown whether grasslands were barriers for tree kangaroo movement. These alpine grasslands cover extensive areas of the high elevation mountains above 3000 m and effectively provide an upper boundary for tree kangaroo distribution. Furthermore, the large expansions of alpine grassland during drought years can help provide some understanding of the effects of climate change on tree kangaroo habitat.

CONCLUSIONS

As an arboreal animal, the Matschie's tree kangaroo is dependent on a complex 3-dimensional forest structure. To truly understand the resource needs and behaviors of the Matschie's tree kangaroo, it is necessary to build on traditional 2D home range and resource selection studies and include vertical movements and temporal patterns to identify resource needs, threats, and potential barriers to movement (Powell and Mitchell, 2012). However, animal home range sizes that are small relative to the spatial resolution of the satellite imagery, presence-only location data from collars, and complex forest structure limited the ability to make specific correlations between location and habitat even from such relatively high resolution imagery.

The results of this study support the hypothesis that tree kangaroos are not habitat generalists and had some locations (emergent canopy) frequently visited while grasslands were rarely traversed. Whether these patterns of movement are driven more by forest species composition or more by forest structure remains undetermined. The destruction of complex forest at higher elevations from climate change-associated fires and frosts, and at lower elevations from population expansion and swidden agriculture, poise substantial threats if indeed the complex structure of primary cloud forests is necessary for *D. matschiei*.

Despite the low sample size, each individual animal exhibited interesting differences in movement pattern, velocity, and habitat use. *D. matschiei* appear to be very sensitive to the vertical structure of their habitat, with no evidence of individual animals crossing forest clearings, which has significant implications however movement data were of insufficient resolution to see if they similarly avoid forest gaps. The far larger 100% MCP sizes from the VHF data than the GPS data add evidence to the argument that VHF data overestimates the area in which individual animals are found due to sampling error or disturbance by trackers.

As with many rapidly developing technologies, using custom GPS collars and UAS for mapping proved challenging to implement effectively in the remote and difficult field locations. However, the ability to obtain previously unachievable insights into animal behavior and ecology proves the need to continue applying novel techniques to solve outstanding challenges. Though the push is to continually apply the most advanced tools available to a research question, research design and intentional application of these techniques is essential in being able to provide valid and valuable information. This proof of concept paves the way to advancing the understanding of behavior from high resolution motion logging and machine learning to understand 3d movement and behavior patterns in future research.

ACKNOWLEDGMENTS

This research was supported by funding from The National Geographic Society, GEF 5 (through UNDP), Conservation International, Ministry of Environment of the Government of Germany through the German Development Bank (KfW), Gay Jensen and Bob Plotnick Conservation and Research Leadership gift, other private donors and family foundations. We would like to thank the hard work and dedication of many people including Doug Bonham for developing the GPS collars with us; Daniel Okena, Nicholas Wari, Stanley Gesang, and the entire YUS research team; Field vet team Erika Crook, Trish Watson, Carol Esson; JCU Andrew Krockenberger; the YUS landowners; Rick Passaro; TKCP staff; Falk Huettmann; and Joe Pontecorvo.

REFERENCES

Ancrenaz, M., Dabek, L., O'Neil, S., 2007. The costs of exclusion: recognizing a role for local communities in biodiversity conservation. PLoS Biol. 5 (11), e289.

Anderson, K., Gaston, K.J., 2013. Lightweight unmanned aerial vehicles will revolutionize spatial ecology. Front. Ecol. Environ. 11 (3), 138–146.

Betts, H.D., Brown, L.J., Stewart, G.H., 2005. Forest canopy gap detection and characterisation by the use of high-resolution digital elevation models. N. Z. J. Ecol. 95–103.

Betz, W., 2001. Matschie's tree kangaroo (Marsupialia: Macropodidae, *Dendrolagus matschiei*) in Papua New Guinea: estimates of population density and landowner accounts of food plants and natural history. DissertationUniversity of Southampton, Southampton, England.

Brown, D.D., Kays, R., Wikelski, M., Wilson, R., Klimley, A.P., 2013. Observing the unwatchable through acceleration logging of animal behavior. Anim. Biotelemetry 1 (1), 20.

Burt, W.H., 1943. Territoriality and home range concepts as applied to mammals. J. Mammal. 24 (3), 346–352.

Byers, J.B., 2020. Investigating the Spatial Behavior and Habitat Use Of The Matschie's Tree-Kangaroo (*Dendrolagus matschiei*) Using GPS Collars And Unmanned Aircraft Systems (UAS). Master's thesisUniversity of Montana.

Cagnacci, F., Boitani, L., Powell, R.A., Boyce, M.S., 2010. Animal ecology meets GPS-based radiotelemetry: a perfect storm of opportunities and challenges. Philos. Trans. R. Soc. B 365, 2157–2162.

Chambers, J.Q., Asner, G.P., Morton, D.C., Anderson, L.O., Saatchi, S.S., Espírito-Santo, F.D., Palace, M., Souza Jr., C., 2007. Regional ecosystem structure and function: ecological insights from remote sensing of tropical forests. Trends Ecol. Evol. 22 (8), 414–423.

Coombes, K.E., 2005. The Ecology and Habitat Utilization of Lumholtz's Tree Kangaroos, *Dendrolagus lumholtzi* (Marsupialia: Macropodidae), on the Atherton Tablelands, Far North Queensland. PhD DissertationJames Cook University.

Davies, A.B., Ancrenaz, M., Oram, F., Asner, G.P., 2017. Canopy structure drives orangutan habitat selection in disturbed Bornean forests. Proc. Natl. Acad. Sci. 114 (31), 8307–8312.

Flannery, T.F., 1995. Mammals of New Guinea. Cornell University Press, New York.

Fleming, C.H., Fagan, W.F., Mueller, T., Olson, K.A., Leimgruber, P., Calabrese, J.M., 2015. Rigorous home range estimation with movement data: a new autocorrelated kernel density estimator. Ecology 96 (5), 1182–1188.

Goetz, S.J., Steinberg, D., Betts, M.G., Holmes, R.T., Doran, P.J., Dubayah, R., Hofton, M., 2010. Lidar remote sensing variables predict breeding habitat of a Neotropical migrant bird. Ecology 91 (6), 1569–1576.

Hebblewhite, M., Haydon, D.T., 2010. Distinguishing technology from biology: a critical review of the use of GPS telemetry data in ecology. Philos. Trans. Roy. Soc. B: Biol. Sci. 365, 2303–2312.

Isenberg, M., 2017. Integrating external ground points in forests to improve DTM from Dense-Matching Photogrammtry, Rapid Lasso GmbH. Available from: https://rapidlasso.com/2017/06/13/integrating-external-ground-points-in-forests-to-improve-dtm-from-dense-matching-photogrammetry. [16 July 2020].

Jensen, R., Fazang, K., 2005. General Collection of Flora in Wasaunon—Using 32 Nearest Neighbor Method. Forest Research Institute Unpublished.

Kie, J.G., Matthiopoulos, J., Fieberg, J., Powell, R.A., Cagnacci, F., Mitchell, M.S., Gaillard, J., Moorcroft, P.R., 2010. The home-range concept: are traditional estimators still relevant with modern telemetry technology? Philos. Trans. Roy. Soc. Lond. B: Biol. Sci. 365 (1550), 2221–2231.

Koh, L.P., Wich, S.A., 2012. Dawn of drone ecology: low-cost autonomous aerial vehicles for conservation. Trop. Conserv. Sci. 5 (2), 121–132.

Lichti, N.I., Swihart, R.K., 2011. Estimating utilization distributions with kernel versus local convex hull methods. J. Wildl. Manag. 75, 413–422. https://doi.org/10.1002/jwmg.48.

Lyons, A.J., Turner, W.C., Getz, W.M., 2013. Home range plus: a space-time characterization of movement over real landscapes. Movement Ecol. 1 (1), 2.

McGreevy, T.J., Dabek, L., Gomez-Chiarri, M., Husband, T.P., 2009. Genetic diversity in captive and wild Matschie's tree kangaroo (*Dendrolagus matschiei*) from Huon Peninsula, Papua New Guinea, based on mtDNA control region sequences. Zoo Biol. 28 (3), 183–196.

McLean, K.A., Trainor, A.M., Asner, G.P., Crofoot, M.C., Hopkins, M.E., Campbell, C.J., Martin, R.E., Knapp, D.E., Jansen, P.A., 2016. Movement patterns of three arboreal primates in a Neotropical moist forest explained by LiDAR-estimated canopy structure. Landsc. Ecol. 31 (8), 1849–1862.

Mohan, M., Silva, C.A., Klauberg, C., Jat, P., Catts, G., Cardil, A., Hudak, A.T., Dia, M., 2017. Individual tree detection from unmanned aerial vehicle (UAV) derived canopy height model in an open canopy mixed conifer Forest. Forests 8 (9), 340.

National Geographic, 2009. Adorable Tree Kangaroos Fitted With Tiny Video Cameras. Available from: https://www.youtube.com/watch?v=5WAiBl_b2cE. [16 July 2020].

Newell, G.R., 1999. Home range and habitat use by Lumholtz's tree-kangaroo (*Dendrolagus lumholtzi*) within a rainforest fragment in North Queensland. Wildl. Res. 26, 129–145.

Noonan, M.J., Tucker, M.A., Fleming, C.H., Akre, T.S., Alberts, S.C., Ali, A.H., Altmann, J., Antunes, P.C., Belant, J.L., Beyer, D., Blaum, N., 2019. A comprehensive analysis of autocorrelation and bias in home range estimation. Ecol. Monogr. 89 (2), e01344.

Porolak, G., 2008. Home Range of the Huon Tree Kangaroo, *Dendrolagus matschiei*, in Cloud Forest on the Huon Peninsula, Papua New Guinea. Doctoral dissertationJames Cook University.

Porolak, G., Dabek, L., Krockenberger, A.K., 2014. Spatial requirements of free-ranging Huon tree kangaroos, *Dendrolagus matschiei* (Macropodidae), in upper montane Forest. PLoS One. 9(3), e91870.

Powell, R.A., Mitchell, M.S., 2012. What is a home range? J. Mammal. 93 (4), 948–958.

Procter-Grey, E., 1985. The Behavior and Ecology of Lumholt's Tree Kangaroo, *Dendrolagus lumholtzi* (Marsupalia:Macropodidae). PhD thesisHarvard University.

Pugh, J.A., 2003. Identification of Huon Tree Kangaroo (*Dendrolagus matschiei*) Habitat in Papua New Guinea through Integration of Remote Sensing and Field Observations. Master's ThesisUniversity of Rhode Island.

Riley, S.J., DeGloria, S.D., Elliot, R., 1999. Index that quantifies topographic heterogeneity. Intermount. J. Sci. 5 (1–4), 23–27.

Ropert-Coudert, Y., Wilson, R.P., 2005. Trends and perspectives in animal-attached remote sensing. Front. Ecol. Environ. 3 (8), 437–444.

Stabach, J.A., 2005. Utilizing Remote Sensing Technologies to Identify Matschie's Tree Kangaroo (*Dendrolagus matschiei*) Habitat. PhD dissertationUniversity of Rhode Island.

Stabach, J.A., Dabek, L., Jensen, R., Wang, Y.Q., 2009. Discrimination of dominant Forest types for Matschie's tree kangaroo conservation in Papua New Guinea using high-resolution remote sensing data. Int. J. Remote Sens. 30 (2), 405–422.

Stabach, J.A., Dabek, L., Jensen, R., Porolak, G., Wang, Y.Q., 2012. Utilization of remote sensing technologies for Matschie's tree kangaroo conservation and planning in Papua New Guinea. In: Remote Sensing of Protected Lands. Taylor & Francis Publishers.

Wells, Z., Dabek, L., Kula, G., 2013. Establishing a conservation area in Papua New Guinea—Lessons learned from the YUS conservation area. In: Beehler, B.M., Kirkman, A.J. (Eds.), Lessons Learned from the Field: Achieving Conservation Success in Papua New Guinea. Conservation International, Woodland Park Zoo, Tree Kangaroo Conservation Program.

CHAPTER

24

Investigating Matschie's Tree Kangaroos With 'Modern' Methods: Digital Workflows, Big Data Project Infrastructure, and Mandated Approaches for a Holistic Conservation Governance

Falk Huettmann

EWHALE Lab – Biology & Wildlife Department, Institute of Arctic Biology, University of Alaska Fairbanks (UAF), Fairbanks, AK, United States

OUTLINE

Tree Kangaroos
https://doi.org/10.1016/B978-0-12-814675-0.00015-4

INTRODUCTION

Expeditions and field work have been the method of choice for tackling questions of exploration and conservation concern in remote regions for millennia. Those works are to progress society and global well-being (Brockway, 2002; Brown, 2017; Revkin, 2004). Famous examples can be found with historical explorations like James Cook, Charles Darwin, the Russian World Expeditions, James Hooker, U.S. Exploring Expedition, Lewis and Clarke expedition or modern ones like those by The Smithsonian Institute/US, Jane Goodall, Jacques Cousteau, George Schaller and even space explorations (Table 24.1; Huettmann, 2020a for a modern example in Asia). While new information was obtained and science-cultural advances were made from all those expeditions – usually of global relevance – virtually all of those explorations to this very day still share the common problem of public raw data presentation and lack of modern data delivery; many questions remain (Brockway, 2002; Drayton, 2005 for misuse of such information and efforts; Revkin, 2004).

The data sharing efforts are important, and they tend to shape the conservation management and subsequent fate of entire continents and the world. One example is Antarctica and the Antarctic Treaty System (ATS) with mandatory research data-sharing as a global good (De Broyer et al., 2014; National Science Foundation, 2020). For how this affects decisions and policy, see Huettmann (2012) and Huettmann et al. (2015) for the Ross Sea and the largest remote fisheries and Marine Protected Area (MPA) in the world subsequently influencing global climate and conservation aspects.

TABLE 24.1 List of data details for some famous expeditions and field work of relevance and impact for Papua New Guinea.

Name of expedition	Focus	Citation	Comment on data availability	Workflow
James Cook	Circumnavigating the world; exploration	Diamond (1999)	Hardcopy, incomplete, little digital and online	None
Charles Darwin	Circumnavigating the world; natural history	Darwin (1989)	Hardcopy, some digital and online	None
Russian World Expeditions	Russian Far East and connecting routes (Alaska, Papua New Guinea, India, Antarctica etc.) and associated research	Ivashintsov (1980), Bulkeley (2014), Brown (2017)	Hardcopy, very little digital	None
The Smithsonian Institute, American Museum of Natural History, Australian Museum	Biological Warfare in the wider Pacific region	MacLeod (2001), Rauzon (2016)	Classified	None
	Papua New Guinea and natural history collections	Webb (1995), Stone and Berryman (2012), Taylor (2016), LeRoy and Diamond (2017)	Incomplete, vast delay, hardcopy, very little online available	None

A good role model and policy framework for modern global biodiversity issues was set up in 1992 with the Convention on Biological Diversity (RIO convention; CBD, 2020a) and its Global Biodiversity Information Facility (GBIF) web portal and Darwin Core data format (CBD, 2020b; GBIF, 2020). It served as a type of data open access to the global community (Ohse et al., 2009; Nemitz et al., 2012), now also done with open access open source R software tools (rgbif; Chamberlain et al., 2020). Using the rgbif package in R, one can access the programmatic interface to the Web Service methods provided by GBIF. This package gives direct access to data from GBIF (GBIF, 2020) via their Application Programming Interface (API).

Workflows are an important concept allowing to collect standardized field data and document each step from data collection to data cleaning, analysis, and inference. Ideally, this is done in a digital form allowing for repeatable and transparent steps (Huettmann, 2005, 2009; Zuckerberg et al., 2011). However, specific and standardized workflows are still difficult to come by for expedition work (Table 24.1) while complexities of funding, political problems, and stochastic events such as hurricanes, earthquakes, civil wars, and pandemics interfere with the in-time delivery of field expedition data (Brown, 2017; see Carlson, 2011 for missing International Polar Year IPY data and World Data Centers; see Revkin, 2004 for expeditions on rubber and cinchona tree exploration as well as for extensive and precious field data lost due to fire in herbaria; Webb, 1995 for an example of publication delay). Nations and their institutions run quickly into inherent complexities, legacies, and problems when dealing with expeditions and data, and good outcomes. Conservation is usually time-critical to be effective (Rosales, 2008; Silvy, 2012). Such a related science is to be transparent and repeatable, specifically when carried out in the public realm and when high impact and progress of the wider public good is to be achieved (see Brockway, 2002 for a negative example and minimal data shared in an Open Access framework with the global public). While technical advances for in-time online delivery and predictive modeling stand as accepted standards from such efforts (see Huettmann, 2005 for ISO metadata and repositories), those principles are widely heralded (International Science Council, 2020) but in reality are hardly achieved (secret data still rule in academic and industrial research) (Brown, 2020; Carlson, 2011).

Matschie's tree kangaroo (also called Huon tree kangaroo, *Dendrolagus matschiei*; Taxonomic Serial Number TSN 552712; itis.gov) is an endemic species of the Huon Peninsula, located in the Morobe Province of Papua New Guinea. It is found in a remote and vast, ancient wilderness landscape that is difficult to access (Flannery, 1995a,b; Flannery et al., 1996). The species was first described in 1907 by the German naturalist Paul Matschie. Subsequently it was collected by several international expeditions, and specifically studied for the last 30 years, namely by Dabek (1994), Flannery (1995a,b), Flannery et al. (1996), Porolak (2008), and Porolak et al. (2014); an overview can be viewed on Animal Diversity Web (2020). This species is of wider conservation concern and is listed as Endangered on the IUCN Red List (Ziembicki and Porolak, 2016).

Many expeditions have been made to PNG and some of those focus and report on tree kangaroos (Flannery, 1995a,b; Flannery et al., 1996; Martin, 2005). However, only a few of them have delivered their data and made them available for public use, or even for purposes of conservation management. The vast majority of those data are not available or even useful in a meaningful way (Huettmann, 2020b). Lack of accessibility to data is a massive and generic problem with publicly funded expeditions or those supported by a national trust resource while ancient habitats and co-evolved lifestyles are getting lost and transitioning toward the global urban during the Anthropocene (Crutzen, 2006). Whereas vast expedition expenditures are made, the actual deliveries – data – are frequently not shared and information is not used effectively and thus the destructive globalization and framework moves forward as part of an imbalance between the Anthropocene and its conservation management (for Papua New Guinea: Chan, 2006; Henton and Flower, 2007; Kulick, 2019; for parts of Asia: Regmi and Huettmann, 2020).

RESEARCH STEPS AND WORKFLOW PROCESS

Using Matschie's tree kangaroo research as an example, the following steps show how progress can be achieved under the wider global framework and with modern, latest, and best available digital options at hand showing required research steps and a workflow:

Step 1: Robust Field Data Collections in the Field: Starting with the X, Y, and Z

As typically found in wildlife research (Silvy, 2012), the data collection needs a research design. For Matschie's tree kangaroo, this tends to involve very remote field work in conditions difficult or even hostile (Chapter 23; Diamond, 1999; Flannery, 1998; Mack, 2014). This can be due to steep terrain, ruggedness, remoteness, weather, and difficulties when acting in a foreign culture and rainforest environment with life-threatening diseases (West, 2006). Even more so then, the data collected and obtained are very precious and should be meaningful and precise – explicit in time and space with a good description – and be worked up in-time and made available to a global audience. In the most basic way, this includes using an analog notebook and a global positioning system (GPS) with a time stamp, ideally accompanied by specific photos. Such data can describe presence or absence of a species; ideally, abundance within a given area; often indirect measures are taken like documenting grazing signs, feces, or scratch marks (Fig. 24.1; Chapter 26). Care is to be taken that all of this can actually be achieved as initially planned, e.g. gear not broken, batteries available, and no lost equipment (none of those are trivial and have ended many field projects). Such types of raw data for Matschie's tree kangaroos can be seen and are described in Huettmann (2020b).

Step 2: In-time Description of Collected Field Data: Adding ISO-compliant Metadata in (Self-archiving) Repositories

Field data require a work-up which usually refers to data compilation, data cleaning and data digitization (Silvy, 2012). Once the work-up has been completed, it is a relatively easy but mandatory task to technically describe the digital data and their scientific content, e.g. taxonomy used, and statistical methods employed. That is achieved with metadata (data that describe data) in a standardized form (Huettmann, 2005), namely a set of described fields (e.g. Federal Geographic Data Committee FGDC) in a digital format for online delivery (e.g. using Extended Mark-up Language XML). This is covered with the International Organization for Standardization (ISO) codes, namely 19,115 (FGDC, 2020) and associated ones that remain a key issue for 'findability' of species taxonomies and key words for the public (Huettmann, 2005, 2009). Those ISO-compliant metadata can then be posted in metadata repositories, e.g. with governmental online portals or libraries; several of those metadata standards can be found and followed (Huettmann, 2005, 2009; for online editors see USGS Metavist website (USGS, 2020); see also National Center for Ecological Analysis and Synthesis (NCEAS, 2020) for MORPHO as the metadata editor with the National Science Foundation). While specific metadata standards differ, they tend to agree on a basic internationally available minimum set. Such metadata can be seen at University of Alaska Fairbanks (UAF) Scholarworks (forthcoming) or by request from the author.

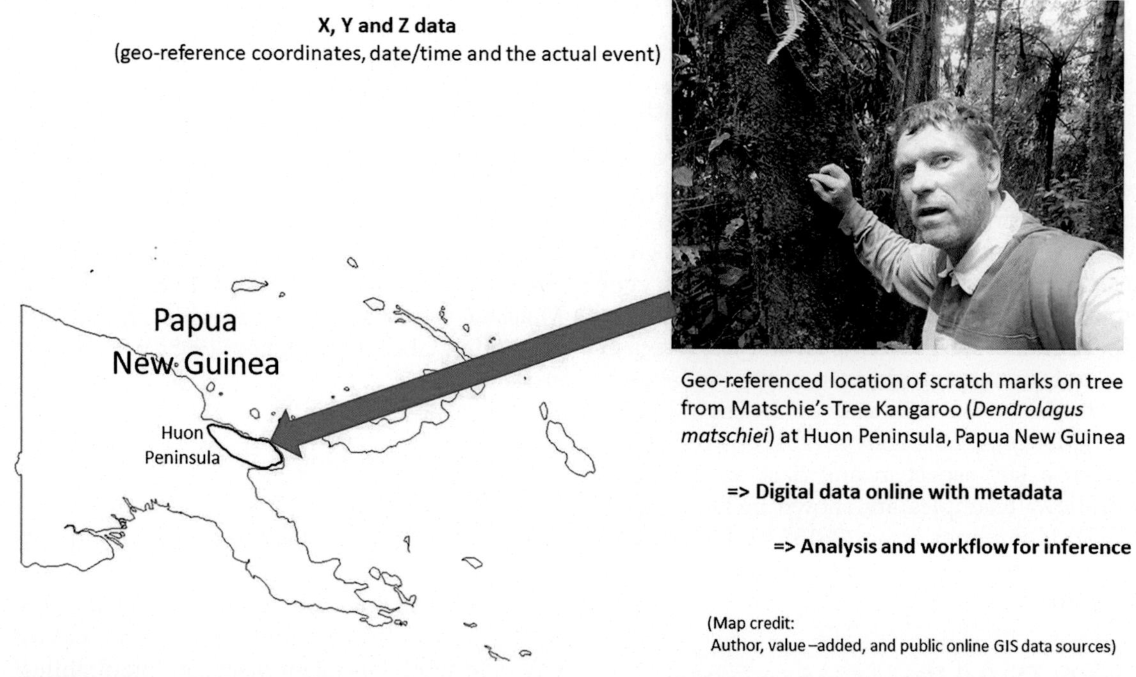

FIGURE. 24.1 Geo-referenced location of scratch marks on a tree from Matschie's tree kangaroo (*Dendrolagus matschiei*) at Huon Peninsula, Papua New Guinea.

Step 3: Mapping Field Data With Open Source GIS

"Without geography you are nowhere". Arguably, virtually all data carry a spatial component and are hardly meaningful without spatial information (Fortin and Dale, 2005; see Diniz-Filho et al., 2003 for autocorrelation; Magness et al., 2010 for policy; Schneider, 2009 for scale). That is also true for DNA data, stable isotopes, and most other 'lab data' often perceived – wrongly – as stand-alone events and in isolation. This considers that including spatial context is a mandatory step for looking at data – any data (Silvy, 2012) – when assessing them and for truly understanding them in the analysis and inference (Cushman and Huettmann, 2010; Drew et al., 2011). This is also part of any ecological understanding of data, as spatial aspects are part of the discipline from the very beginning (Fortin and Dale, 2005).

Step 4: Data Mining of Field Data: Finding and Learning Signals

Once digital – data are entered and available online – it should be exposed to a pre-analysis and get data mined as a test/trial (as promoted and shown in Humphries et al., 2018). This has many advantages as it sets a pilot test run for data validity and serves as a rapid assessment (Huettmann and Schmid, 2014; Kandel et al., 2015), later to be fine-tuned and for specific questions over time. Data mining can be done with algorithms of AI or machine learning (Mueller and Massaron, 2016; Hastie et al., 2016). Ideally, such algorithms can already learn the data (where data then become virtually obsolete and all get captured as an AI/machine learning algorithm with very high accuracy; see Huettmann et al., 2018 for mimicking data). If the data cannot be fully learned, at least signal extraction is a very powerful tool for rough data, and later to be model-predicted for generalizable inference (Humphries et al., 2018).

Step 5: Predictions with AI/ML Algorithms

From the above steps, it is easy to move into predictions (Breiman, 2001). A key advantage is that predictions – e.g. from an algorithm that was able to learn and mimic the data with good accuracy (Humphries et al., 2018) – one can generalize, and test the outcome quantitatively to other situations, e.g. for impact studies, forecasting, or to compare and confront the model with other real world data and evidences (Hilborn and Mangel, 1997; Silvy, 2012). Predictions remain a key aspect in any data analysis and workflow. Examples are shown in Drew et al. (2011), Humphries et al. (2018), and then with (Huettmann, 2020b) for Matschie's tree kangaroos.

Step 6: Digital Project Infrastructure

Virtually any project needs gear and tools (Silvy, 2012). In the year 2020 onwards, this has primarily a digital aspect because 'tools' tend to be fully digital, online, and include several data sets. But this detail is commonly overlooked, underestimated, underfunded, and lacks expertise and long-term consistency (Huettmann, 2015; Zuckerberg et al., 2011; see Costello et al., 2014 for financial sustainability; Carlson, 2011 for a reality perspective). Beyond presence data (Huettmann and Schmid, 2014; De Broyer et al., 2014; Kandel et al., 2015; Dornelas et al., 2018; Elith et al., 2020) a basic example consists of habitat map data for species and their habitats and climate data, as well as socioeconomic data (Sriram and Huettmann, 2017). An example for the Matschie's tree kangaroo (Huettmann, 2020b) is presented for free download at UAF Scholarworks (forthcoming) or upon request to the author consisting of over 100 GIS layers for Papua New Guinea. Another aspect is the actual software needed to run data visualization, data analysis, assessment, and reporting. Such data and software repositories can now be found online, e.g. with github.com in R and Python code.

For Papua New Guinea, such digital infrastructures are dispersed and often found outside of PNG, and located in Australia, the United States (USA) or the European Union (EU) and the United Kingdom (UK). See Table 24.2 for some whereabouts of crucial PNG data. For Matschie's tree kangaroo, specific online data are widely absent (Flannery, 1995a,b; Flannery et al., 1996).

Step 7: Toward Doing Everything in 'The Cloud'

Except for the data collection for fieldwork (Step 1), all the steps (Steps 2–6) can, and actually should, be done ideally in 'The Cloud'. This includes gathering already existing data before any field work has taken place and maintaining these data in 'The Cloud' along with data collected during and certainly afterwards the field work is done (Huettmann, 2015). While having a local back-up is important, 'The Cloud' has many advantages when it comes to global access, computing power, speed, output diversity, impact, storage, even security, and longevity (Humphries and Huettmann, 2018). Thus far, 'The Cloud' is a diverse safety measure but local data copies still should be kept, if at all possible (Zuckerberg et al., 2011). A new angle comes along with online repositories for data and code, and when commercial services are able to help (e.g. the Amazon Cloud, 2020; Microsoft AZURE, 2020).

Step 8: Assembling an In-time Workflow

Currently, there are few resources and software to achieve these steps, all in a good workflow, from field work to data cleaning, rapid assessments, data delivery to publication. However, there is a clear push to achieve an efficient

TABLE 24.2 Data of relevance for Papua New Guinea. CGIR – Consortium of International Agricultural Research Centers; USA – United States of America; UN – United Nations; CGIAR – Consortium of International Agricultural Research Centers.

Data set name	Content	URL/citation	Location	Comment
Worldclim	Climate and altitude	www.worldclim.org	USA	A quasi-standard for ecological models
DIVA-GIS	National infrastructure	http://www.diva-gis.org/	International (Several CGIAR Institutes)	Free and open GIS with data
ETOPO1	Altitude	https://ngdc.noaa.gov/mgg/global/global.html	USA	A USA government data project
104 GIS layers	Compilation of climate, topography and socio-economic layers	https://scholarworks.alaska.edu/handle/11122/7151 (compiled by author)	USA	Many sources; unpublished manuscript appendix
Google Earth Engine	Earth science data (Remote Sensing etc.)	https://earthengine.google.com/	USA	A commercial data compilation and data viewer
GBIF	Biodiversity (museum specimen etc.)	GBIF.org	UN	The global biodiversity data standard
Bishop Museum Hawaii	Specimen	bishopmuseum.org/	USA	A long-standing Research stakeholder for the region
Sydney Museum	Specimen	australianmuseum.net.au/	Australia	A major long-term repository for Papua New Guinea

workflow, and research contracts should be set up, and provide, target workflow items as milestones so that at a later stage, those can be merged and re-done, all as one singular workflow benefitting the species of study and its habitat. In laboratory work, this is already a certain standard, such as DNA, stable isotope, and industrial product assessments, where double-development and blind-testing is essential, usually legally binding and 'best professional' practice. In many research works and funding bodies, such reasoning – as stated above for transparent standardized digital workflows available for the public – remains widely absent, certainly in wildlife research, with Matschie's tree kangaroos, and with the IUCN website (IUCN, 2020; Kandel et al., 2015; Zhang et al., 2019a,b).

Step 9: Scaling It Up – Generic, Large-Scale, and Global Applications for Testing and Generalization

Arguably, data and findings are not just to fit local situations, but they are to generalize wider; that is where models come in handy because – if done right – they allow to be scaled up (Drew et al., 2011; Humphries et al., 2018). The above outlined steps feed directly into such wider generalization, specifically when a research design exists where training and alternative testing data were collected. When a generalization gets confronted with other evidence and reality (Hilborn and Mangel, 1997; Humphries et al., 2018), it can be assessed and/or improved; just like a regular hypothesis test formulated by K. Popper's falsification theory as the modern scientific standard (Wilkinson, 2013; for applications see Breiman, 2001; Cushman and Huettmann, 2010; Humphries et al., 2018; Silvy, 2012). If the generalization fails, then the previous steps are to be checked and re-done; information can be gained. A workflow provides an 'easy workbench' – a testing ground – for doing so; re-runs can be done within seconds in a transparent and repeatable fashion. It is a great way for scaling it up and extending. Examples can be seen with Kandel et al. (2015) and now are also forthcoming with Matschie's tree kangaroo to indicate how such concepts apply overall for the wider Huon peninsula and PNG (Huettmann, 2020b).

PROBLEMS AND HURDLES TO OVERCOME

Without open access and digital data sharing, modern approaches to science are limited. Thus, a more holistic conservation science-based management cannot be achieved, nor can those telecoupled systems be managed well (Liu et al., 2018). As discussed earlier, just a few Matschie's tree kangaroo data are shared, and in the public set, the bulk in GBIF are hardly usable. That is an obvious and significant problem for research done in Papua New Guinea as a whole and for its major players (Table 24.3).

The term 'data' here refers to valid taxonomies and referencing field survey events in space and time (such as date – day, month, and year – and time of day). A geo-referencing with decimal coordinates in latitude, longitude, and elevation (meters above sea level) should easily be possible these days and thus is a common expectation.

Another crucial aspect to deal with is authenticity of the data and detectability (Silvy, 2012). These are not really addressed in GBIF and its data scheme (Darwin Core, 2020) though, as the global standard. Instead, GBIF currently just

TABLE 24.3 A selection of projects and research with very little data shared for public consumption and progress in Papua New Guinea.

Citation	Topic	Comment
Flannery (1995a,b, 1996) Flannery et al. (1996) Martin (2005)	Mammals and tree kangaroos	Over a decade of field work and research
Mack (2014)	Bird research in Papua New Guinea	Over a decade of field work and research
Mayr and Diamond (2001)	Bird collections for Melanesia	Many decades of bird surveys and collections
West (2006)	Anthropology	Many field survey and conservation projects

serves, and double-serves data but does not quality-screen much of its data. The consequence is that such data and global efforts become less trusted, and thus, less used. Related to this are the needed expertise and budgets assigned and distributed to administer the data holdings and GIS data layers (Costello et al., 2014), including software updates and longevity. An agreed upon quality flag of data, e.g. fit for use, remains absent in virtually all IUCN data as well as GBIF data. Finally, it comes home to (moral) value and ethics (e.g. what are data used for?), just as it was expressed by Bandura (2007) and Huettmann (2015). Like with most IUCN data, for Matschie's tree kangaroo, we see none of those aspects discussed yet.

A REALITY OUTLOOK FOR WORKFLOWS AND CONSERVATION OF SPECIES LIKE MATSCHIE'S TREE KANGAROO

Despite vast public data holdings in GBIF and by many institutions able to inform policy, expeditions are still the *'bread and butter'* for science-based conservation management (Silvy, 2012; West, 2006); often they include specimen collection and other intrusive methods. The new digital opportunity pushes beyond that for online in-time digital data mining and AI/Machine Learning prediction-based concepts. This actually becomes the method of choice for good conservation management and in times of globalization and the associated Anthropocene (Drew et al., 2011; Huettmann, 2015; Humphries et al., 2018). The described workflow certainly should be done for any expedition inquiry first and can be done in parallel and becomes the method of choice after an expedition and to inform policy (Magness et al., 2010).

Clearly, we are not 'there' yet. Thus far, globalization has not been kind to wilderness regions and endemic species (see Huettmann, 2018 for decay of the ecological niche for mammals world-wide). That conservation crisis is true globally and includes the Pacific Rim (Huettmann, 2014; Czech and Krausman, 2001 for Hawaii and California) as well as for Papua New Guinea and parts of northern Australia where tree kangaroos live (Flannery, 1995a,b; Flannery et al., 1996; Martin, 2005). Still, the Huon Peninsula remains one of the least affected wilderness areas in the world – from reef to ridge – but currently is under threat, 'even' beyond climate change. Mining and development schemes have started to perforate PNG landscapes and watersheds and those stakeholders keep pushing for more (West, 2006; see Chan, 2006 for legacy of mining).

Tree kangaroos – two species in Australia and 12 species in New Guinea – make for a peculiar case within that conservation matrix because most of the macropodid (kangaroo) species overall are actually located in Australia, and their endemic conservation status -including digital data for the public – overall speaks for the reality check of the taxa (Belanger, 2018). In the meantime, Matschie's tree kangaroo remains widely understudied and misjudged, but large range declines have been suggested for this species [Huettmann, 2020b] and many threats are reported, e.g. stray dogs. While actual population estimates remain unknown, and could be higher than assumed, the future outlook for this species remains in dire straits. This requires a massive change in business taking the needs of Matschie's tree kangaroos fully into account and can be better achieved using modern and best-available tools for a more holistic approach safeguarding endemic species.

ACKNOWLEDGMENTS

FH is grateful to the late L. Spears and R. O'Connor, as well as A.W. (Tony) Diamond, T. Lock and D. Klein for data sharing visions, discussions and actions leading to views and work presented here. FH is also extremely grateful to 'the rangers' providing the bulk of field detection data about Matschie's tree kangaroos in Papua New Guinea; and acknowledge other data contributors in GBIF, as well as S. Sriram for her large global GIS layer compilation indeed. Further, H. Alvarez Berrios, H. Hera and her entire team were essential for support during FH's field campaigns. One must acknowledge -and will never forget – how M. Dunleavy supported UAF and the wildlife department. Further, the funders (GEF/UNDP) are acknowledged via the Woodland Park Zoo in Seattle for running and supporting this project for a long time. The members of the Tree Kangaroo Conservation Program field office, as well as the villages of Sapmanga, Gomdam and in YUS are very kindly thanked, specifically for field work campaigns and the mega-transect that allowed for many aspects of this work.

REFERENCES

Amazon Cloud, 2020. Available from: https://aws.amazon.com/. (2 June 2020).

Animal Diversity Web, 2020. *Dendrolagus matschiei*, Huon Tree Kangaroo. Available from: https://animaldiversity.org/accounts/Dendrolagus_matschiei/ (2 June 2020).

Bandura, A., 2007. Impeding ecological sustainability through selective moral disengagement. Int. J. Innov. Sustain. Dev. 2, 8–35.

Belanger, P., 2018. Extraction Empire: Undermining the Systems, States, and Scales of Canada's Global Resource Empire, 2017—1217. MIT Press, Cambridge, MA.

Breiman, L., 2001. Statistical modeling: the two cultures (with comments and a rejoinder by the author). Stat. Sci. 16, 199–231.

Brockway, L.H., 2002. Science and Colonial Expansion. The Role of the British Royal Botanic Gardens. Yale University Press, United States.

Brown, S., 2017. Island of the Blue Foxes: Disaster and Triumph on the World's Greatest Scientific Expedition (A Merloyd Lawrence Book). Da Capo Press, New York, NY.

Brown, K., 2020. The Big Secret in the Academy Is That Most Research Is Secret: The Dangerous Rift Between Open and Classified Research. Available from: https://www.aaup.org/article/big-secret-academy-most-research-secret#.X2Tg4WhKhPY.

Bulkeley, R., 2014. Bellingshausen and the Russian Antarctic Expedition, 1819-2. Palgrave MacMillan, New York.

Carlson, D., 2011. A lesson in sharing. Nature 469, 293.

Chamberlain, S., Oldoni, D., Barve, V., Desmet, P., Laurens, G., Mcglinn, D., Ram, K., 2020. Package 'rgbif', Interface to the Global 'Biodiversity' Information Facility API. R Package Version 3.0.0. Available from: https://cran.r-project.org/web/packages/rgbif/rgbif.pdf (2 June 2020).

Chan, J., 2006. Playing the Game: Life and Politics in Papua New Guinea. University of Queensland Press, Brisbane.

Convention on Biological Diversity (CBD), 2020a. The Rio Conventions. Available from:https://www.cbd.int/rio/ (20 July 2020).

Convention on Biological Diversity (CBD), 2020b. Global Biodiversity Information Facility. Available from: https://www.cbd.int/cooperation/csp/gbif.shtml (20 July 2020).

Costello, M.J., Appeltans, W., Bailly, N., Berendsohn, W.G., de Yong, Y., Edwards, M., Froese, R., Huettmann, F., Los, W., Mees, J., 2014. Strategies for the sustainability of online open-access biodiversity databases. Biol. Conserv. 173, 155–165.

Crutzen, P.J., 2006. The "Anthropocene" In: Ehlers, E., Krafft, T. (Eds.), Earth System Science in the Anthropocene. Springer, Berlin, Heidelberg.

Cushman, S., Huettmann, F. (Eds.), 2010. Spatial Complexity, Informatics and Wildlife Conservation. Springer, Tokyo.

Czech, B., Krausman, P., 2001. The Endangered Species Act: History, Conservation, Biology, and Public Policy. John Hopkins University Press, Baltimore, MD.

Dabek, L., 1994. Reproductive Biology and Behavior of Captive Female Matschie's Tree Kangaroos, *Dendrolagus matschiei*. (Thesis)University of Washington, Seattle, WA.

Darwin, C., 1989. The Voyage of the Beagle. Penguin Books, New York.

Darwin Core, 2020. Available from: https://dwc.tdwg.org/. (20 July 2020).

De Broyer, C., Koubbi, P. (Eds.), with Griffiths, H., Danis, B, David, B., Grant, S., Gutt, J., Held, C., Hosie, G., Huettmann, F., Post, A., Ropert-Coudert, Y., van den Putte, A., 2014. The CAML/SCAR-MarBIN Biogeographic Atlas of the Southern Ocean. Scientific Committee on Antarctic Research (SCAR), Cambridge. Available from: https://www.scar.org/library/scar-publications/occasional-publications/3501-biogeographic-atlas-of-the-southern-ocean-selected-chapters/ (2 June 2020).

Diamond, J., 1999. Guns, Germs and Steel: The Fate of Human Societies. Norton Company, New York.

Diniz-Filho, J.A.F., Bini, L.M., Hawkins, B.A., 2003. Spatial autocorrelation and red herrings in geographical ecology. Glob. Ecol. Biogeogr. 12, 53–64. https://doi.org/10.1046/j.1466-822X.2003.00322.x.

Dornelas, M., Antão, L.H., Moyes, F., et al., 2018. BioTIME: a database of biodiversity time series for the Anthropocene. Glob. Ecol. Biogeogr. 27, 760–786. Available from: https://doi.org/10.1111/geb.12729.

Drayton, R., 2005. Nature's Government: Science, Imperial Britain and the 'Improvement' of the World. Orient Longman, London.

Drew, C.A., Wiersma, Y.F., Huettmann, F. (Eds.), 2011. Predictive Species and Habitat Modeling in Landscape Ecology. Springer, New York.

Elith, J., Graham, C.H., Valavi, R., Abegg, M., Bruce, C., Ford, A., Guisan, A., Hijmans, R.J., Huettmann, F.,

Lohmann, L., Loiselle, B., Moritz, C., Overton, J., Peterson, A.T., Phillips, S., Richardson, K., Williams, S.E., Wiser, S.K., Wohlgemuth, T., Zimmermann, N.E., 2020. Presence-only and presence-absence data for comparing species distribution modeling methods. J. Biomed. Inform. 15, 69–80.

Federal Geographic Data Committee (FGDC), 2020. Geospatial Metadata. Available from: https://www.fgdc.gov/metadata/iso-standards (20 July 2020).

Flannery, T.F., 1995a. The Mammals of New Guinea, second ed. Reed Books, Sydney.

Flannery, T.F., 1995b. Mammals of the South-West Pacific and Moluccan Islands. Comstock, Cornell, Ithaca, New York.

Flannery, T.F., 1998. Throwim Way Leg. The Text Publishing Company, Melbourne.

Flannery, T.F., Martin, R.M., Szalay, A., 1996. Tree Kangaroos—A Curious Natural History. Reed Books Australia, Melbourne, Victoria.

Fortin, M.-J., Dale, M.R.T., 2005. Spatial Analysis: A Guide for Ecologists. Cambridge University Press, Cambridge.

Global Biodiversity Information Facility (GBIF), 2020. Available from: gbif.org. (2 June 2020).

Hastie, T., Tibshirani, R., Friedman, J., 2016. The Elements of Statistical Learning: Data Mining, Inference, and Prediction. Springer, New York.

Henton, D., Flower, A., 2007. Mount Kare Gold Rush: Papua New Guinea, 1988–1994. Mt Kare Gold Rush, Cotton Tree, Queensland.

Hilborn, R., Mangel, M., 1997. The Ecological Detective: Confronting Models With Data. Princeton University Press, Princeton, NJ.

Huettmann, F., 2005. Databases and science-based management in the context of wildlife and habitat: towards a certified ISO standard for objective decision-making for the global community by using the Internet. J. Wildl. Manag. 69 (20), 466–472.

Huettmann, F., 2009. The global need for, and appreciation of, high-quality metadata in biodiversity work. In: - Spehn, E., Koerner, C. (Eds.), Data Mining for Global Trends in Mountain Biodiversity. CRC Press, Taylor & Francis, Boca Raton, FL, pp. 25–28.

Huettmann, F. (Ed.), 2012. Protection of the Three Poles. Springer, Tokyo, p. 337. Available from: https://www.springer.com/gp/book/9784431540052.

Huettmann, F., 2014. Economic growth and wildlife conservation in the north Pacific Rim, highlighting Alaska and the Russian Far East. In: Gates, E., Trauger, D., Czech, B. (Eds.), Peak Oil, Economic Growth, and Wildlife Conservation. Springer, New York, NY, pp. 133–156.

Huettmann, F., 2015. On the relevance and moral impediment of digital data management, data sharing, and public open access and open source code in (tropical) research: the Rio convention revisited towards mega science and best professional research practices. In: Huettmann, F. (Ed.), Central American Biodiversity: Conservation, Ecology, and a Sustainable Future. Springer, New York, pp. 391–418.

Huettmann, F., 2018. Climate change effects on terrestrial mammals: a review of global impacts of ecological niche decay in selected regions of high mammal importance. In: Dellasala, D., Goldstein, M.I. (Eds.), Encyclopedia of the Anthropocene. In: vol. 2. Elsevier, Waltham, MA, pp. 123–130.

Huettmann, F., 2020a. A governance analysis of the snow leopard, its habitat, and data: who owns charismatic animals and who drives and uses the agenda for what? In: - Regmi, G.R., Huettmann, F. (Eds.), Hindu Kush-Himalaya Watersheds Downhill: Landscape Ecology and Conservation Perspectives. Springer Dordrecht, Holland, pp. 459–472 (Chapter 23).

Huettmann, F., 2020b. Tree Kangaroo Project. An Open Access GIS data analysis using machine learning.

Huettmann, F., Schmid, M., 2014. Publicly available open access data and machine learning model-predictions applied with open source GIS for the entire Antarctic Ocean: a first meta-analysis and synthesis from 53 charismatic species. In: Veress, B., Szigethy, J. (Eds.), Horizons in Earth Science Research. In: vol. 11. pp. 24–34 (+ data publications in dSPACE with UAF library) (Chapter 3).

Huettmann, F., Schmid, M.S., Humphries, G.R.W., 2015. A first overview of open access digital data for the Ross Sea: complexities, ethics, and management opportunities. Hydrobiologia 761, 97–119. Available from: https://doi.org/10.1007/s10750-015-2520-x.

Huettmann, F., Craig, E.H., Herrick, K.A., Baltensperger, A.P., Humphries, G.R.W., Lieske, D.J., Miller, K., Mullet, T.C., Oppel, S., Resendiz, C., Rutzen, I., Schmid, M.S., Suwal, M.K., Young, B.D., 2018. Use of machine learning (ML) for predicting and analyzing ecological and 'presence only' data: an overview of applications and a good outlook. In: - Humphries, G., Magness, D.R., Huettmann, F. (Eds.), Machine Learning for Ecology and Sustainable Natural Resource Management. Springer, Cham, pp. 27–61.

Humphries, G.R.W., Huettmann, F., 2018. Machine learning and 'The Cloud' for natural resource applications: autonomous online robots driving sustainable conservation management worldwide? In: Humphries, G., Magness, D.R., Huettmann, F. (Eds.), Machine Learning for Ecology and Sustainable Natural Resource Management. Springer, Cham, pp. 353–377.

Humphries, G., Magness, D.R., Huettmann, F. (Eds.), 2018. Machine Learning for Ecology and Sustainable Natural Resource Management. Springer, Cham.

International Science Council, 2020. Open-Access-to-Scientific-Data-and-Literature. Available from: https://council.science/ (21 July 2020).

International Union for Conservation of Nature (IUCN), 2020. Available from: IUCN.org. (27 July 2020).

Ivashintsov, N.A., 1980. Russian Round-The-World Voyages, 1803–1849: With a Summary of Later Voyages to 1867 (Materials for the Study of Alaska History). University of Alaska Press, Fairbanks, AK.

Kandel, K., Huettmann, F., Suwal, M.K., Regmi, G.R., Nijman, V., Nekaris, K.A.I., Lama, S.T., Thapa, A., Sharma, H.P., Subedi, T.R., 2015. Rapid multi-nation distribution assessment of a charismatic conservation species using open access ensemble model GIS predictions: red panda (*Ailurus fulgens*) in the Hindu-Kush Himalaya region. Biol. Conserv. 181, 150–161.

Kulick, D., 2019. A Death in the Rainforest: How a Language and a Way of Life Came to an End in Papua New Guinea. Algonquin Books, Toronto.

LeRoy, M., Diamond, J., 2017. Rollo Beck's collection of birds in Northwest New Guinea. Am. Mus. Novit. 1–36. Available from: https://www.amnh.org/research/research-library/library-news/papua-new-guinea-expedition.

Liu, J., Dou, Y., Batistella, M., Challies, E., Connor, T., Friis, C., Millington, J.D.A., Parish, E., Romulo, C.L., Bicudo Silva, R.F., Triezenberg, H., Yang, H., Zhao, Z., Zimmerer, K.S., Huettmann, F., Treglia, M.L., Basher, Z., Chung, M.G., Herzberger, A., Lenschow, A., Mechiche-Alami, A., Newig, J., Roch, J., Sun, J., 2018. Spillover systems in a telecoupled Anthropocene: typology, methods, and governance for global sustainability. Environ. Sustain. 33, 58–69. https://doi.org/10.1016/j.cosust.2018.04.009.

Mack, A., 2014. Searching for Pekpek: Cassowaries and Conservation in the New Guinea Rainforest. Cassowary Conservation and Publishing, New Florence, PA.

MacLeod, R., 2001. "Strictly for the birds": science, the military, and the Smithsonian's Pacific Ocean Biological Survey Program, 1963-1970. J. Hist. Biol. 34, 315–352.

Magness, D., Morton, J.M., Huettmann, F., 2010. How spatial information contributes to the management and conservation of animals and habitats. In: Cushman, S., Huettmann, F. (Eds.), Spatial Complexity, Informatics and Wildlife Conservation. Springer, Tokyo, pp. 429–444 (Chapter 23).

Martin, R., 2005. Tree Kangaroos of Australia and New Guinea. CSIRO, Victoria.

Mayr, E., Diamond, J., 2001. The Birds of Northern Melanesia: Speciation, Ecology, and Biogeography. Oxford University Press.

Microsoft AZURE, 2020. Available from: https://azure.microsoft.com/. (2 June 2020).

Mueller, J.P., Massaron, L., 2016. Machine Learning for Dummies. Wiley & Sons, New York.

National Center for Ecological Analysis and Synthesis (NCEAS), 2020. Morpho. Available from: https://github.com/NCEAS/morpho (20 July 2020).

National Science Foundation, 2020. The Antarctic Treaty. Available from: https://www.nsf.gov/geo/opp/antarct/anttrty.jsp (2 May 2020).

Nemitz, D., Huettmann, F., Spehn, E.M., Dickoré, W.B., 2012. Mining the Himalayan uplands plant database for a conservation baseline using the public GMBA webportal. In: - Huettmann, F. (Ed.), Protection of the Three Poles. Springer, Tokyo, pp. 135–158.

Ohse, B., Huettmann, F., Ickert-Bond, S., Juday, G., 2009. Modeling the distribution of white spruce (*Picea glauca*) for Alaska with high accuracy: an open access role-model for predicting tree species in last remaining wilderness areas. Polar Biol. 32, 1717–1724.

Porolak, G.B., 2008. Home Range of the Huon Tree Kangaroo, *Dendrolagus matschiei*, in Cloud Forest on the Huon Peninsula, Papua New Guinea. (Unpublished M.Sc. thesis)James Cook University, Australia. Available from: https://researchonline.jcu.edu.au/29818/1/29818_Porolak_2008_thesis.pdf.

Porolak, G., Dabek, L., Krockenberger, A.K., 2014. Spatial requirements of free-ranging Huon tree kangaroos, *Dendrolagus matschiei* (Macropodidae), in upper montane forest. PloS One. 9(3), e91870https://doi.org/10.1371/journal.pone.0091870.

Rauzon, M., 2016. Isles of Amnesia: The History, Geography, and Restoration of America's Forgotten Pacific Islands. Latitude 20 Publishers, Honolulu, HI.

Regmi, G.R., Huettmann, F. (Eds.), 2020. Hindu Kush-Himalaya Watersheds Downhill: Landscape Ecology and Conservation Perspectives. Springer Dordrecht, Holland.

Revkin, A., 2004. The Burning Season: The Murder of Chico Mendes and the Fight for the Amazon Rain Forest. Island Press, New York.

Rosales, J., 2008. Economic growth, climate change, biodiversity loss: distributive justice for the global north and south. Conserv. Biol. 22, 1409–1417. https://doi.org/10.1111/j.1523-1739.2008.01091.x.

Schneider, D., 2009. Quantitative Ecology: Measurements, Models and Scaling. Academic Press, Elsevier, New York.

Silvy, N.J., 2012. The Wildlife Techniques Manual—Management, 2 volumesseventh ed. The Johns Hopkins University Press, Baltimore, MD 414 pp.

Sriram, S., Huettmann, F., 2017. A Global Model of Predicted Peregrine Falcon (*Falco peregrinus*) Distribution With Open Source GIS Code and 104 Open Access Layers for Use by the Global Public. (unpublished) Available from: https://doi.org/10.5194/essd-2016-65 https://www.earth-syst-sci-data-discuss.net/essd-2016-65/ (29 May 2020).

Stone, C., Berryman, J., 2012. A Squatter Went to Sea: The Story of Sir William Macleay's New Guinea Expedition (1875) and his Life in Sydney (1957). The Robert Menzies Collection: A Living Library. Sydney, Australia.

Taylor, P.M., 2016. Assembling, Assessing and Annotating the Source Materials for the Study of the 1926 Expedition. Available from: https://www.researchgate.net/

publication/331976682_Assembling_Assessing_and_Annotating_the_Source_Materials_for_the_Study_of_the_1926_Expedition.

USGS, 2020. Metadata Wizard 2.0. https://www.usgs.gov/software/metadata-wizard-20 (20th July 2020).

Webb, V.-L., 1995. Photographs of Papua New Guinea: American Expeditions 1928-29. Pacific Artspp. 72–81.

West, P., 2006. Conservation Is Our Government Now: The Politics of Ecology in Papua New Guinea. Duke University Press, Durham, NC.

Wilkinson, M., 2013. Testing the null hypothesis: the forgotten legacy of Karl Popper? J. Sports Sci. 31, 919–920.

Zhang, L., Huettmann, F., Liu, C., Sun, P., Yu, Z., Zhang, X., Mi, C., 2019a. The use of classification and regression algorithms using the random forests method with presence-only data to model species' distribution. MethodsX 6, 2281–2292. Available from: https://www.sciencedirect.com/science/article/pii/S2215016119302596.

Zhang, L., Huettmann, F., Liu, C., Sun, P., Yu, Z., Zhang, X., Mi, C., 2019b. Classification and regression with random forests as a standard method for presence-only data SDMs: a future conservation example using China tree species. Ecol. Inform. 52, 46–56. https://doi.org/10.1016/j.ecoinf.2019.05.003.

Ziembicki, M., Porolak, G., 2016. *Dendrolagus matschiei*. The IUCN Red List of Threatened Species 2016: e. T6433A21956650. Available from: https://doi.org/10.2305/IUCN.UK.2016-2.RLTS.T6433A21956650.en (28 June 2019).

Zuckerberg, B., Huettmann, F., Friar, J., 2011. Proper data management as a scientific foundation for reliable species distribution modeling. In: Drew, C.A., Wiersma, Y., Huettmann, F. (Eds.), Predictive Species and Habitat Modeling in Landscape Ecology. Springer, New York, pp. 45–70 (Chapter 3).

FURTHER READING

Menzies, J.I., 1991. Handbook of New Guinea Marsupials and Monotremes. Kristen Press, Inc., Madang

CHAPTER

25

Veterinary Techniques for the Assessment of Health in Wild Tree Kangaroos

Erika (Travis) Crook[a], *Carol Esson*[b], *and Patricia Watson*[c]

[a]Utah's Hogle Zoo, Salt Lake City, UT, United States

[b]Ulysses Veterinary Clinic, Stratford, QLD, Australia

[c]Redmond Fall City Animal Hospital, Redmond, WA, United States

OUTLINE

INTRODUCTION

In order to conserve the endangered Matschie's tree kangaroo (*Dendrolagus matschiei*) in Papua New Guinea (PNG), an understanding of their basic biology, health parameters, and physical requirements is imperative. The local people of Papua New Guinea have been living alongside wild tree kangaroos for centuries and are able to provide valuable insights into tree kangaroo natural history as well as the endemic tree kangaroo population and health trends. Conducting scientific health assessments on the tree kangaroos complements this local traditional knowledge. Evaluating wild animals in their native habitat also provides critical information that benefits the species living in managed care (e.g., zoos, sanctuaries). Ideally the environment provided in managed care would thus simulate the habitat, vegetation, diet, and thermal gradients observed in the wild.

Tree Kangaroos
https://doi.org/10.1016/B978-0-12-814675-0.00001-4

Matschie's tree kangaroos are found only on the Huon Peninsula of PNG in the montane rainforest from 1000 to 3300 m above sea level (Ziembicki and Porolak, 2016; Chapter 1). According to the IUCN Red List (Ziembicki and Porolak, 2016) it is estimated that the wild population on the Huon Peninsula consists of approximately 2500 mature individuals. More accurate estimates of total population are being studied currently (Dabek, personal communication). The Huon Peninsula has an average annual rainfall of approximately 2.5 m with air temperatures ranging from 5°C to 30°C (Porolak et al., 2014). The tree kangaroos spend 90% of their time in the montane forest at an average height of 18–20 m where the canopy height averages 28–30 m. The tree kangaroos spend 10% of the time on the ground per studies at one field site (Porolak et al., 2014). Habitat use and behavior is an area of continued study (Chapter 23).

Health assessments of wild Matschie's tree kangaroos on the Huon Peninsula started in 2004 as part of the research initiative of the Tree Kangaroo Conservation Program (TKCP) (Chapters 13, 16, 22, 23). During field studies, the animals are caught to have radio or GPS collars placed, exchanged, or removed for studies on home range and activity. A veterinarian is always part of the research team to ensure proper handling and animal welfare for the animals. An added benefit is that in conjunction with the radiotelemetry studies at the time of capture, biological samples can be collected and a health assessment can be performed.

TREE KANGAROO HEALTH ASSESSMENTS

Capture Methods

Over the years, field techniques for capturing, anesthetizing, and examining tree kangaroos have evolved to be as stress free as possible while still allowing the research team to gather vital information on home ranges and forest usage. The tree kangaroo field research team consists of international researchers, veterinarian, sometimes a veterinarian technician and zookeeper, local field assistants, and traditional hunters. The field team also consists of local conservation rangers, guides, and villagers that help transport equipment and food to a temporary field camp within the forest where tree kangaroos are found (Fig. 25.1). The field work usually involves hiking for 1–2 days from a village in order to reach the remote forest location. A working space of a table and benches are made using branches, vines and ferns from the forest and tarps are set up to protect from the daily rain (Fig. 25.2).

Local assistants, hunters, and trackers divide into teams and search in the forest for the elusive tree kangaroos (Fig. 25.3). Weather conditions are taken into account prior to searching for tree kangaroos. The clouded forest is prone to large periods of rain especially in the afternoons. It is too dangerous to catch animals in heavy rain, both for the people and the animals as wet trees increase the risk of falling for the tree kangaroos. Time of day is also taken into account so that there is plenty of time to release the animal on the day of capture. If no animal is found by 2:00 p.m., the search is stopped and resumed the next day.

Tree kangaroos have two main types of defense and escape, which are to go up higher into the canopy to get away and hide, or to leap down to the ground and hop/run away. Early on, hunting dogs were used to try to tree the animals for capture. This technique was not successful as the tree kangaroos either escaped or the dogs caught them on the ground and harmed them. The current capture method is to clear vegetation from around the base of the tree where a tree kangaroo has been spotted. Then one or two hunters climb adjacent trees and encourage the animal to jump to the ground (Fig. 25.4).

The tree kangaroo may stretch out its appendages to create some drag that helps to slow its descent. When it lands on the ground it takes a few seconds to recover before moving, which allows the hunter to catch the tree kangaroo at the base of the tail and to place it in a burlap sack

FIGURE 25.1 Example of a field camp at Wasaunon, located in the Sarawaget Ranges on the north coast of the Huon Peninsula, Papua New Guinea. Researcher tent area as well as a covered veterinary exam area in the background. *Source: TKCP.*

(Fig. 25.5). This method proves safe and is the preferred method by the hunters.

Depending on the work that needs to be conducted on the tree kangaroo, there are several options for what happens next. If the main goal is to change or remove a radio-collar, then this procedure can be performed at the site of capture without sedation if the animal is relatively calm. The body of the animal remains in the burlap sack while allowing access to the animal's head. Covering the animal's eyes during manual restraint can be calming. If more extensive health assessments are planned and the field veterinarian has accompanied the capture team, then procedures can be done immediately at the site of capture. If the procedures are to be done at the camp, then the trackers will hike back to camp with the tree kangaroo in a burlap sack. Marsupials are prone to capture myopathy which can result in severe muscle damage from extreme exertion or stress (Vogelnest, 2015). Therefore myopathy has to be taken into consideration when determining the methods used for health assessment and radio-collar placement. The time taken to capture the animal and subsequent handling has to be monitored closely and kept to a minimum to reduce stress.

Sedation Techniques

Prior to delivering a sedative, the tree kangaroo is weighed in the burlap sack using a handheld spring scale (Fig. 25.6).

FIGURE 25.2 Veterinary equipment and sample table. The table is from natural forest products while the tarp is to protect from the daily rain. *Source: TKCP.*

In the initial field seasons, to reduce stress on the tree kangaroos while collecting biological samples, an inhalant anesthetic agent, isoflurane, was administered to the tree kangaroos via a facemask. This system used ambient air (instead of oxygen) to deliver the isoflurane gas. Because tree kangaroos live at high altitude with less oxygen in ambient air, it was determined that an injectable protocol was safer for sedation.

The first injectable protocol was Tiletamine + Zolazepam (Telazol® or Zoletil®) 1.7–2.1 mg/kg administered intramuscularly into the quadriceps muscle through the burlap sack. This resulted in heavy sedation. One individual required a supplement of Ketamine (Ketaset®) 2.5 mg/kg intramuscularly. The authors modified the protocol for the subsequent field season to Tiletamine + Zolazepam (1.55 mg/kg) plus Ketamine (1.2 mg/kg) in order to administer

FIGURE 25.3 Trackers showing Dr. Lisa Dabek a tree kangaroo high up in the forest canopy. *Source: TKCP.*

lower doses of the longer acting Tiletamine +Zolazepam. The protocol has now evolved to a benzodiazepine that exerts anxiolytic, sedative, muscle-relaxant, and amnestic effects. Diazepam (Pamlin) at a dose of 1.0 mg/kg intramuscularly through the burlap sack provides adequate sedation to allow safe handling of the tree kangaroo for radio-collar placement, body measurements, and blood sampling. Using a sedative, instead of a true anesthetic, minimizes the risk even further in this remote region.

Health Assessment, Data, and Sample Collection

The tree kangaroo is kept in the burlap sack after the sedation is administered. After the injection, the tree kangaroos will sit in a hunched

FIGURE 25.4 Local YUS hunter/TKCP field assistant climbing a tree adjacent to a tree kangaroo. The hunter encourages the tree kangaroo to jump down into an area of cleared vegetation where it can be restrained by the base of its tail before being placed in a burlap sack. *Source: TKCP.*

FIGURE 25.5 Captured tree kangaroo placed in a burlap sack. *Source: TKCP.*

FIGURE 25.6 Weighing a tree kangaroo in a burlap sack with a handheld spring scale prior to sedation. *Source: TKCP.*

over position as the sedation takes effect. The tree kangaroo is visually monitored every 10 minutes through the opening of the sack. The level of sedation is determined by assessing the alertness and response of the animal to gentle physical stimulation (Fig. 25.7).

Once the animal is calm enough to allow safe handling, usually within 10 minutes of the administration of the sedative, the animal is removed from the sack and placed on the exam table. Physiological parameters of heart rate, respiration rate, and body temperature are measured every 15 minutes while the animal is sedated. All animals have had voluntary respirations without respiratory depression. As the sedation is performed at a high elevation area, ambient temperature can drop below 15°C, it is therefore imperative to monitor the animal's body temperature as sedation can impair the animal's ability to thermoregulate. Hot water bottles are placed alongside the sedated animal to aid in supporting body temperature.

After the initial physiological measurements have been performed to ensure the animal is stable, the tree kangaroo is scanned for a microchip (transponder) in case it had been previously captured. If no microchip is found, one is inserted subcutaneously between the shoulder blades. Photographs of the animal's unique facial markings are taken as an additional means of identification; Chapter 2). A full physical examination is then performed (Fig. 25.8).

The body weight of the animal is complemented with a body condition score assessment determined by palpation of the muscle mass over the scapula and hips. A body condition score from 1 to 9 is assigned using the husbandry guide developed by the Association of Zoos and Aquariums Tree Kangaroo Species Survival Plan® (AZA TKSSP, 2017; Chapter 20). The eyes are

FIGURE 25.7 Field veterinarian assessing the level of sedation and auscultating the sedated tree kangaroo prior to removal from the burlap sack. *Source: TKCP.*

FIGURE 25.8 Local trackers, hunters, field assistants, and rangers watching the field veterinarian work on a sedated tree kangaroo. *Source: TKCP.*

examined for injury or other defects such as cataracts, and the ears for signs of wounds or external parasites. The oral cavity and teeth are examined and used for age determination depending on the amount of wear on the premolars and molars. The pelage is examined for rub marks or fur loss. The skin is examined for irritations, wounds, and external parasites (ectoparasites). Any ectoparasites found are collected and stored in 70% ethanol. A pouch check is also performed and pouch joeys have been seen on several occasions (Fig. 25.9).

Morphometric measurements consist of head length from occipital crest to nose tip, head width across the zygomatic arch, neck circumference for preparation of the radio-collar, crown-rump length from the occipital crest to the tail base, and tail length from tail base to the tail tip (Fig. 25.10).

Hind limb measurements include hind foot length from heel to tip of foot pad, width across the section near the claws, and the narrowest width near the heel. The hind leg is measured from the ankle to the knee, and from the knee to the hip. The forelimb measurements include the point of the shoulder to the elbow, elbow to the wrist, and front paw from the wrist to the beginning of digits, and width across the widest part of the paw. These morphometric measurement data are being analyzed and prepared for a future publication by one of this chapter's co-authors (C. Esson).

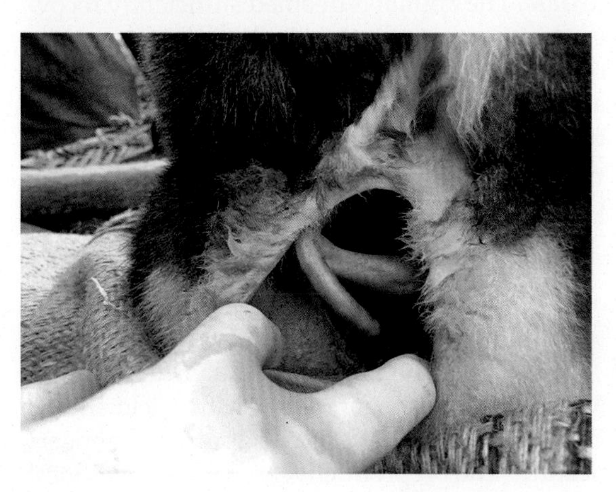

FIGURE 25.9 Matschie's tree kangaroo joey in the mother's pouch. *Source: Rob Liddell.*

Blood samples are collected from the ventral tail vein at the base of the tail using a 22 g needle and 6 ml syringe. Blood is placed into 2 ml red top tubes with clot activator for serum (Interpath services, Australia), whole blood into 2 ml EDTA purple top tubes (Interpath services, Australia), and 1 ml lithium heparin green top

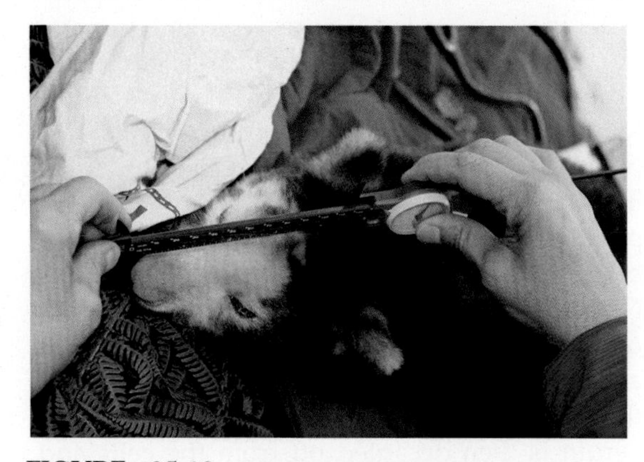

FIGURE 25.10 Morphometric measurements being taken on a sedated tree kangaroo. *Source: Doug Bonham.*

FIGURE 25.11 Radio-collared tree kangaroo recovering smoothly from the sedative. Being monitored during the recovery to determine when to return to the forest. *Source: Rob Liddell.*

tubes (Zebravet, Australia). A hand-held Abaxis i-STAT analyzer (Abaxis, Inc., Union City, CA USA, and REM Systems Pty, Ltd., North Ryde, Australia) is used in the field to analyze whole blood samples to provide immediate chemistry and blood gas values. Blood smears are made, air dried, and then fixed in Diff Kwik Fixative (Fronine Laboratory Supplies, Riverstone, NSW Australia) for future laboratory microscopic evaluation looking at red blood cell morphology, white blood cell count estimate, and differentials. The presence/absence of hemoparasites is also recorded. To preserve genetic material, whole blood is stored in EDTA, ethanol, a drop onto FTA cards, and a drop onto 903 Protein Saver cards (GE Healthcare, Piscataway, NJ, USA, and UK Limited, Little Chalfont, Buckinghamshire, UK) and stored at ambient temperature (Smith and Burgoyne, 2004). Whole blood is also placed on Advantec Nobuto strips (Toyo Roshi Kaisha, Ltd., Tokyo, Japan; distributor Advantec MFS Inc.) for future pathogen analyses. The serum tubes stand upright until the serum separates from the red blood cells. Serum is then decanted off from the cells and stored in sterile cryovials, double bagged and kept cool in the river at camp. Serum can be kept at 4–8°C for up to 7 days before loss of protein structure occurs (Chase et al., 2012). Since field-trips have a longer duration than a week, Nobuto strips are also used as a safeguard for the preservation of whole blood and serum. Storage methods will be compared when the samples are analyzed for antibodies to several pathogens. These pathogens will include *Toxoplasma gondii*, *Coxiella burnetii*, *Leptospira* spp. and *Mycoplasma* spp. If feces are present in the holding burlap sack, they are preserved in ethanol for future genetic, parasite, or feeding ecology studies. Feces are also placed in paper envelopes with silica gel for drying for DNA samples. Hair samples are plucked from the base of the tail and abdomen. Hair is preserved in ethanol for further genetic studies and also kept in paper envelopes for future cortisol studies.

During one of the most recent field trips, radiographs and ultrasound were performed on the wild tree kangaroos by Dr. Robert Liddell (Chapter 15). Advancements in solar powered equipment allowed these diagnostic modalities to be used in the field.

Recovery and Release

At the end of the sampling, the tree kangaroos are administered a non-steroidal anti-inflammatory (meloxicam, 0.2 mg/kg subcutaneously) and vitamin E (16 iu/kg intramuscularly) as a precaution for capture myopathy. During some field seasons, subcutaneous fluids of 50 ml sodium chloride has been administered to support the cardiovascular health of the animals under longer acting sedatives. The tree kangaroo is replaced into the burlap sack and kept warm with hot water bottles during recovery from sedation. Recovery takes approximately four hours when Telazol or Telazol +Ketamine is used. Recovery takes approximately two hours when Diazepam/Pamlin is used. Benzodiazepines (Diazepam/Pamlin) can be antagonized (reversed) with flumazenil given as an intramuscular or intravenous injection but has not been necessary in field work to date. The animal is deemed sufficiently recovered from the sedation and ready for release once it is alert and has regained its reflexes and strength (Fig. 25.11). The animal is returned to the tree where it was caught, and released while being observed to ensure it is able to climb the tree competently (Fig. 25.12).

RESULTS

From our forays into the mountain homes of the Matschie's tree kangaroos at various field sites in Papua New Guinea, the population appears healthy and reproductively active. The local people report that the populations seem to be increasing due to the protected areas established

FIGURE 25.12 Releasing a newly radio-collared tree kangaroo back to the same tree where it was captured. *Source: TKCP.*

by the Tree Kangaroo Conservation Program (Chapter 10). Health assessment data from one field season has been published (Travis et al., 2012). During that June 2007 field season in Wasaunon, the following physiologic values were obtained from sedated tree kangaroos (Table 25.1): mean cloacal temperature was 36.9 ± 0.8°C, mean heartbeat per minute was 131 ± 17, and mean respiration per minute was 80 ± 36. The mean adult male body weight (kg) was 6.7 kg ± 1.0 and the mean adult female body weight was 7.3 kg ± 0.3.

During all field seasons, the general body condition score for the wild tree kangaroos was 4/9, as there is no obvious accumulation of body fat, and the ribs, hips, and spine are somewhat palpable covered by lean strong muscles. That score per the SSP guide is "slightly under-conditioned," but in a field setting that is considered a normal and expected body condition score. Oral cavity evaluations revealed teeth with patchy dark staining on the enamel but no gingivitis or tartar. The dark color is likely from the tannins in the foods ingested.

External parasites have been observed on the wild tree kangaroos. *Trombiculidae* mites in the inguinal and scrotal area and *Ixodidae* family ticks on the ear margins or under the chin have been documented. No ticks were found during initial field studies but seem to be increasing in number in the last four years. However, the number of ticks collected is small therefore the apparent increase cannot be validated statistically.

Data on heavy metal, trace nutrient, electrolyte, vitamin, and mineral analysis from one June 2007 field season of wild and captive Matschie's tree kangaroos are presented in Tables 25.2 and 25.3. During that same field season, cloacal samples collected with a synthetic tip swab and preserved in sterile saline were all negative for *Mycobacterium avium* via real-time polymerase chain reaction (PCR) (Travis et al., 2012). During that same season, some free-ranging and captive individuals showed evidence of endoparasites; positive fecal exams revealed *Eimeria* spp. oocysts and strongyle spp. type ova (Travis et al., 2012).

TABLE 25.1 Physiological values from sedated wild Matschie's tree kangaroos (*Dendrolagus matschiei*) from Wasaunon field site, Papua New Guinea, June 2007 (Travis et al., 2012).

Parameter	Mean ± SD[a]	Range	*n*
Cloacal temperature (°C)[b]	36.9 ± 0.8	35.8–37.7	6
Heartbeats per minute[b]	131 ± 17	112–160	6
Respirations per minute[b]	80 ± 36	40–130	6
Adult male body weight (kg)	6.7 ± 1.0	6.0–8.2	4
Adult female body weight (kg)	7.3 ± 0.3	7.0–7.5	3[c]
Environmental temperature (°C)	13.7 ± 0.9	12.8–15.0	5

[a] *SD, standard deviation.*
[b] *These values are derived from the first data points recorded once anesthetized.*
[c] *Two of the female weights were collected on non sedated individuals.*

TABLE 25.2 Vitamin, mineral, trace nutrient, and electrolyte analysis on serum from Papua New Guinea wild and captive Matschie's tree kangaroos (*Dendrolagus matschiei*), June 2007 (Travis et al., 2012).

	Wild Tree Kangaroo[a]			Captive Tree Kangaroo[b]			
Parameter	Mean ± SD[d]	Range	*n*	Mean ± SD	Range	*n*	*P*-value[c]
Vitamin E (μg/mL)	8.5 ± 1.7	6.7–10.3	4	4.8 ± 0.9	3.8–6.3	5	0.014
Vitamin A (ng/mL)	123.8 ± 37.7	91.0–178.0	4	155.0 ± 47.3	102.0–225.0	5	
Beta carotene (μg/mL)	2.7 ± 0.1	2.6–2.7	4	2.8 ± 0.1	2.7–2.9	5	0.014
Bicarbonate (mmol/L)	14.0 ± 4.2	11.0–17.0	2	8.7 ± 1.2	8.0–10.0	3	
Calcium (mg/dL)	8.35 ± 0.07	8.3–8.4	2	8.77 ± 0.12	8.7–8.9	3	
Phosphorus (mg/dL)[e]	6.6 ± 2.1	5.1–8.0	2	6.0 ± 2.7	3.6–9.0	4	
Iron (μg/dL)	160.5 ± 38.9	133.0–188.0	2	181.7 ± 61.1	142–252	3	
Magnesium (mg/dL)	3.6 ± 0.7	3.1–4.1	2	2.8 ± 0.2	2.6–3.0	4	
Sodium (mmol/L)	132.5 ± 2.1	131–134	2	134.5 ± 1.7	132–136	4	
Potassium (mmol/L)[e]	6.4 ± 0.2	6.2–6.5	2	6.9 ± 1.3	5.0–7.8	4	
Chloride (mmol/L)	97.0 ± 1.4	96.0–98.0	2	93.8 ± 2.6	90.0–96.0	4	
Manganese (ng/mL)	1.4 ± 0.7	1.0–2.5	4	1.7 ± 0.8	0.6–2.6	4	
Cobalt (ng/mL)	0.4 ± 0.2	0.3–0.7	4	0.6 ± 0.3	0.2–0.9	4	
Copper (μg/mL)	0.1 ± 0.02	0.08–0.12	4	0.16 ± 0.05	0.12–0.23	4	0.021
Selenium (ng/mL)	7.6 ± 2.8	4.0–10.0	4	12.1 ± 1.6	10.0–13.0	4	0.043
Molybdenum (ng/mL)	1.1 ± 0.6	0.4–1.8	4	21.5 ± 3.5	16.5–24.6	4	0.021

[a] *Wasaunon Field Site.*
[b] *Rainforest Habitat, University of Technology, Lae.*
[c] *P-value listed if a significant difference* <0.05 *was detected.*
[d] *SD, standard deviation.*
[e] *Values may have been affected by sample handling.*

TABLE 25.3 Heavy metal analysis on whole blood from Papua New Guinea wild and captive Matschie's tree kangaroos (*Dendrolagus matschiei*), June 2007 (Travis et al., 2012).

	Wild Tree Kangaroos[a]			Captive Tree Kangaroos[b]			
Parameter (ng/ml)	**Mean ± SD[d]**	**Range**	***n***	**Mean ± SD**	**Range**	***n***	***P*-value[c]**
Selenium	34.7 ± 15.6	18.0–61.0	6	115.1 ± 33.1	78.0–155.0	5	0.006
Lead[e]	1.3 ± 0.4	<1–2.0	6	4.6 ± 1.5	2.6-6.4	5	0.006
Arsenic[f]	<1	n/a	6	42.6 ± 40.2	5.0-101.0	5	0.003
Cadmium	<1	n/a	6	<1	n/a	5	n/a
Thallium	<1	n/a	6	<1	n/a	5	n/a
Mercury	<5	n/a	6	<5	n/a	5	n/a

[a] *Wasaunon Field Site.*
[b] *Rainforest Habitat, University of Technology, Lae.*
[c] *P-value listed if a significant difference of <0.05 was detected.*
[d] *SD, standard deviation.*
[e] *One wild tree kangaroo had a result of <1 ng/ml which was assigned a value of 0.99 ng/mL for statistical analysis.*
[f] *Data points reported by the laboratory as <1 ng/mL for wild tree kangaroos were assigned a value of 0.99 ng/mL for statistical analysis.*

Data and biological samples from six subsequent field trips (2014–2019) are currently being analyzed. Since juveniles and adults of both sexes have been examined, morphometric data between sex and age classes are being evaluated. This may help elucidate sexual dimorphism if present in addition to provide more data of body measurements for future reference material.

The results of the blood samples measured via the i-STAT machine are being integrated into reference ranges for hematology and biochemistry values. Genetic analysis (DNA) will be performed on blood samples in order to identify relationships among the resident animals. This will increase our knowledge on home range interactions between animals and dispersal of offspring. Blood samples will be examined for several potentially zoonotic pathogens (pathogens that can transmit from animals to humans). As these animals are hunted for food in non-protected areas, it is important for the local people to be made aware of any potentially harmful pathogens the tree kangaroos may be carrying.

Health of the tree kangaroos and impacts of capture and effects of wearing the radio-collars are being investigated. Cortisol, a stress hormone, can be measured in hair samples and can provide information on stress levels before and after capture (Heimbürge et al., 2019). This information will help guide or possible modify future capture and sampling techniques.

Fecal samples contain much information. They continue to be analyzed for endoparasite load and type. Cortisol in fecal samples will be compared with the cortisol hair analyses. DNA for genetic studies extracted from the fecal samples stored in ethanol will provide comparative data to the DNA extracted from the blood samples.

CONCLUSIONS

Baseline reference data for the Matschie's tree kangaroo are being developed from biological samples and health assessments performed in the field. The goal is to develop a greater understanding of how Matschie's tree kangaroos interact with their environment and to provide knowledge to aid in the conservation of this unique species. Currently, the tree kangaroo populations of the Huon Peninsula appear healthy with no threatening disease issues. The populations may even be rebounding in areas that are being conserved. However, continued monitoring of this population is recommended. As population density increases, the risk of pathogen transfer also increases; disease not observed previously may become a concern. Another reason to study the wild tree kangaroos is to provide biological and ecological information that benefits the tree kangaroos living under managed care (e.g., zoos, sanctuaries) (Chapters 20 and 22). The people of Papua New Guinea take great pride in their tree kangaroos and the Tree Kangaroo Conservation Program has assisted in increasing awareness and resources to help keep the forests and the people healthy. This benefits the tree kangaroos and all the species that call the forest their home.

REFERENCES

American Association of Zoos and Aquariums (AZA) Tree Kangaroo Species Survival Plan® (TKSSP), 2017. Tree Kangaroo Body Condition Score Chart. Available from: https://nagonline.net/wp-content/uploads/2017/03/Tree-Kangaroo-BCS-Poster.pdf. [29 June 2020].

Chase, B.A., Johnston, S.A., Legutki, J.B., 2012. Evaluation of biological sample preparation for immunosignature-based diagnostics. Clin. Vaccine Immunol. 19 (3), 352–358.

Heimbürge, S., Kanitz, E., Otten, W., 2019. The use of hair cortisol for the assessment of stress in animals. Gen. Comp. Endocrinol. 270, 10–17.

Porolak, G., Dabek, L., Krockenberger A.K., 2014. Spatial requirements of free-ranging Huon tree kangaroos, *Dendrolagus matschiei* (Macropodidae), in upper montane forest. PLoS One 9(3): e91870. Available from: <https://researchonline.jcu.edu.au/33165/>

Smith, L.M., Burgoyne, L.A., 2004. Collecting, archiving, and processing DNA from wildlife samples using FTA® databasing paper. BMC Ecol. 4, 4. https://doi.org/10.1186/1472-6785-4-4.

Travis, E.K., Watson, P., Dabek, L., 2012. Health assessment of free-ranging and captive Matschie's tree kangaroos (*Dendrolagus matschiei*) in Papua New Guinea. J. Zoo Wildl. Med. 43, 1–9.

Vogelnest, L., 2015. Marsupialia (Marsupials). In: Miller, R.E., Fowler, M.E. (Eds.), Fowler's Zoo and Wild Animal Medicine. Elsevier Saunders, St. Louis, MO, pp. 255–274.

Ziembicki, M., Porolak, G. 2016. *Dendrolagus matschiei*. The IUCN Red List of Threatened Species 2016: e. T6433A21956650. Available from: <https://doi.org/10.2305/IUCN.UK.2016-2.RLTS.T6433A21956650.en> [12 November 2019].

CHAPTER

26

Using Non-Invasive Techniques to Study Tree Kangaroos

Sigrid Heise-Pavlov[a], *Thomas J. McGreevy, Jr.*[b], *and Simon Burchill*[c]

[a]Centre for Rainforest Studies at The School for Field Studies, Yungaburra, QLD, Australia
[b]Department of Natural Resources Science, University of Rhode Island, Kingston, RI, United States
[c]Tree Kangaroo and Mammal Group Inc, Atherton, QLD, Australia

OUTLINE

INTRODUCTION

Effective species conservation requires knowledge of a species' distribution, its habitat requirements (including shelter, access to water and food, and migration), its population dynamics (mortality and population growth) and behaviors (including intra- and interspecific interactions) (Conde et al., 2016; Johnson et al., 2018). Acquisition of this knowledge is often

Tree Kangaroos
https://doi.org/10.1016/B978-0-12-814675-0.00013-0

problematic, particularly when the species inhabits less-accessible habitats, is difficult to find (cryptic), has low population densities, or is hard to keep in captivity to be studied (Bland et al., 2015).

Various methods to gain this knowledge became very popular in the past and are outlined in textbooks on ecological methods (Pearl, 2000). Examples are capture-mark-recapture methods to study home ranges and migratory behavior of species, physiological examinations of anesthetized (darted) individuals, and behavioral ecology studies on captive individuals. These methods have been widely used for the study of tree kangaroos (Coombes, 2005; Martin, 2005; Newell, 1999; Procter-Gray, 1985). Over time it became clear that these methods have impacts on an individual's health and its behaviors and therefore may result in unrepresentative data on a species' ecology, physiology, and behavior (Gompper et al., 2006). Radio collars, when placed on a mammal, can cause impaired movements (Banks et al., 1975), can hinder digging and climbing capabilities (Corner and Pearson, 1972; Moorhouse and MacDonald, 2005), and can cause skin irritation and hair loss (Bartholomew, 1967). However, major advances in GPS and radio telemetry techniques have reduced these shortcomings. Still, the application of radio collars can produce skewed information on a species interactions and home range as the stress associated with the capture and marking of an animal can change its orientation and movement through its habitat (Berteaux et al., 1994; Daly et al., 1992; Harris et al., 1990; White and Garrott, 1990) although modern devices are small and effects on an animal's movement could be considered negligible.

The increased understanding of the limitations of these invasive methods to produce reliable information on a species resulted in the development and application of non-invasive methods over the last decades. This trend in ecological studies was facilitated by advancements in genetics, electronics and micro-technology (Marvin et al., 2016; Niemi and McDonald, 2004). Non-invasive methods allow the collection of data on a species with a minimum or no effect on the studied individuals as the direct contact/interaction between the animals and the human investigator is reduced or completely excluded. Similar to invasive methods, the effect of non-invasive methods on studied animals follows a gradient, since some methods still require a type of animal-human interaction. Potential negative effects of the application of non-invasive methods to a studied species can therefore not be excluded. Thus, the suitability of many such methods for different species still remains to be evaluated and may become obvious only after a long-term use of these methods (Nuske et al., 2014).

This chapter will describe various non-invasive methods that have been applied to the study of tree kangaroos. It attempts to evaluate these methods by considering their advantages and disadvantages in collecting reliable data on the ecology and behavior of tree kangaroo species. Some of these methods are still under development and shall be described here to encourage further refinement and testing. The chapter also will list some potential methods in researching tree kangaroo species which have not been developed yet but might be worth considering.

GENETIC METHODS

The genetic analysis of non-invasively collected samples from wildlife dates back to the early 1990s. The first application was the analysis of deoxyribose nucleic acid (DNA) extracted from Pyrenean brown bear (*Ursus arctos*) hair samples by Taberlet and Bouvet (1992). Shortly after, similar techniques were used to analyze DNA extracted from European brown bear (*U. a. arctos*) fecal samples by Höss et al. (1992). Intestinal cells are sloughed off as the fecal material travels through an animal's digestive tract and become the source of DNA from the animal. In a given mammalian cell, there is DNA from both the nuclear genome and mitochondrial genome. Nuclear DNA (nDNA) is found in the nucleus of a cell, inherited from both parents, and contains the information about all the animal's chromosomes. Mitochondrial DNA (mtDNA) is found in the mitochondrion, represents an intact record of maternally inherited DNA, and typically contains about 20,000 base pairs of information. In a cell there are usually two copies of the nDNA genome, but there can be hundreds to thousands of mitochondria each with a copy of the animal's mitochondrial genome (Birky et al., 1989). Thus, the first genetic studies using fecal samples focused on the analysis of mtDNA because of its higher quantity in a DNA extraction.

The analysis of mtDNA extracted from fecal samples can be used to identify the species, quantify a population's mitochondrial genetic diversity (Waits and Paetkau, 2005), and facilitate phylogeographic analyses to determine how a population's mitochondrial lineages are geographically distributed (Avise et al., 1987). More extensive genetic analyses can be conducted on nuclear DNA including estimating a population's nuclear and immuno-genetic diversity, identifying an individual from fecal samples, and determining the animal's gender and parentage. Additional analyses can determine an estimation of the size and density of a population, estimate the effective population size, test for hybridization between closely related species, characterize population structure, and determine conservation units (Frankham et al., 2004; Waits and Paetkau, 2005). When DNA is extracted from a fecal sample, it also includes DNA from the organisms the animal consumed, bacterial DNA, and parasite DNA. This can allow for the identification of the plants that a herbivore consumed or the prey species that a carnivore consumed (De Barba et al., 2014; Valentini et al., 2009). The analysis of bacterial and parasite DNA allows a researcher to investigate the pathogens an animal has been exposed to and their microbiome (Srivathsan et al., 2016; Zhu et al., 2011).

The use of fecal DNA as a non-invasive method has advantages with regard to the reduction of time, costs and stress that are involved in locating and handling an animal compared to obtaining a tissue sample (Kilpatrick et al., 2013). The use of conservation dogs to locate samples in the field has recently increased and has greatly improved researchers' ability to locate samples, determine the species that created the sample, and determine a species' distribution (Orkin et al., 2016; Smith et al., 2005; Vynne et al., 2011).

Another recently introduced non-invasive DNA-based method is the use of environmental DNA (eDNA) (Bohmann et al., 2014) that includes invertebrate-derived DNA (iDNA), which is genetic material ingested by invertebrates such as leeches, midges, ticks, and mosquitos. The use of iDNA has been tested and successfully resulted in demonstrating the existence of data-deficient cryptic species in inaccessible habitats (Schnell et al., 2012). A refined method in detecting and monitoring vertebrates using iDNA from leeches encourages further studies and lends support to the potential future application of this method in studying tree kangaroos in dense rainforests of Australia and New Guinea (Schnell et al., 2018).

Regardless of the sources of DNA material obtained, genetic methods are limited by possible contamination and the aging of collected material that affect the purity of the extracted DNA material from collected samples. Non-invasively collected samples are often scarce and the extracted DNA is typically of low quality and quantity, resulting in an increased risk of genotyping errors leading to false identification of individuals (Piggott and Taylor, 2003). Furthermore, sample age and season of collection of fecal material can negatively affect amplification rates and reliability of microsatellite genotypes amplified from fecal DNA (Piggott, 2004). Thus, sampling fresh material from which DNA can be extracted may be a limiting factor in the application of genetic methods to tree kangaroo species that live in a highly humid environment in which DNA material rapidly decays. The method to preserve the samples during collection and storage, the DNA extraction technique, and reducing the time from collection to extraction all would need to be optimized to successfully use nDNA markers on DNA extracted from fecal samples.

Another drawback for the application of genetic methods is that extracted DNA is often fragmented in small pieces. However, the development of next-generation sequencing technologies has greatly increased one's ability to develop nDNA markers that amplify small pieces of DNA (e.g., single nucleotide polymorphisms [SNP] markers). SNP markers could readily be developed for tree kangaroos to greatly increase the amount of genetic information obtained from non-invasively collected samples and address the remaining evolutionary and ecological questions.

The application of genetic methods in the analysis of non-invasively collected material from tree kangaroos has increased our knowledge on the genetic diversity of the species and their phylogenetic relationships, provided a more reliable species identification in areas with sympatric species, and contributed to a better understanding of the distribution of tree kangaroo species. One of the first studies that applied genetic methods to collected fecal samples from tree kangaroos was done by Bowyer et al. (2002) on the genetic diversity of the Lumholtz's tree kangaroo (*Dendrolagus lumholtzi*) (LTK). This analysis revealed relatively low levels of genetic diversity in this species, which has been attributed to the Pleistocene climatic fluctuation that resulted in large-scale rainforest contractions that have imposed an ancient population bottleneck on the ancestral *D. lumholtzi* population. Further genetic studies by Bowyer et al. (2003) looked at the phylogenetic relationships between the two Australian species of tree kangaroos—Lumholtz's tree kangaroo and Bennett's tree kangaroo (*D. bennettianus*) and the Australian and New Guinea tree kangaroo species using mtDNA.

After McGreevy Jr. et al. (2009, 2010a,b, 2011, 2012, 2016) refined the applied genetic methods, more questions on genetic diversity and phylogenetic relationships of New Guinea tree kangaroos could be answered. Microsatellite markers were developed and tested for their efficacy using both tissue and fecal samples from Matschie's tree kangaroos (*Dendrolagus matschiei*) maintained in zoos (McGreevy Jr. et al., 2010b). To expedite the species identification in the wild based on collected tree kangaroo fecal samples from the Huon Peninsula, McGreevy Jr. et al. (2010a) developed a polymerase chain reaction (PCR) method to distinguish fecal DNA from Matschie's tree kangaroos from two other sympatric species (the New Guinea pademelon [*Thylogale browni*] and the small dorcopsis [*Dorcopsulus vanheurni*]). The number of tree kangaroo taxa that had DNA extracted from fecal samples was expanded by McGreevy Jr. et al. (2012) and provided additional insight into the taxonomy of the genus.

The most comprehensive phylogeny of tree kangaroos was recently conducted by Eldridge et al. (2018). Using mtDNA and nDNA, this study confirms the presence of a paraphyletic ancestral long-footed and a derived monophyletic short-footed group, and also identifies six major lineages (Chapters 1, 2). Despite these recent applications of DNA methods to extant tree kangaroo species, the species versus subspecies status of numerous taxa within the groups still remains unclear. The analysis of non-invasively collected samples could potentially identify putatively new taxa, which would then need confirmation through the more traditional analysis of voucher specimens (Pleijel et al., 2008; Yates, 1985). The continued molecular genetic analysis of tree kangaroos also could shed light on the process of adaptive radiation and speciation.

DNA extraction methods and storage solutions have improved over the past decade and population estimates based on analyzing DNA from fecal samples are now more common. Genetic methods may therefore enable us in the future to better assess population structures and diet of tree kangaroos. This will ultimately facilitate the planning of conservation actions and can determine the efficacy of ongoing conservation efforts, such as the Tree Kangaroo Conservation Program (Chapter 16; TKCP, 2019).

NON-INVASIVE ECOLOGICAL FIELD METHODS BASED ON TREE KANGAROO SIGNS

The use of physical signs, left behind by animals, can be a cost efficient and reliable non-invasive method to study the distribution, habitat use, and behavior of vertebrates (Triggs, 2004). Tree kangaroos leave behind feces, scratch marks, and hairs. Being folivores, tree kangaroos consume a large amount of foliage but also utilize other plant material, such as ferns, lichens, mosses, flowers, buds, and fruits (Anton, 2014; Coombes, 2005; Martin, 2005; Chapter 2). This results in high defecation rates, despite their low metabolic rates. Scratch marks are mainly left behind by tree kangaroos when they descend from a tree using their strong claws to maintain contact with the tree when gliding down in a posterior position (Martin, 2005). Scratch marks from tree kangaroos consist of three parallel 6–10-cm-long linear imprints on the bark of trees. Hairs of tree kangaroos are mainly found in and on the surface of feces resulting from the swallowing of hairs during intense grooming and subsequent ingestion. While feces (sometimes in conjunction with hairs) have been used for studies on diet and habitat use of tree kangaroos, scratch marks have been included in studies on habitat use and distribution of tree kangaroos.

The Use of Feces in Ecological Research on Tree Kangaroos

The identification of incompletely digested food particles in feces presents a non-invasive technique that has been used in studies of many mammal species, specifically in herbivores (Davis et al., 2015; Dickman, 1995; Reynolds and Aebischer, 1991). Leaf and stem epidermis and cuticles remain often undigested within or on the surface of the feces and can be extracted and identified based on unique cuticular pattern of material from different plant species (Storr, 1961). The technique allows qualitative assessments of the composition of the diet. For example, Jones (2000) proved that feces from captive Matschie's tree kangaroos contain fragments from all plant species the individuals were feeding from. Based on this assessment, Coombes (2005) used this technique to analyze the diet of Lumholtz's tree kangaroos (LTK) in the wild and to detect potential preferences of individuals for certain plant species. The presence of cuticles from petioles and young leaves from some species confirms the observed preference for young plant material by LTKs. Information about the preferred plant species is valuable for revegetation and conservation planning with focus on tree kangaroos.

The technique of using indigested plant material from feces to identify consumed plants is time consuming and prone to errors in identification when plant material can't be extracted appropriately. Coombes (2005) recommends using this technique to confirm observations of the consumption of plant species by animals in the wild. The technique is not suitable to make assessments on the proportion of plant species consumed (Ellis et al., 2014) as differences in the digestion of plant material lead to mispresentation of material from different plant species in feces. Furthermore, retention times of plant material in feces differ between zero and 13 days (Jones, 2000) making the collection of fresh feces necessary. This is problematic given the humid environment of tree kangaroo habitats.

The presence of feces within potential tree kangaroo habitat has been used to obtain information on a species' distribution and habitat use (Fig. 26.1). Coombes (2005) used fecal counts in conjunction with home range studies on radio-collared individual LTKs to examine the intensity of habitat use outside of individuals' home ranges. A positive relationship between the number of feces and the occurrence of some tree species indicated variations in the preference of some individual tree kangaroos for certain food plants.

Similarly, Lofty (2017) used the number of feces within 1 m radius around the base of 30 trees at various spots with different undergrowth densities in rainforests and restoration sites on the Atherton Tablelands in Northeastern Australia to determine whether habitat use by LTKs was affected by undergrowth density. A general linear model showed that none of the measured variables of undergrowth density influenced the number of feces at these spots, suggesting that LTKs are able to use habitat with different undergrowth densities. This supports the high habitat versatility of this tree kangaroo species.

The presence of feces from tree kangaroos in various habitats can be used to assist in constructing the distributional range of the species.

FIGURE 26.1 Fresh Lumholtz's tree kangaroo feces (measurements: 2.5 cm × 1.5 cm). *Source: S. Heise-Pavlov.*

Betz (2001) conducted distance transect sampling on the Huon Peninsula in New Guinea to estimate the density of Matschie's tree kangaroos using fecal pellets, which was an important first step in measuring density. Unfortunately there was a high rate of misidentification of dung with forest wallaby (Dorcopsulus vanheurni) and New Guinea pademelon (THylogale brownii). The quality of the nDNA of nDNA extracted from the fecal samples was too poor for reliable individual identification.

Attempts to use fecal counts to assess the density of tree kangaroos were conducted for LTKs on the Atherton Tablelands in Australia. A comparison between abundance estimates based on fecal counts with those derived from spotlighting showed no close relationship (Kanowski, personal communication, 2017). Problems with detection of feces, their aging, and variations in defecation rates contribute to the difficulty of using feces for abundance estimates in tree kangaroos.

The probability of detection of feces is based on key parameters such as fecal pellet production, detection probability, and decay rate. Production rates of feces are difficult to estimate in wild populations and are likely to vary with gender, age, and reproductive stage of the producer, but also with season. A highly variable defecation rate between 60 and 127 feces within 24 hours is reported for captive LTKs by Heise-Pavlov and Meade (2012). Phillips (personal communication, 2018) could not find a significant difference in defecation rates between male and female captive LTKs but confirmed the high variation in the production of feces in this species with between 29 and 134 feces per day. A significant decrease in both consumption rates and defecation rates during the warmer months was found in captive LTKs (Tazawa, 2012).

Dense undergrowth in rainforests and the activity of coprophagous invertebrates reduce the detectability of feces, which can result in the negation of the presence of the species in an area (i.e., false negative error) (Royle and Link, 2006). The activity of dung beetles (*Coleoptera, Scarabaeidae*) can profoundly affect detection rates. An up to 100% disappearance of fresh feces from LTK habitat was reported by Heise-Pavlov and Meade (2012) for an upland rainforest in Northeastern Queensland, Australia, where up to 50 dung beetles were captured in pitfalls over two nights. Vernes et al. (2005) describe high dung beetle species richness for wet sclerophyll habitats in the Wet Tropics of Australia. Moist conditions support dung beetle activity (Grimbacher et al., 2006; Hill, 1993). Consumption rates for feces vary depending on seasons with highest dung beetle activities during tropical wet seasons making this season unsuitable for methods that depend on the detection of fecal material.

Seasons also affect decay rates of fecal material. Experiments on decay rates of feces from LTKs in Northern Australia estimated that 50% of deposited material decays within 5.1 and 7.2 days (Phillips, personal communication, 2018). However, in an experiment that Coombes (2005) conducted during the dry season from August to September (without any rain) within the Wet Tropics of Australia, feces took up to 2 months to decay. Detailed studies on decomposition of feces from LTKs under different climatic conditions were done by Heise-Pavlov and Meade (2012) and revealed that the appearance of feces (size, shape) changes considerably depending on the prevailing weather conditions and may result in misidentification, particular with feces from sympatric species such as pademelons (*Thylogale stigmatica*) and brushtail possums (*Trichosurus vulpecula*). To avoid misidentification, only fresh feces should be used.

The Use of Hairs in Ecological Research on Tree Kangaroos

Feces found in tree kangaroo habitat can belong to sympatric species. For the LTK, pademelons, black wallabies (*Wallabia bicolor*) and possums (e.g., *Trichosurus vulpecula*) can inhabit the same habitat and produce feces similar in size and shape to those from LTKs. This can make the species identification of feces difficult. An attempt was therefore made to include hairs in the identification process to increase the reliability of information on the origin of collected feces (Cammisa, 2013; Evans, 2012; Richardson, 2012). Hairs were extracted using different methods such as dissection of feces under dry conditions, in ethanol, under heat in a drying oven, or using static devices (Evans, 2012; Richardson, 2012). Various traits of the extracted hairs (such as cuticle scale pattern, medulla lattice pattern and cross section) were subsequently investigated by applying techniques described in the interactive computer key, Hair ID by Triggs and Brunner (2002). Results show that none of the investigated hair traits was usable for a clear distinction between feces from these sympatric species (Evans, 2012). This might be attributable to changes of hair traits due to degradation during the extraction process or during digestion (Evans, 2012). In conclusion, the inclusion of hairs in species identification of feces does not prevent misidentification of feces produced by sympatric species (Cammisa, 2013; Lobert et al., 2001).

The Use of Scratch Marks in Ecological Research on Tree Kangaroos

The distribution of scratch marks (Fig. 26.2) on trees was used by Meeks (2009) and Chan (2008) to examine preferences of LTKs for trees with certain traits for climbing and for certain plant species for food. Trees with scratch marks from LTKs had a significantly smaller Diameter at Breast Height (DBH) than those without scratch marks (average DBH of 19.8 cm ± 1.01 cm standard error vs 25.9 cm ± 1.07 cm). Furthermore, trees with scratch marks did not always belong to known food species for LTKs, which suggests that LTKs may use trees with suitable traits (such as smaller DBH) as "highways into the canopy" to reach foliage of preferred tree species.

While the use of scratch marks on tree trunks for studying distribution, habitat use, and diet of tree kangaroos can be beneficial, this method is limited for several reasons: (1) the visibility of scratch marks is reduced on trees with rough bark and thus they may go undetected; (2) scratch marks can last for a long time, negating assumptions on the current use of a habitat; and (3) the longevity of scratch marks varies and depends on bark healing processes which differ between tree species resulting in a skewed distribution of trees with visible scratch marks. However, scratch marks should not be ignored as a tool to gain information on tree kangaroos as long as their limitations are recognized, and they are combined with other reliable, ecological field methods.

The Combined Use of Feces and Scratch Marks in Ecological Research on Tree Kangaroos

The combined presence of scratch marks and feces was used to examine whether antipredatory behavior of LTKs leads to the preference for certain microhabitat features by LTKs (Heise-Pavlov et al., 2011). Since LTKs descend from the canopy when threatened, to flee on the forest floor (Procter-Gray and Ganslosser, 1986), it was assumed that they prefer trees with structural features that allow a fast descent to the ground. Trees with signs from LTKs (scratch marks, feces, or both) were examined with respect to their DBH and degree of obstruction, such as the presence of vines, epiphytes, and the proximity of branches from neighboring trees. Trees with smaller trunks (average DBH around 20 cm), trees with no epiphytes, and trees with less obstructions from neighboring trees, shrubs, and lianas within 0.5 m of the tree trunk were indeed found to be associated with signs of scratch marks, feces, or both (Heise-Pavlov et al., 2011). This suggests that LTKs may prefer trees with these features since they facilitate the

FIGURE 26.2 Scratch marks made by Lumholtz's tree kangaroos on tree trunks. *Source: S. Heise-Pavlov.*

vertical movement of individuals within a rainforest habitat.

Scratch marks and feces have also been used to determine the intensity of habitat use of LTKs in Northeastern Australia. This was based on a modified Spot Assessment Technique (SAT) that Phillips and Callaghan (2011) developed to study tree preferences and activity levels of koalas (*Phascolarctos cinereus*) in different habitats. The original SAT method consists of a point-based tree sampling methodology that utilizes the presence/absence of koala fecal pellets within a 1 m radius around the base of trees within sampling plots that contain a minimum of 30 trees. Due to highly variable decay rates of tree kangaroo feces in the humid Wet Tropics of Northeastern Australia, several attempts have been made to test the applicability of the SAT method for assessing habitat utilization of LTKs, and to modify it accordingly. First, no correlation was found between the results of the SAT method and results of direct searches for LTKs within a habitat (Goetz, 2009). Second, the modification of the range of tree trunk sizes to be incorporated in the SAT method and the inclusion of a visibility scale for scratch marks as a surrogate for their age also did not result in a correlation between areas that were identified by SAT as of "high activity" and areas in which tree kangaroos had frequently been observed

(Smetana, 2016). Third, a study by Kremp (2017) showed that from all measured variables in the modified SAT method of Smetana (2016), the amount of feces was the most significant factor in distinguishing between habitats of high and low use by LTKs. This study also revealed that the distribution of feces across trees of different DBH size classes potentially can assist in identifying areas with resident LTKs, opposed to areas which are only occasionally visited by LTKs. By adding distribution of feces to the modified SAT method, the validity of SAT for assessing the degree at which LTKs use a habitat seems to increase. Future studies need to test this assumption.

In summary, challenges of detecting tree kangaroo feces in dense rainforest undergrowth which are amplified by their rapid decomposition under humid conditions, their misidentification with feces from sympatric species, and their consumption by dung beetles reduce the reliability of feces for ecological research on tree kangaroos. Therefore, a cautionary approach should be applied when interpreting data obtained from fecal analyses alone. Attempts to combine fecal counts with other tree kangaroo signs, such as scratch marks on trees, appear promising for the assessment of the intensity at which a habitat is used by tree kangaroos, but also show limitations which are linked with the above listed restrictions in the detectability of feces and the longevity and visibility of scratch marks.

THE APPLICATION OF ADVANCED TECHNOLOGY TO NON-INVASIVE METHODS OF RESEARCH ON TREE KANGAROOS

Recent advancements in technology have supported ecological field research by providing equipment that allows for better detection and observation of animals. While technology of radio-telemetry still requires the capture and handling of individuals to fit them with a transmitter, modern technology aligns with the desire of ecologists to minimize stress to an animal while maximizing obtained data on an animal's ecology and behavior (Marvin et al., 2016). The application of remote sensing technology in ecological research opens up avenues to study species that are difficult to detect and to observe, such as small and/or arboreal species (Chapter 23; Claridge et al., 2005; Mills et al., 2016). Thus, during the last decades, the use of remote sensing and digital technologies such as camera traps, thermal imagery, and photography in combination with advanced software have been applied to tree kangaroo research.

Remote Sensing Cameras

Remote sensing cameras have been used in capture-recapture (CR) or distance sampling (DS) methods (Rowcliffe et al., 2008; for overviews of methods: Borchers et al., 2015; Buckland et al., 2001). In CR, a series of detectors (e.g., traps or cameras) is installed at multiple sampling locations. The resulting "capture history" at the locations at which each uniquely identified animal was detected is used to estimate the probability of detection, and hence accounts for undetected animals (Borchers et al., 2015). Recent advancements of the CR method to spatially explicit capture-recapture (SECR) methods (Borchers and Efford, 2008; Efford, 2004; Royle et al., 2013) now enable the estimation of animal densities by using distances between detectors at which animals are (and are not) detected to estimate a distance-based detection probability surface. While SECR does not include distances to animal locations, the DS method estimates the detection probability surface by using observed distances of animals from detection points (Borchers et al., 2015). DS requires only a single survey occasion and can therefore account for animals missed.

Camera trapping was extensively used in research and conservation of the Tenkile tree kangaroo (*Dendrolagus scottae*) in Papua New

Guinea (Thomas, 2014) by the Tenkile Conservation Alliance (TCA) (Chapter 22) for studying and monitoring Tenkile tree kangaroo. A detailed description can be found on the TCA website (TCA, 2020).

Remote sensing cameras have been intensively used to study the use of a restored riparian habitat by LTK on the Atherton Tablelands in Australia (Burchill, personal communication). Cameras were placed at locations at which tree kangaroo signs were detected. These signs included scratch marks on trees, feces, but also evidence of feeding activities such as chewed plant parts on the ground, or favorite species such as umbrella trees defoliated or with only fresh new growth. Apart from locations with tree kangaroo signs, locations at which tree kangaroos have been previously observed were included in the selection of spots for the installation of cameras. Cameras were also installed near animal tracks since it appeared that LTKs may use the same tracks on a regular basis.

Although the pre-selection of locations for the installation of cameras based on signs of tree kangaroo activities proved to be useful, it was advantageous to initially set cameras to view a wider range of an area instead of focusing on a detected activity sign. This increases the chance to pinpoint animals to the detected signs and to assess the suitability of these signs for capturing tree kangaroo images using a camera. It also allows obtaining a better understanding of tree kangaroo movements and behaviors. Subsequently cameras can be installed closer to detected activity points to further study tree kangaroo's habitat use.

For the installation of cameras, available knowledge of the ecology and behaviors of tree kangaroos was taken into consideration. For example, recent research on the climbing abilities of LTKs revealed that individuals often select trees with smaller DBHs to reach the canopies of larger trees (Chan, 2008; Heise-Pavlov, 2017). Cameras were therefore installed in a way that they focus on trunks of trees with smaller DBH. However, the technical limits of a camera determine their distance to the focal point. Furthermore, the cardinal direction at which the cameras are installed can influence the quality of the photos taken at certain times of day. If the camera is directly facing east or west in more open forest or young revegetation plots, photos taken in early morning or late afternoon, respectively, may be of low quality due to the direct focus towards the sun ("white photos"). This may result in a loss of 1–2 hours of photo opportunities.

The acquired images allowed some conclusions on the use of certain areas by tree kangaroos in this riparian restored habitat (Burchill, personal communication). The analysis of areas with tree kangaroo movement activity by Ciarolo (2020) showed that the species' use of restored habitats seems to be more determined by the presence of certain microhabitat features, such as trees of certain growth forms and branching characteristics, the presence of pioneer species and vines for climbing, than the age of a restored habitat. However, more data are needed to confirm the significance of this conclusion.

Since most remote sensing cameras also record the times at which images are taken, information on activity cycles of animals can be obtained. A study by Semper (2020), using images of LTK movements in the aforementioned restored habitat, confirmed a cathemeral activity pattern of LTKs. However, an observed trend towards a crepuscular activity pattern may suggest a certain plasticity in the species' behavioral ecology when using restored habitats. A rather crepuscular activity pattern in LTKs' movements was also observed in LTK mothers with their joeys at this restored habitat. Most of observed LTK mothers at this site spent most of the day with their joeys stationary in one tree but descended and moved on the ground to another site around dusk (Heise-Pavlov, personal observation, 2014) (Fig. 26.3).

FIGURE 26.3 Mother Lumholtz's tree kangaroo with juvenile descending from a tree within a restored riparian forest. *Source: S. Burchill.*

Images can further be used to retrieve information on habitat features that have been used by LTK individuals (Fig. 26.4), to study interactions between individuals and, in conjunction with methods of individual recognition, to obtain information on home range sizes and abundances (Mendoza et al., 2011).

Remote motion sensing cameras present a valuable tool for the study of tree kangaroos, specifically with focus on their movements which they frequently do on the forest floor. Tree kangaroos do not seem to display novelty shyness (Heise-Pavlov, 2016), so that the installation of a camera in their habitat is unlikely to adversely affect their behavior. A developed algorithm to estimate the number of animals in an area using distances between detection points and distances between detection points and detected animals, facilitates population density estimates without disturbances to the animals (Thomas et al., 2010).

Motion sensing cameras can be very sensitive to small movements and therefore may capture a lot of images that are triggered by moving leaves, twigs, or other undergrowth. This can be a very limiting factor to their use in dense rainforests since heavy rain can also trigger the capture of images. As a result, the analyses of a high quantity of images can be very time-consuming and only a small proportion of images may be useful for the study. High quality weather-protected cameras with a range of settings are recommended to obtain good-quality photos that allow answering various questions on the behavioral ecology of tree kangaroos. Preliminary surveys of the study site with a small number of cameras can assist in selecting promising detection spots, to get experienced in

FIGURE 26.4 Retrieving measurements of habitat features that have been used by Lumholtz's tree kangaroos (see Fig. 26.3). *Source: S. Burchill.*

camera mounting and to assess the usefulness of the camera type for the study.

Thermal Imaging

Thermal imaging techniques support safe and non-invasive measurements and the acquisition of results that cannot be obtained by any other method (Cilulko et al., 2013). The techniques are based on the detection of temperature distribution patterns on the surface of bodies. In ecological studies it can be used to locate individuals in their habitats, to determine the size of wildlife populations, to study responses of animals to environmental factors, and to analyze thermoregulation in animals (McCafferty, 2007).

Thermal imaging devices have proved useful in the detection and observation of tree kangaroos, which are hard to spot (Underwood and Gillanders, personal communication, 2017). Animals appear as warm spots against a dark, cool background in the thermogram. Underwood (personal communication, 2017) reported a significant increase in the total number of sightings of arboreal mammals observed through the hand-held thermal imager compared to the number of sightings when using a spotlight only.

Thermal imaging has its limitations, which are particularly apparent in warm tropical rainforests, the habitats of tree kangaroos. The method is most effective in detecting animals in lowland deciduous forests and under dry conditions since water and high moisture hinders the detection of the warmth of a body (Cilulko et al., 2013). Dense vegetation also has been found to limit the effectiveness of thermal imaging devices (Butler et al., 2006; Ditchkoff et al., 2005). Thus, although thermal imaging devices have been used to detect tree kangaroos in Australian rainforests, the suitability of this method for ecological studies of tree kangaroos is limited; and thus, this method has not yet been applied to scientific studies on tree kangaroos.

Visual Methods of Individual Recognition in Tree Kangaroos

The ability to distinguish individuals of the same species is crucial for effective conservation because it not only allows researchers to assess population densities, but also to gain information on territoriality, habitat requirements, conspecific and heterospecific interactions, and movement patterns (Crouse et al., 2015; Sutherland, 2008). The challenge is to find an effective method that will maintain the balance between cost effectiveness, time efficiency, and overall ethical standards for the subjects' welfare. Currently, most available identification methods, both direct and indirect, do not effectively manage all of these factors equally. However, the development of advanced photographic equipment, digital image capture methods including the use of smartphones, and identification software provide a range of techniques that can facilitate non-invasive identification of individual wildlife.

Since tree kangaroos are known to be very susceptible to stress (Coombes and Shima, personal communication, 2016), there is a growing interest to replace the use of radio collars to observe and follow tree kangaroos by non-invasive methods to identify individuals in Australia. Preliminary studies have been recently conducted on the suitability of photos for individual recognition of tree kangaroos (McMahill, 2016; Chapter 19). Identifying individuals using facial features has been used in studies of sea turtles (Carter et al., 2014), chimpanzees (Yale Environment, 2011), and cassowaries (Crome and Moore, 1990).

The characteristic black facial mask of LTKs seems to differ between individuals with respect to its extent and shape (Andersen and Cianelli, personal communication, 2014) and therefore triggered research on using some of its traits to distinguish between individuals (Fig. 26.5).

In a series of studies, photos from faces of different LTK individuals were investigated to develop a method of individual recognition based on photos. The first study examined the suitability of three facial features for individual recognition: the percentage of the extent of the facial mask in various sections of the face, ratios of measurements of the facial mask, and the presence and location of specific facial markings

FIGURE 26.5 (A–C) Three faces of Lumholtz's tree kangaroo individuals showing the characteristic black facial mask. *Source: S. Burchill, P. Heise-Pavlov.*

(Azcue, 2012). The results showed that ratio measurements (e.g., Nose-Eye to Nose-Ear Ratio) contribute the most to a distinction between the sampled LTK individuals.

Douglas (2015) applied the software Wild ID and its algorithms to LTK facial photos to identify possible matches between photos which were then confirmed by the user (software created by Bolger, 2011). The results showed that Wild ID software was successful in matching photos from the same individual. The application of Wild ID to photos from LTKs is cost-effective and would be beneficial in studies on movements of individual LTKs across landscapes (Douglas, 2015). Elnadouri (2016) demonstrated the utility of the Wild ID software in identifying residential LTKs at a restoration site by a 93.2% user verification of the photo matches that were proposed by the software.

The application of the photo identifying methods is influenced by the quality of the photos. Photo quality plays a higher role when Wild ID software is applied than for the identification based on ratio measurements. Photos of low quality are likely to influence the Wild ID software program's ability to identify matches of photos originating from the same individual (Elnadouri, 2016).

McMahill (2016) considered factors such as user friendliness and time and cost efficiency to determine the most reliable method for individual LTK identification based on facial images. Using distance and similarity indices and non-parametric multidimensional scaling, McMahill (2016) found that cropping photos to a certain size and using easy obtainable ratio measurements proved to be the most reliable and efficient method in identifying individual LTKs with an 83% success rate. It remains to be investigated whether the described methods are suitable for other tree kangaroo species with less profound facial features. However, the selection of ratio measurements can be adjusted to the different species of tree kangaroo.

The first application of the developed methods for individual recognition of LTKs was done in a study on habitat use of the species in a riparian restoration area on the Atherton Tablelands in Northeastern Australia (Heise-Pavlov et al., 2018). Based on the ratio method, the application of Wild ID software and observer verifications, home ranges of four females could be calculated and were compared with those reported from animals in rainforest fragments in the same area (Heise-Pavlov et al., 2018).

CITIZEN SCIENCE IN TREE KANGAROO RESEARCH

Integrating public outreach into research, or "citizen science," can provide many benefits to research, such as reducing costs, increasing available resources, and broadening the feasible geographic scale of projects (Conrad and Hilchey, 2011; Dickinson et al., 2010). Numerous studies have evaluated the value of citizen science and found that the public can provide useable, valid data (Holt et al., 2013; Kremen et al., 2011; Lukyanenko et al., 2016), but that this depends on the development of a rigorous sampling protocol and training of citizen scientists (Kobori et al., 2016).

Citizen science has become a particularly valuable tool for the study of cryptic species as their detection is usually time and cost consuming (Lunney and Matthews, 2001). Due to limitations in the detection of tree kangaroos high in dense canopies of rainforests, it is therefore not surprising that the involvement of the public in studies on the distribution and habitat use of tree kangaroos has been considered (Fig. 26.6).

During a community survey of the distribution of LTKs on the Atherton Tablelands in 1998 and 1999 (Kanowski et al., 2001), information on tree kangaroo sightings was collected from residents using questionnaires that were

FIGURE 26.6 Citizen scientists observing Lumholtz's tree kangaroo and conducting fecal counts at tree bases. *Source: S. Heise-Pavlov.*

posted, and via interviews of residents of the Atherton Tablelands. This information gathering was accompanied by a publicity campaign to support the correct identification of the target species. Obtained data were verified through in-person and phone interviews with record providers by asking them for a detailed description of the sighted animals (Kanowski et al., 2001). Records were then vetted by an experienced panel of local residents and ecologists. Misidentification of sighted animals could not be completely excluded (Kanowski et al., 2001) but was minimized by limiting the inclusion of records to high-elevation areas in order to exclude sightings of the sympatric species, the Bennett's tree kangaroo. Based on the recorded sighting data, a distribution map of LTKs on the Atherton Tablelands was generated (Kanowski et al., 2001), which also allowed the derivation of attributes that characterize LTK suitable habitat.

Hauser and Heise-Pavlov (2017) used records from this community survey, data from the Queensland's Parks and Wildlife Service, and more recent community-based sighting records to evaluate the validity of data from the community for studies on the distribution and habitat use of this species. By comparing maps generated from community records with those generated from data obtained from scientific studies, the authors demonstrate the value of community based collected incidental data even though incidental data were collected in a non-standardized way (Hauser and Heise-Pavlov, 2017). In a similar study Heise-Pavlov and Gillanders (2016) show how incidental sighting records can be utilized to assess the use of a highly fragmented landscape by LTKs.

Detailed observations of tree kangaroos by community members also assisted in the study of behavioral traits of LTKs. This included research on conspecific interactions and home range sharing in LTKs (Heise-Pavlov et al., 2018) and studies on mother-joey relationships in this species, which largely depended on observations of rehabilitated and released animals by wildlife caretakers (Chapters 6 and 7; Schiffer, 2018; Osterman, 2018). Citizen scientists will also provide a valuable resource to study the movements and habitat use of tree kangaroos at a landscape-scale by applying the above-mentioned photographic methods of identification of individual tree kangaroos.

Research and conservation of tree kangaroos in New Guinea depends largely on the involvement of local communities and landowners (Dabek, 2014; Chapters 17, 23, 24). The establishment of the YUS Conservation Area, Tree Kangaroo Conservation Program, and the Tenkile Conservation Alliance program are showcases for species conservation with indigenous people (Tree Kangaroo Conservation Program, 2018; Huon Tree Kangaroo Conservation case study, 2013; Chapters 10, 12, 13, 16, 22). Through conservation awareness programs, ranger training, and scientific data collections, indigenous people learn to value the protection of these species and their habitats (Porolak, 2008). Indigenous knowledge of habitat use by local tree kangaroos and observations of tree kangaroos by community members proved to be invaluable information sources to gain insight into dietary requirements of tree kangaroo species (Porolak, 2008; Chapter 22).

The presented cases of the application of citizen science to tree kangaroo research and conservation demonstrate how broad-scale data can be obtained in a cost-effective manner. However, to ensure a high reliability of collected data, involved citizen scientists need to be trained and guided appropriately. The provision of feedback and a sense of being part of a conservation program have greatly facilitated the participation of community members in the described programs. Data from untrained community members should be subject to rigorous verification processes (Kanowski et al., 2001).

CONCLUSIONS

Due to the cryptic life of tree kangaroo species in the high canopies of tropical rainforests, the study of their ecology and behavior to inform conservation measures relies more and more on the application of non-invasive methods. Advancements in extracting, amplifying, and identifying DNA from tree kangaroo biological material, but also from eDNA has greatly contributed to our knowledge on the genetic diversity of tree kangaroo species, their phylogenetic relationships, and their distributional ranges.

Signs associated with tree kangaroo presence, such as feces, scratch marks, and hairs, have increasingly been used in field studies on distribution, habitat use, food preferences, and behavior of tree kangaroos to assist in habitat protection and restoration. However, they are largely unsuitable for abundance estimates of tree kangaroo populations. This is due to variations in the detectability and longevity of these signs that place limits to their use in ecological field research. These limitations can be partly accounted for by using signs in combination and incorporating knowledge on their decay rates.

Remote-sensing cameras and thermal imagery are examples of how advanced technologies can provide data on movements, home range sizes, and behaviors of tree kangaroos that would otherwise only be obtainable via invasive methods, such as radio-collaring of individuals. However, the application of these methods requires a certain degree of knowledge about the likely locations of tree kangaroos within a habitat. These methods have also proved to be suitable for estimating the number of animals in an area by using algorithms that include distances between detection points, and distances between detection points and detected animals. Improvements in the capacities of photographic equipment and the development of recognition software have resulted in attempts to distinguish between individual tree kangaroos. Developed methods have been tested in captive and wild tree kangaroos and will facilitate our research on tree kangaroo movements and habitat use.

The range of non-invasive methods that have been developed and used in research on tree kangaroos invokes their application in citizen-science projects that will further enhance and support our research on tree kangaroos. Reported cases have shown that involving community members as well as wildlife caretakers in research on tree kangaroos is cost-effective and can boost our data on the ecology and behavior of these species which assists their conservation.

The reported examples clearly show that research and conservation of tree kangaroos will benefit from further developments of non-invasive field methods that are applicable to these species.

ACKNOWLEDGMENTS

The authors thank Rand Herron and numerous University of Rhode Island undergraduate student volunteers for processing fecal samples. The authors also thank the villages of Keweng, Teptep, Yawan, Towet, and all other villages in the Yopno-Urawa-Som Local Level Government in Papua New Guinea (PNG), the Tree Kangaroo Conservation Program field teams, PNG Department of Environment and Conservation, PNG National Research Institute for sample collection and logistics.

Many students of the Centre for Rainforest Studies at The School for Field Studies in Australia contributed their enthusiasm and many hours of field and laboratory work to research non-invasive methods to study Lumholtz's tree kangaroos on the Atherton Tablelands. This work was supported logistically by The School for Field Studies and its staff who accompanied the students in the field.

Wildlife rehabilitators Margit Cianelli and Karin Semmler collected feces from tree kangaroos, possums and pademelons without which studies on the use of feces in researching Lumholtz's tree kangaroos would not have been possible. We are thankful to Steve Phillips who introduced the SAT method to us. Many people provided photos of Lumholtz's tree kangaroos to develop methods for individual recognition in this species, including Margit Cianelli, Peter Heise-Pavlov, Kristen Bell, Amanda Freeman, and staff of David Fleay Wildlife Park, Dreamworld Zoo, and the Wildlife Habitat in Port Douglas.

REFERENCES

Anton, T., 2014. To eat or not to eat: a question of diet, detoxification mechanisms, and survival of the Lumholtz's tree kangaroo (*Dendrolagus lumholtzi*). Unpublished Directed Research Report, Centre for Rainforest Studies, The School for Field Studies.

Avise, J.C., Arnold, J., Ball, R.M., Bermingham, E., Lamb, T., Neigel, J.E., Reeb, C.A., Saunders, N.C., 1987. Intraspecific phylogeography: the bridge between population genetics and systematics. Annu. Rev. Ecol. Syst. 18, 489–522.

Azcue, I., 2012. Studying cute faces: an investigation into using facial recognition to distinguish between individual Lumholtz's tree-kangaroos (*Dendrolagus lumholtzi*). Unpublished Directed Research Report, Centre for Rainforest Studies, The School for Field Studies.

Banks, E.M., Brooks, R.J., Schnell, J., 1975. A radiotracking study of home range and activity of the brown lemming (*Lemmus trimucronatus*). J. Mammal. 56, 888–901.

Bartholomew, R.M., 1967. A study of the winter activities of bobwhites through the use of radiotelemetry. Occas. Pap. C.C. Adams Cent. Ecol. Std. 17, 1–25.

Berteaux, D., Duhamel, R., Bergeron, J.-M., 1994. Can radio collars affect dominance relationships in *Microtus*? Can. J. Zool. 72, 785–789.

Betz, W., 2001. Matschie's Tree Kangaroo (Marsupialia: Macropodidae, *Dendrolagus matschiei*) in Papua New Guinea: Estimates of Population Density and Landowner Accounts of Food Plants and Natural History (Masters thesis). University of Southampton, Southampton, England.

Birky, C.W., Fuerst, P., Maruyama, T., 1989. Organelle gene diversity under migration, mutation, and drift: equilibrium expectations, approach to equilibrium, effects of heteroplasmic cells, and comparison to nuclear genes. Genetics 121, 613–627.

Bland, L.M., Collen, B., Orme, C.D.L., Bielby, J., 2015. Predicting the conservation status of data-deficient species. Conserv. Biol. 29, 250–259.

Bohmann, K., Evans, A., Gilbert, M.T.P., Carvalho, G.R., Creer, S., Knapp, M., Douglas, W.Y., De Bruyn, M., 2014. Environmental DNA for wildlife biology and biodiversity monitoring. Trends Ecol. Evol. 29, 358–367.

Bolger, D., 2011. Wild ID 1.0. Computer Software. Vers 1.0. Darthmouth University. Available from: http://wildid.teamnetwork.org/index.jsp. (12 April 2015).

Borchers, D.L., Efford, M.G., 2008. Spatially explicit maximum likelihood methods for capture-recapture studies. Biometrics 64, 377–385.

Borchers, D.L., Stevenson, B., Kidney, D., Thomas, L., Marques, T.A., 2015. A unifying model for capture-recapture and distance sampling surveys of wildlife populations. J. Am. Stat. Assoc. 110 (509), 195–204.

Bowyer, J., Newell, G., Eldridge, M.B., 2002. Genetic effects of habitat contraction on Lumholtz's tree-kangaroo (*Dendrolagus lumholtzi*) in the Australian wet tropics. Conserv. Genet. 3, 59–67.

Bowyer, J.C., Newell, G.R., Metcalfe, C.J., Eldridge, M.D.B., 2003. Tree-kangaroos *Dendrolagus* in Australia: are *D. lumholtzi* and *D. bennettianus* sister taxa? Aust. Zool. 32, 207–213.

Buckland, S.T., Anderson, D., Burnham, K., Laake, J., Thomas, L., Borchers, D., 2001. Introduction to Distance Sampling: Estimating Abundance of Biological Populations. Oxford University Press, Oxford.

Butler, D.A., Ballard, W.B., Haskell, S.P., Wallace, M.C., 2006. Limitations of thermal infrared imaging for locating neonatal deer in semiarid shrub communities. Wildl. Soc. Bull. 34 (5), 1458–1462.

Cammisa, N., 2013. Assessing the suitability of an indirect method to study Lumholtz's tree-kangaroos, *Dendrolagus lumholtzi*. Unpublished Directed Research Report, Centre for Rainforest Studies, The School for Field Studies.

Carter, S., Ian, J., Bell, P., Miller, J.J., Gash, P.P., 2014. Automated marine turtle photograph identification using artificial neural networks, with application to Green Turtles. J. Exp. Mar. Biol. Ecol. 452, 105–110.

Chan, C., 2008. Features of plant species preferred by the Lumholtz's tree-kangaroo (*Dendrolagus lumholtzi*) and observations on territory markings in Northeast Queensland Australia. Unpublished Directed Research Report, Centre for Rainforest Studies, The School for Field Studies.

Ciarolo, T., 2020. Identifying Lumholtz's tree-kangaroo (*Dendrolagus lumholtzi*) habitat hotspots within replanted riparian forest on the Atherton Tablelands. Unpublished Directed Research Report, Centre for Rainforest Studies, The School for Field Studies.

Cilulko, J., Janiszewski, P., Bogdaszewski, M., Szczygielska, E., 2013. Infrared thermal imaging in studies of wild animals. Eur. J. Wildl. Res. 59, 17–23.

Claridge, A.W., Mifsud, G., Dawson, J., Saxon, M.J., 2005. Use of infrared digital cameras to investigate the behaviour of cryptic species. Wildl. Res. 31, 645–650.

Conde, D.A., Colchero, F., Silva, R., Syed, H., Jongejans, E., Jouvet, L., Baden, M., Meiri, S., Gaillard, J.-M., Chamberlain, S., 2016. Exploring data gaps at the species level: starting with demographic knowledge. In: Annual Conference of the Taxonomic Databases Working Group (TDWG). Available from: https://mbgocs.mobot.org/index.php/tdwg/tdwg2016/paper/view/1145. (6 May 2020).

Conrad, C.C., Hilchey, K.G., 2011. A review of citizen science and community-based environmental monitoring: issues and opportunities. Environ. Monit. Assess. 176, 273–291.

Coombes, K., 2005. The Ecology and Habitat Utilisation of the Lumholtz's Tree-Kangaroo, *Dendrolagus lumholtzi* (Marsupialia: Macropodidae), on the Atherton Tablelands, Far North Queensland (Ph.D. thesis). James Cook University, Cairns.

Corner, G.W., Pearson, E.W., 1972. A miniature 30-MHz collar transmitter for small mammals. J. Wildl. Manag. 36, 657–661.

Crome, F.H.J., Moore, L.A., 1990. Cassowaries in North-Eastern Queensland—report of a survey and a review and assessment of their status and conservation and management needs. Wildl. Res. 17, 369–385.

Crouse, D., Richardson, Z., Jain, A., Tecot, A., Baden, A., Jacobs, R., 2015. Lemur face recognition: tracking a threatened species and individuals with minimal impact. MSU technical report. MSU-CSE-15-8.

Dabek, L., 2014. Tree-Kangaroo Conservation Program in Papua New Guinea: an overview. Mammal Mail 15 (1), 2–4.

Daly, M., Wilson, M.I., Behrends, P.R., Jacobs, L.F., 1992. Sexually differentiated effects of radio transmitters on predation risk and behaviour in kangaroo rats *Dipodomys merriami*. Can. J. Zool. 70, 1851–1855.

Davis, N.E., Forsyth, D.M., Triggs, B., Pascoe, C., Benshemesh, J., Robley, A., Lawrence, J., Ritchie, E.G., Nimmo, D.G., Lumsden, L.F., 2015. Interspecific and geographic variation in the diets of sympatric carnivores: dingoes/wild dogs and red foxes in south-eastern Australia. PLoS ONE 10, e0120975.

De Barba, M., Miquel, C., Boyer, F., Mercier, C., Rioux, D., Coissac, E., Taberlet, P., 2014. DNA metabarcoding multiplexing and validation of data accuracy for diet assessment: application to omnivorous diet. Mol. Ecol. Resour. 14 (2), 306–323.

Dickinson, J.L., Zuckerberg, B., Bonter, D.N., 2010. Citizen science as an ecological research tool: challenges and benefits. Annu. Rev. Ecol. Evol. Syst. 41, 149–172.

Dickman, C., 1995. Diets and habitat preferences of three species of crocidurine shrews in arid southern Africa. J. Zool. 237, 499–514.

Ditchkoff, S.S., Raglin, J.B., Smith, J.M., Collier, B.A., 2005. From the field: capture of white-tailed deer fawns using thermal imaging technology. Wildl. Soc. Bull. 33 (3), 1164–1168.

Douglas, C., 2015. Using facial recognition software to identify individual Lumholtz's tree kangaroos (*Dendrolagus lumholtzi*). Unpublished Directed Research Report, Centre for Rainforest Studies, The School for Field Studies.

Efford, M.G., 2004. Density estimation in live-trapping studies. Oikos 106, 598–610.

Eldridge, M.D., Potter, S., Helgen, K.M., Sinaga, M.H., Aplin, K.P., Flannery, T.F., Johnson, R.N., 2018. Phylogenetic analysis of the tree-kangaroos (*Dendrolagus*) reveals multiple divergent lineages within New Guinea. Mol. Phylogenet. Evol. 127, 589–599.

Ellis, W., Attard, R., Johnston, S., Theileman, P., McKinnon, A., Booth, D., 2014. Faecal particle size and tooth wear of the koala (*Phascolarctos cinereus*). Aust. Mammal. 36, 90–94.

Elnadouri, A., 2016. Facial recognition: discrete technique for identifying and tracking individual Lumholtz's tree-kangaroo in fragmented habitats (*Dendrolagus lumholtzi*). Unpublished Directed Research Report, Centre for Rainforest Studies, The School for Field Studies.

Evans, R., 2012. Using hair to identify Lumholtz's tree-kangaroo (*Dendrolagus lumholtzi*) feces. Unpublished Directed Research Report, Centre for Rainforest Studies, The School for Field Studies.

Frankham, R., Ballou, J.D., Briscoe, D.A., 2004. Introduction to Conservation Genetics. Cambridge University Press, Cambridge.

Goetz, M., 2009. Defining activity levels and habitat use of *Dendrolagus lumholtzi* on the fragmented landscape of the Atherton Tablelands in Far North Queensland, Australia. Unpublished Directed Research Report, Centre for Rainforest Studies, The School for Field Studies.

Gompper, M.E., Kays, R.W., Ray, J.C., Lapoint, S.D., Bogan, D.A., Cryan, J.R., 2006. A comparison of noninvasive techniques to survey carnivore communities in northeastern North America. Wildl. Soc. Bull. 34, 1142–1151.

Grimbacher, P.S., Catterall, C.P., Kitching, R.L., 2006. Beetle species' responses suggest that microclimate mediates fragmentation effects in tropical Australian rainforest. Austral Ecol. 31, 458–470.

Harris, S., Cresswell, W., Forde, P., Trewhella, W., Woollard, T., Wray, S., 1990. Home-range analysis using radio-tracking data–a review of problems and techniques particularly as applied to the study of mammals. Mammal Rev. 20, 97–123.

Hauser, W., Heise-Pavlov, S., 2017. Can incidental sighting data be used to elucidate habitat preferences and areas of suitable habitat for cryptic species? Integr. Zool. 12 (3), 186–197.

Heise-Pavlov, S., 2016. Evolutionary aspects of the use of predator odors in antipredator behaviors of Lumholtz's tree-kangaroos (*Dendrolagus lumholtzi*). In: Schulte, B., Goodwin, T. (Eds.), Chemical Signals in Vertebrates. vol. 13. Springer, New York, pp. 261–280.

Heise-Pavlov, S., 2017. Current knowledge of the behavioural ecology of Lumholtz's tree-kangaroo (*Dendrolagus lumholtzi*). Pac. Conserv. Biol. 23, 231–239.

Heise-Pavlov, S., Gillanders, A., 2016. Exploring the use of a fragmented landscape by a large arboreal marsupial using incidental sighting records from community members. Pac. Conserv. Biol. 22, 386–398.

Heise-Pavlov, S.R., Meade, R., 2012. Improving reliability of scat counts for abundance and distribution estimations of Lumholtz's Tree-kangaroo (*Dendrolagus lumholtzi*) in its rainforest habitats. Pac. Conserv. Biol. 18 (3), 153–163.

Heise-Pavlov, S.R., Jackrel, S.L., Meeks, S., 2011. Conservation of a rare arboreal mammal: habitat preferences of the Lumholtz's tree-kangaroo, *Dendrolagus lumholtzi*. Aust. Mammal. 33, 5–12.

Heise-Pavlov, S., Rhinier, J., Burchill, S., 2018. The use of a replanted riparian habitat by the Lumholtz's tree-kangaroo (*Dendrolagus lumholtzi*). Ecol. Manag. Restor. 19, 76–80.

Hill, C.J., 1993. The species composition and seasonality of an assemblage of tropical Australian dung beetles (*Coleóptera: Scarabaeidae: Scarabaeinae*). Aust. Entomol. 20, 121–126.

Holt, B.G., Rioja-Nieto, R., Aaron MacNeil, M., Lupton, J., Rahbek, C., 2013. Comparing diversity data collected using a protocol designed for volunteers with results from a professional alternative. Methods Ecol. Evol. 4, 383–392.

Höss, M., Kohn, M., Pääbo, S., 1992. Excrement analysis by PCR. Nature 359, 199.

Huon Tree-kangaroo Conservation case study, 2013. Available from: www.speciesconservation.org/case-studies-projects/huon-tree-kangaroo/4711. (5 August 2018).

Johnson, K., Baker, A., Buley, K., Carrillo, L., Gibson, R., Gillespie, G.R., Lacy, R.C., Zippel, K., 2018. A process for assessing and prioritizing species conservation needs: going beyond the Red List. Oryx 54 (1), 125–132.

Jones, K.M.W., 2000. Tree-Kangaroo (*Dendrolagus* spp.) Diet: Faecal Analysis as a Technique to Determine Food Plants and Feeding Patterns (Honours thesis). University of Adelaide, South Australia.

Kanowski, J., Felderhof, L., Newell, G., Parker, T., Schmidt, C., Stirn, B., Wilson, R., Winter, J.W., 2001. Community survey of the distribution of Lumholtz's Tree-kangaroo on the Atherton Tablelands, north-east Queensland. Pac. Conserv. Biol. 7, 79–86.

Kilpatrick, H.J., Goodie, T.J., Kovach, A.I., 2013. Comparison of live-trapping and noninvasive genetic sampling to assess patch occupancy by New England cottontail (*Sylvilagus transitionalis*) rabbits. Wildl. Soc. Bull. 37 (4), 901–905.

Kobori, H., Dickinson, J.L., Washitani, I., Sakurai, R., Amano, T., Komatsu, N., Kitamura, W., Takagawa, S., Koyama, K., Ogawara, T., Miller-Rushing, A.J., 2016. Citizen science: a new approach to advance ecology, education, and conservation. Ecol. Res. 31, 1–19.

Kremen, C., Ullman, K.S., Thorp, R.W., 2011. Evaluating the quality of citizen-scientist data on pollinator communities. Conserv. Biol. 25, 607–617.

Kremp, M., 2017. Revising the spot assessment technique to better evaluate habitat use and population density of Lumholtz's tree-kangaroos (*Dendrolagus lumholtzi*). Unpublished Directed Research Report, Centre for Rainforest Studies, The School for Field Studies.

Lobert, B., Lumsden, L., Brunner, H., Triggs, B., 2001. An assessment of the accuracy and reliability of hair identification of south-east Australian mammals. Wildl. Res. 28, 637–641.

Lofty, G., 2017. Through thick and thin: the effects of understory density on the Lumholtz tree-kangaroo's (*Dendrolagus lumholtzi*) habitat use in the Atherton Tablelands. Unpublished Directed Research Report, Centre for Rainforest Studies, The School for Field Studies.

Lukyanenko, R., Parsons, J., Wiersma, Y.F., 2016. Emerging problems of data quality in citizen science. Conserv. Biol. 30, 447–449.

Lunney, D., Matthews, A., 2001. The contribution of the community to defining the distribution of a vulnerable species, the spotted-tailed quoll, *Dasyurus maculatus*. Wildl. Res. 28, 537–545.

Martin, R., 2005. Tree-Kangaroos of Australia and New Guinea. The University of Chicago Press, Chicago.

Marvin, D.C., Koh, L.P., Lynam, A.J., Wich, S., Davies, A.B., Krishnamurthy, R., Stokes, E., Starkey, R., Asner, G.P., 2016. Integrating technologies for scalable ecology and conservation. Global Ecol. Conserv. 7, 262–275.

McCafferty, D.J., 2007. The value of infrared thermography for research on mammals: previous applications and future directions. Mammal Rev. 37 (3), 207–223.

McGreevy Jr., T.J., Dabek, L., Gomez-Chiarri, M., Husband, T.P., 2009. Genetic diversity in captive and wild Matschie's tree kangaroo (*Dendrolagus matschiei*) from Huon Peninsula, Papua New Guinea, based on mtDNA control region sequences. Zoo Biol. 28, 183–196.

McGreevy Jr., T.J., Dabek, L., Husband, T.P., 2010a. A multiplex PCR assay to distinguish among three sympatric marsupial taxa from Huon Peninsula, Papua New Guinea, using the mitochondrial control region gene. Mol. Ecol. Resour. 10, 397–400.

McGreevy Jr., T.J., Dabek, L., Husband, T.P., 2010b. Microsatellite marker development and Mendelian inheritance analysis in the Matschie's tree kangaroo (*Dendrolagus matschiei*). J. Hered. 101, 113–118.

McGreevy Jr., T.J., Dabek, L., Husband, T.P., 2011. Genetic evaluation of Association of Zoos and Aquariums Matschie's tree kangaroo (*Dendrolagus matschiei*) captive breeding program. Zoo Biol. 30, 636–646.

McGreevy Jr., T.J., Dabek, L., Husband, T.P., 2012. Tree kangaroo molecular systematics based on partial cytochrome b sequences: are Matschie's tree kangaroo (*Dendrolagus matschiei*) and Goodfellow's tree kangaroo (*Dendrolagus goodfellowi*) sister taxa? Aust. Mammal. 34, 18–28.

McGreevy Jr., T.J., Dabek, L., Husband, T.P., 2016. Comparative mtDNA analysis of three sympatric macropodid from a conservation area on the Huon Peninsula, Papua New Guinea. Mitochondrial DNA 27, 2673–2678. https://doi.org/10.3109/19401736.2015.1022761.

McMahill, M., 2016. The matching game: analysis of four facial recognition methods on the Lumholtz's tree-kangaroo (*Dendrolagus lumholtzi*). Unpublished Directed Research Report, Centre for Rainforest Studies, The School for Field Studies.

Meeks, S., 2009. Tree preference and microhabitat selection of the Lumholtz tree-kangaroo (*Dendrolagus lumholtzi*). Unpublished Directed Research Report, Centre for Rainforest Studies, The School for Field Studies.

Mendoza, E., Martineau, P.R., Brenner, E., Dirzo, R., 2011. A novel method to improve individual animal identification based on camera-trapping data. J. Wildl. Manag. 75, 973–979.

Mills, C.A., Godley, B.J., Hodgson, D.J., 2016. Take only photographs, leave only footprints: novel applications of non-invasive survey methods for rapid detection of small, arboreal animals. PLoS ONE 11, e0146142.

Moorhouse, T.P., MacDonald, D.W., 2005. Indirect negative impacts of radio-collaring: sex ratio variation in water voles. J. Appl. Ecol. 42, 91–98.

Newell, G.R., 1999. Home range and habitat use by Lumholtz's tree-kangaroo (*Dendrolagus lumholtzi*) within a rainforest fragment in north Queensland. Wildl. Res. 26, 129–145.

Niemi, G.J., McDonald, E., 2004. Application of ecological indicators. Annu. Rev. Ecol. Evol. Syst. 35, 89–111.

Nuske, S., Fisher, D., Seddon, J., 2014. Common species affects the utility of non-invasive genetic monitoring of a cryptic endangered mammal: the bridled nailtail wallaby. Austral Ecol. 39, 633–642.

Orkin, J.D., Yang, Y., Yang, C., Yu, D.W., Jiang, X., 2016. Cost effective scat detection dogs: unleashing a powerful new tool for international mammalian conservation biology. Sci. Rep. 6. art. no. 34758.

Osterman, A., 2018. Scent marking behaviour in a young rehabilitated male Lumholtz's tree-kangaroo (*Dendrolagus lumholtzi*). Unpublished Directed Research Report, Centre for Rainforest Studies, The School for Field Studies.

Pearl, M.C., 2000. Research Techniques in Animal Ecology. Columbia University Press, New York.

Phillips, S., Callaghan, J., 2011. The Spot Assessment Technique: a tool for determining localised levels of habitat use by koalas *Phascolarctos cinereus*. Aust. Zool. 35, 774–780.

Piggott, M.P., 2004. Effect of sample age and season of collection on the reliability of microsatellite genotyping of faecal DNA. Wildl. Res. 31, 485–493.

Piggott, M.P., Taylor, A.C., 2003. Remote collection of animal DNA and its applications in conservation management and understanding the population biology of rare and cryptic species. Wildl. Res. 30, 1–13.

Pleijel, F., Jondelius, U., Norlinder, E., Nygren, A., Oxelman, B., Schander, C., Sundberg, P., Thollesson, M., 2008. Phylogenies without roots? A plea for the use of vouchers in molecular phylogenetic studies. Mol. Phylogenet. Evol. 48, 369–371. Available from: https://www.researchgate.net/publication/5428762_Phylogenies_without_roots_A_plea_for_the_use_of_vouchers_in_molecular_studies. (30 June 2020).

Porolak, G., 2008. Home Range of the Huon Tree-Kangaroo, *Dendrolagus matschiei*, in Cloud Forest on the Huon Peninsula, Papua New Guinea (Master thesis). James Cook University, Cairns.

Procter-Gray, E., 1985. The Behavior and Ecology of Lumholtz's Tree-Kangaroo, (*Dendrolagus lumholtzi*; Marsupialia: Macropodidae) (Ph.D. thesis). Harvard University, Cambridge, MA.

Procter-Gray, E., Ganslosser, U., 1986. The individual behaviors of Lumholtz's tree-kangaroo: repertoire and taxonomic implications. J. Mammal. 67, 343–352.

Reynolds, J.C., Aebischer, N.J., 1991. Comparison and quantification of carnivore diet by faecal analysis: a critique, with recommendations, based on a study of the fox *Vulpes vulpes*. Mammal Rev. 21, 97–122.

Richardson, W., 2012. Methods of hair extraction and identification from scat samples of Lumholtz's tree-kangaroo (*Dendrolagus lumholtzi*). Unpublished Directed Research Report, Centre for Rainforest Studies, The School for Field Studies.

Rowcliffe, J.M., Field, J., Turvey, S.T., Carbone, C., 2008. Estimating animal density using camera traps without the need for individual recognition. J. Appl. Ecol. 45, 1228–1236.

Royle, J.A., Link, W.A., 2006. Generalized site occupancy models allowing for false positive and false negative errors. Ecology 87, 835–841.

Royle, J.A., Chandler, R.B., Sollman, R., Gardner, B., 2013. Spatial Capture-Recapture. Academic Press, Boston.

Schiffer, M., 2018. Come back Mommy!: a case study on mothering behavior in rehabilitated Lumholtz's tree kangaroo (*Dendrolagus lumholtzi*). Unpublished Directed Research Report, Centre for Rainforest Studies, The School for Field Studies.

Schnell, I.B., Thomsen, P.F., Wilkinson, N., Rasmussen, M., Jensen, L.R., Willerslev, E., Bertelsen, M.F., Gilbert, M. T., 2012. Screening mammal biodiversity using DNA from leeches. Curr. Biol. 22, 262–263.

Schnell, I.B., Bohmann, K., Schultze, S.E., Richter, S.R., Murray, D.C., Sinding, M.H.S., Bass, D., Cadle, J.E., Campbell, M.J., Dolch, R., Edwards, D.P., Gray, T.N.E., Hansen, T., Hoa, A.N.Q., Noer, C., Heise-Pavlov, S., Sander Pedersen, A.F., Ramamonjisoa, J.C., Siddall, M.E., Tilker, A., Traeholt, C., Wilkinson, N., Woodcock, P., Yu, D.W., Bertelsen, M.F., Bunce, M., Gilbert, M.T.P., 2018. Debugging diversity-a pan-continental exploration of the potential of terrestrial blood-feeding leeches as a vertebrate monitoring tool. Mol. Ecol. Resour. 18, 1282–1298.

Semper, C., 2020. Investigating activity patterns of Lumholtz's tree kangaroos (*Dendrolagus lumholtzi*) in a restored riparian area. Unpublished Directed Research Report, Centre for Rainforest Studies, The School for Field Studies.

Smetana, M., 2016. Habitat utilization of a wildlife corridor by Lumholtz's tree-kangaroos (*Dendrolaus lumholtzi*) in the Atherton Tablelands of Queensland, Australia. Unpublished Directed Research Report, Centre for Rainforest Studies, The School for Field Studies.

Smith, D.A., Ralls, K., Cypher, B.L., Maldonado, J.E., 2005. Assessment of scat-detection dog surveys to determine kit fox distribution. Wildl. Soc. Bull. 33, 897–904.

Srivathsan, A., Ang, A., Vogler, A.P., Meier, R., 2016. Faecal metagenomics for the simultaneous assessment of diet, parasites, and population genetics of an understudied primate. Front. Zool. 13 (1), 17.

Storr, G.M., 1961. Microscopic analysis of faeces, a technique for ascertaining the diet of herbivorous mammals. Aust. J. Biol. Sci. 14, 157–164.

Sutherland, W.J., 2008. The Conservation Handbook: Research, Management, and Policy. John Wiley & Sons, Blackwell Publishing, Oxford.

Taberlet, P., Bouvet, J., 1992. Bear conservation genetics. Nature 358, 197.

Tazawa, M.M., 2012. Seasonal variation in consumption and defecation rates of Lumholtz tree-kangaroo (*Dendrolagus lumholtzi*). Unpublished Directed Research Report, Centre for Rainforest Studies, The School for Field Studies.

Tenkile Conservation Alliance, 2020. Objective 4: conduct research - component 1 (distance sampling) and component 2 (camera traps). Available from: https://www.tenkile.com/conduct-research.html. (7 June 2020).

Thomas, J., 2014. Fauna survey by camera trapping in the Torricelli Mountain Range, Papua New Guinea. In: Camera Trapping: Wildlife Management and Research. CSIRO Publishing, Collingwood, Australia, pp. 69–76.

Thomas, L., Buckland, S.T., Rexstad, E.A., Laake, J.L., Strindberg, S., Hedley, S.L., Bishop, J.R., Marques, T.A., Burnham, K.P., 2010. Distance software: design and analysis of distance sampling surveys for estimating population size. J. Appl. Ecol. 47, 5–14.

Tree Kangaroo Conservation Program, 2018. Tree-kangaroo Conservation Program: Annual Report 2018. Available from: www.zoo.org/document.doc?id=2537. (15 June 2019).

Tree Kangaroo Conservation Program, 2019. Available from: https://www.zoo.org/tkcp.

Triggs, B., 2004. Tracks, Scats, and Other Traces: A Field Guide to Australian Mammals. Oxford University Press, Oxford.

Triggs, B., Brunner, H., 2002. Hair ID. Ecobyte Party Ltd., CSIRO Publishing, Queensland.
Valentini, A., Miquel, C., Nawaz, M.A., Bellemain, E., Coissac, E., Pompanon, F., Gielly, L., Cruaud, C., Nascetti, G., Wincker, P., Swenson, J.E., Taberlet, P., 2009. New perspectives in diet analysis based on DNA barcoding and parallel pyrosequencing: the trnL approach. Mol. Ecol. Resour. 9 (1), 51–60.
Vernes, K., Pope, L.C., Hill, C.J., Bärlocher, F., 2005. Seasonality, dung specificity, and competition in dung beetle assemblages in the Australian Wet Tropics, north-eastern Australia. Trop. Ecol. 21, 1–8.
Vynne, C., Skalski, J.R., Machado, R.B., Groom, M.J., Jácomo, A.T.A., Marinho Filho, J., Neto, M.B.R., Pomilla, C., Silveira, L., Smith, H., Wasser, S.K., 2011. Effectiveness of scat detection dogs in determining species presence in a tropical savanna landscape. Conserv. Biol. 25 (1), 154–162.
Waits, L.P., Paetkau, D., 2005. Noninvasive genetic sampling tools for wildlife biologists: a review of applications and recommendations for accurate data collection. J. Wildl. Manag. 69, 1419–1433.
White, G.C., Garrott, R.A., 1990. Analysis of Wildlife Radio-Tracking Data. Academic Press, Elsevier, Toronto.
Yale Environment, 2011. Facial recognition software used in research of apes, elephants. E360.yale.edu.
Yates, T.L., 1985. The role of voucher specimens in mammal collections: characterization and funding responsibilities. Acta Zool. Fenn. 170, 81–82. Available from: http://www.ibiologia.unam.mx/pdf/Yates1985.pdf. (29 June 2020).
Zhu, L., Wu, Q., Dai, J., Zhang, S., Wei, F., 2011. Evidence of cellulose metabolism by the giant panda gut microbiome. Proc. Natl. Acad. Sci. U. S. A. 108 (43), 17714–17719.

SECTION VII

THE FUTURE OF TREE KANGAROOS

CHAPTER

27

The Future of Tree Kangaroo Conservation and Science

Lisa Dabek[a] *and Peter Valentine*[b]

[a]Tree Kangaroo Conservation Program, Woodland Park Zoo, Seattle, WA, United States
[b]College of Science and Engineering, James Cook University, Townsville, QLD, Australia

As of this year 2020, 12 of the 14 species of tree kangaroo remain in the threatened categories of Vulnerable, Endangered, or Critically Endangered on the IUCN Red List of Threatened Species. It is crucial that we understand these animals enough to conserve them. Knowledge about the world's tree kangaroos has expanded quite dramatically in the last two decades since the first book on tree kangaroos was published (Flannery et al., 1996). That is partly due to the application of modern scientific techniques to better understand biology, behavior, and taxonomy. But this has been uneven across the fourteen species that we identify in this volume. For most of these species, there has been little progress in knowledge, not least because they are so challenging to study in the wild. This is partly a consequence of the environments where many species survive—remote places with formidable access constraints, usually quite distant from roads and settlements. To be a tree kangaroo science and conservation field researcher takes a special quality of energetic enthusiasm and robust field skills. It may be partly because of this challenge, that we still do not have a thorough assessment of the entire tree kangaroo group and the population and habitat use of each species remains poorly understood. There is scope for significant field studies to address this gap.

In a few sites across the tree kangaroo distribution, there have been substantial investments in research and conservation action. In Australia, the two endemic tree kangaroo species have been quite well studied across a range of research questions and the outcomes have proved very helpful in establishing key conservation baselines and effective longer term management targets. While both species are considered near threatened, due largely to habitat loss, much is known about how to ensure long term protection and significant work has been undertaken to assist their survival, as presented in Chapters 1, 2, 3, 5, 6, 7, and 8.

Tree Kangaroos
https://doi.org/10.1016/B978-0-12-814675-0.00024-5

Questions do remain about the potential impacts on each species from climate change and there still are opportunities for more work in this area. Habitat protection is relatively secure where it falls within the Wet Tropics World Heritage Area. However, remaining fragments of rainforest that provide important habitat create additional challenges from urban settlement and its consequences, notably poorly managed domestic dogs and road kills.

The tree kangaroos in New Guinea face a more uncertain future (Beehler and Laman, 2020; Chapter 4). The relentless expansion of logging, much of it illegal, and the transformation of rainforests into oil palm and other plantations, not only impact directly on tree kangaroos, but further expands the network of roads and urban settlements that can prove devastating for local populations. Hunting remains a significant threat, especially in the absence of community-based conservation programs and alternative options. In Indonesian New Guinea, there is limited investment in protected areas across the habitats for some species and for others, low population densities, rugged landscapes and remote access have limited the necessary research. The case of the Wondiwoi tree kangaroo (*Dendrolagus mayri*) is instructive. First described from a single specimen collected in 1928 (Chapter 1), it was not recorded for the next 90 years until Michael Smith, undertaking botanical field work in the Wondiwoi Peninsula, was able to produce the first photograph of a living animal (Fig. 27.1). While this has raised hopes that this species, previously considered most likely extinct, continues to survive in these remote areas, it also highlights the prospect that other species in similarly remote areas are awaiting rediscovery and protection.

FIGURE 27.1 The first photograph of a live Wondiwoi Tree Kangaroo, taken in 2018 by Michael Smith. What other novel observations are out in the cloud forests of New Guinea? *Source: Michael Smith.*

There has been some limited research into tree kangaroos within Indonesia (Flannery et al., 1996) (Chapter 17), but the scope exists for a much greater investment in what is already an urgent need. International NGOs have been supportive but there is not yet a strong local center for such research. The ongoing political conflict between Indonesian authorities and Papuan people does not provide a positive environment for community-based conservation programs and continues to disrupt conservation efforts. Even the management of protected areas is often undertaken remotely so that on-ground skills (especially monitoring and research) are limited. There is room for a stronger alliance among Government programs of conservation and research for tree kangaroos, local universities, and the local communities of Papuan people who are the custodians of significant traditional ecological knowledge. The essential role of protected areas in providing for the long-term conservation of tree kangaroos is outlined in Chapter 9 with specific guidance for Papua, West Papua, and Papua New Guinea (PNG). There are some excellent models of workable solutions available from PNG (Chapters 10, 11, 12, 13, and 16).

The eastern portion of New Guinea, the independent country of Papua New Guinea, has a very different land management regime mainly because of the land tenure system. Local communities are able to fully participate in conservation programs as stewards of their own land. It is this capacity of Papua New Guinea society that has formed the foundation of several community-based programs with positive outcomes for tree kangaroos. Future conservation success will still require expanded protected areas and associated income and community programs as the template suggests in Chapter 9. But no program is likely to succeed without considering the links between livelihood support and conservation outcomes (Chapter 14) or linking human health and wildlife health (Chapter 15). The examples from the Tree Kangaroo Conservation Program—TKCP in the YUS Conservation Area and from the Tenkile Conservation Alliance in the Torricelli Mountains have long shown the critical roles of community in achieving conservation outcomes. A partnership between the local community and international NGOs, together with appropriate engagement with the District, Provincial, and National Governments, has demonstrated success with tree kangaroo conservation a key part of overall outcomes. Such programs require long term investment (many decades) and must maintain continuity among all the partners. Skill development and opportunities for local leadership are essential elements of these programs. As concluded in Chapter 16, "the longevity of tree kangaroo conservation depends upon local stewards and landowners". A focus on indigenous-led conservation and

FIGURE 27.2 Junior Rangers in the YUS conservation program (Huon Peninsula, PNG). Building the inclusion of local young people in the conservation program. *Source: Ryan Hawk.*

collaboration and networks of indigenous approaches to conservation is gaining momentum. It is also essential that community-based programs include young people. By providing better access to education and the identification of various potential roles, there is increased hope for the future. In the YUS TKCP program, for example, the Junior Ranger Program builds conservation of tree kangaroos into everyday life (Fig. 27.2).

There needs to be more investment in increasing the number of PNG and Papuan scientists and conservation practitioners to focus on a broad array of conservation work. There is a focus on trying to create more community-based protected areas that will be managed sustainably as part of the PNG Policy on Protected Areas. Sustainable financing of protected areas must be a part of this initiative.

Successful conservation of tree kangaroos in a changing world requires excellent data for science and management. One element of future needs is an extension of some of the more sophisticated tracking and analysis techniques to gain a greater understanding of each species. In Chapter 7 the importance of including behavioral ecology in conservation programs is explored. Section VI has a number of chapters that demonstrate current technology being applied at some sites with potential for expansion to cover additional species and situations. Increasing use of non-invasive techniques such as camera traps can further add to our understanding of ecology. The use of altitude and motion logging GPS collars and Unmanned Aircraft Systems (UAS) aerial imagery will allow us to understand so much more about the movements and habitat needs of tree kangaroos in the wild (Chapter 23). The use of open source data mining and AI and Machine Learning is the way of the future for answering broad scale questions of tree kangaroo abundance and distribution (Chapter 24). The use of human medical imaging technology in the field will help us to better understand the physiology and health of tree kangaroos (Chapter 15, 25). There needs to be health assessments on more species of tree kangaroos. There must also be research on the effects of climate change on tree kangaroos since the health and altitudinal range of species could be decreasing.

An important unfinished project is to complete the assessment of tree kangaroo phylogeny and help finish the taxonomy of the group. Chapter 1 is based on the latest understanding but it is diminished by the absence of some taxa from the genetic work. Today's technology enables such work to be completed by small, non-invasive samples from the wild, including feces, hairs, and even environmental DNA (eDNA).

Ideally, we would recommend a complete survey and documentation of all 14 species across Australia, PNG, and Indonesian New Guinea, similar to the complete Birds of Paradise survey and documentation by Laman and Scholes (2012).

A significant aspect of this volume is the essential roles that zoos play in contributing to the science and conservation of tree kangaroos (Chapters 18–22). Especially with an elusive and hard to find group of species (Chapter 2), maximizing research studies of zoo animals has enabled the information to be gathered that is not otherwise obtainable. Much of the foundation of what we know about the biology and reproductive behavior of New Guinea species came from the coordinated efforts of zoos. The international network of zoos has taken a strong lead in conservation, research, and management of several species of tree kangaroos. It will be important to secure commitments from all zoos that have tree kangaroos to support conservation in the wild. Papua New Guinea's Port Moresby Nature Park is an excellent example of a local PNG zoo that is involved in supporting conservation and education of tree kangaroos within the region. Globally, zoos have the ability to engage a wider audience to support conservation of tree kangaroos. We must continue to strengthen the international collaboration of zoos, conservation organizations, government, universities, and local communities.

In conclusion, we can be hopeful for this incredible group of species, but it will take the commitment of local people as stewards, the commitment of governments to prioritize the health of wildlife, the commitment and collaboration of local and international conservation practitioners to provide solutions, and the education of the

FIGURE 27.3 The future of tree kangaroos and people. *Source: Ryan Hawk.*

future generations of people who live in these countries (Fig. 27.3).

REFERENCES

Beehler, B., Laman, T., 2020. New Guinea: Nature and Culture of Earth's Grandest Island. Princeton University Press, Princeton, NJ.

Flannery, T., Martin, R.W., Szaley, F., et al., 1996. Tree Kangaroos: A Curious Natural History. Reed Books, Melbourne.

Laman, T., Scholes, E., 2012. Birds of Paradise: Revealing the World's Most Extraordinary Birds. National Geographic.

CHAPTER

28

Tree Kangaroos: Ghosts and Icons of the Rain Forest

Sy Montgomery
Nature Author, Hancock, NH, Unites States

A kangaroo who lives in trees! Who ever heard of such a thing?

Certainly not the kids in Mrs. Camille Bell's class at Spessard Holland Elementary School in Satellite Beach, Florida. Until, that is, one day in 2016, when the students read about the Matschie's Tree Kangaroo in a book in their 5th grade reading program.

Impossibly cute, the Matschie's soft orange and yellow coat, long tail, pink nose, and marsupial "belly pocket" make it seem like a living plush toy—or perhaps something designed by Dr. Suess. "It was such an endearing and precious animal," Mrs. Bell said. "And when the kids learned it was in trouble, it hooked us." From that moment on, she said, "it was tree kangaroo fever!"

At the school where she's taught reading and language arts since 1991, Mrs. Bell keeps a plaque on her desk. It says this: "The future of the world is in my classroom today." What happened next proves the saying is true—and proves that tree kangaroos can inspire even children who live half a planet away from them to help protect these compelling animals and their cloud forest home.

"My children got into this conversation in class," she explained. "They said, we're only ten years old, but we want to help! How do we start?"

First, the class decided, they had to educate others. They started with the 500 students and 40 staff at their own school. They took a poll. Nobody had ever heard of this animal. Nobody knew that it only lived in one small area of the world—the cloud forest of the Huon Peninsula of Papua New Guinea. Nobody knew that its survival was threatened. No one had heard the story of how Dr. Lisa Dabek fought, against all odds, to successfully track the 'roos through the cloud forest for the first time and probe the secrets of their lives. Nobody knew how crucial it was to enlist the help of local villagers in the effort to save the 'roos and their forest. Nobody knew it could be done at all.

The students wanted to spread the word. So, the kids made the Matschie's tree kangaroo the focus of an intensive research project. Each 5th grader created his or her own Powerpoint talk—and each child presented his or her findings, one-on-one, to students and staff in all the other grades in the building.

Tree Kangaroos
https://doi.org/10.1016/B978-0-12-814675-0.00031-2

But they didn't stop there.

"They wrote essays on the 'roo," their teacher said. "They wrote poetry on the 'roo." (One child called the 'roos in her poem "the chubkins of the trees....they look like moss, they smell sweet like orchids"). They created artwork. Two children wrote a song, and a third set it to music from a Harry Potter movie. (A sampling of the lyrics: "I'm a Matschie's Tree Kangaroo/Help my dream come true/Please try to love us/Be kind, we are the best....")

The children mounted an evening of entertainment and education about the tree kangaroo. They put on the show for the Parent-Teacher Organization's Santa Evening Social.

But even that didn't seem like enough.

The kids knew that Dr. Dabek needed money to continue her research. And the Tree Kangaroo Conservation Program (TKCP) needed money to expand its conservation and education efforts throughout all the areas the tree kangaroos live.

Mrs. Bell remembers, "Then a little girl named Jeanette said, 'Why can't we do something where we make things—and sell them to raise money to help the tree kangaroo project.' And then the whole class erupted!"

Hands shot up. "I know how to do weaves!" offered one girl. "I know how to make rock art!" said another. Every kid had something to offer: "I can make emoji pencils!" "I can make bead animals!"

"It just blew up!" Mrs. Bell said. The tree kangaroos took over the curriculum. Now the animals extended their influence beyond the language arts and biology, into sophisticated math.

"We started making charts of who knew how to make what. We put together a marketing plan," Mrs. Bell explained. Included in the plan were decorated, informational "Save the 'Roo jugs," where people could donate coins, which the kids placed at community centers like the bagel shop, the YMCA, vet clinics. Three kids, with their parents' permission, went door-to-door soliciting donations during the halftime break of a TV football game. The kids figured the fans would be happy and feeling generous. (They were right.)

Other kids ran a hot chocolate stand out of school. Another manned a stand serving lemonade. One particularly shrewd boy ran a combination lemonade and potato chip stand—because the young entrepreneur figured if his customers ate the potato chips, they would buy more lemonade.

But the heart of the fund-raising plan was the Straw Market.

In the school auditorium, the kids decided to create a replica of the famous straw markets of the Bahamas, where vendors offer tourists handmade straw crafts. Here children would sell their wares to the other children and staff of the school. Everything, they decided, would cost $1. They set out to create their inventory and worked at it for three months.

They made animal magnets out of clothes pins. They made flower head bands. They made Chinese lanterns out of construction paper. They made "stress balls" out of rubber balloons filled with sand. Every child, the marketing plan decreed, had to make at least 30 items for the customers to be satisfied. The team of boys and girls who decided to make rings and bracelets made 1000.

"To say these children were over the top in love with these 'roos would be an understatement," said Mrs. Bell. "They were making these things in my room, after school. They got parents involved. We collected materials for our arts and crafts from donors—so everything we'd sell would be pure 100 percent profit for the tree kangaroos."

On seven, eight-foot-long tables draped with green cloth and fringed with plastic grass, the children hawked their wares and worked the cash register, as class after class came and went, for one period after another. The market ran for three days. One of the products, the stress balls, sold out the first day.

The culmination of the year-long unit on tree kangaroos was a Skype interview with Dr. Lisa

Dabek herself. Even after all their research, the children still had many questions. How did she set up her relationships with the people in Papua New Guinea? How did she procure the tree kangaroos for her zoo? Did she get sick or injured during her research in Papua New Guinea?

On the appointed day, each child approached the microphone hooked up to the computer to ask the researcher his or her question. The moment the kids saw Dr. Dabek's image on the screen in the school auditorium, they burst into applause. Dr. Dabek graciously answered all their questions from her office, occasionally using a plushie stuffed tree kangaroo to illustrate her points when needed.

But after their interview was concluded, the kids had something to tell Dr. Dabek, too. They showed her some of their crafts and images of the straw market. And then two of the students, fifth graders Kaden Johnson and Reece St. Pierre, held up a giant check for the money they had raised for her project: $2174.36.

"I was so moved by what these kids did," said Dr. Dabek. "I was completely surprised and overwhelmed."

Mrs. Bell's 5th graders were particularly industrious, but they are far from the only kids who have been touched by the tree kangaroos. How can any child resist such an animal? "The tree kangaroo is marketably adorable," wrote the Bulletin of the Center for Children's Books.

As of last year, there were more than 125,000 copies in print of the book Mrs. Bell's students read in class, *Quest for the Tree Kangaroo*, which I wrote. (For comparison: the average children's book from a major publisher sells 3000 copies in its lifetime, according to Publisher's Weekly.) Many of these copies reside in school and community libraries, where the book owes its popularity, in some part, to librarians' careful attention to reviews, honors, and awards. The book has won more than a dozen awards and honors. It was named one of the top books of the year for children by The Washington Post, the American Society for the Protection of Animals, and the Sibert awards committee of the American Library Association. It was honored by both the National Science Teachers Association and the National Council of Teachers of English. But why did it win these honors? Why do librarians buy the book for their shelves, and why do kids take it out to read? This is due entirely to the charisma of the tree kangaroo, the lure of its exotic habitat, and the inspiration of its champion, Dr. Dabek.

Both the book's publisher, Houghton Mifflin, and the Tree Kangaroo Conservation Program have developed classroom study guides to help teachers use the book in the classroom. Lisa's work with these animals was also the subject of an article I wrote for Ranger Rick's Nature magazine, with a circulation of 525,000. The magazine figures that an additional 200,000 children see the magazine each year by viewing used copies.

So, it's little wonder that both Dr. Dabek and I have heard from hundreds of students in classrooms around the country. They've Skyped and corresponded and arranged personal classroom visits and school wide assemblies. Kids and teachers have sent us emails, drawings, letters, and questions. Children have made posters and mobiles of tree kangaroos for their schools and libraries. One of the recurring sentences in the many letters from children that Dr. Dabek receives is this:

"I want to be just like you."

In the tidy, handmade village of Yawan in the highlands of Papua New Guinea's Huon peninsula, it was Traditional Dress day. All the kids were wearing beautiful barkcloth clothes or grass skirts (Fig. 28.1). After their teacher, Pekison Kusso, blew the conch shell to call the children to the classroom, the

FIGURE 28.1 YUS students in traditional dress. *Source: TKCP.*

youngsters in his class got to work. Today's focus was conservation—a subject Pekison considers just as important as reading or math. If people don't learn to conserve, he told me through a translator, we could lose everything—the tree kangaroos, the forests, even the vines people use to build their homes. "But because we teach conservation," he said, "children will grow up finding it easy to conserve in the future."

Seven-year-old Joel was drawing a tree kangaroo at his desk. "When I see a tree kangaroo, I'm excited!" the little boy told me. "Even seeing a picture of one makes me happy!" he said (Fig. 28.2). Nine-year-old Ali said that he thinks tree kangaroos are so beautiful that no one should hunt them. And then one of Dr. Dabek's visiting assistants told the boy an important fact: "These kangaroos are found only here. Nowhere else. Not anywhere else in Papua New Guinea. Not in Africa or Europe or America!"

The boy's dark eyes widened. What did Ali think of that?

"It's good it's found only here," he decided. "Our place has something special. Something we can be proud of."

Even in areas which have voluntarily set aside land to protect tree kangaroo habitat, known as the YUS Conservation Area, the people "are not fully aware of how special their place is to people outside YUS borders," said Anne O'Dea. Under contract with the Tree Kangaroo Conservation Program (TKCP), she and partner Steve Winderlich, of Windydea Professional Services and Consultants, are working to change that. Two years ago, they helped TKCP staff develop and implement a program called Junior Rangers aimed specifically at the young people who live in the villages dotted through Matschie tree kangaroo habitat. Windydea and TKCP staff teach local volunteer teachers how to run the program. Starting with

FIGURE 28.2 Tree kangaroo drawing from YUS student in Yawan Village. *Source: TKCP.*

children as young as three through adults older than 30, the program aims to show young people how to garden more sustainably, read maps, identify plants, animals, habitats, and threats, and the tools and techniques that scientists use to do field research and monitoring (Fig. 28.3).

"The very first lesson is called 'YUS is Special'," explained environmental consultant O'Dea. "The youth lesson is designed to connect the junior rangers with other people from TKCP, government, etcetera, to see why YUS is special to them." The lesson helps the junior rangers see YUS from perspectives both inside and outside of YUS borders. It includes the view of a child on the other side of the world: "My name is Kristi. I am 7 years old and I live in America. I love seeing the Tree Kangaroos in Woodland Park Zoo. I care that YUS is protected because one day I want to see them in the wild." And the Junior Rangers program also features the book, Quest for the Tree Kangaroo. The children of YUS are delighted to find

FIGURE 28.3 Classroom in Isan Village in YUS. *Source: TKCP.*

in its pages, photos of their neighbors—including children. They are thrilled to discover that, in a country on the other side of the world, fellow kids are learning about Papua New Guinea's native animals.

"It's coming full circle!" observed Dr. Dabek.

"I want to tell you about a very special place I visited a few years ago."

In 2017, for the March for Science in Boston—an Earth Day event which drew 70,000 demonstrators, the largest of the more than 600 marches in support of science across the US and around the world—I was honored to be invited to speak. I was particularly glad that my audience would be what I considered the most important people at the march. I was asked to address the children at the march.

Of course, wise people understand that children are our future. "I really believe the future of conservation is with kids," Dr. Dabek said. "The more kids around the world understand the importance of protecting plants and animals, the better off we'll be." Her focus on the future is rightly echoed by TKCP Conservation Strategies Manager Karau Kuna: "Longterm wildlife conservation requires local communities to understand the direct link between protecting their forests and protecting their future," he said. That's why TKCP so values the focus on teaching conservation: "Programs like Junior Rangers help teach our young people to look after the environment better," said Ranger Nelson Teut, "and become the leaders of tomorrow."

But children are not just the leaders of tomorrow. They are the leaders of today. Even though they cannot drive or vote, even though they don't earn a salary, kids are powerful agents of change in their homes, in their schools, in the community. They profoundly influence their parents' choices in terms of what items they buy, what foods the family eats, what charities they support, whether they recycle, where vacations are spent, and so on. Like Mrs. Bell's students at Spessard Holland Elementary School, children can be teachers, too. One respected educator, Dr. Cindy Thomashow (now co-director of the graduate program in Urban Environmental Education at IslandWood in partnership with Antioch University in Seattle) found in her work that parents report learning more about the environment from their own children than from TV, radio, magazines, and newspapers combined.

So, as I was speaking before the kids crowding around the Parkman Bandstand on the Boston Common at the March for Science, I chose to feature an experience with which I knew children would powerfully connect: my expedition with Dr. Dabek to study the tree kangaroos.

"It was on New Guinea—a place that has been called a Stone Age Island.....

I set the scene: Papua New Guinea has been described as "a lost world," "a land that time forgot". New Guinea was mostly unexplored by outsiders till the middle of the 20th century, for a number of excellent reasons: Tangled jungles. Steep mountains. Erupting volcanoes. Aggressive crocodiles. Poisonous snakes. Tropical diseases. In short, this is the sort of place a kid with a taste for adventure would dream about. Especially since, because of New Guinea's long isolation, animals survive here like no others on the planet...from dinosaur-like giant cassowaries, to kangaroos who live in trees.

"I joined a team of researchers trying to do what others had told them was impossible," I told the kids. "We wanted to capture and radio collar a particular kind of tree kangaroo—the Matchie's tree kangaroo. It's about the size of a big cat, with orange and yellow fur and a sweet pink nose. It spends most of its time 80 feet high in the trees. It eats orchids. Nobody had done this before."

As teachers and parents well know, it can be difficult to keep kids' attention under even the best of circumstances. But on this day, the temperature in the 40s, there was a knifing wind, and a misty rain was falling.

Yet hearing about the tree kangaroos, these kids, shivering under umbrellas, were riveted.

"To get to the animals we had to hike along some of the toughest trails I've ever done," I continued. "For three days, we struggled up steep slopes slippery with sucking mud. We were up so high—10,000 ft—we were literally in the clouds. The air was thin and hard to breathe. The second day of hiking, I quietly wondered whether I was having a nine-hour heart attack.

"Leading our group was my friend, Dr. Lisa Dabek," I continued. "And it's her story I want to share with you today, because this scientist is as amazing as the Matchie's tree kangaroos she studies...."

I told the kids about Lisa's childhood asthma—it was so bad she had to give away her pet cat, Twinkles. I told them about how she overcame it in order to study animals. I told them how meeting a tree kangaroo at a zoo changed her life forever—how she was inspired to help them in the wild. I told them how she spent weeks searching for them in the cloud forest—and how after five weeks in the field, she saw only two. And then she didn't see another for *seven years*. But Lisa kept trying.

It's easy to see how scientists can get discouraged. Even on the expedition I joined to research my book, at the end of every day, our group was exhausted, bruised from falling, and wet and muddy. Once we ran out of food and had to eat ferns. Other things went wrong. Equipment didn't work in the rain. Our satellite phone failed. We worried about the stinging nettles, and also about the leeches—because they can get in your *eye*.

FIGURE 28.4 Lisa Dabek and young Matschie's tree kangaroo at Wasaunon Field Research site in YUS. *Source: Jonathan Byers.*

But, as I told the kids, all our discomfort vanished the minute we saw our first wild Matschie's tree kangaroo. "He was even more adorable than we imagined," I said. "His colors were crazy-brilliant. His nose was the cutest, softest pink. He looked like a big stuffed animal. And his fur was as soft as a cloud. How do I know? Because I touched it—because we captured, radio collared, and tracked *four* tree kangaroos on that trip (Fig. 28.4). And today, Lisa has data on many dozens of tree kangaroos, data that are helping to save this species and the cloud forest upon which these and so many other wonderful animals—*and* the wise, local people—depend."

"This story," I assured the crowd, "reminds us the power of what science can do. Science can change the world."

"Doing science isn't always easy," I reminded the kids. "But we're not going to quit. Like Lisa, we're not going to back down!"

Both in and out of classrooms, both in Northern and Southern hemispheres, the beautiful, unlikely-seeming tree kangaroos of Papua New Guinea continue to inspire both children and adults (Fig. 28.5). No wonder: They're such beautiful, appealing animals; their habitat is both exotic and Edenic. But perhaps the greatest appeal of their story is that it illustrates what may well be the most compelling promise of our time: that yes, we *can* find the way to save the tree kangaroos and their forests—and thus our oceans, and our air, and the climate of our beautiful Earth.

FIGURE 28.5 Papa Manauno gently holding a tree kangaroo in his forest.

Index

Note: Page numbers followed by *f* indicate figures and *t* indicate tables.

J

K

L

M

T

U

V

Printed in the United States
By Bookmasters